Sir Francis Ronalds
Father of the Electric Telegraph

Sir Francis Ronalds
Father of the Electric Telegraph

Beverley F. Ronalds

Imperial College Press

ICP

Published by

Imperial College Press
57 Shelton Street
Covent Garden
London WC2H 9HE

Distributed by

World Scientific Publishing Co. Pte. Ltd.
5 Toh Tuck Link, Singapore 596224
USA office: 27 Warren Street, Suite 401-402, Hackensack, NJ 07601
UK office: 57 Shelton Street, Covent Garden, London WC2H 9HE

Library of Congress Cataloging-in-Publication Data
Names: Ronalds, Beverley Frances.
Title: Sir Francis Ronalds : father of the electric telegraph / Beverley Frances Ronalds
    (formerly CSIRO, Australia & The University of Western Australia, Australia).
Description: Abingdon, Oxon, UK ; [Hackensack] New Jersey : Imperial College Press, [2016]
Identifiers: LCCN 2015049360 | ISBN 9781783269174 (hc : alk. paper)
Subjects: LCSH: Ronalds, Francis, Sir, 1788–1873. | Telegraph--Great Britain--History. |
    Inventors--Great Britain--Biography. | Meteorologists--Great Britain--Biography.
Classification: LCC TK5243.R66 R664 2016 | DDC 621.383092--dc23
LC record available at http://lccn.loc.gov/2015049360

British Library Cataloguing-in-Publication Data
A catalogue record for this book is available from the British Library.

Cover image: Portrait of Sir Francis Ronalds painted by his nephew Hugh Carter (1867).
© National Portrait Gallery, London

Desk Editors: Kalpana Bharanikumar/Mary Simpson

Typeset by Stallion Press
Email: enquiries@stallionpress.com

In memory of Dad, and my cousin Elsie

CONTENTS

PREFACE

Sources and Acknowledgements

The Ronalds family has always been passionate about its history. The branches that reached the USA, Canada, Australia and New Zealand joined the few who remained in Britain in maintaining genealogies and preserving mementos. My father's interest was sparked when his Uncle Oscar returned from a trip to England in 1926 armed with a beautifully drawn family tree. Dad's research culminated in the booklets *Cousins by the Dozens* (1966) and *The Ronalds Family of Australia* (1985). On his death my first cousin Elsie Day took the baton, building a skeleton of the entire family back five or more generations and fleshing out fascinating detail of their lives. Her early efforts were in the microfiche hoods of Melbourne, but in more recent years she was able to employ the power of the internet. Other relatives have written books about family members, published old letters and diaries or put extensive family trees on the web. For a small and rather ordinary family, the level of interest is remarkable.

I am most fortunate in having these past efforts as the foundation for my own work, as well as being able to explore libraries, archives and museums in Europe, North America and Australasia — and meeting my distant cousins there.

The volume of original material available is perhaps surprising. There are two major sources — the Ronalds Archive at the Institution of Engineering and Technology (IET) in London, England and the Harris/Ronalds Collection in London, Ontario. The former, although science-based, also includes personal papers as well as the Ronalds Library. The Ontario collection at Western University (WU) and Eldon House contains hundreds of letters from the extended family, together with diaries and other belongings that provide invaluable insight into how they lived. A third collection of Ronalds Papers, including Sir Francis Ronalds' very helpful autobiographical letter, is held at University College London (UCL). Further letters to, from and about Sir Francis and his family, as well as relevant personal manuscripts, survive at the Royal Society; the Universities of Cambridge, Leeds, Texas and Harvard; the British Library; the Science Museum in South Kensington; Wellcome Library; Society of Antiquaries; Dr Williams's Library; National Portrait Gallery; Nichols Archive Project; National Cooperative Archive, Manchester; Herbert Museum, Coventry; Bakken Museum, Minneapolis; Wheeler Gift Collection; Chicago History Museum; Newberry Library, Chicago; Historical Societies of Massachusetts, Indiana and Pennsylvania; Origins of Cyberspace Library; State Libraries of NSW, South Australia and Victoria; Ballarat Archives Centre; Alexander Turnbull Library in Wellington, New Zealand; Staats- und Universitätsbibliothek in Hamburg; and in private collections. Brief excerpts from nearly 500 of the most relevant letters and manuscripts are quoted in the text.

Primary sources pertaining to Sir Francis' scientific endeavours survive in several archives. Meteorological data are retained at the National Meteorological Archive at Exeter as well as at the IET. His inventions may be viewed at the Science Museums in London, Edinburgh, Oxford and Cambridge, and also Kelmscott House in Hammersmith. There are early scientific photographs at the IET, Cambridge and Oxford Universities, the Meteorological Archive and the British Library.

Sir Francis' rarer printed works are held in the IET; British Library; National Art Library; Royal Society; Wellcome Library; Science Museum; Royal Institute of British Architects; Society of Antiquaries; Bodleian Library; Cambridge University Libraries; National Libraries of Wales and Scotland; Bibliothèque nationale de France; Société polymathique du Morbihan; Université de Strasbourg; Université d'Aix-Marseille; Accademia Galileiana; Library of Congress; New York Public Library; Metropolitan Museum of Art; Yale University; University of St. Thomas and elsewhere. I also recognise with pleasure the various digital libraries by which 18th and 19th century books, journals and newspapers are now so readily available.

Useful detail on the family's homes and surroundings is found in manorial documents, lease agreements, tithe maps and rate books at the London Metropolitan Archives (LMA), and at Local History Centres and Record Offices at Chiswick; Islington; Hounslow; Hammersmith; Croydon; Brighton; Derby; Lichfield; Leicester; Edwards County; and Ballarat. Artwork, including portraits, is held at Eldon House, WU, the IET, National Portrait Gallery, Kelmscott House and in the extended Ronalds family. Original genealogical data exist in Ontario and UCL as well as with the family.

Further information pertaining to Sir Francis' story was obtained from the British Museum; National Archives at Kew; London Fire Brigade Museum;

Museum of Science and Industry, Manchester; Fox Talbot Museum; Hastings Museum & Art Gallery; Guernsey Museum; Museum of the Isles, Skye; Musée de Préhistoire de Carnac; Lynn Woodwork Museum; Royal Institution; Royal Astronomical Society; Académie des sciences; Institution of Civil Engineers; RSA; Lambeth Palace Library; British Science Association; Royal Archives; King's College London; University of Cambridge, Institute of Astronomy; University of Westminster Archives; London Metropolitan University; the Correspondence of William Henry Fox Talbot project; Royal Horticultural Society; Royal Botanic Gardens Kew; Society of Ornamental Turners; Fishmongers' Company; Drapers' Company; Flyfishers' Club; Registrar General for Scotland; High Council of Clan Donald; Chatsworth; Bonhams; Battle Town Council; American Library Association; National Library of New Zealand; National Library of Australia; State Library of Western Australia; Central Highlands Regional Library, Victoria; and Environment Canada.

I warmly thank staff at all of these entities, and others I have visited, for their time and assistance — and particularly the archive personnel at the IET and in London Ontario. Thanks too to Prof Michael Twyman; Prof Martin Kemp; Prof Larry Schaaf; Prof Johannes Volmer; Prof Ruth Watts; Dr Giovanni Bonello; Dr Jacqueline Latham; Dr Alison Morrison-Low; Dr Gerald Kutney; Revd Glenn Coggins; Revd David Beazley; The Ven Stephan Welch; Mark Beswick; Rebecca Storr; Carolyn Hammond; Tara Wittmann; Terry and Brenda Fieldhouse; Jon Cable; Barb and Bob Newell; Lillia D'Agostino; Cecile Arbus; Julian Pooley; John Cloake; Robbie Brothers; Helen Elletson; Elena Payami; Mark Hurn; Dr Sharon Howard; Jeremy Norman; Rachel Howell; Mick Hall; Dr Nobuyuki Kawano; Michael Taffe; Wendy Skilbeck; Jennie de Protani; Jennifer Burrell; Lorraine Powell; Dr Erica Smyth; Steve Uphill; John Birchall; my editor Dr Lawrence Osborn; and the Imperial College Press editorial team. Experts who very kindly commented on particular sections of the work are:

Dr John Carras, Honorary Fellow, CSIRO, Australia,
Dr Serge Cassen, Directeur de recherches CNRS, Université de Nantes, France,
John Edwards, http://www.ornamentalturning.co.uk,
Prof Giles Harrison, Department of Meteorology, University of Reading, UK,
Roger Watson, Curator, Fox Talbot Museum, UK.

Acknowledgement is due too to my near and distant cousins for their generosity in sharing information, photographs (and even DNA). Finally, I thank science and engineering friends who advised on several inventions, John Carras, Colin Brice and Dale Twycross who kindly reviewed the draft, others who have endured several years of my one-track conversation on Sir Francis Ronalds, and my wonderful husband Ian who has relived each of my ancestor's achievements with me.

Editorial Notes

As with many families of this time, particular Christian names recur frequently, which may be helpful in genealogy but is frustrating in storytelling. As the eldest son, Sir Francis Ronalds acquired his father's name, who had in turn received a family name on his mother's side. Five of Sir Francis' nephews were christened Francis and the name continued to pass down the generations to the degree that I, my father, my nephew and great-niece all have the middle initial 'F'. In this book Sir Francis is referred to as Frank, the name used by his family, in the early part of the story to distinguish him from his father and other relatives.

An even more popular name is Hugh. Frank's grandfather Hugh Ronalds was almost certainly given a family name. It then passed to his son, both families of grandchildren, and seven of his eight sets of great-grandchildren. It too has continued in use and at least one descendent named Hugh Ronalds is alive today. Various simple descriptors are used before or after the name to help identify particular people in the text. Frank's grandfather is called 'Old Hugh', and his son Hugh Jnr or (Frank's) Uncle Hugh, depending on the context. Similarly, Frank's maternal grandfather is William Field Snr. His first cousin Dr Henry Ronalds (a physician) and nephew Dr Edmund Ronalds (with a PhD) are always given their title.

A few editorial changes have been made to quotes from handwritten documents to assist reading. The 'long s' in use at the time (for example, 'afsist') has been replaced with today's 'short s'. Spelling of certain specific or unusual words like proper nouns or those in another language have been adjusted to the modern version, although alternative spellings used reasonably consistently are retained. All abbreviations, capitalisation and underlining are also retained. Some supplementary punctuation has been included: for example, the closing parenthesis, which was omitted frequently in Frank's writing. Editorial entries are as follows:

[blank]   Space left in original text for later insertion or to retain anonymity
[missing]   Original text no longer visible
[?]   Word(s) uncertain or not deciphered
[sic]   Erroneous or dated spelling retained (first use only)

FOREWORD

Sir Francis Ronalds was a pioneer: an inventor; a communicator; and a custodian of knowledge. He is celebrated in the Institution of Engineering and Technology (IET) as the first person to develop a working electric telegraph in the UK and, through his bequest of the Ronalds Library, as the founder of the IET's Library and Archives collections. He is also a great example of an innovator who followed his own path and offered his inventions for public benefit.

Sir Francis built his electric telegraph in 1816, and offered it to the UK Government without attempting to patent his invention. He was ahead of his time — the Admiralty turned this new technology down as being 'wholly unnecessary' and not a substitute for tried-and-tested semaphore. The development of telegraphy and the global telecommunications network is fundamental to the history of the IET, which was founded as the Society of Telegraph Engineers in 1871. The IET also continues to support innovation and open source technology — today, we have over 167,000 members worldwide, who are developing the solutions to help engineer a better world that will reach far into the future.

The Ronalds Library was bequeathed to the Society of Telegraph Engineers following Sir Francis' death in 1873. It was an international collection of considerable significance, covering the wider developments of engineering, particularly telecommunications, in the UK and Europe. The IET has built on this bequest, and that first library of 2,000 books has grown to a collection of over 5 linear kilometres of physical collections and a wealth of digital resources. In 2006 the quality of the IET's Library and Archives collections was recognised with the award of Designation status — described by the Arts Council as collections that 'deepen our understanding of the world and what it means to be human'. The IET's collections, knowledge networks and information services continue to develop and are expanding into a global, digital information network. I like to think that Sir Francis would approve of the vast collection of knowledge that has grown out of his original bequest and remains a key source of information for many around the globe.

The world we live in today is very different from the one that Sir Francis knew, but he had the vision to see the potential of mass communication and to make his ideas and resources freely available to humanity. The IET is very grateful for the part that Sir Francis has played in our history as an Institution and as a pioneer for the technology that has become such an integral and essential part of everyday life.

Naomi Climer
2016 President
Institution of Engineering and Technology

ABOUT THE AUTHOR

During her engineering career, Dr Beverley Frances Ronalds served as Group Executive at CSIRO (Australia's national science laboratory), the Woodside Chair of Oil and Gas Engineering at The University of Western Australia, and as a Lecturer at Imperial College London. She also helped design petroleum production facilities for the North Sea, the Australian North West Shelf and the Gulf of Mexico. She has been listed regularly as one of Australia's most influential engineers and served on various Councils advising State Premiers and State and Commonwealth Ministers. She is a Fellow of the Australian Academy of Technological Sciences and Engineering, the Institution of Civil Engineers, the Institution of Engineers Australia and the Australian Institute of Company Directors.

Now retired, Beverley has greatly enjoyed getting to know her great-great-great-uncle Sir Francis Ronalds, who, like her father, was a creative and modest engineer and an inspiration.

Chapter 1

INTRODUCTION

Who invented the telegraph? If you are American, you will answer 'Morse'. A British person might say 'Wheatstone', and Europeans will probably have another view again. In fact, a man named Ronalds was knighted as 'the original inventor of the electric telegraph'. He had his strong constitution to thank for the accolade; he was 82 years of age and had witnessed several decades of the credit being given to others — just as it is now.

Most people questioned about venerable Observatories in London will think of the Royal Observatory, Greenwich. In the mid-19th century, another observatory was established on the west side of London near Kew. Before long it had surpassed Greenwich in the quality of its instruments and become the national and international centre of the new science of systematic meteorology and weather forecasting. This was achieved despite active undermining by the Director at Greenwich. The Head of this new observatory was — as you might guess — the same man Ronalds.

And that's not all. Through his life, Ronalds documented close to 200 inventions. Some of these are still in use today and others had a significant influence on modern living, yet his association with them is unknown. Why? Much of the reason relates to his nature. He did not seek recognition for his work, preferring to develop further ideas rather than tell people what he had already done. He had no wish to be seen as a leader in the scientific community.

The bicentenary of his most noticed invention (the telegraph in 1816) is a good time to learn more about this very modest man and explore his large portfolio of achievements.

Sir Francis Ronalds FRS was born on 21 February 1788 at the family cheese-monger business in the City of London. His was an archetypal family of the

Unitarian faith, having a strong and talented mother who provided good education for her children and encouraged them to follow their dreams (Chapter 4). Youngest brother Alfred achieved fame with his book on fly fishing — which ran to 12 editions — before migrating to Australia. Another brother Hugh helped found a community in the American Midwest, while sister Emily epitomised the family's active interest in social reform.

Ronalds was apprenticed when he turned 14 and 4 years later was running the family business (Chapter 2). An indication of the size of the enterprise is given by his early sale of shares in 12 of his trading ships. With his reserved nature, however, he did not enjoy life as a merchant and was pleased to be able to hand over to his brother Edmund in 1814.

Within weeks of his 'retirement', he was writing scientific papers on his favourite subject of electricity. He was perhaps the first to delineate for a battery the parameters known now as electromotive force and current, and gave the initial description of the effects of induction in distorting signals in long cables. Fundamental science was of less interest, however, than practical application and Ronalds can be considered to be the first electrical engineer. Within two years he had invented the battery-operated clock, improved electric generators and effective modes of electrical insulation (Chapter 6). He mused of his electric telegraph: 'Let us have *electrical conversazione offices*, communicating with each other all over the kingdom...give me *materiel* enough, and I will electrify the world' (Chapter 7). The rejection of his telegraph by the Admiralty in 1816 as being 'wholly unnecessary' was to become infamous in his lifetime.

The period 1818–20 was his 'Grand Tour' to Europe and the Near East (Chapter 8). He met numerous people of interest, including those who brought the statue of Ramesses II and the Elgin Marbles to England and who helped unravel the mysteries of Egyptian hieroglyphics. His own scientific activities included climbing both Mounts Vesuvius and Etna during eruptions and taking atmospheric electricity readings at the craters. He returned to Italy in 1823–24, probably to pursue his idea of importing sulphur into the UK, which was on the brink of becoming a pivotal element of the industrial revolution (Chapter 9).

Ronalds next turned his attention to innovations of a mechanical nature. He created a successful business selling his patented instruments for drawing accurately in perspective (Chapter 10). Surveying the Neolithic stones at Carnac in France with his friend Dr Alexander Blair, they published precise drawings of the ruins using his instruments, which have earned him the descriptor 'archaeologist' (Chapter 11). He additionally developed the ubiquitous portable tripod stand with three pairs of hinged legs, an attachment for the lathe that helped start the mass production of furniture and, as part of a comprehensive risk management strategy for fire, a Fire Finder that was reinvented in the USA in the 20th century and also remains in use (Chapter 12).

In 1842, he began his role as Honorary Director of the Kew Observatory. He shaped its successful mission and fought the battles necessary for it to survive its early years and go on to become arguably the world's premier meteorological and geomagnetic observatory (Chapter 13). With the most comprehensive atmospheric electricity measurements to date, and the provision of his equipment to other observatories around the world, Ronalds strove to delineate the global interrelationships between atmospheric electric potential, air–earth currents and geomagnetism (Chapter 14). He additionally made important advances in the new discovery of photography, including the first 'movie camera' to capture the continuous modulation of physical parameters (Chapter 15). His photo-barograph aided weather forecasting until the last decades of the 20th century.

After his second retirement at age 65, he spent nine years on the Continent completing an electrical library and continuing to watch the roll-out of the telegraph around the world. Following his death on 8 August 1873, the Ronalds Library was bequeathed to the new Society of Telegraph Engineers, helping to ensure the future of what became the Institution of Electrical Engineers and is now the Institution of Engineering and Technology (Chapter 16). Cambridge University Press reprinted his library catalogue in 2013. In all, he fitted many noteworthy accomplishments into his life.

The layout of this biography reflects both time and theme: the order of the topics is approximately chronological but each is considered in detail in a particular chapter. Chapters vary in style as a result. Chapters 2–4 introduce Ronalds' personal attributes and the influence of his large and supportive family. The focus of Chapters 5, 7, 11 and 13 is the human side of science — skills, behaviours and interactions with colleagues. Chapters 6, 10, 12, 14 and 15 cover many of his inventions and are primarily technical in nature, although also considering related business matters. International travels are described in Chapters 8, 9 and 16. His activities that brought the greatest ramifications (the telegraph and the establishment of Kew Observatory) are found in Chapters 7 and 13 respectively. Each chapter stands essentially alone to facilitate ready reference to particular topics.

We start by exploring life for Ronalds in the cheesemonger business.

Chapter 2

FOUNDED ON CHEESE

Ronalds' birth at the family's Thames Street cheesemonger warehouse, his apprenticeship and a sketch of his activities over a decade of running a large business, including the risks he faced, what he learnt and why he 'retired' from business life.

Sir Francis Ronalds' earliest surroundings were the City of London, with a backdrop of the Thames and the 600-year-old London Bridge with its 18 irregular arches. They were also the scene of his apprenticeship, and while still a teenager (as he later reminisced): 'the very severe task of managing a business which then returned at least 150.000£ per annum devolved upon my over-loaded shoulders'.[C375,1] Frank, to use his familiar name, did not enjoy his years in merchant business and was very glad to be able to leave in 1814 to pursue scientific interests.

His maternal ancestors — the Field family — had been merchants for about five generations when Frank entered the business. According to Frederick Pierce's *Field Genealogy* (1901),[B117] the first English 'Field' had come across from France with William the Conqueror. In the long line since then the standout was John Field (*c*.1520–87). He was an astronomer who published the first English calendar based on Nicolaus Copernicus' model of the solar system — with the Sun, rather than the Earth, being the centre of rotation. He was gaoled for his efforts by Queen Mary.

Another John Field, the astronomer's great- or perhaps great-great-grandson, was a successful London cheesemonger in the latter part of the 17th century.[A165] His side of the family were also gentlemen farmers at Cockernhoe in Hertfordshire.

---

[1]References starting with a letter are listed in the Bibliography.

The business was later in the hands of Isaac and then Nathaniel Field, who passed it before his death in 1755 to his nephew (and Frank's grandfather) William Field Snr.[W3] William Snr set up in partnership with Nathaniel Hardcastle (the husband of his sister Elizabeth) as *Hardcastle & Field Merchants*. Their premises were at 8 Old Swan Stairs, at the river end of Swan Lane, which led south from 104 Upper Thames Street.[W4,W7] William Snr's older brother John meanwhile took over their father's business as a seedsman, situated in Lower Thames Street.[J4]

Their addresses were at that time the heart of the City of London. Upper and Lower Thames Street met at Old London Bridge, then the only bridge entry into the City. Fishmongers' Hall, home of the Worshipful Company of Fishmongers, was situated close to the busy intersection. Thames Street was aptly named: it had once been a waterside path but, starting with the Romans, land reclamation over the centuries meant that it now ran parallel to and a little north of the river. The river was the main means of communication with the City and access to it was provided by numerous small alleyways like Swan Lane. It was in this little area that some of the first deaths from the Plague had been recorded in 1665.[B36] The Great Fire of 1666 also broke out just metres away, with Fishmongers' Hall being the first major building to be consumed.

Succession planning, and providing for offspring, continued to be important for the family. As Nathaniel and Elizabeth had no children, they introduced their nephew Joseph Hardcastle into the business. William Snr and his wife Anna née Bayley helped to establish each of their children in the City. William Snr apprenticed his sons William Jnr, George Snr and Charles over the period 1782–1801.[J4] Daughters Emily and Charlotte married respectively William Venning and Thomas Gibson, who set up *Venning & Gibson Silk Manufacturers* in Milk Street. Another daughter, Ann, married James Tatlock whose silk broker business was also in Milk Street (Table 2.1). Importantly for this story, their second daughter Jane married one Francis Ronalds in 1785.

Francis was the son of 'Old Hugh' Ronalds and Mary née Clarke of Brentford. His older brothers took over the family's nursery and the evidence points to Francis serving his apprenticeship with his Clarke relations, who were silkmen. The Ronalds and Field families would have had linkages through both silk and seed. That in silk was particularly strong, with Francis' ribbon-weaving business in Wood Street being just around the corner from James Tatlock — and James being one of the witnesses at Francis and Jane's wedding.[J102]

The indefatigable William Snr quickly set up business with his new son-in-law. *Field & Ronalds Cheesemongers* was established at 109 Upper Thames Street, just around the corner from Old Swan Stairs, and Francis and Jane were living at the property when Frank was born there in 1788.

Table 2.1:   Descendants of William FIELD and Catherine née Stackhouse: Paternal Grandparents of Frank's Mother.

1725–1812	1761–1863	1786–1895
Children	**Grandchildren**	**Great-grandchildren**
John *m* M Robinson *seedsman*	J Elizabeth *m* John HEATH *manufacturer*	Edmund *hosier*
William (Snr) *m* Anna Bayley	Ann *m* James TATLOCK *silk broker*	George Thomas *silk brokers*
cheesemongers	William	
	Jane *m* Francis RONALDS *cheesemongers*	Jane (Jnr) **Francis (Frank)** William Edmund Hugh Mary Ann Emily Charlotte Charles Nathaniel Alfred Maria
	William (Jnr) *m* Mary Payne	Mary William Emily Caroline Clara Hardcastle *ship's chief mate*
	Frederick	
	Emily *m* William VENNING	
merchants	George (Snr) *m* Ann Bolland *silk manufacturers*	George (Jnr) *hop merchant* Eliza Frederick Catherine (Kate) Anna Arthur *wine merchant*
	Stackhouse Charlotte *m* Thomas GIBSON	Thomas Field *silk manufacturer*
	Nathaniel Maria Caroline	
	Charles *m* Ann Slatter	Charles Ann
Elizabeth *m* Nathaniel HARDCASTLE	*Joseph HARDCASTLE*	
Catherine		
Sybilla *m* Thomas FIELD *stationer*	Katherine *m* William COTES *silk manufacturer*	Elizabeth Maria William Nathaniel Hannah Mary
	Sybilla	

At William Snr's purchase of the freehold in 1787, the title deed described 109 Upper Thames Street as:

> All those 3 messuages or tenements thrtofore *[sic]* erected & built…whereon formerly stood four messes *[sic]* or tenements…burnt down by the dreadful fire in London which happened in the year of our Lord 1666…All that messuage or tenement…in the tenure or occupation of the s^d Wm Field…known by the name or sign of the Wheatsheaf… Together with the Counting house (theretofore a Yard)…And all those new brick buildings…lately erected & built by the s^d Wm Field…& used by him as Warehouses [J12]

The building's frontage along the south side of Thames Street was 'on the ground or Shop floor 19 feet & 1 inch of assize little more or less' and '26 feet & 3 inches' on the upper floors where it spanned across Wheatsheaf Alley. It extended back '100 feet & 1 inch' towards the Thames (Figs. 2.1 and 2.2). William Snr paid £700 for the property. On his death in 1812, Jane purchased it from her brother (William Snr's heir) for the generous sum of £3,450. It was sold again for £3,400 when Jane died in 1852.[J12]

The Fields also acquired the property next door on the west side of Wheatsheaf Alley, extending from Thames Street down to the river and including a wharf with 'crane-room and counting-house'.[B36] The wide-ranging business interests of the Field family at this address included being tallow chandlers — selling candles made from processed animal fat.

The property behind 109 Thames Street on the river belonged to John Garratt, a wholesale dealer in tea and hops. As Lord Mayor of London he laid the first stone for the 'new' London Bridge in June 1825.[B161] He sold the premises in 1828 and it became the new location of The Shades, a celebrated tavern since about 1750 at the south-east corner of Fishmongers' Hall:

> Casks of wine were placed against the walls, from which pints, half-pints, &c., were drawn, as called for drinkers sitting there. At the end of the room…was placed on a slab a fine old Cheshire cheese *[perhaps provided by Field & Ronalds?]* and loaves of bread, from which wine customers took as much as they pleased gratuitously [2]

Finally, the east boundary of 109 Upper Thames Street adjoined Fishmongers' Hall. The Fishmongers' Company examined all fish coming into London. In Frank's time the Hall, reopened in 1671 after the Great Fire, was a stately red brick building overlooking the river (Fig. 2.3). The Thames Street façade comprised a central entrance ornamented with the Company's arms, with three houses on either side that were tenanted to merchants (Fig. 2.4). After the new bridge was completed, the Hall was replaced with the Greek-revival structure that stands today.

The intermingling odours of cheese, ham, tallow, fish, wine, tea, hops and the raw sewage flowing into the Thames, on top of the other usual London smells of the time — including horses and the smoke of coal fires — must have made a pungent cocktail.

---

[2] *Mercury* Hobart 18/1/1864 2g.

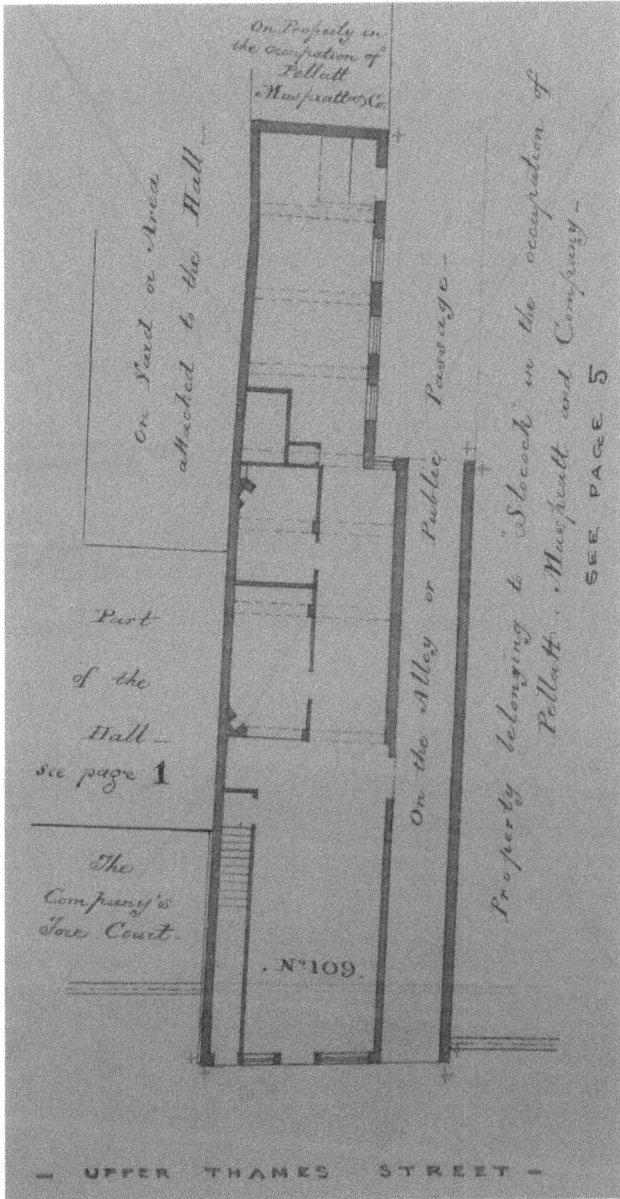

Fig. 2.1:   Plan of 109 Upper Thames Street (*circa* mid-19C). With kind permission of Fishmongers' Company LMA CLC/L/FE/G/045/MS21538/001.

Before long, the next generation began to enter the business. On 15 April 1802, seven weeks after his 14th birthday, Frank began his formal seven-year apprentice-ship to his father.[14] His brother Edmund later joined him. Francis died, however, when Frank was 18. Jane decided to keep the business and her father William Snr took the opportunity to officially retire — he was now 77 years of age. It fell to Frank

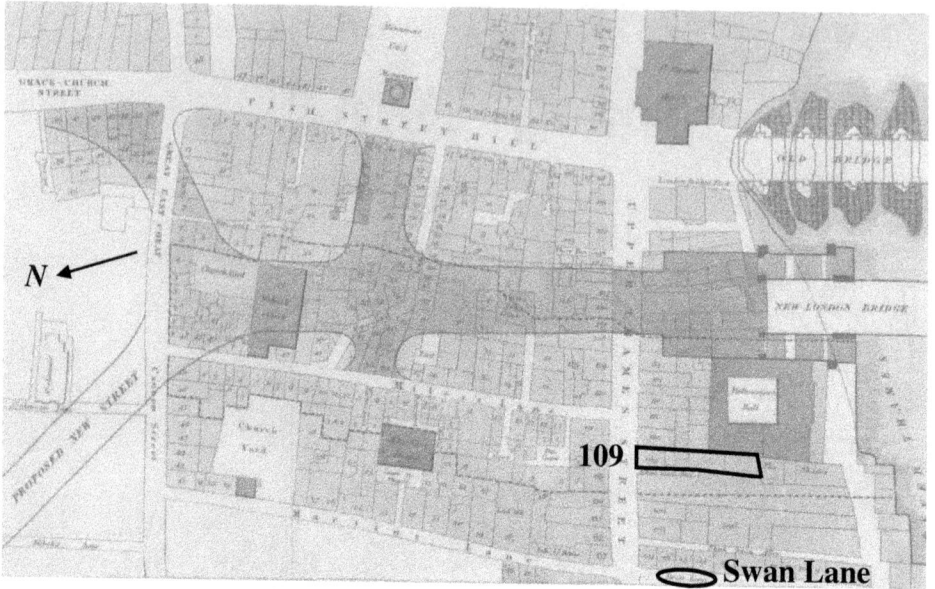

Fig. 2.2:   Plan of St Michael's Parish showing Swan Lane, 109 Upper Thames Street and location of the new London Bridge. Ref. B36.

Fig. 2.3:   View across the Thames showing the new bridge construction (1825). LMA 28095.

A Old Swan Stairs, *B* Garratt's house (backing onto 109 Upper Thames St), *C* Fishmongers' Hall, *D* Monument to the Great Fire of London.

Fig. 2.4:   View east down Upper Thames St from near No. 109, showing Fishmongers' entrance on right and the new bridge approach (1831). LMA 4802.

to run the company and to teach Edmund their father's and grandfather's ways. At this point he had four years of business experience, and Edmund just one. With Jane having nine other children at home and elderly family members to care for, and Frank being underage, she took on a partner to advise him in company affairs. The partner was Michael Coventry and the new entity was *Ronalds & Coventry.*

Frank stayed in the cheesemonger business for a total of 12 years, nearly 8 of which he was effectively the manager. He and Jane 'retired'[3] the day after his 26th birthday, handing the company over to Edmund who was then 23 years old. Edmund retained the business for another 38 years until Jane died.

Hardcastle & Field Merchants, *Field & Ronalds Cheesemongers* and their successors were large and geographically diverse concerns. Cheese and butter were major staples at that time, even more so than today, and London's population was growing rapidly. The *London Tradesman* described the job in 1747:

He is only a Retailer of Cheese, Butter, Eggs, Bacon and Hams: His Skill consists in the Knowledge of the Prices and Properties of these kind of Goods. It is pretty precarious, and liable to a great many Accidents; their Cheese lose in their Weight,

[3] *London Gazette* 23/4/1814 864b.

their Hams stink, and their Bacon rusts, notwithstanding all the Care they are able to take; were it not for such Accidents as these, their Trade would be very profitable.[B3]

Wholesaling was another important aspect of the business for many Cheesemongers. Agricultural historian George Fussell has quoted a passage recorded in 1770:

> The most considerable cheesemongers in London have formed themselves into a club. They are owners of about 16 ships, which are employed between London, Chester and Liverpool…They have factors in Cheshire who buy up the Cheese for them and lodge it in their Warehouses at Chester. At their weekly Meetings they settle what Quantities each shall have brought up to Market. By this Means the Market is fed in such a Manner, that they command whatever price they please.[A146]

Another view from the period was that because of the London monopoly:

> Farmers in Cheshire are obliged to sell their cheese to the factors or brokers of the cheesemongers at their own price, upon trust, with very long credit.[B177]

Some 20th century commentators have argued that this approach enhanced the reliability of food delivery to the capital. While the cartel certainly appeared disadvantageous to the farmers, the cheesemongers were also accepting much of the risk of storage, transport and sale of the goods.

'Clubs' were common in those days. One example is the Fishmongers' Company right next door. The London cheesemongers never formed a comparable Company, although a Cheesemongers Benevolent Institution was still in existence in 1848, of which Edmund was a Steward.[4] The cheesemongers also worked effectively as a group to lobby Parliament. They were certainly a powerful and wealthy merchant body and their businesses were perhaps a little too close-knit even for those times.

Both Nathaniels (Field and Hardcastle) and William Snr all made mention of their ships in their wills. In 1786 it became a legal requirement to register ships with the Customs officers in their home port, with boat ownership divided into 64 shares. Rupert Jarvis' analysis of early shipping data for the Port of London suggests that William Snr was the largest ship owner amongst the Thames Street cheesemongers.[A160] It is apparent from 12 ships in which Frank had an interest in 1809 that ownership could be spread thinly and that the cheesemongers were trading up both the east and west coasts of England. Cheese and butter were also transported to and from London by canal, river and wagon.[A146]

Some of the family's ships also worked internationally. Frank's uncle Charles Field was based in Russia: he married in St Petersburg in 1805 and the couple's first child was born there. Such trade was hampered after 1806 (the very year that Frank took over his father's company) when Napoleon embargoed British trade with the rest of Europe.

---

[4] *The Times* 15/6/1848 1e.

Years later, Edmund was importing produce from as far afield as Canada: one lot being 230 Fk^ns *[firkins]* of Butter, about 7 Tons, shipped from Montreal upon which the Duty the Freight Charges & Comm^n amount to between 11/ *[shillings]* & 12/ p 112^ls *[a British hundredweight or cwt]* & the Butter sold @ 56/ leaving 41/ p Cw^t as net proceeds for the Canadian Merchant.^C111

In London, deals were made at the Royal Exchange, which was located half a kilometre north of the Thames Street premises. The Exchange had a large chiming clock to summon merchants, at which times the central courtyard was crowded with British and foreign traders, brokers, ships' masters and others with commercial interests. A business colleague said that Joseph Hardcastle 'used to stand on 'Change, the centre of the most cheerful group there, very refined in his mirth, but very witty.'^B79

Business was also done in more informal settings. William Herbert reminisced in his *History and Antiquities of the Parish and Church of St. Michael, Crooked Lane* (1833) that 'Fishmongers' wharf had been for many years the little '*Change*' of the oldest and most respectable inhabitants of St.Michael's and the adjoining parishes',^B36 and went on to mention William Field Snr, George Field Snr's father-in-law and others.

The Thames Street premises were the partnership's headquarters as well as the logistics, storage and distribution hub for the operation, and had a shop for local sales. There were also residential areas on the site. It was Francis and Jane's family home for just a short while, although Francis would have continued to stay there when he had late business in the City, as would have Frank during his apprenticeship. Servants and staff also resided at the property. Staff numbers are difficult to ascertain — Francis' will makes mention of 'each and every of my Clerks and other Servants'^W6 while other documents note the roles of porter, carman and warehouseman as well as kitchen staff, the regional agents and the masters and crews of the numerous ships.

Ships in the Cheese Trade — By WINSTANLEY and SON, at Garraway's, THIS DAY, April 6, at 12 o clock, by order of the Executors of Mr. FRAS. RONALDS, deceased.

SHARES in the following regular Cheese Ships, trading from Liverpool and Hull to London — One Sixteenth of the Mars, One Sixteenth of the Active, One Sixteenth of the Bradford, One Sixteenth of the Hope, One Sixteenth of the Providence, One Sixteenth of the Nantwich, One Sixteenth of the Diligence, One Sixteenth of the Success, One Sixteenth of the Fortitude, One Thirty Second of the Zealous, One Thirty Second of the Ceres, and One Thirty Second of the Britannia. Printed particulars may be had at Lloyd's Coffee house, at Garraway's and of Winstanley and Son, Paternoster-row.

The Times 6/4/1809 4b

It is possible to glean a sense of the business ethos of the interrelated companies. The older partners were certainly shrewd and talented businessmen and risk management was always an essential part of business. It seems probable on balance that they acted close to but not beyond the limits of acceptable practice for the period. Joseph Hardcastle was just 13 years of age when he joined his uncle's family and would have imbibed the senior partners' values. His daughter Emma wrote of him:

> he maintained a character for spotless integrity and unsullied honour. To him, from the very outset, belonged the reputation of the English merchant of the old school... He was remarkable for a happy combination of prudence and decision. No important proceeding in business was adopted until it had been maturely pondered. But when his mind was once made up, he acted with promptitude and energy, and then awaited the event with unruffled tranquillity...how cautiously he shunned the entanglements of dangerous speculation [B79]

Frank similarly reminisced on his father's commercial philosophy:

> you must concentrate your forces, you must have one main object in view and pursue that object steadily and vigorously. You must have a system & (if times are not so much altered as to render my experience quite out of date) you must at least found your system upon the old one, on that of our Father, on a careful & judicious attention to the real interests of your country connections together with a vigilant matching of markets. It always appeared to have been his constant practice to keep his country trade steadily in view whilst he missed few opportunities which offered & which he could avail himself of, to take a manifestly good thing. He never went much out of his way to speculate but laid in wait like a Tiger & if he made an unsuccessful pounce his country trade served him for a devilish good hiding bush.[C38]

Speculation was part of the business of being a merchant for Frank and Edmund — as long as it aligned with business strategy. Writing to Edmund in 1819 about their brother Hugh's bold decision to build a tannery in the remote Illinois prairies, Frank noted: 'I may be too cautious perhaps, I look at these kind of projects in a very different way from a mercantile speculation, particularly a speculation in Butter & Bacon, or Broad Cloth, or Silks or Sugars. Don't feel at all offended'.[C21]

It seems that Francis and his sons were a little more opportunistic than Joseph. William Snr's sons were perhaps even more adventurous and in June 1804 William Jnr and George Snr were declared bankrupt. Such events were highly public affairs and separate notices for each of the two brothers appeared immediately in the *London Gazette*[5] and elsewhere. A second instalment followed three weeks later:

---

[5] *London Gazette* 9/6/1804 722.

The Creditors who have proved their Debts under a Commission of Bankrupt awarded and issued forth against William Field...and...George Field...are desired to meet the Assignees of the said Bankrupts' Estates and Effects...in order to assent to or dissent from the said Assignees selling or disposing of by public Sale...all or any Part of the Bankrupts' Stock in Trade, Household Furniture, Lease or Leases, Goods, Chattels, and Effects to which they, or either of them may be entitled...and also to their paying the Wages of the different Clerks and Servants of the said Bankrupts, or such Part thereof, as they may judge expedient...[6]

Regular updates continued. Before the end of the year significant dividends had been paid to the creditors and the final full dividends were paid in July 1806.

In fact it was not unusual to suffer financial difficulties, even if it was frowned upon and risked debtors' prison. Although the Bank of England had been founded at the end of the 17th century and the first merchant bank appeared in London in the 1760s, business funding through such institutions was not commonplace. It was generally necessary to rely on family members and business associates for financial support to manage cash flow.

Frank's uncle James Tatlock was helped out by William Snr when he was bankrupted in 1811. James' son Thomas, who was also a silk broker, repeated the act 15 years later. Thomas' aunt Emily Venning, in distributing her assets across her siblings and their families in her will in 1834, noted regarding him that 'if the said Thomas Tatlock shall at any time be found bankrupt...then...immediately thenceforth the life interest of him the said Thomas Tatlock of and in the said trust promises shall cease and be void...as if the said Thomas Tatlock were dead'.[W8] Evidently other members of the family were kinder. *Field & Ronalds* put the 150-ton Brig *Jupiter* up for sale in late 1805 (Fig. 2.5), which might have been to assist William Jnr and George Snr with their dividend payments. Jane helped out her two brothers again in 1820. Francis' will indicates that he had lent money to his brother Silvanus, his ribbon-weaver relations in Coventry, and his brothers-in-law at *Venning & Gibson Silk Manufacturers*. William Snr also supported *Venning & Gibson* as well as *Ronalds & Coventry* in the 1805–10 period. Years later, Jane loaned Edmund £12,000 during a general downturn in the economy.

Nonetheless, Frank's first close encounter with insolvency, through his uncles and next-door neighbours William Jnr and George Snr, would have been ringing in his ears when he took over his father's business later that same year. For someone who always lacked confidence, his new responsibilities must have been truly daunting.

---

[6]*London Gazette* 30/6/1804 816.

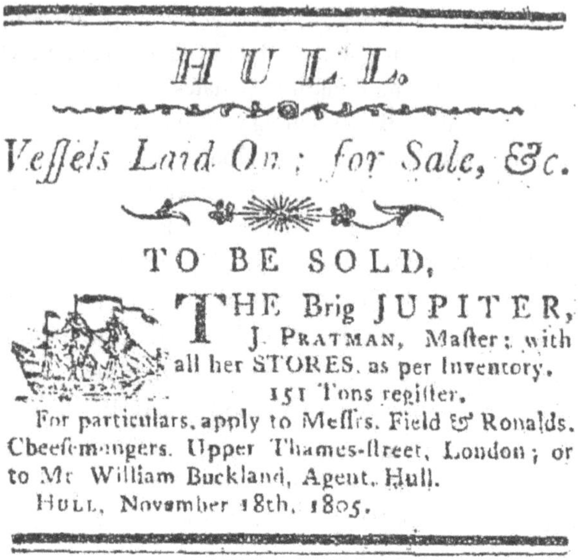

HULL.

Veſſels Laid On; for Sale, &c.

TO BE SOLD,

THE Brig JUPITER, J. PRATMAN, Maſter; with all her STORES, as per Inventory, 151 Tons regiſter.

For particulars, apply to Meſſrs. Field & Ronalds, Cheeſemongers, Upper Thames-ſtreet, London; or to Mr William Buckland, Agent, Hull.

HULL, November 18th, 1805.

Fig. 2.5: Sale advertisement for Brig *Jupiter*. *Hull Packet* 26/11/1805.

He soon also had to deal with the other side of bankruptcy — bad debts.[7] Making matters even more difficult was that he lacked respect for his business partner Michael Coventry; he used uncharacteristically disparaging language when referring to him throughout his life. The primary problem seems to have been their quite different business philosophies and risk appetites. Reminiscing 50 years later about his early managerial experience, Frank wrote: 'The first year terminated very successfully but the proceedings of a speculating Quaker (who long after ruined himself…& who was, quasi, trust *[sic]* upon me as a partner) together with a disgust of trading habits…soon caused me to steer a different cource *[sic]*.'[C375] It was probably Coventry's speculation that resulted in *Ronalds & Coventry* needing to borrow money from William Snr. It was also at this time that the company sold shares in 12 ships. Of course it is possible that he was apportioning undue blame to his partner for a financial set-back they incurred together, although it would be contrary to his normal humility. Either way, he would have hated admitting the early failure to his mother and grandfather.

He believed that with Coventry he had 'to contend with selfish & shuffling tricks'.[C38] Coventry finally departed in 1820 after the business had suffered another misfortune. Frank wrote to his family at the time:

> one of my opinions which I have urged a hundred times has at last been almost proved to be correct viz that Master Michael *[Coventry]* would never do either

[7] *London Gazette* 6/11/1813 2190-1.

himself or anyone with whom he was connected much good. I don't crow, I even pity him...He must return to his old trade of Bag-lumping I suppose, the only one he was ever fit for...

I am glad you have got rid of Michael <u>principally</u> because I know his doctrine to be contrary to mine, or rather because I know him to have <u>no doctrine</u>, no system, no plan, it is better to have some plan altho' a bad one than none because you can improve it always until it becomes a good one.[C38]

Risks with strategy, cash flow, transport, product quality, marketing and politics were not the only concerns for the business. Another perennial threat to operations was fire, and the premises were gutted in January 1820. It was fortunate that Edmund, his wife Eliza and their seven-month-old son were then living in Islington rather than at the property. *The Times* carried the story:

About half past one o'clock yesterday morning, an alarming fire broke out on... Thames-street, which rapidly communicated to adjoining houses...the flames then took the direction of the premises of Messrs. Ronalds and Coventry, the whole of which, together with two houses in Swan-lane...were entirely consumed. The roof of Fishmongers'-hall sustained considerable damage, and great apprehensions were entertained for the safety of this elegant edifice...The loss of the property is estimated at 200,000l.[8]

The fire had started near the corner of Upper Thames Street and Swan Lane and continued down the lane as far as No. 13. It also spread east along Thames Street through No. 106 (property of George Field Snr's father-in-law), No. 107–08 (the Fields) to No. 109 (Edmund) — with Garratt's property behind it — and finally the Hall.

The impact on the extended family was far-reaching. It is not certain whether the properties were insured, although the Ronalds business philosophy suggests that precautions of this type would have been taken. The first fire insurance companies had formed after the Great Fire. *Hardcastle & Field* held insurance in the 1790s,[J5] and Edmund and Garratt purchased cover for their shared facility at Old Swan Wharf in 1821–22 while their properties were being rebuilt after the fire.[J29]

Crime was a further issue that impacted upon the business. Through today's eyes the cases are appalling, although they were of serious concern to management at the time. In 1786 Mark Powell, a servant at *Hardcastle & Field*, was found guilty of forging a draft on the company to the value of £450.14s.[9] Sentenced to death, he was hanged at the gallows outside Newgate prison eight weeks after the trial. Public executions continued until 1868.

---

[8] *The Times* 24/1/1820 3a.
[9] 31/5/1786, available at http://www.oldbaileyonline.org. Version 7.2. (Accessed on 20/7/2015).

Frank appeared at the Old Bailey in 1811 concerning a more common offence: JOHN BIGGS was indicted for feloniously stealing, on the 16th of April, a cheese, value 6*s*. the property of Jane Ronalds, Francis Ronalds, and Michael Mills Coventry.

FRANCIS RONALDS. I am a cheesemonger; my partner's names are Jane Ronalds, and Michael Mills Coventry. I can only speak to the property.

STEPHEN JORDAN. I am a porter to Mr. Ronalds. On the 16th of April, about four o'clock in the afternoon, I saw the prisoner passing by the warehouse in Upper Thames-street several times; a pile of cheese stood about seven or eight yards from the door. He came into the warehouse, he took one of them and made off with it. I immediately pursued him and took him in the street, with the property on him. He begged of me to let him go and he would never do the like again. This is the cheese, it is my master's cheese.

Prisoner's Defence. Necessity made me do it.

GUILTY, aged 75

Confined Fourteen Days in Newgate, and Whipped in Jail.[10]

In one of two incidents Edmund dealt with in 1827, 'HENRY WATTS and JOHN MOORE were indicted for stealing, on the 28th of February, 1 wooden firkin, value 6*d*., and 50lbs. of butter, value 35*s*.'[11] Watts, aged 20, and Moore, 22, were found guilty and sentenced to Transportation for seven years. Watts sailed to

---

St. Luke's HOSPITAL for LUNATICS.

THE ANNUAL DINNER of the GOVERNORS of this CHARITY will be at London Tavern in Bifhopfgate-ftreet, THIS DAY, the 14th inftant.

STEWARDS.

The Rev. Sir Wm. Cheere, Bart.	Thomas Higgins, Efq.
Cornelius Denne Efq.	Mr. Wm. Hobfon
George Dance Efq.	James E. L. Nealfon Efq.
Sam. Robert Gauffen Efq.	Mr. Francis Ronalds
The Rev. Dr. Glaffe	Mr. Ifaac Smith.

Dinner will be on Table precifely at Four o'Clock; and no Servants except thofe of the Officers and Stewards will be admitted.

JOHN WEBSTER, Secretary.

The Times 14/7/1790 1c

---

[10] 29/5/1811, available at http://www.oldbaileyonline.org. Version 7.2. (Accessed on 20/7/2015).
[11] 5/4/1827, http://www.oldbaileyonline.org. Version 7.2. (Accessed on 20/7/2015).

Van Diemen's Land seven months later with 129 other convicts but it is unclear what happened to Moore.

In addition to their business activities, the family participated in the broader affairs of the City of London in various ways. Although *Field & Ronalds* was right next door to Fishmongers', the family were members of the Drapers' Company, while their ancestors had belonged to the Tallow Chandlers' Company, another of London's Livery Companies.[A165] The Companies descend from the medieval guilds set up to regulate the practice of particular trades. Then as now they had voting rights for the City of London Corporation and also led education and charity initiatives as well as social and cultural events.

William Snr and his brother were apprenticed through Drapers', as was their father before them. Francis paid a fee of 46s.8d on 1 April 1802 to become a Freeman by redemption so that he could formally apprentice Frank.[J6] William Snr had in the meantime been elected Master of the Company in both 1790 and 1801.[J4] After serving a second time, he was presented with a silver waiter engraved with both the Fields' and Drapers' coats of arms.[W10] Nathaniel Field had similarly been Master of the Tallow Chandlers in 1746–47.[A165] Edmund followed in his grandfather's and great-great uncle's footsteps, being elected Master in 1864 and 1865.[J4] He was the only one of Frank's siblings to continue their association with the Drapers' Company beyond the registration of their apprenticeship and, after four generations, links with the Company ceased.

Edmund had developed into a confident and respected merchant. On various occasions when another trader was declared bankrupt, he was assigned their personal estate and goods as trustee. Edmund also took an interest in the political affairs of the City. He spoke at a meeting of 'the inhabitants of Bridge Ward' held at Fishmongers' Hall in 1822 where he 'approved of Mr Alderman Garratt's conduct in investigating the expenses of the Ward, and endeavouring to reduce their taxation'.[12] The story passed down to Edmund's granddaughter that his wife Eliza could take much of the credit for his success:

> Grandfather was no sort of business man himself but nothing ever went wrong when he took Grandmother's advice. She used to sit beside him when he dined at 4 o'clock on returning from the City, filling up his glass till the daily bottle of port was empty.[C419]
> No wonder he was gouty![C421]

Frank learnt a great deal in his years as a cheesemonger, which he utilised in his personal activities and in several later business settings. *Field & Ronalds* subscribed to *Jones's English System of Book-keeping, by Single or Double Entry*,

---

[12]*The Times* 14/11/1822 2e.

published in 1796, and he used these methods through his life. On his Grand Tour, he demonstrated both an interest in business, and a good appreciation of it. In Constantinople, he noticed:

> The Bazaars…seem to contain every thing which can be desired by the inhabitants in the utmost profusion, & what seems to me to be always a mark of great extent of trade in a place is that the principal dealers in each particular article herd together in particular quarters or divisions. Thus you see a bazaar of Tobacco, another of Tobacco pipes, others of Coffee, of Fruits &c &c.[J21]

The advantages of concentrating related businesses geographically in 'clusters' were popularized 170 years later by Harvard University Professor Michael Porter in *The Competitive Advantage of Nations* (1990).[B170]

Earlier in Frank's trip: 'I could find nothing to admire very much at Lyons (because I could not gain admission to the silk manufactories)'.[J17] Lyon had long been renowned for its silk industry, which would have been of significant interest to him with his family's textile activities. He also slipped occasionally into the language of business. Musing about attractive women in Paris: 'the pretty ones [are] more modest because they discover by experience the value of modesty. The english pretty women are naturally modest but both seem to understand the market price of their Commodities tolerably well.'[C19]

He understood the principles of merchant business well, but it did not suit him personally. He had a retiring demeanour and lacked the influencing skills necessary in deal-making and other challenging business situations. He certainly would not have been able to stand up to his older partner Coventry and he was probably overly cautious after his early experiences with him. The exploitation that often accompanied successful business would also have been distasteful to him. He disliked the routine of business life too and prized the freedom to be able to explore his many ideas as he wished and in his own time. He harboured a hope that some of his ideas might help improve society.

Frank was in fact involved in a variety of business activities over much of his life but, after *Ronalds & Coventry*, these were of his own choosing. He developed the concept of a sulphur import business from Sicily and later manufactured and marketed his patented tracing instruments successfully for nearly two decades. His commercial skills were also invaluable in re-establishing and running the Kew Observatory.

What happened to other members of this merchant family in succeeding generations? Paradoxically, it seems that the more successful a business, the less likely it was that offspring would continue it. The children had sufficient education, money and status to follow other paths. For some it was desirable to be a 'gentleman' rather than to work, while others used their privileged position to pursue scientific, artistic or humanitarian goals. Daughters, if they wished, could remain independent and enjoy a comfortable life without marriage. Edmund was the only one of Francis and Jane's seven sons to spend his working life as a merchant — and he did not encourage his children to continue as cheesemongers. Two centuries after John Field came down to London, the prime Thamesside real estate had been sold and the Field/Ronalds family was no longer in the merchant business.

Chapter 3

SCENES IN THE STORY

A thumbnail sketch of key backdrops to Ronalds' life — his ancestry, homes, lifestyle and the lifelong influence and support of his extended family.

Frank's story has numerous scene changes, just as it has many characters. His mother Jane set up home at nine different places in the years after she was married. It was quite common for merchants in the City of London to acquire a 'country house' as their family and wealth grew. Jane did not settle in one such place, but lived to the north, south and west of London as well as in the City itself (Fig. 3.1). Frank took each of these moves with her, apart from the first one before he was born. Most of the locations Jane chose were nearby to either her relatives or those of her husband Francis. As a result, the extended family retained close linkages and had important influences on each other through their lives.

Staying at the seaside was a popular pastime, for pleasure or when convalescing after an illness. Jane and her family would rent a house for a month at Brighton or Hastings on the south coast perhaps twice a year. Jane did not retire to Brighton like her brother George Snr, but her daughter Emily did and Frank himself spent his last years just a short distance from Hastings. Frank had other travels too, both for business and pleasure. He often stayed in the City on business and in numerous trips he explored much of England as well as the continent.

Jane and Frank were not the only members of the extended family to enjoy a change of scenery. Jane's paternal forebears had come down to London from Hertfordshire. Frank's paternal grandparents, 'Old Hugh' and Mary, moved from Scotland and Coventry respectively to settle in Brentford. Frank's brother Alfred migrated to Australia and another brother Hugh set up home in the USA, as did a first cousin and a nephew (Table 3.1). Five of his brother Edmund's children went to New Zealand, and two of his first cousin Dr Henry Ronalds' sons to Canada

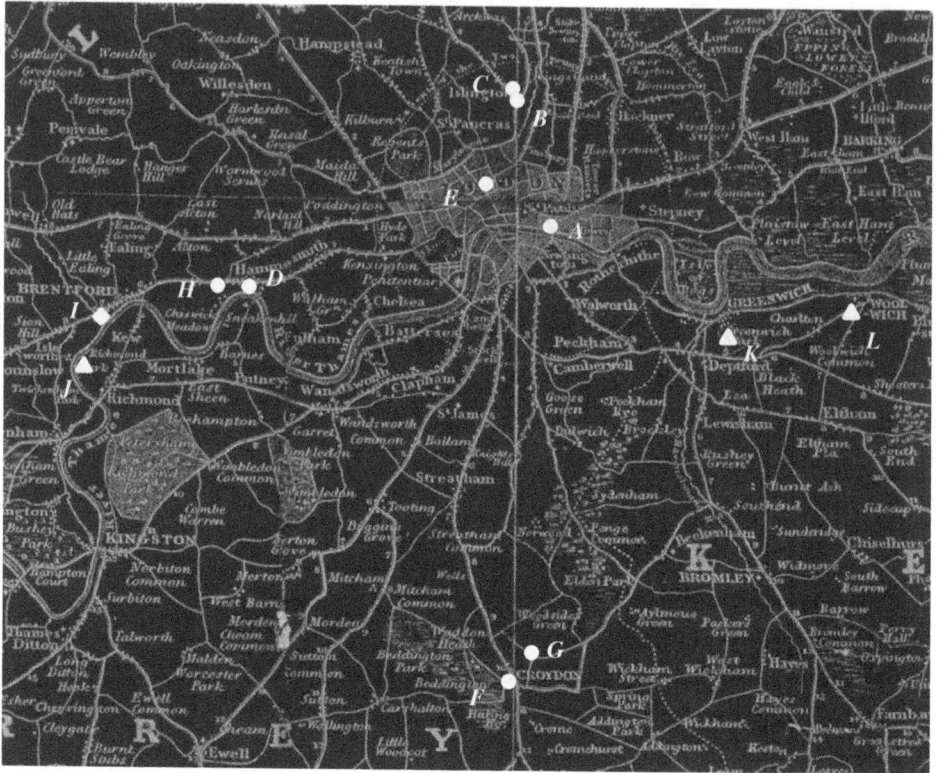

Fig. 3.1: Map of London and surrounds showing Ronalds homes and other important locations. Modified from LMA 30497 (1833).

A City of London, B Canonbury, C Highbury, D Hammersmith, E Bloomsbury, F Central Croydon, G Croydon environs, H Chiswick, I Brentford, J Kew Observatory (Chapter 13), K Royal Observatory, Greenwich (Chapter 13), L Royal Artillery Barracks, Woolwich (Chapter 13).

(Fig. 3.2). The exodus to the New World was so complete that this branch of the Ronalds name probably disappeared from the UK four generations after Old Hugh adopted it.

The intention of this chapter is to outline Frank's key physical and social surroundings. We start with important locations in the Ronalds family's history — Scotland, Coventry and Brentford — and then visit each of his homes. His overseas travels are detailed later in the book, while the emigration of his brothers and their children is summarised in the next chapter.

3.1 Ronalds Scottish Ancestry

Frank's paternal grandfather has long been known as 'Old Hugh' by the Ronalds family and regarded as their 'founder'. It turns out that he was born at Kilmarnock on 7 February 1725, with his baptismal record noting carefully: 'the child being begotten and born two months before the tyme'.[J2] His recently-wed parents were John Ronald (no 's') and Jean née Fergushill of Kilmaurs.[J2]

Kilmarnock lies in the Scottish lowlands, 35 km SSW of Glasgow. From Kilmarnock it is 3 km to Kilmaurs and then 2 km to Rowallan Castle,[B30] where John was gardener.[J2] The estate exists today as a golf course. The Fergushill family had long lived in the region and held a barony until heiress Jeane Fergushill died in 1658.

After Old Hugh had mastered the family profession, he moved south. This was not unusual — Scottish gardeners were highly regarded in England at the time. His close friend William Aiton Snr also migrated and went on to establish the future Kew Botanic Gardens for the royal family. Henry Jnr (Fig. 3.3), son of Frank's cousin Dr Henry, hinted at another reason for him leaving Scotland:

> Hugh Ronald
> Born in Scotland near Kilmarnock in Ayrshire in the year of our Lord one thousand seven hundred & twenty six on coming to England somewhere about the time of the rebellion he added the letter S to the original name Ronald thereby changing the name to Ronalds.[J91]

More remarkable than the move itself was that Old Hugh had the means to establish his own nursery in England. Another sign of the family's position was that his will included the gift of 'my red Cornelian seal with my Coat of Arms and my Father's two razors' and 'my white cornelian seal with a cypher and crest' to his two eldest sons.[W5] Henry Jnr's record of the armorial bearings notes their source to be 'an engraving which was taken from the old Family seal by my cousin Francis Ronalds [Frank]'.[J91] These arms belong to the MacDonalds of Clanranald, a highland clan that strongly supported Bonnie Prince Charlie in the ill-fated Jacobite uprising of 1745. 'Donald Mcdonald of Maydert, Captain of Clanranald'[J1] matriculated a coat of arms in 1672, surmounted with a crest of a castle with a single tower on a torse (or wreath). A later Captain and Chief, Reginald George Macdonald of Clanranald, registered revised arms in 1810 with a crest including a triple-towered castle.

Frank used both of these coats of arms, as did the rest of the family.[A196] A copper bookplate (Fig. 3.4) survives in the Ronalds Archive at the Institution of Engineering and Technology (IET) that features a single-towered castle; he would have engraved and printed it based on Old Hugh's seal when he started collecting books on electricity. Even after his knighthood he did not establish his own armorial bearings, but continued to use the new

Table 3.1:   Descendants of 'Old Hugh' RONALDS and Mary née Clarke: Frank's Paternal Grandparents.

1755–1852	1785–1895	1811–1938	
Children	**Grandchildren**	**Great-grandchildren**	**Destination**
Mary			
Jane			
Henry			
Hugh (Jnr) *m* Elizabeth Clarke	Hugh *m* Elizabeth Barclay		
	Mary		
	Elizabeth (Betsey)		
	Francis		USA
	(Dr) Henry *m* Elizabeth Lucy Robertson	Henry (Jnr)	CANADA
		Lucy	
		Hugh	
		Eleanor	
		William	CANADA
		Francis	
		Mary	
		John	
		Malcolm	
		Eleanor	
	John		
	Silvanus		
	Matthew		
	George		
	Robert		
John			
Francis *m* Jane Field	Jane (Jnr) *m* James MONTGOMREY (Snr)	James (Jnr)	
		Ronald	
		Jane	
		Susanna	
		Francis	
		Hugh	
		Gertrude	
		Malcolm	USA
		Elizabeth	
	Francis (Frank)		
	William		
	Edmund *m* Eliza J Anderson	(Dr) Edmund	
		Eleanor	
		Mary Ann	
		Eliza	
		Charlotte	

(Continued)

Table 3.1: *(Continued)*

1755–1852	1785–1895	1811–1938	
Children	**Grandchildren**	**Great-grandchildren**	**Destination**
		Marion	NEW ZEALAND
		Janet	ALGERIA
		Francis	NEW ZEALAND
		Hugh	
		James	AUSTRALIA
		Ellen	
		Eliza	NEW ZEALAND
	Hugh *m* Mary (Kate) Flower	Mary	USA
		Hugh	USA
		Emily	USA
		Jane	USA
		Francis	USA
		Richard	USA
	Mary Ann *m* Peter MARTINEAU	Francis	
		Catherine	
		Mary Anne	
		Sarah	
		Hugh	
	Emily		
	Charlotte		
	Charles *m* Catherine Fisher	William	
		Julia	
	Nathaniel		
	Alfred *m* Margaret Bond	Maria	AUSTRALIA
		Alfred	
		Francis	AUSTRALIA
		Nathaniel	AUSTRALIA
		John	AUSTRALIA
		Charles	AUSTRALIA
		Margaret	AUSTRALIA
		Hugh	
	m Mary Ann Harlow	Alexander	AUSTRALIA
		Mary	AUSTRALIA
		Eliza	AUSTRALIA
		Julia	AUSTRALIA
	Maria *m* Samuel CARTER	Alexander	
		Hugh	
		John	
		Jane	
Silvanus *m* Charlotte Cuncliffe			

Fig. 3.2:   Frank's cousin Dr Henry Ronalds (1790–1847) aged 51; portion of oil on canvas by Charles Langley (1842). Eldon House, Tara Wittmann photography.

Fig. 3.3:   Dr Henry's son Henry Ronalds Jnr (1821–63) aged 21; portion of oil on canvas by Charles Langley (1842). Eldon House, Tara Wittmann photography.

Fig. 3.4: Frank's early copper bookplate with single-towered castle. IET 1.11.

Clanranald coat of arms. His portrait in the National Portrait Gallery depicts these arms (Fig. 3.5), and books in the Ronalds Library are labelled with either the entire arms (Fig. 3.6) or the triple-towered crest. The Ronalds genealogy has also been preserved carefully by all the different branches of the family. Each version starts with Old Hugh and his wife Mary, with nothing earlier apart from the Clanranald arms.[J91–J93,J98,J99]

It was incorrect for the family to use these arms — they did not have the surname Macdonald for a start — but doing so illustrated their universal interest in preserving what they understood to be their heritage. Such errors were quite common in those times as there was imperfect knowledge of the rules of heraldry. Rather than record that he was a clansman, it had perhaps been Old Hugh's intention to show his allegiance to Clanranald, having participated in the rebellion — it has been a family tradition ever since that he did so. Alternatively he and his family might simply have taken an unconventional interpretation of Clanranald (or 'Clan Ronald'[J93] as Frank's sister Emily recorded it) based on his name and Scottish origin.

Fig. 3.5: Frank aged nearly 80, with Clanranald coat of arms in corner; old photo of Hugh Carter's oil painting (1867). Ref. T17.

Today's evidence suggests that the Ronald family were not highlanders but had resided in Ayrshire for some time. DNA testing of a descendent of Frank's brother Alfred, down the male line, indicates kinship with the Cunninghams, who were an ancient family in the area. A John Ronald had been gardener to John Cunninghame at an estate having interrelated ownership with Rowallan.[W1] George Black's *Surnames of Scotland*[B145] advises that the appellation Rannalsoune had existed historically in Ayrshire. The name Hugh Ronald was not unique in the area in the family's time. Robbie Burn's 1780 poem 'The Ronalds of the Bennels' concerns a land-owner near Kilmarnock, whose son was christened Hugh. Finally, John's brother was another Hugh Ronald — he was gardener to renowned merchant and MP Daniel Campbell at his properties near Glasgow.[W1]

Of course it is also feasible that the DNA results reflect an interloper and the family's ancestors were indeed from Clanranald. We will probably never know the full story.

Fig. 3.6:   Frank's later bookplate depicting Clanranald coat of arms. IET Ronalds Library.

3.2   Coventry, and the Clarke and Carter Families

We travel now to Coventry, 150 km north west of London, and the heritage of Frank's paternal grandmother Mary née Clarke.[A196,A197] Mary's family had property in Hill Street near the St John the Baptist church in Coventry[W2] and was prominent in the running of the city. Her brother and two of her nephews all served as Mayor.[B168] They intermarried with another important family, the Carters — Samuel Carter Snr was Gaoler and his brother was Undersheriff (chief operating officer for the city); they held these influential positions for several decades.[B34,B155] The Carters' father was also Mayor on two occasions.

There was considerable interplay between the Ronalds, Clarke and Carter families in Coventry and London over several generations. Mary Ronalds née Clarke, her eldest son and one of her grandsons all lived their last years in

Fig. 3.7:   Almost certainly Frank's portrait of his cousin Betsey Ronalds (1788–1854) (*c.*1830); drawn with his perspective tracing instrument. WU X1593.

Coventry. Dr Henry's 20-year-old daughter wrote excitedly to her Aunt Betsey (Fig. 3.7) when the family arrived there in 1842:

> You cannot think my dear Aunt how very fashionable we are…We have already received our invitation to the Drapers' ball, a very aristocratic affair I believe and I understand all the young ladies are busy thinking of their dresses which are to be quite grand…we shall not want gaiety [C109]

Mary's second son Hugh Jnr and his wife Elizabeth née Clarke were cousins.[192] Her fourth son Francis (Frank's father) was apprenticed to his maternal relatives from the mid-1770s and then formed the partnership *Clarke & Ronalds Ribbon Weavers* in the City of London. Even closer to Frank, his sister Maria married the Gaoler's son and namesake. This Samuel Carter was elected to Parliament for the seat of Coventry in 1868.

Now that we have been introduced to Frank's more important elders and their abodes, we will explore what was the extended family's hub in his time.

3.3 Brentford: Heartland of the Ronalds Family

Our next stop is Brentford, which is situated on the north bank of the River Thames, 15 km west of the City of London. Although Frank never lived there, it

Fig. 3.8: Lamberts — Ronalds family home at Brentford (*c*.1888). Chiswick Local History Centre.

was an important place for him. His grandparents had established themselves there on their marriage in 1754. Old Hugh died a few weeks before Frank was born, but he would have remembered his grandmother who lived until he was eleven. Brentford remained the heartland of the family for 130 years.

Old Hugh, Mary and their family lived at 'Lamberts' (Fig. 3.8), which was situated on the south side of the High Street adjoining the vicarage of the Church of St Lawrence.[A197] Faulkner's *History and Antiquities of Brentford, Ealing, & Chiswick* (1845) noted that the house was 'undoubtedly of the Elizabethan period'.[B55] A 1889 article in Kemp's *Sketcher and Theleme* elaborated that the 'barge or vergeboard of the gable' was 'one of the finest specimens in England'.[A131]

Brentford turned out to be the ideal place for Old Hugh and Mary to have set up their home and nursery.[B157] Opposite the town, on the south bank of the Thames and easily accessible by ferry or bridge, was the royal palace and the newly created Botanic Garden at Kew. Brentford, in addition, was 'a place of great trade, being one of the greatest thoroughfares in the kingdom; stage-carriages passing every half hour in the day. There are also regular market-boats every tide'.[B8]

Over time, Ronalds' nursery blossomed to become the 'best known'[B164] in the area, with even wider global reach than the family's cheesemonger business.[A197] The *Gardeners' Chronicle* later reminisced that it 'acquired a great and, we might say, European reputation. Then it was one of the very best fruit nurseries in the kingdom, and young men sought admission there with as great desire as they would in these days to Kew *[Gardens]*.[A123] At peak, nearly 50 acres were under cultivation at six sites.

It was very soon after the Colony in Australia was settled in 1788 that Sir Joseph Banks first approached Frank's uncle, Hugh Jnr, to supply plants and seeds for its new inhabitants. Banks had accompanied James Cook on the voyage that 'discovered' the east coast of Australia in 1770. His orders seem to cover almost every type of vegetable, fruit, herb and flower that can readily be imagined, together with trees, shrubs, fodder plants, nuts, tobacco and numerous sorts of grapevine.[J106] He configured a plant cabin for the deck of the various boats and gardeners tended them on their long voyage.[A197]

Hugh Jnr (Fig. 3.9) was described in one of his obituaries as having 'a very extensive knowledge of the different branches of science connected with Horticulture'.[A45] He was an early Fellow and office-bearer of what became the

Fig. 3.9:   Frank's uncle Hugh Ronalds Jnr (1760–1833) aged about 50; oil on ivory miniature (*c*.1810). Eldon House.

Royal Horticultural Society. In 1831 he published his 'superb'[A45] book *Pyrus Malus Brentfordiensis*,[B35] with 'pyrus malus' being the species name for apples assigned by Carl Linnaeus (1707–78), the father of modern taxonomy. The illustrations were made by his daughter Betsey.[A198]

Two of Old Hugh and Mary's children had had offspring: Hugh Jnr and Frank's father Francis. It was Francis' branch that continued the international horticultural tradition into the late 20th century. He and his wife Jane shared a fondness for horticulture and established a large garden at their new home at Highbury. The first gift Frank sent home to his mother on his Grand Tour was seeds that might not be available in England. He was being characteristically self-effacing when he wrote: 'The Seeds I sent you are the very commonest kind I suppose, they will come up in the shape of lupins & Larks ferns, don't laugh at my Panorama of the "Beauties of Nature".'[C22] Jane also exchanged unusual plant seeds with family friend and likely relative Edward Clive Bayley, a merchant in St Petersburg.[C86]

Frank's brother Hugh, who settled in the USA, had 'two favourite pursuits, horticulture and literature'.[B103] Another brother Alfred, and his son, both established large nurseries in Australia. Overall, there were at least seven generations of Ronalds horticulturalists from their beginnings in late 17th century Scotland.[A197]

Despite emigration, Brentford continued to be a centre of focus for both Francis' and Hugh Jnr's families. Francis and Jane were married and later buried at St Lawrence's church. Frank inherited a freehold property from his father that was located on the north side of Brentford High Street adjacent to the bridge across the Brent. His older sister Jane Jnr wed a Brentford man by the name of James Montgomrey (Snr) and lived her married life there. Their descendants continued the family timber merchant business at 'Montgomrey Wharf' into the 1900s. Lamberts endured until soon after Hugh Jnr's youngest child Robert (Fig. 3.10) died there, an octogenarian bachelor, in 1880.[A196,1] Frank, his parents and siblings, had meanwhile enjoyed various homes, and we visit these properties next (Table 3.2).

3.4 Islington

Frank grew up in Islington, where his family resided for 24 years. Parents Jane and Francis had lived in the City of London only briefly after their marriage. Their first child Jane Jnr was born close to *Clarke & Ronalds Ribbon Weavers* and their second home was the premises of *Field & Ronalds Cheesemongers*. Around the time his brother William was born, and Frank was a year old, the family moved again. They resided initially in Canonbury Place, where William sadly died, but they welcomed further children Edmund, Hugh, Mary Ann and Emily. At

---

[1]Contents of the old Ronalds family home are now in the Harris/Ronalds Collection in London, Ontario.

Fig. 3.10:   Frank's cousin Robert Ronalds (1799–1880) aged about 70; oil on canvas (*c.*1870). Eldon House, Terry Fieldhouse photography.

Highbury Terrace the family was rounded out with the arrival of Charlotte, Charles, Nathaniel, Alfred and Maria.

Frank and his siblings would have had very happy childhoods in Islington. They lived close to their relations on the Field side of the family (Fig. 3.11). They enjoyed a large garden, and the boys would have developed their skill in bird-shooting in the fields surrounding the village. Having an elevated situation, the area was known for its healthy climate and 'bracing air'.[B40] Even today there are good views from Canonbury Tower and the upper floors of Highbury Terrace. Islington was also convenient, lying just 5 km north of the City, and Francis probably walked or rode his horse into business each day.

Perhaps the first family members to live in Islington were Jane's parents William Snr and Anna Field and their younger children, who moved there soon after Jane and Francis' wedding in 1785. An aunt and an uncle soon joined them. Three of Jane's married siblings as well as two of her first cousins later settled in the village with their families. Their proximity would have been of great comfort to her and the children when Francis died in 1806. Frank's cousins would have looked up to him as one of the oldest of their group and he would similarly have

Table 3.2: Residences of Frank and his Mother Jane.

Family Homes

Dates	Years	Town	Address	Acres
1785–87	2	London	Coleman Street	
1787–89	2	London	109 Upper Thames Street	
1789–96	7	Islington	11 Canonbury Place	
1796–13	17	Islington	1 Highbury Terrace	5
1813–20	7	Hammersmith	26 Upper Mall	1
1820–22	2	Bloomsbury	40 Queen Square	
1823–25	3	Croydon	Old Town	
1825–33	8	Croydon	Heath Lodge	26
1833–52	19	Chiswick	2 Chiswick Lane	1

Frank Elsewhere

Dates	Years	Town	Address
		Not born	
1818–20	2	*Travels — Europe, Eastern Mediterranean*	
1823–24	1	*Travels — Italy*	
1852–53	1	Charing Cross	Golden Cross Hotel
1853–53		Oxford	Observatory
1853–62	9	*Travels — Europe*	
1863–73	10	Battle	9 St Mary's Villas

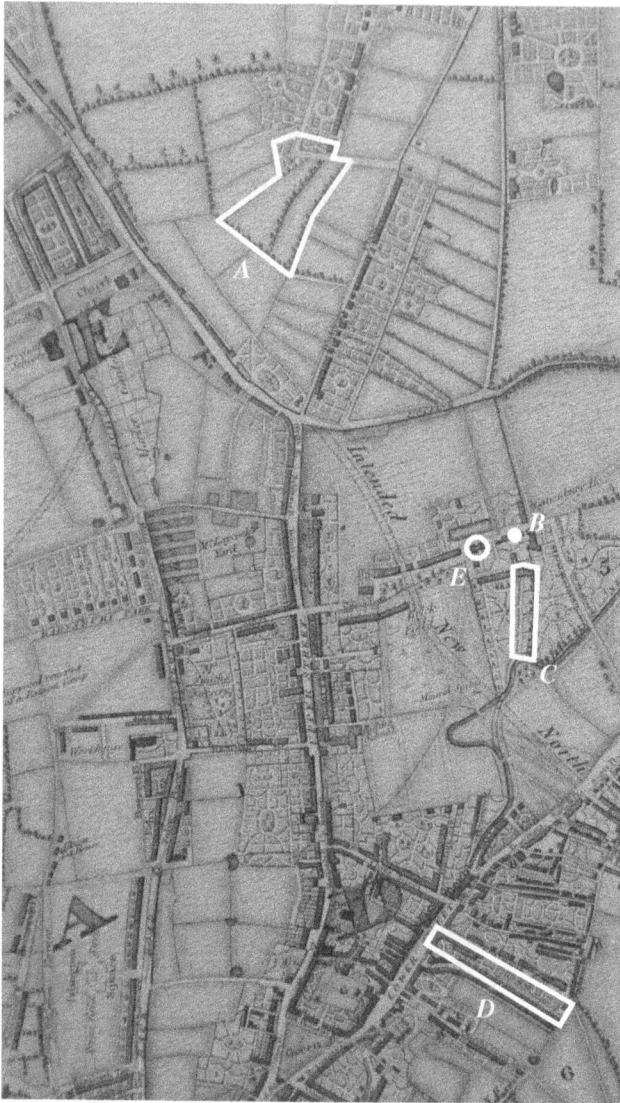

Fig. 3.11:   Map of Islington (*c.*1822) showing family homes.

A 1 Highbury Terrace, *B* 11 Canonbury Place, *C* 6 Canonbury Place (William and Anna Field), *D* Islington Prebend (Elizabeth Hardcastle), *E* Canonbury Tower.

idolised his youngest uncle Charles Field who was just eight years his senior. They were all good friends as adults, with Charles accompanying Frank on his trip to Carnac in France.

The family played an active role in the affairs of the village. When the war against France resumed in 1803 with Napoleon's threatened invasion of the UK, volunteer forces were set up across the country. Francis, together with Jane's brothers-in-law William Venning and Thomas Gibson, were three of the eight captains in the Loyal Islington Volunteer Infantry, which was funded by a local subscription.[2] Their uniform was 'a scarlet jacket turned up with black, light blue pantaloons, short gaiters and beaver caps'.[B40] Francis' brother Hugh Jnr was on the organising committee for a similar corps formed in Brentford[B55] and the latter's teenage son Hugh was an ensign.

William Snr and Anna had purchased 6 Canonbury Place, which was described in their lease agreement as being 'part of the Capital Messuage or Great Mansionhouse called Canbury *[sic]* House'.[J10] The property comprises the southern end of the east range of the manor house, with the seven-storey Canonbury Tower standing diagonally opposite at the north west corner of a courtyard garden. On the far side of the tower was a large fishing pond (Fig. 3.12). Canonbury Manor had been rebuilt in the 16th century, and William Snr and Anna's property still contains elements from this period. Particulars of the house and garden and the Fields' furniture and fittings survive in their auction notice, held at Columbia University.[J11] To the south of the manor were large 'pleasure grounds'[J8] extending down to the New River; maps from the period show these to be manicured gardens. Two of the octagonal summer houses at the corners of the manor's walls

Fig. 3.12:   Pond at Canonbury Tower (1827); 11 Canonbury Place is near the centre of the picture. Islington Local History Centre.

---

[2]*London Gazette* 10/9/1803 1196.

Fig. 3.13:   View north from New River (1824); 6 Canonbury Place is partially visible behind the property's summerhouse on extreme right. Islington Local History Centre.

remain today, one of which lay near the south east end of William Snr and Anna's grounds (Fig. 3.13). The house — a Grade II* listed building — is now an academy, although the former grounds have been built upon.

William Snr's sister Elizabeth Hardcastle moved up to Islington from the City shortly after the death of her husband Nathaniel. Her 'large and handsome' 'Mansion-house'[B40] lay in the Islington Prebend (which belonged to St Paul's Cathedral)[J9] on what is now Essex Road.

Jane and Francis' first Islington address was more modest than those of her senior family members, who had been very successful in business together. It was at 11 Canonbury Place, just around the corner from William Snr and Anna's, where the children would have spent much of their time. It was rebuilt in the mid-19th century.

Their second property, a kilometre north of Canonbury at 1 Highbury Terrace, was more substantial. It was one of a pair of double-fronted villas with coach houses and stables built in the early 1790s. As well as gardens extending to the mews behind, they also owned meadows and gardens to the side of the house, giving a total of five acres of land.[J8] In front of the house lay another large meadow, now called Highbury Fields (Fig. 3.14).

Fig. 3.14:   Highbury Terrace from Highbury Fields; No. 1 is on the far left. Islington Local History Centre.

The track along the south side of the house became a throughway in the 1880s and was named Ronalds Road in honour of Frank (Fig. 3.15). In 2013, a plaque was unveiled on the façade to commemorate his early science activities there. As Frank had quipped:

> chemistry was my chief amusement & my most memorable performance…was the blowing up of a large hydrogen gazometer *[sic]* in the breakfast room of No 1 Highbury Terrace. The immediate consequence of which clever achievement was the transfer of my studio to a small cock-loft over the Coach-house.[C375]

The home is now used for retirement housing.

In 1813, after 17 years at Highbury, Jane decided to move her family to Hammersmith. Her role as oldest surviving daughter in caring for the previous generation had come to an end the previous year with the deaths of her father William Snr and Aunt Elizabeth Hardcastle. In Hammersmith she would be closer to her daughter Jane Jnr, who had moved to Brentford after her marriage in 1810 and already had two sons. Jane kept one Highbury Terrace for the rest of her life but never returned to live there.

Not surprisingly, given their happy childhood memories, several of her children came back to Islington after they married. Mary Ann lived at 16 Highbury Terrace for over 60 years; her father-in-law Peter Martineau Snr had purchased 6 Canonbury Place after William Snr's death. Edmund and his new wife Eliza moved to Canonbury Lane in 1819. Frank joked to Emily at the time: 'I am glad he seems comfortably fixed with a house, I suppose he pokes the fire with great importance

Fig. 3.15:   Facade of 1 Highbury Terrace in 2014 with Ronalds Road sign on the left and green commemorative plaque to Frank.

and shoves round his port exactly in time & tune and remarks very scientifically upon it.'[C22] Edmund and Eliza returned to Canonbury in the 1830s and resided in a substantial stand-alone house.

3.5 Hammersmith

The hamlet of Hammersmith was 10 km from the City of London and 5 km east by road (or one loop of the Thames downstream) from Brentford. Jane chose a home on the Upper Mall, right on the north bank of the river (Fig. 3.16). Today it is the best known of all of the family's properties. It was described in the 1915 *Survey of London* as a 'very charming house'[B132] and another occupier, William Morris, wrote that 'the situation is certainly the prettiest in London'.[C409] Morris, a famous textile designer, poet, author and socialist, later named the property Kelmscott House.

The house had been built in the late 1780s and is a substantial edifice. There are five bays along the three main floors and also a basement for the servants and a fifth level with dormers. To the rear was a 180 m long garden perambulated by a gravel walkway and described by Morris as 'really most beautiful'.[C409] It has since been halved in length by the construction of the Great West Road. The front is still

Fig. 3.16:   View across the Thames to Hammersmith Upper Mall in 2012; white rectangular plaque honouring Frank's telegraph is visible near the top of the former stable on the far left.

framed by a curve of the river wall bowing into the Thames, a sign of the importance of the property. The building is today Grade II* listed.

At Hammersmith the family was able to enjoy more time with their Ronalds cousins. Frank's brother Charles wrote to Robert when both lads were 14:

> we shall go into the house at Hammersmith at michælmas *[29 September]*, when if you come home *[from boarding school]* we can work hard at the ship…You can get the mast as I think she had better have but one. She is a great deal too *p*eep *[the 'd' in deep being written upside down for emphasis]* for her size which am afraid cannot be well altered but she will carry 12 or more cannons very well which I will get.[C2]

The house holds importance for several members of the family. Supplementing Charles' plans for sailing model ships on the river, it is easy to imagine Alfred developing his interest in fly fishing there. Frank was also a fisherman; he acquired a boat and made more than one analogy between fishing rods and the atmospheric electricity conductor he developed at Hammersmith, as described in Sec. 6.4:

> the light *[conductor]* rods…and their lanterns are, in fact, but little more cumbersome than a long fishing rod, and its line.[C328]

It was from this idyllic setting that Hugh decided to leave to build a pioneer settlement on the prairies of Illinois. The family's best-known activity at Hammersmith however was the large-scale demonstration of Frank's electric telegraph. He had quickly 'fitted up...a more elegant hay-loft'[C375] above the adjoining stable to continue his experiments. He also had his own study, which would have been his other favourite place in the house. Jane showed even greater understanding in allowing him to dig a trench almost the full length of her beloved garden to install the wire for his telegraph in 1816.

Mary Ann and Edmund were both married in Hammersmith, and Frank embarked on his Grand Tour shortly after Hugh emigrated. It was perhaps the family's shrinking (if not the mutilated garden!) that determined Jane on another change of scenery after seven years. In a summary of letters received from his family in Sicily, Frank noted:

> My Mother.
> Mentions her idea of moving, undetermined whether to chuse [sic] Town or Country or between both.[J23]

He did not mind and wrote to Charlotte:

> I can only say I wish to have no other influence in the matter than in what respects the general harmony of my "dear family". You know my opinion about expence [sic], I have discussed it in your ears too much perhaps, believe me. I really don't regard <u>myself</u> in the matter at all [C24]

Charlotte would have been worried about his electrical apparatus, but his comments show that he had no plans to continue his telegraph experiments with the many metres of buried cable. He did care about his workshop equipment, however, writing to his mother:

> Be so good as the winter is coming on "<u>in your pants</u>" to let my lathe and tools <u>particularly</u> the slide rest be sent to Holzappfel's[*-3] [sic] in Cockspur Street [at Trafalgar Square] for him to keep for me, they will I fear be entirely ruined else.[C25]

On their departure from Hammersmith, he also enquired, 'Did you have my shelves & glass cases which contained the apparatus removed, they cost a good deal & if left should be valued as fix [missing].'[C38]

A 20th century owner, Marion Helena Stephenson, bequeathed Kelmscott House to the William Morris Society, with the contents of a museum located in the coach house and basement. A section of Frank's insulated telegraph wire restored from the garden is held there, along with Frank's bust, and there is a tablet on the façade commemorating his telegraph. How the plaque and bust got there is described in the final chapter.

---

[3] Asterisk * indicates brief biography of the person is provided in the Appendix.

3.6 Bloomsbury

When Frank returned home in late 1820 from his Grand Tour, the family had settled into Queen Square in the heart of Bloomsbury. Jane had chosen to come into town. Charles ran his business as a solicitor here, and it would also have been convenient for Nathaniel who was now in partnership with Edmund at Upper Thames Street. The square had been constructed in the early 18th century; their home was situated on the south east corner and was again five storeys including dormers and basement. The middle of the square formed a large garden (Fig. 3.17).

Evidently Queen Square did not suit Jane because the family stayed only two years. She would have regretted not having her own garden. Sixty years later, the house became part of the Italian Hospital and was replaced with a purpose-built building at the end of the century. It is now part of Great Ormond Street Hospital.

3.7 Croydon

The next move was to Croydon, a town 14 km south of the City. Rate books show that Jane's first home there was a substantial property in the Old Town comprising

Fig. 3.17: View south down Queen Square (1812); Jane's house No. 40 is in the centre distance on the street corner. LMA 8236.

Fig. 3.18:   Frank's perspective sketch of the family's first home at Croydon (*c*.1825); the woman seated in the foreground would be one of his sisters. IET 1.7.40.

offices, a garden and land as well as the house (Fig. 3.18). It seems likely that it was associated with the Old Palace, which had been a summer home for the Archbishops of Canterbury since the 13th century. The palace had been sold in 1780 and for a time became private residences. The principal buildings are now Grade I listed and serve as a school.

As with Jane's other location choices, there was good road access to the centre of London from Croydon. Close friend Sir Frederick Henniker* must still have lamented Frank going so far afield, because Frank wrote to him in November 1822:

> You are very good to say that you hope I have not left London. I will not be so egotistic as to express any regrets to any of my friends on the occasion but I do not intend to neglect [?] opportunities of continuing my London acquaintance.[C48]

Frank had multiple business interests at this time that took him to Birmingham and even abroad as well as to London. He gave a London address in the preface to his book on the electric telegraph and stayed frequently at the London Coffee House on Ludgate Hill when in the city for business or for social activities like the opera. Joseph de Gourbillon*, his travelling companion in Italy, denoted coffee houses as 'a place to receive persons who never visit the public-house'.[B14] Frank also spent eight months in Italy on a business trip in 1823–24. On his return he leased a property in Upper Church Street, just around the corner from the Old Palace. This was almost certainly a business address for him as he had started to manufacture drawing instruments.

Jane moved house again in 1825 at what would have been a sad and lonely time for her. Charlotte had died a few months earlier. Emily was in America visiting Hugh, Nathaniel would have been residing in the city most of the time and Charles had just married. Frank took Maria to Paris for six weeks, probably soon after Emily returned home, to help them recover from the loss of their favourite sister. Alfred would still have been at home, but he had moved up to Lichfield (not far from Coventry) by 1829 and wed in 1831.

Jane chose a large estate called Heath Lodge. The entrance was at what is now the intersection of Lower Addiscombe and Morland Roads, 2 km north east of the Old Palace in Croydon. A sweeping drive led past the large fish pond, complete with island, up to the main house and garden with offices, stables, stockyards and lawn behind (Fig. 3.19). Further on was a cottage with kitchen garden and an orchard. The whole was surrounded by meadows to give a total of 26 acres.

HEATH LODGE, CROYDON

Fig. 3.19: View over the pond to the main house at Heath Lodge (1833); island with trees is on the far right. Courtesy Museum of Croydon M/98/31.

It might have reminded Jane of happier times in Islington, and it turned out to be a forerunner of the estates several of her younger relatives later acquired in the southern counties.

Jane enjoyed Heath Lodge for eight years. The two-storey house, which had been built in the early 1790s, was demolished in the mid-20th century.

3.8 Chiswick

Jane's final move was to Chiswick at the age of 66, where she spent the last 19 years of her life. The three-storey house was attached to an acre of gardens. Built in the 18th century, it is Grade II listed and now part of a convent (Fig. 3.20). Brentford, Hammersmith and the intermediate Chiswick were all linked by the High Road into London that ran adjacent to her home at the top of Chiswick Lane — Jane was back close to family at Brentford and friends at Hammersmith.

The Nichols* family in Hammersmith were welcome visitors, as documented in numerous entries in daughter Mary's diary: 'Harriett and I drank tea with Mrs. Ronalds at Chiswick Met the Misses and Masters Montgomery *[sic]*, 2 Misses Geary, and Miss Watson.'[133] Jane would have played a significant role in caring for her younger Montgomrey grandchildren. Her eldest daughter Jane Montgomrey

Fig. 3.20:   No. 2 Chiswick Lane in 2012.

had died in 1827 and her husband struggled increasingly after her passing: mention was made in family letters of his need for 'healing of a wounded spirit', 'he is certainly a true subject for pity and regret'.[C72] In 1837 the courts determined him to be 'of unsound mind'.

Another grandchild, Charles' daughter Julia, also received special attention — as a baby she had lost both her mother and her only sibling. The Brentford cousins tell us:

> My Aunt with Frank, M[r] Martineau, Charles and Julia dined with us yesterday…they have left Julia to spend a few days with Mary *[her second cousin of similar age]* and we are to take her back and dine at Chiswick on Wednesday. Tomorrow she is going to have a little party.[C95]

A few years later when Jane was aged 76:

> Aunt Ronalds is very well. She has just sent Julia off to school again & on Saturday last walk'd over to see us & was not quite sure whether she should walk back again to Chiswick.[C110]

There was also loss and celebration in the immediate family. Jane's son Nathaniel had died soon after she moved and, like Charlotte, was laid to rest in the family vault at St Lawrence's church. Her 'baby' Maria was married shortly afterwards. This left just Frank, Emily and Julia living with Jane in her last years, together with a cook, housemaid and footman.

A typical family holiday at the seaside at this time is recorded by diarist Henry Crabb Robinson*, who was also spending time at Brighton. On 20 November 1841:

> In the evening I went by self-invitation to M[rs] Martineau *[Mary Ann]* & M[rs] Ronalds *[Jane's rented house]* — M[r] M: and M[r] R: (the son) *[almost certainly Frank]* were there and we had an agreeable chat de omnibus *[wide-ranging]* — And also a rubber of Whist[J13]

He also recorded numerous family get-togethers at the Martineaus' Highbury home:

> I went to dine at Peter Martineau's where was a family party of Fields, Ronaldsons *[sic]*, Wild*[e]*s *[Peter's brother-in-law]* & Martineaus. And this was on the whole as agreeable as such parties generally are[4]

The three properties Jane still owned — at Upper Thames Street, Highbury Terrace as well as Chiswick Lane — were all sold after her death in 1852. Frank, being of age 64, used the change to start his gradual retirement from the nearby Kew Observatory (Fig. 3.1), which he had been running for a decade. He stayed variously at the Golden Cross Hotel at Charing Cross, with Uncle Charles Field in St Marylebone (both in London), and in Oxford (on observatory business) for a year before embarking on a nine-year sojourn to France, Switzerland and Italy.

---

[4]Ref. J13 8/12/1842.

3.9 Battle

On returning to England in 1863, Frank settled at the village of Battle, situated 8 km inland from the coast at Hastings. He lived in (but did not buy) a relatively modest semi-detached house with a small garden on a new estate (Fig. 3.21) — St Mary's Villas were conveniently located close to the railway station that had opened in 1852. He had one servant.

Once again, family was nearby. Maria and Samuel Carter owned a large house and farm just half a kilometre further out of the village.[B84] There was a 'Field colony at Hastings',[B117] while Emily spent her last years in Brighton, in the same street as four other Field cousins she had grown up with in Islington. Brighton is 45 km from Battle and was by then readily accessible by rail via Hastings. Charles had died and his daughter Julia lived with Frank in his last years. Alfred had also passed away after migrating to Australia. Mary Ann remained in Highbury and Edmund was just a few kilometres up the road from her in Highgate. The family continued to support each other until the end.

Fig. 3.21:   Frank's last home at 9 St Mary's Villas, Battle (on right), in 2012.

Chapter 4

FRANK AND HIS FAMILY

Description of a Unitarian family — their upbringing, values and interests and the pivotal role of Ronalds' mother in assisting her children's disparate and sometimes unconventional endeavours. Ronalds' physical appearance and personality traits and an introduction to each of his siblings.

Frank's father Francis Ronalds was aged 23 and Jane Field just 18 when they married at St Lawrence's church in 1785. Jane bore 12 children in less than 18 years, with Francis dying when the youngest was 2 and Jane had just turned 40 (Figs. 4.1–4.2).

We start this chapter with some comments on various facets of the family's situation — their religion, education, political and social attitudes, and wealth — before outlining what life at home was like for the siblings. Next follows a snapshot of Frank's personal characteristics and finally brief sketches of each of his brothers and sisters and their own families. Family letters and journals written by Frank on his Grand Tour are a key source for these descriptions, which are therefore rooted in the period 1818–20 when the offspring were young adults (Table 4.1) — other surviving personal correspondence is much less 'frank'. Jane was her children's most important influence up to that time and consequently we also gain considerable insight into her talents and personal attributes.

4.1 Religion

Frank's family and many of their friends were Protestant Dissenters. Their religion influenced all aspects of their lives: where their birth was registered, how they were educated, jobs they could hold, whom they married and where they were buried.

Fig. 4.1:   Frank's mother Jane Ronalds née Field (1766–1852); reproduction of miniature. Sarah Crowley, descendent.

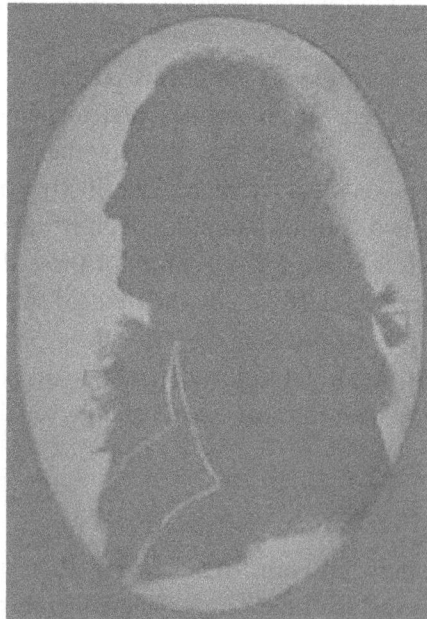

Fig. 4.2:   Frank's father Francis Ronalds (1761–1806); reproduction of silhouette. Sarah Crowley, descendent.

Table 4.1:   Ronalds Family: Particulars at Time of Frank's 1818–20 Grand Tour.

Name *Nickname*	Age (1818)	Abode	Spouse	Children	Comments
Mother *M^{rs} M*	52	Hammersmith → Bloomsbury			
Siblings			**Spouse**	**Children**	**Comments**
Jane (Jnr)	32	Brentford	James Montgomrey	1811 James 1812 Ronald 1814 Jane 1815 Susanna 1817 Francis 1819 Hugh	1810 Married
Francis *Frank*	30				On tour
Edmund *Ned*	28	Canonbury	Eliza Anderson	1819 Edmund	Cheesemonger 1818 Married
Hugh	26	Illinois, USA	Kate Flower		Tanner 1818 Emigrated to America 1820 Married
Mary Ann *Brăss*	24	Highbury	Peter Martineau *Squirrel*	1819 Francis	1816 Married
Emily	22	Family home			
Charlotte *Whig*	21	Family home			
Charles *Vicar*	20	Family home			Solicitor
Nathaniel *Nat*	18	Family home			Apprentice Cheesemonger
Alfred	16	?			Apprenticed
Maria *Muffin Woman*	14	Family home			
Pet					
Dido		Family home			Dog (female)

Specifically, the Ronaldses adopted the Unitarian faith, which gained some popularity amongst the commercial and industrial classes in the later 18th century, although remaining illegal until 1813 (when Frank was 25). Unitarianism is so named because of the belief in one God rather than the Trinity of the Father, the Son and the Holy Spirit. It combines scripture with rational thought and scientific enquiry in understanding the world and God, rather than having a basis in ritual and mystery. There is also significant emphasis on broad and practical education, personal improvement, liberty and equality and duty to society. The extended Ronalds family's activities and interests in many ways epitomise these tenets. The obituary of William Snr's first cousin John Field, for example, noted him to be 'an example of sincere religion without enthusiasm, economy and frugality devoid of meanness and parsimony, diligence and success in the affairs of life untainted with avarice, and every where accompanied with the strictest integrity and uprightness.'[A1]

Frank explained Unitarianism in his typical practical style while studying the art in Vicenza:

> [Jacobo] Bassano is a Unitarian Painter. He makes Mary a good wholesome Dame, Joseph a country lad[?], & Jesus a chubby little fellow always ready for the bottle & the society of this holy family generally consists in a conclave of Cows & Pigs... Bassano studied nature so long and so very successfully that he did not care to leave natural for supernatural objects. Truth for Mystery. However he was perhaps rather too fond of Cows & Pigs even for the age he lived in.[J22]

His father demonstrated a similar 'down-to-earth' philosophy — Francis' will started:

> I desire to be buried in a vault [the existing family vault] in the church yard of New Brentford in the County of Middlesex and that my funeral may be devout but without any ostentations pomp or superfluous Expense [W6]

Frank went one step further, writing in Naples:

> Ancient Tombs. It is surely a much more decent and polite method of taking leave of our fellow creatures to be burnt than to be encumbering the earth with one's nasty carcass after death. I desire if should die without expressing it in my will that my executor will have me burnt.[J18]

This desire seems to have been forgotten when he passed away 54 years later.

The family's elders — the Fields as well as the Clarkes in Coventry — were all nonconformists. Old Hugh Ronalds was a leading member of the Society of Protestant Dissenters in Brentford.[J3] His name is on the 999-year lease for their chapel on what is now Boston Manor Road, where he contributed to the building fund and purchased two pews. His children and grandchildren in Brentford were also trustees. Robert recorded that his grandfather 'brought with him all the zeal and persuasions belonging to the Kirk [Church], which to Scotch people was their whole life and soul at that day.'[J98]

The births of Frank and his siblings were registered in Dr Williams's Library and they did not have Anglican baptisms. They went on to marry Protestant Dissenters, with the exception of Alfred's first wife Margaret, who was Catholic.

The Ronalds family's links to St Lawrence's in Brentford are explained by the requirement for weddings to be performed in an Anglican church and because many dissenting places of worship had no graveyard. Other family members like William Field Snr and Frank's baby brother William were buried at Bunhill Fields, an unconsecrated burial ground a little north of the City of London. Hugh Jnr provided landscape design and many trees for the new Kensal Green cemetery — the first commercial cemetery catering for both Anglicans and Dissenters.[A181] As late as 1860, when Alfred died in Ballarat, the Church of England minister in the settlement declined to read a burial service, resulting in a flurry of letters for and against his decision to the newspapers.

Nonconformists could not graduate from the two English Universities, Oxford and Cambridge, and were barred from public office. Frank's slightly younger cousin Dr Henry was the first family member to obtain a university education, which he achieved by studying in Edinburgh. University College London (UCL) was founded in 1826 as the first secular university in England. Dr Williams's Library is now housed at the adjacent University Hall, which was built as a memorial for an 1844 Act of Parliament that enhanced legal toleration of Unitarians; the reform was spearheaded by Frank's third cousin and family friend Edwin Wilkins Field and his statue is in the Library.

Comments made by Frank at various places he visited further illustrate his religious upbringing, including his interest in exploring other beliefs and the natural questioning of his own. In Padua:

> How many idle vagabonds in these parts are paid ten times as much *[as a University professor]* for wearing a gown and living at their ease because they profess to teach a quantity of dogmatical nonsense called Religion & Theology.[J22]

Rome:

> I saw a Cardinal kneel down close by the side of dirty fellow whom very few of the lowest Rank of Methodist Parsons in England would venture to approach within a yard of.[C21]

After witnessing the miracle of a saint's blood liquefying in Naples:

> I fell insensibly into a comparison of Unitarianism & Catholicism but I actually came to the conclusion that the former is the best belief because if at any time one should find oneself cheated by the Book, and the Parsons, it is a less bore to be cheated in an affair relating to reason than in one relating to imagination & enthusiasm.[J18]

For him, the most incredible part of the miracle was its acceptance:

> the wide spread belief in it amongst all ranks did not appear perhaps to him *[travel writer John Eustace]* a catholic half so miraculous as it does to me a Unitarian born, bred and educated and cushioned[?] with the ridiculous prejudices of my Country.[J18]

In Constantinople:

> I have felt very <u>lone</u> all the time I have been in Turkey in <u>religious</u> affairs…If it were not for sociabilities' sake I would not <u>profess</u> any particular creed whatever. I fell into some comparison between Unitarianism & mohammedism & I was not surprised to find that we agreed on a great many points [J21]

while what he saw in Jerusalem disgusted him:

> It don't much matter whether the right kind of respect is inspired or not to the priests, indeed as they themselves with their gorgeous robes of sanctity form a part of this theatrical representation…perhaps they are as well pleased with admiration at the splendour of the chapel as by any other sentiment of their flocks…talk of the treasures & magnificence of their Churches and you will see their eyes glisten with delight. [J20]

> The Holy week may be passed in Jerusalem with no small gratification of curiosity in witnessing the various fooleries, contentions, and habits of the Pilgrims &c &c — but it is mixed with some pain to see the Tomb of the prince of peace made the very spot chosen for exhibiting the most bitter & <u>malicious</u> sectarianism that ever disgraced Christianity. [J25]

4.2 Education

We know that Frank attended two schools. The first was Revd Dr Nathaniel Phillips' establishment in Walthamstow, located about 6 km from Islington, and it was followed by Revd Eliezer Cogan's boarding school at Cheshunt, further up the Lea Valley and about 18 km from home. Cogan's daughter later married Frank's first cousin Thomas Field Gibson. [B105] The other boys had a similar essentially Unitarian education. Charles attended Revd John Potticary's boarding school in Blackheath [C2] and the Brentford cousins boarded in the Uxbridge Academy run by Revd Thomas Beasley.

There is considerable evidence that they learnt the classics, French, literature, art, history, geography, music, dance, as well as some science and mathematics and skills for their future livelihood. [B105] Hugh Jnr wrote to son Robert at age 14: 'I send the Instruments as you desire, which hope you will make good use of. I am glad Mr Beasley is going to try you in Mensuration *[measurement]*'. [C3] Revd Cogan was highly regarded as a Greek scholar. Dr Henry wrote his dissertation for his medical degree in Latin [B11] — copies are held at the Wellcome Trust and the British Library. Robert wrote letters home in French to assist with his studies, [C1] before translating part of Alain Lesage's novel *Histoire de Gil Blas de Santillane*. He also transcribed his favourite poetry into his journal, while Dr Henry's son Henry Jnr went one step further and composed verse. It is remarkable how much could be picked up before their apprenticeship commenced.

Unfortunately there is little record of the girls' education beyond that it worked. Dr Henry's future wife attended boarding school from the age of six and had additional tuition by 'Masters'[J7] in music, drawing and French. She was both orphaned and an only child; it is more likely that Jane had a governess for her brood. Always interested to learn new things, Frank's sisters later supplemented their French with Italian or German.[C106]

Girls were also taught the skills to run a large household, including budgeting and accounts, purchasing and staff management. When Henry Jnr decided to migrate to Ontario at age 21, it was his Aunt Betsey who provided much of the sound personal and business advice:

> I should hope you will not expend all the money that you have rec[d] but keep some of it in the Bank & receive the interest of it, there will be a great deal of pleasure in saving up a little...let every body know that you have got a start from home & that you are come to make yourself a respectable Citizen...Keep all Flatterers & Sychophants *[sic]* at arms length...Industry is the parent of wealth & Idleness of Poverty. But kindness is always felt & esteemed a kind thought[C122]

4.3 Political and Societal Attitudes

The long political and social persecution suffered by the Dissenters ensured that they were comfortable questioning the status quo and seeking change. Emily stands out within the Ronalds family in this regard, forming friendships with early socialists Robert Owen,[B123] Fanny Wright[B176] and James Pierrepont Greaves[B180] that her siblings shared.[C59–C61] She helped plan and fund Fanny's experimental community for the emancipation of slaves near 'the little town of Memphis'[A163] in 1825 — 40 years before America abolished slavery. Anti-slavery and socialism were also important themes in her brother Hugh's settlement in Illinois.

The siblings had a role model for their pioneering work nearby in Joseph Hardcastle, partner in the Fields' cheesemonger business. Joseph, together with renowned abolitionists William Wilberforce and Thomas Clarkson, had become a Director in 1791 of an entity established for the resettlement in Africa of former American slaves; the approach was closely aligned with his efforts in founding the London Missionary Society.[B79] Meetings were held at the Old Swan Stairs property at Thames Street. Fanny's approach was rather more radical and included the renunciation of religion and marriage and the promotion of 'the physical amalgamation of the two colors'.[1]

---

[1]Explanatory Notes respecting the Nature and Object of the Institution of Nashoba and of the Principles upon which it is founded. Addressed to the Friends of Human Improvement, in all Countries and of all Nations, *New Harmony Gazette*, 30/1/1828, 6/2/1828, 13/2/1828.

We see a hint of Frank's politics in 1820 when a wave of revolutions was in progress on the Continent. He was delayed in Germany on his way home from Italy, where secret societies called the Carbonari had been instrumental in an uprising in Naples against the king:

> I have reason to believe from the many questions put to me at the Police offices, the strict examination of my papers &c &c that I am suspected of belonging to the Society of Carbonari & that my letters were <u>treacherously</u> opened and detained by the Ministers of Police. No matter. If they open <u>this</u> they may learn that they are acting upon principles well calculated to add one more to the rapidly increasing and <u>inextinguishable</u> set of people, one who never thought that the contribution of his write, his feather in the scale, could help the downfall of governments whose <u>unworthy</u> proceedings every day more & more prove that no <u>foreign aid</u> is wanting to <u>spur on</u> this <u>desirable event</u>…You must excuse a little politicks *[sic]* this once. I never talk politicks except when I am obliged and then they say I talk treason.[C43]

He is caught at a bad time here — he was a liberal rather than an extremist in his politics.

4.4 Wealth and Finances

Francis' death in 1806, although very sad, did not bring the hardship that might be anticipated — the family was well off and, even more importantly, Jane was completely competent to bring up her children on her own.

It is clear from Francis' will that he held Jane's abilities in the highest regard — he knew she could take on all aspects of the family's affairs and gave her the choice as to how to do it:

> I recommend but do not enjoin my said wife to carry on the cheese trade (but not the Tallow trade) during such time and to such extent after my decease for her own use and benefit as she shall think fit and with a view if she shall also think fit and proper but not otherwise of introducing my two oldest or any other or others of my son or sons into the said trade and business when they shall become capable of exercising and carrying on the said trade [W6]

In addition to the business, he bequeathed her the family home and its contents and £14,000 in any form she wished. The approach is very different from other wills of this time in the family where the widow was provided with a stipulated annual allowance from the estate by the trustees so that expenses would be covered without her needing to worry about investment strategy.

The remainder of Francis' estate was for the children, expressly allowing for the possibility that Jane might be pregnant at his death. The funds would be invested initially, with the interest covering their maintenance and education, and the capital transferred as each child came of age. Jane was the first-named executor and carried through these arrangements. Passing the bulk of the estate directly to offspring was usual at the time as a woman's assets were relinquished by law to her husband on marriage; this approach thus guarded against the possibility of the widow remarrying and the children not being supported adequately by their step-father. It was also normal in the Ronalds' family, if not elsewhere, for assets to be distributed essentially uniformly among the offspring rather than to the oldest son or sons. The estate proved to be the major livelihood for nearly all of the off-spring — Frank's only other income sources were anything he took from his years running the family business, his earnings from sales of several inventions and the small government pension he received from age 64.

Some indication of the buying power of £14,000 may be gleaned from domestic budgets published in the period. Maria Rundell's *New System of Practical Domestic Economy* gave suggested budget breakdowns for various annual incomes. She recommended an annual budget of £250 for 'A Gentleman, his Lady, Three Children, and a Maid-Servant',[B31] inclusive of housing and education costs. If the income was £1,000 per annum, the same family could sustain six servants — including a coachman and footman for the four-wheeled carriage and pair of horses. Francis and Jane lived very comfortably, but not extravagantly — they seemed not to have a carriage at this time, for example. Jane now had more than enough money to raise her standard of living should she wish without any supplementary income from the cheesemonger business.

It is not known how many servants Jane employed while most of her children were still at home. Perhaps the most formal recognition of the valued service the servants gave is in Uncle Charles Field's will, which stipulated on a sliding scale the size of bequest to be given to each, depending on the overall length of time they had spent with him.[W10] If a servant had been in his employ for 21 years, the payment would be £105. Any legacy duty payable for these gifts was to come from his estate and each servant was to be supplied by his executors with suitable mourning attire. He and his sister shared three servants.

Jane's own will, written when she was aged 82, is another illustration of her commercial skills. It is a very long and complex document. In it she provides for each child still living or with surviving offspring, but with the amounts differing according to the extent she had financed 'their establishment in life',[W9] marital arrangements and other personal considerations. Each bequest involved both a lump sum and a proportion of the residual estate to facilitate these variations and to distinguish for her married daughters the split between a personal payment and a contribution to the trust established as part of their marriage settlement. The form of the bequest also differed according to whether the receiver was in England or overseas and their fiscal acumen. The document shows an astonishing grasp of her many financial dealings through her 45 years of widowhood.

Jane would have negotiated her children's marriage contracts, which were significant affairs — we know she contributed £3,500 to Maria and Samuel Carter's settlement.[W9] Again, their purpose was to ensure that the wife and children would be provided for.

Jane's energy and competence were also happily applied to more menial tasks like home improvement. Frank illustrates this side of her character when he was at the ruins of the Parthenon in Athens some years after the sculptures had been removed by Lord Elgin:

> How would Mrs M [his nickname for his mother] long to have a thorough rout here! Down with all the nasty little modern ruins, away with every thing that one can remove except the antiques themselves. If Lord Elgin had possessed real taste in lieu of a covetous spirit he would have done just the reverse of what he has, he would have removed the rubbish and left the antiquities.[J21]

4.5 Family Life

Jane's first priority of course after Francis' death was to nurture her large, young family. She was evidently successful: the sentiment of Crabb Robinson's* summation of her daughter Mary Ann as 'not a handsome woman but a well bred person'[2] would have extended to all of the siblings. There are numerous examples of her care and concern for her children throughout their lives. When Frank was aged 59 and she was going on 81, she wrote to him at Oxford: 'I hope to see you before we return having taken it [1 Royal Crescent Brighton where she was holidaying] for a month…do let me hear from you in a post or here, whether you have quite got rid of your cold'.[C195]

Home-life was stimulating. As well as frequenting the opera and ballet and going to balls, the family sang and played music together. Much of the conversation was centred on recent reading, issues of the day and cultural and philosophical questions. Both girls and boys were encouraged to be informed, to think deeply and to articulate opinions on what was happening around them — even gossip.

While Frank was on his Continental travels, rumours were circulating about King George IV's wife and she was brought to trial in 1820 citing adultery. It created great excitement — circulation of *The Times* more than doubled and Crabb Robinson devoted some pages to it in his diary. Frank conducted a thorough though informal survey of the views of people in places where Caroline and her friend Bartolomeo Bergami had visited — he knew it would be of interest to the family. After relaying various specific comments he had heard, he summed up in

---

[2]Ref. J13 4/12/1822.

a slightly more generous way than had Crabb Robinson: 'I believe I may say that the general opinion in these parts seems very much against her innocence & also very much against the prosecution'.[J22] He was excited to report a little later that he stayed in the same suite they had used in Karlsruhe; he made a drawing of the rooms including where Bergami's and Caroline's beds were positioned.

He also wrote home on more intellectual topics:

We don't express contempt half so forcibly by ridicule as the french. We are much too grave on ridiculous subjects. Must one acknowledge that their Ridicule is sometimes more pungent than our Sarcasm? Settle this point some Sunday after dinner. The comparative effects of Ridicule and Sarcasm (I can't find a better word, I mean a sort of light but somewhat severe criticism) for Public opinion (not on the individual) is the case to be determined you know. Compare Pope *[(1688–1744), English poet and master of satire]* with Moliere *[(1622–73), French playwright, actor and master of comedy]* if you think the comparison just.[C19]

To Charlotte:

Lamb has I suppose been elected for Westminster *[at a by-election in 1819]*, what was the parliamentary junto? How do opinions go about Burdett now? *[He defeated Lamb in the 1820 general election]*. What have been the principal subjects of debate this Season? How goes trade?[C23]

And to Emily:

Give me a little litterary *[sic]* News, what have you new in the novel line. Hey, they are not going to turn me out of the Book Club for absence I hope.[C22]

Surrounded by accomplished females at home, Frank greatly respected women and enjoyed their friendship. He used 'man or woman'[C43] and 'his or her'[J22] in his writing at a time when most people would not be inclusive. While in the Arab world he wrote: 'the want of conversable beings but particularly of the society of women folks for so long a time had begun to render me stupid & barbarous'.[J21] His spirited mental banter with Charlotte in Italy[J18] illustrates the intellectual parity in his relationships with his sisters. This did not stop him teasing them. Writing to Mary Ann:

You are very impudent and very bold. How dare you venture to assert that women's understandings would *[missing]* the world better than men's, recollect it is peculiar to the english ladies to possess strong understandings and they are but a small part of god almighty's creation. But perhaps you mean "the world would be better off" their understandings than on them.[C21]

Jane remained the head of the family throughout her life. The position that suited Frank, as the senior male, was that of counsellor. When he embarked on his Grand Tour, he asked Charles (the oldest of the brothers still at home) to deputise for him, probably to encourage him to mature: 'I hope Charles has filled my pastoral Chair this evening with becoming dignity.'[C18] There are examples in Frank's letters of advice given to most of his siblings; it was sometimes given gently and sometimes with directness, but he was always concerned lest it might cause offence. He also assisted his sisters, in particular, in various activities such as banking. Another part of his role in the family was in facilitating a circle of friends and acquaintances; his sisters Mary Ann and Maria both met their husbands through him.

Unmarried children continued to live at the family home unless work took them elsewhere. Frank spent much of his life with his mother and their relationship was particularly close — they had complementary skills and strong mutual respect. Jane encouraged his pursuit of science even when it interfered with her home and garden and he, in turn, used his creative abilities to make life more comfortable for her. Surviving correspondence and diaries show that they entertained together: he participated in visits she made and received, and his friends commonly gave their respects to Jane in their letters to him. Jane would always have listened to his advice. A probably typical example was in 1819 when he wrote to her and 'Recommended the allowance for the Children's board &c to be increased.'[123]

With 11 children surviving to adulthood, it is possible to identify some general trends in the family. Six of the siblings were fortunate in inheriting Jane's longevity (she lived to the age of 85); the others more closely followed Francis who died at 45. The cause of death for Charles and Alfred was recorded as apoplexy (stroke), which was quite likely also their father's affliction. Most of the children demonstrated a keen sense of adventure — of wishing to push boundaries and experience new things — an attitude that Jane must have fostered. Emily and Hugh had the same serious outlook on life. All the girls were strong-willed and self-assured, particularly in comparison with Frank, but not to the extent of entering the public arena. The older boys had greater contact with their father, and Edmund and Hugh echoed him in embracing business life with confidence and contributing to community affairs, although they perhaps lacked his commercial nous. Their younger brothers Nathaniel and Charles seemed more inclined to view their inheritance as the means for an easy life. Frank and Alfred had a different driver: both were highly creative in a variety of areas and their wealth underpinned these scientific and artistic endeavours.

All credit to Jane that she nurtured these diverse traits and encouraged her children personally and financially to pursue their dreams — perhaps she recognised some of her own inclinations in them. It is through her support that Frank and Alfred remain known today. In an era when boys automatically followed their fathers into their business, and girls were groomed for marriage and motherhood, the freedom Jane gave her children is remarkable. She is in many ways the hero of this story.

4.6 Frank the Person

A sense of Frank's physical appearance in adulthood is given in his passports.[J28] Translating from the French, he was 175 cm (5 feet 9 inches) tall, with brown eyes, a long face, high forehead, round chin, long (or big) nose and an average mouth. At age 37, he had black hair, dark eyebrows, brown whiskers and a brown complexion. Nine years later, his hair and eyebrows were reported as being brown (literally 'chestnut') and his complexion 'ordinaire'. When he was 30, he was denoted as having a flushed complexion:

> un teint coloré! Surely this can't be the case…I certainly felt as if I had a teint coloré a little before my examination, for at the <u>visit</u> of an English Dandy's trunk the officer taking the tag of the lace of a pair of stays between his thumb and fore finger drew them forth in a most whimsical manner and amused the circumstanding Monsieurs highly asking at the same time if it were chargeable, but this could be only un teint coloré pour le moment.[C18]

Most members of the family shared similar physical features. Thick, dark hair was prevalent. Surviving portraits and miniatures also indicate a rather long face with a prominent nose and thin mouth (Fig. 4.3).

Although Frank walked significant distances, his lifestyle in other respects was not very healthy. His eyesight deteriorated and he became stooped quite early in life from constant poring over books and intricate instruments by candlelight. His mentor and friend Edward Sabine* teased him when they were both in their

Fig. 4.3:   Frank's self-caricature doodle (*c.*1830). IET 1.6.127.

seventies: 'I was not a little surprised to learn from your brother in law *[Samuel Carter]* that I am your senior in years! I hope to see that y^u preserve a due measure of juvenility*[?]* in your looks.'[C380] Even at age 30 when he embarked on his Grand Tour, Frank mocked himself as being 'an invalid'[C20] with 'a train of old batchelor like *[sic]* grievances'.[J17] Jane required detailed updates on his wellbeing, which he gave in each letter. A typical bulletin reads:

> My Health is tolerably good. The Rheumatism gives me a twinge now and then but I always find that these twinges are the immediate result of some particular exposure to cold draught of air or such like accident so that I begin to laugh at it. The state of my Waistcoat is certainly much improved also upon the whole but now and then I am obliged to make wry faces.[C21]

His sisters Mary Ann and Charlotte also suffered rheumatism.

These communiqués were not sufficient for Jane, however, and she invited Frank's travelling companion over for dinner and interrogation on his return to England:

> I am glad Mr Hooper was particular in his account of my health, my reason for not saying so much about it was that I thought his account would be more correct than mine, being untinctured with Hypochondria. However this does not prevent me from seeing that it is improved…I take better care of myself because I have nobody to take care of me.[C22]

Fortunately he became significantly more robust during his travels; he later denoted this period as 'when I thought myself very ill'.[J22] He lived to the same ripe old age as his mother.

The word that perhaps best encapsulates Frank is 'modesty'. His own character, his opinions of other people and things, and even his pragmatic approach to life were all influenced strongly by an ethos of humility and restraint. His parents had a similar philosophy and it was also representative of his religion. He additionally appreciated patience and personal pride, probably because he did not always achieve these himself. Writing home while on his Grand Tour:

> How does my Vicar *[Charles, the anointed family advisor and keeper of 'morals'*[C43]*]* go on with the fulfilment of his duty both in the pulpit and out of it. I hope he is not becoming puffed up with pride and vain glory in holding so important a charge. Does he set an example of patience & charity & modesty? Let him recollect his day of reckoning must come[C22]

> Let your views always aspire Charlotte, be proud, I don't mean conceited or vain or ostentatious, be proud and modest too.[C24]

His preference for streamlined simplicity extended to buildings:

> Nothing now existing can inspire greater Respect for the ancient state of Architecture than such an object *[the Teatro Olimpico in Vicenza]*; here one sees all that we can imagine to be necessary to the purpose intended, all and nothing more than is exactly fitting; and how much more chastity of design & ornament, true symmetry & exact

proportion than <u>we</u> have any <u>original</u> conception of. The fruits of modern genius are ill begotten Bastards compared with these, whenever we deviate from such models we blunder.[J22]

His unpretentious and understated communication was another manifestation of this attitude. He enjoyed silliness: 'Altogether this has been a very pleasant week, vive la bagatelle *[long live triviality]*'.[C19] His writing is sprinkled with self-deprecation and humour to counter any possible impression of self-importance from his more intellectual observations. In contrast, he strongly disliked the exaggeration he found in travel books, for example:

> If you <u>want</u> to "give way to your imagination" or to indulge any enthusiasm at the first view of any famous place or thing, never read an enthusiastical *[sic]* account of it previously but above all never read such a bombastic, sentimental, gasconading stale fellow as Chateaubriand *[writer and diplomat]* whose account of Jerusalem is but a bad Romance.[J20]

Another related aspect of his character was prudence, which he demonstrated particularly regarding money. Communicating on business matters with Edmund (who acted with alacrity):

> You all laugh at my old batchelor like & cautious way of thinking, and are inclined to despise perhaps those very notions which you would yourselves adopt sometimes if they did not come recommended by me of whom you laugh…but…my pride is not at all wounded by your good tempered ridicule.[C38]

Regarding his own finances:

> If I go into Sicily which is not at all improbable I shall perhaps want more money. I have got £100 still but you know my caution and you know how unpleasant it would be to remain without enough[C21]

Although he lived well and bought innumerable devices for his science, he was always careful with money and maintained detailed accounts.

Frank enjoyed receiving compliments from family and colleagues, which boosted his self-esteem, although he protested against them in his self-effacing style:

> And pray how dare you *[Mary Ann]* attempt to <u>nick</u> me into a correspondence by flattery, don't you know and see that I am proof against such temptation, I only write this once to you to say I shan't write any more.[C21]

> I could not persuade myself to go to Vienna, the road home looked so much smoother; & then those pretty letters but I'll tell you what, if you *[Emily]* cram your letters with so much flattery you will really make me desire to have no more of them[C43]

Confidence and experience in dealing with people was a real weakness for him, particularly in challenging circumstances — he had little or no inkling how he might influence others' thoughts or actions. His comment about one of his travelling companions, a numismatist and antiquarian whom he nicknamed

'Coins', suggests recognition of this naivety: Coins 'appeared a mere child in every thing beside antiques but I liked him because he was as original & inoffensive as a child. <u>You</u> *[referring to his armchair in the study]* would have fallen desperately in love with him.'[J17]

Frank liked to give more than he received and hated the thought that his behaviour might give others concern. When he needed to borrow money during his travels, he wrote home regarding the lender: 'If you have any regard for me pray do anything rather than suffer me to remain a moment longer than you can help injured in the opinion of a so generous & open hearted friend.'[C38] This trait would similarly have been the main reason why Frank accompanied at least one of his travelling companions for longer than he should, as described in Sec. 8.5. On the other hand, he admired the ability some people had to be disinterested in others' views of them. In Greece: 'We dined with…the veriest*[?]* frenchman I ever came near, it did one's heart good to observe his perfect self complacency & vivacity, I wish I had a tenth part of his nonchalance.'[J21]

Frank was warm-hearted and kind and enjoyed interacting with people. He was much loved by his family, especially the females, and had numerous close and trusted friends. As an introvert, he preferred to socialise in a small group and was reserved amongst those he did not know well: 'I esteem one or two old friends higher than a large <u>circle</u> of acquaintance'.[C24] Describing a highly enjoyable five-day stagecoach trip from Paris to Lyon with French and Italian fellow-travellers, he ended:

> O but there were a thousand little incidents which kept us all in good humour and full of fun and they appeared to like my Company very well, but I managed to avoid their <u>Kisses</u> at parting and I hope I preserved the proper <u>dignity of an Englishman</u>.[C19]

In Turin, when he visited the site of Giovanni Beccaria's experiments on atmospheric electricity:

> The Frati expressed themselves very glad to see an admirer of their most celebrated associate…They shewed *[sic]* me every thing that remains of the house…and invited me to sup and sleep in order that I might have their view in the morning but I did not stop, which was very foolish.[C19]

In Rome:

> I determined today at dinner to make an effort to go to some of the Concerts and Parties, so with some few husses and haws I <u>plucked up a spirit</u> and <u>scraped</u> with an English gentleman of the name of Stanley who promised to take me to Rufini's*[?]*… They say some of the English walk into these places without introductions as they do into such like parties in London. I am too proud or too modest, I don't know which. Brăss *[Mary Ann]* would say I'm only too lazy[C21]
>
> On Monday I turn my Back upon Rome, I must make my calls (there Charlotte, am I not improved) and get ready tomorrow[C21]

Women to whom he was attracted had certain traits in common. In Lyon he met the Superior of the Convent:

in whom I found a young & beautifull *[sic]* person & a highly fascinating ladylike deportment combined with the most rigid discipline & gravity.[J17]

They met again the next day:

She pleased me more than yesterday, so much grace & true politeness is seldom united as in her case with so much animation.[C19]

In Milan:

I had the honour to accompany a pretty little princess round the Church too, she had very fine large black eyes and was very petite and modest and genteel and all that, but I did not fall in love — no.[C20]

Valletta:

Of other subjects of observation nothing remains worth mentioning except the Ladies. I have certainly seen no women so beautiful since I left England as these I have seen here. Fine penetrating, sparkling, black eyes, beautiful hair & teeth, a fine open confiding expression sometimes united with a jocund smiling hilarity but more frequently with an appealing humility quite distinct from our modesty but almost as engaging.[J19]

He even toyed briefly with the idea of analysing the concept of Italian female attraction 'in a regular, scientific and cool way':[C20]

We must begin with the Air frappante *[striking]*, the touchante *[touching/affecting]* & the brilliante, these three heads must form the first part of the Subject. Secondly tournure *[shape]* & the minor graces. Thirdly form & feature, the fine, the delicate, the insignificant. We shall have little to say on the last head of the discourse viz complexion. Every division comprising many subdivisions must be examined with the most painful research.[J17]

Although he showed a romantic side, Frank's shyness and the high hurdle that was set by the talents of his sisters conspired against him finding a partner. Furthermore, his passion was for science and he would have had to forgo that to supplement his inheritance in supporting a family. At times he became quite melancholy about his bachelor status. His nearest brother Edmund's marriage took place just before Frank embarked on his travels, and he wrote a year later:

I wish Edmund much joy of his Boy and his pretty little Wife & his Boy much happiness in him, it will be very funny for me, to see them all together in a family group.[C25]

He received another jolt when his next brother Hugh wed while he was still away:

Don't talk any more about English beauty & so forth, Edm^d & Hugh are happy, I am an idle vagabond wandering about on the face of the earth as if I had nowhere to lie my head, no faithful bosom, no dimpled cheeks, no lovely ——, but I am getting sentimental [C38]

At another time:

> I suppose Marriages & births are going on at a great rate — remember me to all the old birds and teach the young ones that they have an uncle Frank.[C29]

Travelling through Switzerland in December, he wrote:

> Show me the man or woman who can warm his heart with fine scenery when his or her toes ache with Cold. I mean to make the Course of Switzerland when I have a wife ~~to keep me warm~~ (alas when will that be). How delightfull *[sic]* to —— I won't begin on that subject —— I'll spare you this once.[J22]

Musing to Charlotte on his 31st birthday:

> 32 I think I am and no better off than when I was 30…I wonder what will become of me by next birth day what do you think? Do you think I shall ever be married, it is high time I think of it if I mean to think of it at all, but I have been thinking & thinking so long in fact that I don't think I shall think of it so much than as I do now if I am not married.[C24]

He became a little embarrassed at times about his soft-hearted nature, even when confiding to his mother or sisters:

> n.s. means not sentimental, I shall always put these letters in future after any paragraph, a sentence &c &c where I don't know how to express myself without appearing sentimental, when I can't avoid appearing so without too many words; it will save time [C24]

Frank also cared deeply for animals. He mentions the family dog Dido several times in his letters. At Naples, he visited the *Grotta del Cane*, so named from the practice of sending a dog into the cave to show visitors how the carbon dioxide and water vapour emitted there suffocates an animal. The dog would recover if it was removed from the cave sufficiently quickly. He wrote:

> The poor little skeleton of a dog…excited my pity & having determined to save him one death, poor fellow, I took a good Snuff of the Carbonic acid myself and persuaded my companions to convince themselves of the reality of its existence and of the torture of dying in it by following my example…The argument used by some namely that he is used to it is I take it about as good as that of the Joe Miller's fish woman "Lord Ma'am, the Eels be used to skinning" *[from Ref. B2]*[J18]

In Rome:

> Cruelty to Beasts. I have not yet seen in Italy an instance of that wanton cruelty to horses &c which we ought to be so much ashamed of in England. The italian driver and his beast seem much more sociable here than in England perhaps because they are more nearly associated; barbarians without barbarity.[C21]

In Venice, the city of love:

> It was indispensably necessary to have a little to say to Signore Cupid you know in Venice & I was dreadfully at a loss (having been so much out of practice) to begin my devotions at the little devil's shrine, so had recourse to my snuff box & Sofa

again. How I had at last began, proceeded in & finished this affair cannot be related, I wish it could for I think it would furnish Whig *[Charlotte]* with a tolerably good dish of laughter. At one time I was so perplexed that I had almost determined to skip this duty alltogether *[sic]* & was only resolved to persevere by way of not infringing upon my invariable rule of allways *[sic]* "doing as they do at Rome" [J22]

It turned out his companion was a spaniel.

Frank's inquiring and logical mind is apparent in all his endeavours — scientific, cultural and social. He took numerous travel and historical books in various languages on his travels, which he studied at length. His journals are littered with references to the literature, ranging from ancient Greek and Roman naturalists to quotes from contemporary poetry and opera. He also drew analogies with what he had seen first-hand in Britain. He contrasted the Milan cathedral with that at Salisbury, and likened the irregular landscape near Mt Etna with the undercliff at the Isle of Wight in terms of their geologically recent creation. In the countryside near Constantinople:

The Views now began to assume features very similar to those of Kent and Sussex indeed they reminded me very much of my rides at home [J21]

And in Malta:

I live here just as I should at a good Inn in Canterbury or Bristol except that I pay about half the price. [J19]

Even his unhappy travelling companion in Italy, Joseph de Gourbillon* (see Chapter 8), acknowledged that Frank 'was a very worthy man, and was also extremely well-informed.' [B14] Another companion introduced his cousin to Frank: 'to whom I refer you for much information (if you desire it) respecting many Countries in which M[r] Ronalds & I have lately travelled.' [C39] Frank might not have known in some social situations when to stop sharing what he had seen and discovered.

His studies embraced all aspects of local culture, although there was a hierarchy of interest:

I love Music. I am inspired with delight sometimes in examining the Monuments… But Italy contains much besides Music & Monuments…I think the Pictures are greater things than the Monuments & the Natural Phenomena greater than the Music [C23]

Coming from an artistic family, he enjoyed exploring art and architecture — to a degree. In Rome, he wrote:

I hope I shall "get my business finished here"…about the end of next week. I have gone through all the Antiquities and some of the Pictures very carefully and method-ically (Oh, you systematic blockhead me thinks I hear some Virtuoso squeal out) and I hope to kill the remainder of the pictures by that time. [C21]

as to churches I have tired of them already [C21]

He was characteristically self-effacing about his broad knowledge because he was acutely aware that it was imperfect. Referring to Sicilian farming practices, he wrote:

> I think the chief defect after that of want of active labour is the little regard which they pay to the sorts of Cattle, fruit, Vines &c &c and the little attention paid to the proper renewal and succession of trees &c &c...I have no agricultural science, I never pretended to any, but I should have no common sense if I could not see this.[J18]

On architecture:

> I don't pretend to architectural Science but I think I never saw a more beautiful specimen of the Ionic order or one which exhibits that order in so great perfection as the two standing columns of the Temple of Apollo Didymeus *[in Turkey (Fig. 4.4)]*[J21]

Classical history:

> What can I say of the Hellespont *[also in Turkey]*? I have already I fear tired you by descriptions of scenery & I dare not enter into historic or critical details or classical allusions but I should expose my ignorance[J21]

Religion, *en route* to Jerusalem:

> I met here at the Consul's house a Mr Connor of King's College Oxford engaged in distributing bibles for the Bible society, and made his acquaintance which I have no

Fig. 4.4:   Lithograph of Temple of Apollo Didymus (*c*.1821); drawn on stone by Frank and printed by Hullmandel. IET 1.6.5.

doubt will be agreeable and beneficial to me being wretchedly ignorant of the Bible altho' travelling soon in the holy land.[J21]

The importance of books for Frank is well illustrated by a long outpouring to Charlotte in April 1819 when he realised the books on electricity he had collected in Italy were probably lost. It is an unusual show of emotion and an excruciating one:

My Books, my dear Books, by two short lines of Edmund's <u>letter</u> in which he says you "can hear nothing of the Books which I thought I had sent from Florence" I am deprived at once of sleep, the pleasure of the other letters. I have got up again on purpose to request you to get <u>your things</u> on, send for a place in the stage and go to Edmund's directly and make him understand the extent of my misery in the idea of loosing *[sic]* them. I know it will be difficult and that he may even laugh at my earnestness but <u>you</u> know how much time money and labour I have been at to learn to read German & Italian and when I tell you that it was all for the sake of these very books and some others which I was at infinite pains to collect, that I have faged *[fagged]* my legs off almost wading through the dirty lanes & alleys of the dirtiest towns of this dirty country after them, you will not only use your utmost eloquence with him but you will perhaps use your legs too for my benefit. Besides, they cost me at least 20 pounds which by the by I have frequently dined upon macaroni to save (don't read this out though before Peter *[Martineau]*)...I'd rather return to England with the loss of an ear or an eye than them, in fact I should look like a sow with one ear & I won't return until they are found or replaced. They are rubbish, merest rubbish, but I have made them valuable to me by a silly earnestness and perhaps comparatively speaking a trifling pursuit of a trifling object...In short — (I am getting very long)...I feel their loss as I should feel the loss of a child. Alas my dear Books. Seriously, <u>very seriously</u>. Get Edmund (go with him or get Charles, I know he will do me the kindness) get one of them to call on Zotti the Italian Teacher and use every means in his power to induce him to find them...Engage Peter in my service...Alas, Alas I have forgotten the name of the man at Florence I gave them to... What a fool — a man is to set his heart upon trifles...And so much for my books and the moral of my sad, sad history of grievances.[C24]

The books saga continued through the rest of his journey. It is likely that he only obtained copies of these texts four decades later during his extended stay in Italy.

4.7 Frank's Brothers and Sisters and Their Families

Jane and James MONTGOMREY

As the first daughter, Jane (Jnr) was given her mother's name. Like her sisters, she received from Jane a gift of £100 on her marriage in 1810, which was roughly equivalent to two years' wages for a male teacher.[W9] Her new husband James Montgomrey Snr was a trustee of the Society of Protestant Dissenters in Brentford, and would have been introduced to her by Uncle Hugh the nurseryman.

Frank wrote of her in 1819: 'Bless me what a family she has upon her hands already, it's really a very great change, she must have a great deal of anxiety poor thing ah! We never think enough about parents' anxiety'.[C21] Frank's words were tragically prophetic. Jane Jnr gave birth to 9 children over 12 years, but died when her baby was just three. Her husband was left to watch their daughter die of consumption (tuberculosis) at age 16 — and she was the first of six siblings to do so at quite young ages.[J93] He was unable to cope with the ongoing grief and died insane.

The older children fortunately escaped the disease. Eldest son James Jnr (Fig. 4.5) took on the family timber merchant business; the seven-acre site backed on to the Brent River just off the Thames and had extensive frontage along the Brentford High Street. He was also a Justice of the Peace: in this role he helped regulate local affairs, including trying petty offences, appointing local parish officials and licensing alehouses. In addition he acted as trustee and executor to numerous members of the extended Ronalds family.[A196] Frank requested him to establish and manage a trust to support his niece Julia on his death.[W11]

Fig. 4.5:   Frank's nephew James Montgomrey Jnr (1811–83). WU RC1622.

Edmund and Eliza RONALDS

As Frank's closest brother in age, Edmund grew up with him (Fig. 4.6). Their personal traits were rather different, however, with Frank later telling him: 'I am very glad to hear from you but I don't want to hear anything about Stays yet *[corsets were becoming fashionable for men at the time]*. You are always in such a violent hurry, take things more coolly and you will accomplish matters much better.'[C21] Edmund could also be quite direct in his communication, like their sister Emily. Frank was more guarded (and had little interest in fashion).

Edmund and Eliza would have met in Hammersmith, where both families were residing. Their granddaughter recorded that Eliza 'was a good hard-headed Scotswoman greatly beloved by her sons'[C419] as well as 'a very clever woman and very practical'.[C421] She also had considerable beauty. Frank quipped to sister Charlotte when he learnt that they were expecting their first child: 'I suppose Eliza is very happy, if you should see her remember me to her and make me say what should be said on such an occasion. It's a very good match I suppose, she won't call him *[Edmund]* doddle legs will she?'[C23]

Fig. 4.6:   Frank's brother Edmund (1790–1874). Ronalds family papers, NZ branch.

Eliza and Edmund produced 12 children. Two of their young daughters died within a fortnight of each other in 1829 of scarlet fever. Their eldest son, referred to herein as Dr Edmund, obtained his doctorate in Germany and was appointed Professor at the Queen's College, Galway. He later resigned to run the Bonnington Chemical Works in Scotland, telling Frank: 'I have entirely changed my mode of life & have (with a view to the future of the bairns *[children]*) taken seriously to money grubbing, an occupation sufficiently disgusting', and noting that he was now 'completely ignored, as a tradesman' by 'the savans *[intellectuals]* of Edinburgh'.[C370] With their shared scientific interests, Frank and Dr Edmund were very close.

Edmund and Eliza had by this time run into financial difficulty: their grand-daughter described it as 'the smash'.[C419] He borrowed significant funds from his mother Jane and, very soon after she died in 1852, invested in a silk mill. His idea was to establish his younger sons in business. Depôt Mill was in Derby, employed 34 men, 150 women and 80 children[S2] — and was losing money. Edmund advanced the owners considerable funding and, according to his solicitor in an 1853 court case, they 'made it fly'.[3] Edmund was left to salvage what he could by disposing of the fixed assets at auction.[4] Dr Henry's son Henry Jnr summed up the outcome: 'he must be much reduced in circumstances as two of his daughters have been obliged to go out as Governesses.'[J90]

Five of the children joined Unitarian friends of similar age in New Zealand — Edmund's extremely wealthy brothers-in-law Peter Martineau and Samuel Carter (Table 4.2) provided financial assistance for their travel and establishment there.[J88]

Table 4.2:   Personal Worth of Family Members at Death.

Name	Died	Estate (£)	Comment
Frank	1873	3,000	
Edmund Ronalds	1874	8,000	
Peter Martineau	1869	140,000	
Emily Ronalds	1889	6,381	
Nathaniel Ronalds	1833	800	
Alfred Ronalds	1860	Died intestate with £600 mortgage	
Samuel Carter	1878	90,000	

Source: English Probate Index

---

[3] *Derby Mercury* 27/4/1853 4b.
[4] *Derby Mercury* 9/2/1853 2b.

Life proved to be a significant struggle, as evidenced in their numerous surviving letters. The two girls married and stayed, as did their brother's children, while James died in Australia and the third brother returned home to become a nursery-man. Another of Edmund and Eliza's daughters spent much of her later life in Algiers. Fortunes across the family turned out to be very mixed.

Hugh and Kate RONALDS

Francis formally apprenticed Hugh just 16 days prior to his death in 1806.[J4] Being the third son in the family cheesemonger business did not suit Hugh, however — he wanted to make his own way in the world. His opportunity came at age 25. George Flower and Morris Birkbeck, who were known to the family, had explored the American interior and chosen a suitable prairie in Edwards County, Illinois on which to build a settlement. Almost immediately after the news reached London, Hugh headed off for the Midwest — after filing his first patent 'for certain Improvements in the Art of making Leather'.[P2]

Emigration was appealing for political and economic reasons. The UK was in depression after the long Napoleonic Wars. Dissenting merchants, in particular, disliked the traditional dominance of the aristocracy and the church in shaping the country's institutions and welcomed the principles of liberty and equality embod-ied in the constitution of the United States. The sparse population and the avail-ability of good, cheap land there also meant that knowledge and toil could produce wealth, independent of birth.

Frank was in Egypt around this time and contemplated the advantages of that country as a destination (Egypt was to come under British rule in 1882). As was his nature, he placed greater emphasis on the pragmatics of food and work than on political ideals:

Is there not alas many a poor mechanick [sic] with a capacious belly attatched [sic] to his own proper person & sundry smaller bellys [sic] dependent thereon who would in these sad times rather prefer a good belly full under a bad government to a good government and an empty stomach...

In comparing the advantages of emigration between America & Egypt I think... the latter is far superior to the former in affording a much finer field for the exercise of talent of every kind. The Antiquarian, the Scholar, The man of research in every branch of natural Science, The Mechanic, The merchant, The Manufacturer, may here reap as abundant harvests as the Cultivator, one is not so much tied to the clods.[J20]

Crabb Robinson echoed the last point in summarising Birkbeck's glowing account of the USA in *Notes on a Journey in America* (1818):[B13]

A work which wo[d] obtain greater credit if the writer had not an obvious interest in attracting Colonists to his establishment...His book however satisfies me of what

I had no doubt before…America will not be the choice of a man who has rendered the pleasures of taste & imagination necessary to his comfort.[5]

Crabb Robinson's opinion on the Flowers' decision to emigrate was that 'George has seduced them And his expatriation has been promoted by an unhappy marriage'[6] (George's wife was the only one of the extended family not to travel). Such thoughts would not have soothed Jane Ronalds as her son headed off to their settlement.

For Hugh himself, the decision would have been founded on his belief in hard work and independence. In a letter to Frank in 1823, Hugh called him 'a Batchelor out of business',[C50] before continuing:

I begin to understand now that we cannot go through life without either great occasional meanness or great occasional exertion — besides continual energy and industry…therefore I think it desirable in every one commencing[?] life to chuse the rugged road in order that he may be prepared for all he may meet with [C50]

The first English settlers had set off for the prairie that Birkbeck and Flower selected in March 1818. Sailing from Bristol were 88 emigrants. Most were farm-labourers, mechanics and tradesmen; there were also three women — and Hugh.[B103] After disembarking in Philadelphia, he would have set off on the 1,500 km journey to frontier Illinois on horseback, and then perhaps by 'flat-boat'[B103] down the Ohio River from Pittsburgh, with a final overland section. He arrived a few months before George Flower had returned with his family. He became one of the four initial town-proprietors who each bought a segment of land, chose the name of Albion for the settlement, and funded and oversaw its development.

Forty kilometres south of Albion was another village called Harmony, and both soon gained considerable notoriety. Albion became the crucible of the biggest political issue of the time. American immigration was rising rapidly, encouraged by the very rosy picture painted in Birkbeck's book. He was countered by those who, lamenting the loss of capital and labour from the UK, disparaged life in the New World. Others cited emigration trends in arguing for far-reaching institutional reform, including a Parliament more representative of working men. The debate also raged in America as to the relative merits of settling on the east coast or in the interior.

The controversy generated a number of early British visitors to Albion and much was published about the difficult conditions there. One author wrote: 'I am convinced that any one, who has even a prospect of making a decent livelihood in England, would be a fool and a madman to remove to the Illinois.'[B26] The reports would have done little to allay the family's natural fears for Hugh's welfare. Frank

---

[5]Ref. J13 29/3/1818.
[6]Ref. J13 16/4/1818.

did what he could to keep up their spirits. A nicely worded note to his mother included:

> You have no doubt had the pleasure of hearing from him by this time, but if even you should not have heard…I think you will soon learn to banish anxiety…by experience and knowing for <u>certain</u> that…the chances of his being now <u>unsafe</u> amount to very little.[C22]

With Emily he was more direct but also tried to make her smile:

> I think you are making rather a wrong use of your liveliness of imagination in regard to Hugh. Why <u>feel</u> and much more <u>describe</u> to others who…have not perhaps such a <u>boding unestablished</u> conviction that he is suffering great troubles & privation. Why not rather look to the more comfortable side of the question and suppose that he is sitting smoking *[sic]* his segar *[sic]* or scraping his grumbling friend's guts as he used call his violincello in a nice little cabin with a nice wood fire and a nice piece of wild boar roasting at it (it should be just stunned a little first) or bringing down a good lumping turkey with his rifle or rubbing his hands by the side of his pig sty at the thought of a good Ham or two for the frying pan & a few eggs and plenty of pepper to it. Who has told you he is suffering great troubles & privations? Don't misunderstand me I am not at all disposed to join in any laugh against your sisterly feelings for him and I have no doubt he has had some difficulties to go through poor fellow, but I am inclined to <u>hope</u> he is now well off whilst you seem inclined to <u>fear</u> he is not. You see I cannot write without preaching [C22]

Before long, Hugh proposed marriage to George's sister Kate Flower and built a home for his bride. He continued to downplay their discomforts to his family, although Kate was more open in her letters. In 1827 she described a typical day — she had no servants to assist her, having 'learned to do completely without them',[C57] and was several months pregnant at the time with her fourth child:

> Up at day break — make the fire. Call Kathy *[aged six]* & Hugh *[age four]* & dress Emily *[age two]* — K. sets breakfast. H. picks up wood & chips for the fire. I begin the occupation appointed for the day…one day I wash 2^d Iron 3^d make soap 4th Candles 5th Bake 6th Clean the House. After Breakfast, K. & H. calculate for about half an hour…after which Kathy clears the Breakfast table and puts all the things which have been dirtied the day before into a large tub and it is little Emily's business to wash them *[at age two!]*…about twelve we all sit down, Kate draws writes whilst Hugh reads…I am glad of all the time till evening for needle work all of which I am able to do and have become so expert a tailor…in the evening we go into the garden to gather vegetables & fruit for Papa's supper, little dears have their suppers and go to bed [C57]

Overall, she enjoyed her adopted homeland greatly, describing it as 'a beautiful country, an easy independent way of liberty, plenty of food and fruits'.[C84]

Frank worried about how Hugh would support his family, writing to Edmund:

> Does poor Hugh give any account of the laying out of his money. When you write to him just touch upon this subject…<u>Caution</u> him to lay out as little as he can help

Fig. 4.7:   Frank's portrait of his brother Hugh (1792–1877) at age 42; drawn with his perspective tracing instrument. IET 1.7.70.

in the purchase of experience…Try it on a small scale first. I would have him even look on and learn by other people's experience but I am only repeating what I have said before. I may be too cautious perhaps [C21]

Hugh and Kate needed to turn their hands to various activities to make a living, much as other family members later found in Australia and New Zealand. First they built a tannery. Entries in a neighbour's diary suggest that it provided several benefits to the local community — useful refined products, a means for selling livestock and credit facilities.[A150] Effectively Jane, who provided the investment funding, was assisting the development of the village. Hugh also established stores in Albion and nearby villages and became a wholesale dealer. Crabb Robinson then recorded that he had 'been unfortunate And unable to succeed in trade' and was trying his luck 'as a language master',[7] while Kate wrote of 'my school in the town which adds money a little comfort to our small income'.[C84] Hugh later became a farmer.

---

[7]Ref. J13 30/11/1834.

Hugh, Kate and their children all spent their last days in the United States. The two sons who reached adulthood became physicians. Great-grandsons included a county judge and a history professor, both still in Illinois. Hugh and Kate were able to holiday once in England and several family members visited them in America. Emily was the first in 1824. In 1842, Henry Jnr passed through Albion on his way to set up home in Ontario, Canada. Niece and nephew Sarah and Charles Flower travelled across in 1856, sadly after Kate had died.[B151] Hugh lived to the same good age as Jane and Frank (Fig. 4.7).

Mary Ann and Peter MARTINEAU

Mary Ann and her husband Peter Martineau have the rare honour of being written of in complimentary terms in Crabb Robinson's diaries. He found Mary Ann 'very conversable & sensible'[8] and 'particularly agreeable'[9] while Peter was 'a very gentlemanly man'.[10] Crabb Robinson documented many parties at their home over 40 years, with a typical account being:

> I walked on hastily to Hyghbury *[sic]* Terrace — where I dined with the Martineaus. An Irish Gent (Watton or Waddon or the like) The rest all Ronalds or Fields — An agreeable party enough. In the evening came the Bischofs *[James Bischoff was an author]* & Miss *[Sarah]* Rogers *[sister of poet Samuel Rogers]* &c — I had a pleasant rubber at Whist[11]

The soirées, although often involving family, also included thought-leaders, activists and international visitors, reflecting their interests across science, the arts, politics and religion. They were a popular and outgoing couple at the centre of a diverse social circle.

Peter and his brother David had been schoolmates of Frank and Edmund. Frank and Peter bantered frequently over topics of the moment but their very different personalities could be problematic, as illustrated in Sec. 8.7. Frank nicknamed his sister 'Brăss', noting that 'good Brăss is ductile & malleable',[C21] and probably inferring that these attributes were necessary to live with Peter. The couple naming their first child Francis suggests that Peter admired Frank more than he might have realised.

Mary Ann became interested in Italian novels, poetry and opera when Frank was in Naples, and he recommended a number of books 'if her husband will allow her'[C23] and 'her eyes don't fail yet'.[C23] He also advised regarding her singing:

> Go on with your singing if you have no little cherub to sing to you yet; let me find you much improved. Don't sing so many songs. Sing no very difficult things yet, prefer slow to rapid movements, make yourself a few low notes, don't depend too

---

[8]Ref. J13 22/6/1824.
[9]Ref. J13 21/2/1832.
[10]Ref. J13 4/12/1822.
[11]Ref. J13 17/2/1836.

much upon your higher notes, bawl but don't squawl, keep your temper whilst you practice, and whilst you read these sage admonitions, & recollect that…I am your loving Brother F.R.[C21]

He added to Charlotte:

Look on the top shelf of the shelves on the right hand side of the door in my room for some Books for her. There is I think a <u>small</u> collection of Goldoni's & Metastasio's Plays. She can see how she likes them by these. There is also somewhere the Castle of Otranto in Italian but this is too easy…and she may learn my Air (or duet if she can get any body to sing with her) out of the Matrimonio Segreto of Cimarosa, it is <u>all</u> beautiful…Get her to help you to scold her husband, I suppose amongst all her other requirements she has learnt this, the most <u>usefull</u> *[sic]* of all.[C23]

The Martineaus' principal business was sugar-refining, but Peter also studied beer-making at the Flowers' brewery.[B151] He would have encouraged Hugh's decision to join the Flower family in establishing their American settlement. His and Mary Ann's daughter Sarah later married George's nephew Charles Edward Flower in England. Charles followed in his family's footsteps in running a brewery, and Flowers ale is still sold today. The couple kept a scrapbook with newspaper cuttings about Frank, which is now in the UCL archives, and her diary covering the years 1846–92 has been published.[B151] In it she records her mother Mary Ann's distress as two of her children suffered a progressive paralysis from their teenage years and died in their mid-thirties.

Sarah also described the deaths of her parents. Peter passed away in December 1869:

I had never remembered seeing him in bed — the next morning he became much weaker, but talked to us, and then he asked for my mother and soon after he sank in her arms and gradually died in the most happy, peaceful way [B151]

Mary Ann, nearly 13 years later:

After giving her orders as usual she sent for all the servants and took leave of them, shaking hands and thanking them for all they had done for her saying 'She must go now soon'…and then she died quietly at one o'clock.[B151]

The famous sociologist, author and feminist Harriet Martineau also marked Peter's passing:

My aged cousin, the head of the family, Peter Martineau, died… He was eighty-four years old. He was always good to me, and I feel his departure, though I knew we should never meet again.[B97]

Her brother, the influential Unitarian philosopher Revd Dr James Martineau, conducted the funeral service.

Emily RONALDS

Like her younger sisters, Emily shared her name with a maternal aunt. She was earnest, strong-willed and a little blunt, contrasting with Frank's light-hearted manner, and like her sister Charlotte enjoyed being involved in everything. She was also a reliable support in household management.

Emily first emerges in the letters Frank wrote home from his Grand Tour. She and Charlotte were important correspondents: they wrote more consistently than their brothers and married sisters and provided the political, social and literary updates for which Frank yearned as well as family news. Emily being 22 when he left, the topic of beaus came up in their letters. Two of his little comments were:

Well Emily, who are you thinking or dreaming about? Won't you tell me? Then I must guess and I won't tell you who I guess [C21]

and later:

What's become of a certain Squire — you take don't you? Dismissed, Hey. Yet barring all pother[*12*] as the song goes [C43]

Like Frank, she did not marry.

Emily's trip to the USA at the age of 30 proved to be the turning point in her life. She travelled with George Flower's father and Robert Owen to Albion to visit Hugh and Kate. The journey from Liverpool took 11 weeks and is well documented in the diaries of two other members of the party.[B124,B143] Owen had developed the concept of 'co-operative communities',[B123] where service would be rendered for the good of the whole rather than to the benefit of the owner or for individual gain, and he planned to bring his vision to life at the nearby village of Harmony. Hugh explained the concept:

The Harmonie Community…have just been forming a new Constitution upon the principle of perfect equality — all questions being determined by a majority of members…it is the duty of the foremen to report the daily conduct of each individual in his class to the superintendent — the superintendents report to the assembly — and two thirds of the assembly may expel for bad conduct [C54]

Emily had been in Albion for three months when Fanny Wright arrived. Fanny was already a radical intellectual and social activist: gifted, self-assured and courageous.[B176] Fanny must have found a kindred spirit because just three days later Emily joined her on a six-week trip to New Orleans to meet Fanny's intimate friend General Lafayette (hero of the American and French Revolutions) and inspect slavery at close quarters.[B176] Fanny, Emily and George Flower then planned the development of a cooperative community called Nashoba to assist in assimilating former slaves into society. Fanny purchased several slave families who were to gradually earn their freedom through a system of unified labour.

After a total of five months, Emily suddenly returned home — she had probably just received news of Charlotte's deteriorating health. Buoyed with ideas,

---

[12]From song *The Chapter of Kings* by John Collins.

new confidence and independence from her trip, she soon established an infant school close to the family home in Croydon.[A163] It was just eight years since the first such school in England had opened, influenced by Owen's educational philosophies. Emily also retained her socialist interests. On the day Owen's son arrived in London from Nashoba in 1827, he recorded that he saw 'Miss Ronalds, and several other friends of the cause'[B160] together with his father. Fanny also met up with her 'excellent friend Emily'[C66] in London and they worked together in garnering support for what Crabb Robinson called 'her colony for civilising the blacks'.[13]

Various comments survive from this period regarding Emily's intelligence and fervour. Fanny wrote of her:

She is clever & has I think good feelings — certainly quite liberal views...

she has...taken her stand in favour of human improvement & liberal principles in opposition to old friends & relatives.[A163]

James Pierrepont Greaves, who established a socialist community and school in Surrey,[B180] advised:

She is a thinker, who on some sides thinks with the light — on other sides without it, you will find her open & communicable[C55]

Crabb Robinson met up with her in late 1825:

I dined with Peter Martineau Jr at Hyghbury Terrace — A party & an agreeable one — The most prominent person Miss Ronalds an enthusiast who travelled lately to America...Admires the copper coloured indians & Mr Owen's new principles of Society — This is amusing enough in a person of amiable appearance And who has vivacity & good spirits which will pass for cleverness anywhere in a lady.[14]

And again two years later:

Miss Ronalds was very energetic in favour of the philanthropic plans of her friend Miss Wright. And Greaves was disposed to be eloquent on his infant schools, but these zealots neutralised each other.[15]

Both Fanny and Owen tried to entice Emily back to the USA to live in their socialist communities.[C59,C60] There is no evidence that Emily returned there, although she did continue to travel, making use of her language skills. The Brentford cousins record in 1843:

We dined and slept at Chiswick on Sunday last, our friends there are quite well, my Aunt Ronalds walking about and amusing herself like a young woman and Emily is at home at present but talks of going to Germany again in the Spring. She is quite a rambler.[C118]

She also contemplated travelling to Ontario in 1865 to visit relatives there.

After her mother passed away and the family home was sold, Emily resided on Earlswood Common at Redhill, before moving to Brighton.[C414] She was still

---

[13]Ref. J13 3/11/1827.
[14]Ref. J13 25/11/1825.

full of life in 1884, writing: 'I saw the great exhibition, in a chair which prevented me from seeing many of the worlds wonders'.[C413] Sarah Flower recorded her last visit in late 1889:

> To Brighton to see dear old Aunt Emily. She is very ill and tired out. She died December 4th 1889, and leaves a much honoured name behind her.[B151]

Emily had reached the age of 94 and was the last of the siblings to die.

Charlotte RONALDS

Charlotte was Frank's closest sibling, probably because she was sickly, but they also had much in common. She had the nickname 'Whig' after members of the political party (often wealthy merchants and industrialists) who favoured supremacy of parliament over the monarch, suffrage, abolition of slavery and toleration for Dissenters. Frank had quite a different tone with her, than with her 18-month-older sister Emily, regarding possible suitors:

> Go on like a good girl in the practice of goodness and prudence & modesty and all that and don't think of getting married without my concurrence…
>
> But you'll get married you sly little devil before I get home and tell me nothing about it untill *[sic]* all's settled, and then I shall get mawkish…
>
> Good bye, don't get married, at least if you do look me out a wife (don't you know where to look?)…Good bye once more, don't get married[C23]

Like Frank, Charlotte preferred studying to big parties, and he protected her:

> How do you come off amongst so much music and bon goût *[good taste]* which seems by Emily's & Brăss's account to have so suddenly inspired our worthy & savant family, how do you manage amongst so many great amateurs *[admirers]*? Alas I fear you stand humming and ha-ing & yes-ing & no-ing…whilst you ought to be "giving way to your imagination". Never mind, if you behave well I'll always take your part. We'll be two loggerheads [C23]

He often turned to Whig (more than her older brothers) for matters important to him, including the disappearance of the scientific books he had bought in Florence and ensuring money was transferred to him on his travels: 'I am extremely sorry my pretty whig to bother you with so many and such ennuyants *[tedious]* commissions. But what can I do? It is necessary to confide them to one person and you are the only one whom after mature deliberation I have thought the most fit. Have patience with me'.[C27]

On his travels he developed a storytelling technique of 'conjuring'[J18] Charlotte to join him at a place he enjoyed, and playing out a conversation in which he showed and explained objects of interest to her. These passages would have delighted her and almost made her feel she was there. In them, he depicts her as his equal — inquisitive, intuitive, observant, spirited and happy to speak her mind or, in his words, 'Sly-Boots'.[J18] She is also well read, knowledgeable in science and about foreign countries and good at drawing, although she 'cannot manage figures'.[J18] Both 'hate going over old Houses and Pallaces *[sic]*'.[J18]

Sadly, he only ever travelled abroad with her in his journal. She was the first adult sibling to die, passing away on her 28th birthday in 1825. Some years earlier Frank had written:

> You've been unwell, what has been the matter…I think cupping[*[15]*] might do you good if it is your old complaint. Don't die, that would be as Bad as to get married. Call some D[r], don't be squeamish…Write very particularly how you really are Direct here…I should look more awkward with the loss of my Whig than my Books.[C24]
>
> Farewell my pleasurable little Whig. Vi lascio la mano *[I'll stop writing]*. The further I get from you the more I love you.[C27]

Charles and Catherine RONALDS

Charles was articled to *Swain & Co* solicitors in the City, but established his own legal practice after 'not being taken into the house'.[J23] Although he assisted Frank in patenting his perspective drawing instruments and in other business matters, he was not particularly successful as a solicitor and died relatively poor at the age of 55.

The family of his wife Catherine née Fisher resided at Kensington Palace. Catherine and Charles made it their home for several years until both she and their young son passed away. Daughter Julia then spent most of her time with Jane and Frank at Chiswick, and she also looked after Frank in his old age.

Prior to that sadness, Charles had been the dandy of the family, fashionable, fun-loving and well-connected. Frank wrote home from his Grand Tour:

> Master Charles seems to "pede libero tellus pulsanda" *[dance footloose on the earth]* as actively amongst the quadrille *[dance]* fellows as some other folks amongst some other fellows, but as long as he preserves his pride all'right.[C24]
>
> Tell Charles I'll teach him the true French dandy hork and spit when I come home if it does not go out of fashion first.[C18]

Charles also enjoyed a drink and a wager:

> J. Ashby, W. Sears, W. Tombs, T. Simpson (stroke), and C. Ronalds (steerer), will row E. Hall's crew, of Chelsea, for £5 or £10. Their money will be ready at Mr Kerridge's, the Wilton Arms, Kennerton-street, *[sic]* Wilton-place *[Knightsbridge]*, on Monday evening, from eight to ten o'clock.[16]

Frank looked to Charles in particular to keep him abreast of what was happening in London during his travels:

> I was much obliged to you *[Charles]* for the little scrap of public news you sent me, and as I can get no english papers <u>at all</u> here *[Naples]* I will thank you to go and dine

---

[15] A medical treatment where local suction is created on the skin and the resulting blisters are drained.

[16] *Bell's Life in London* 22/08/1841.

with Peter some day and drink my health over a good bottle of port and <u>sit</u> in judgement together over the events and topics of the day (don't <u>hatch</u> anything to nick me) and give me a well digested report…You can't imagine how it cuts a man to hear news garbled and mangled of his own Country from a foreigner.[C21]

Nathaniel RONALDS

Nathaniel was the least accomplished of the siblings. He was headstrong and Frank enjoyed his company, probably because they were so different.

Nathaniel was formally apprenticed to Edmund at age 14.[J4] Four years later, Edmund wrote to Frank about him and also the attributes of his new horse, prompting Frank to send some advice to Nat:

> Very sorry to see the comparison made between it *[Edmund's horse]* and <u>some</u> people's laziness in mornings. It won't do on the long trot. Nat, don't follow my example, be <u>proud</u>, lazy folks <u>enjoy</u> less repose than active ones. Laziness is as bad as hot weather for the health [C24]

Nathaniel joined the partnership soon afterwards and when he came of age Michael Coventry left the business. Frank wrote to Edmund:

> Nat will now come into play…he will do all in his power to promote the mutual interest and I confidently hope you will both be ultimately compleatly *[sic]* successful & everything be just as it <u>should</u> be [C38]

Evidently the relationship was not completely successful as the partnership was dissolved in 1826, with Edmund continuing the business alone.

A little over a year earlier Nat had been arrested for 'intending to commit a breach of the peace' by duelling with pistols in Regents Park with a Mr George Hind. He participated in a second duel just a few days later in Calais with a Mr Kemp, to clear his name of helping to inform the police of the first duel. Fortunately both were settled without harm. In between the two events, Frank had felt the need to write to the Editor of *The Times* clarifying that he was not the Mr Ronalds referred to in their newspaper report.[17]

During Frank's travels Nathaniel was a good choice to run errands on his behalf:

> Has Nat been so good as to enquire for my books?[C22]

> I sometimes wish I had him *[Nat]* here to take care of me as he did when I went to Brighton once, I think we could agree very well together. If he'd let me snuff the Candles I'd let him carry the purse.[C22]

Nat was Frank's initial contact with home on his 1823–24 trip to France and Italy, although he later turned to the more reliable Emily. Nathaniel died at the age of 32.

---

[17] *The Times* 9/12/1824 3e.

Mr. Nathaniel Ronalds, a young gentleman residing *pro tem.* at the London Coffee-house, Ludgate-hill, and Mr. George Bird Hind, a merchant in the city, were brought up in the afternoon, by Morris and Blackman, on warrants issued upon the oath of Thomas Woods, charging them with intending to commit a breach of the peace, by fighting with pistols.

Mr. Ronalds first presented himself before the Magistrate, followed by Blackman, and said, "I have been taken prisoner by one of your officers, who has a warrant against a person whose name is very much like mine, but is not my name."

Mr. HALLS — You say it is not your name in the warrant; pray what is your name?

The defendant, after much hesitation, mentioned his own name.

Mr. HALLS — inserted the given name in the warrant, and said that was quite sufficient.

Defendant — But I am not to be detained, surely.

Mr. HALLS — Yes, you must indeed, Sir.

In a few moments Mr. Hind made his appearance, and both parties were placed before the Magistrate.

Mr. HALLS said, he supposed, from the warrants before him, that the defendants meditated a duel, and he wished to know from them if that was the case. They could not, as gentlemen, hesitate to answer the question fairly.

Both gentlemen said they did not choose to admit anything. All they chose to say was, that they knew not how the information could have reached the office. Each seemed particularly anxious that he should not be supposed to be the informer, but beyond this they refused to state anything.

Mr. HALLS asked them if they knew each other?

The defendants said they had seen each other before.

Mr. HALLS — Have you quarrelled?

Both defendants were silent.

Mr. HALLS (to Mr. Ronalds) — I see quite enough to convince me that you have quarrelled. Now, to prevent such an absurdity as a duel, have you any objection to shake hands with your opponent?

Mr. Ronalds — I have no desire to do so, Sir.

Mr. HALLS — Will you, Mr Hind, shake hands with Mr. Ronalds?

Mr. Hind — I never saw this person but once, and I am not fond of shaking hands with strangers.

Mr. HALLS said he must have the evidence of the person upon whose information the warrant was granted, and

Thomas Woods was called, who said he had not heard any quarrel between the two defendants himself, but he was informed that they had quarrelled, and he was requested to give information of it.

Mr. Hind — Pray did not the other defendant give you a 10*l.* note to buy pistols?

Witness — Yes, he did.

(Continued)

> *(Continued)*
>
> Mr. Hind — And pray how did you know the spot where we were to meet? (It seems the defendants were stopped on their way to the Regent's Park.) Marylebone is a large place, and how could you be so well acquainted with our route?
>
> Woods said he received the information from a gentleman of the name of Corry. After some further conversation, Mr. HALLS said he had heard quite sufficient, and called upon the defendants to enter into the sureties to keep the peace.
>
> *The Times* 8/12/1824 3c–d

> On Sunday last, a meeting took place at Calais between Mr. Kemp and Mr. Nathaniel Ronalds, the latter gentleman being one of the parties who was compelled to enter into recognizances, as detailed in our Bow-street report of the 8th inst. On the word being given, Mr. Kemp fired his pistol in the air, on which the seconds interfered, and Mr. Kemp having, in the presence of their respective friends, exonerated Mr. Ronalds from all knowledge or participation in the proceedings which led to the interference of the police on the former occasion, matters were amicably adjusted.
>
> *The Times* 14/12/1824 2d

Alfred RONALDS

Frank and Alfred had much in common to bridge their 14-year age gap. Both moved quickly from idea to idea, creating products that endure today, but eschewed publicity for their achievements (Fig. 4.8).

When Alfred returned home after his apprenticeship, industrious Hugh sent some advice via Frank from Albion:

> I'm very sorry to hear that Alfred has no occupation — inaction at his age is I think almost sure to produce rust and bad habits...Were I in his place with my present experience I would make a point of improving his present leisure by learning some mechanical business by which I should at any time be able to earn my living — a species of independence more secure from accident than the largish capital vested in business.[C50]

Alfred in fact used that period to develop skills in lithography and engraving with Frank, as outlined in Sec. 10.7.

By 1829, Alfred was a farmer in Staffordshire and researching his book on fly fishing. Published in 1836, *The Fly-fisher's Entomology* was a great success. American Arnold Gingrich wrote nearly 140 years later in *Fishing in Print — A Guided Tour through Five Centuries of Angling Literature*: 'It's impossible to overstress the importance of Ronalds...it is safe to say that no single book ever had

Fig. 4.8:   Frank's brother Alfred (1802–60) aged about 40; reproduced from a painting. *Fishing Gazette* 20/12/1913 571.

the revolutionary effect on the angling world...of *The Fly-fisher's Entomology*'.[B158] Its uniqueness was in bringing together biological studies with practical techniques and illustrated descriptions of how to make artificial flies. Alfred issued updated versions over the next few years and the twelfth edition appeared in 1921.[B42] It was translated and published in Japanese in 2011 by neurosurgeon Nobuyuki Kawano, and sold out within a year.

Wife Margaret née Bond and several children now in tow, Alfred moved to Wales in 1843 where he sold fishing tackle made with the assistance of their eldest daughter Maria. With his reputation growing, and perhaps because of it, he soon became restless again, and cousin Betsey recorded his thoughts of migrating to Canada.[C144] Margaret's death in 1847 triggered Alfred's decision to move to Australia — taking six of his children with him. He met Mary Ann Harlow on the boat, who produced her first of four more children a few months after marrying him.

Alfred now set up as an engraver, lithographer and printer in Geelong. He was soon lured to the gold rush, but not before creating a medal to commemorate Victoria becoming a separate colony in 1850; it was the first medal struck in

Australasia. Alfred was always an optimist; it was said 'he gave away nineteen medals in order to sell the twentieth',[18] while he wrote home prematurely that he 'was digging for Gold with great Prospects'.[J90] Alfred's last big venture was to set up a nursery in Ballarat to support his large family. Sadly, it did the opposite and he died leaving no will, a £600 mortgage and his youngest daughter Julia just two in a pioneer goldmining settlement. The family lost the nursery and Mary Ann took work as a seamstress.[B165] Her burial plot in Melbourne has no gravestone.

Alfred may well be the most recognised of the Ronalds family today, especially in the fly fishing community and the Victorian Goldfields, but his children were given a rather different start in life from their privileged cousins in the UK. The offspring nonetheless developed with their adopted country and many of Old Hugh's descendants continue to live in Australia.

Maria and Samuel CARTER

Frank made an early mention of Maria when she was 16: 'as to that little muffin Woman if <u>she</u> refused me a kiss when I come home as she did once (I have not forgot it) I'll just squeeze her as flat as a Crumpet'.[C43] In her twenties, she supported Emily's work and appears in letters from Robert Owen as well as Fanny Wright, who described her as 'also unmarried, & very amiable'.[A163] She wed just before her 29th birthday.

Samuel Carter (Fig. 4.9) was an attorney in Birmingham and interrelated with Maria's paternal grandmother in Coventry (Sec. 3.2). His career through the period 1831–68 was inextricably linked with the development of the railways across England — his entry in the *Oxford Dictionary of National Biography* records that in one parliamentary session he had control of 40 bills relating to the two railway companies for which he had been appointed solicitor. His obituary in *The Times* advised that he 'was a man of uncommon shrewdness and ability, a tactician of renown in Parliamentary committee contests, and a doughty politician, yet withal a man of high integrity and gentle and cultivated tastes'[19] while that in the *Daily News* noted that he 'formed many friendships in the world of literature, science, and art'.[20] In later life he was a benefactor to his hometown, including contributing £1,000 towards the new library.

Samuel was a great friend to Frank and a valuable counterpoint to his disregard of business or promotion in relation to his science. Samuel supported the development of Frank's ideas for steam propulsion on the Birmingham canals, spearheaded his knighthood and shaped his collection of electrical books into the

---

[18] *Star* Ballarat 28/4/1860 2b-c.
[19] *The Times* 11/2/1878 5f.
[20] *Daily News* 4/2/1878 2g.

Fig. 4.9:   Frank's portrait of his brother-in-law Samuel Carter (1805–78); drawn with perspective tracing instrument. IET 1.7.46.

'famous' Ronalds Library. The first surviving mention of Maria's assistance with the library is as early as 1820, when Frank wrote home: 'I am very much obliged to Maria for the pain she has taken to make a catalogue of the Books…to see that she writes so pretty & kind. The titles are all perfectly correct'.[C38]

Maria and Samuel's children also strove over many years for greater recognition of Frank's achievements. Son John Carter and son-in-law John Martineau Fletcher were initial trustees of the Ronalds Library while another son Hugh painted his portrait and later donated it to the National Portrait Gallery.

4.8 Concluding Remark

Frank's summary of a packet of letters he received in Sicily dated 28 July 1819 nicely illustrates the differing priorities shared by family members with their faraway son and brother. They range from the business-like, through the frivolous to the sentimental:

Emily.

Acknowledges rec[t] of mine of the 16th June…Affairs of W.F & G.F. *[Frank's uncles William Jnr and George Field Snr]*. *[James]* Montgomrey *[Snr]* extreemly *[sic]* ill with an inflammatory Complaint — Can't attend to his business. M A likely to be confined in 3 M⁰ˢ. *[Mary Ann gave birth to her first child on 10 November 1819].* Lord Byron's Don Juan very indecorous *[Cantos I and II of the poem had just been published].* Zotti dead *[who was to help deliver Frank's books from Florence]…* *[Henry]* Hunt *[Member for Westminster who agitated for radical parliamentary reform]* making another disturbance —

My Mother.

…Intends lending assistance to G.F. Chaˢ & Nat spending a great deal of money. Letter from Hugh, he is much occupied with the new Town.

Chas.

Coats à la francaise, high Collar, not short waisted or broad behind, short skirts, lots of padding in the breast, collar to come down low in front and stick out well & cut short before — Light blue trousers made <u>full</u> and cut out to fit the boot. Trousers going out, pantaloons made full & Hessian Boots. Has had my boat painted outside —

Charlotte

Wants me to come home ——[J23]

Life in the Ronalds family must have been a great deal of fun.

Chapter 5

A LIFE OF SCIENCE

Ronalds' passion for science in all its breadth; his motivations, research methods and skills, and how modesty and lack of confidence undermined his scientific career. A summary list of 180 inventions and the arrangement of their discussion in later chapters.

Science was Sir Francis Ronalds' first love. More specifically, he strove to understand elements of the physical world and apply his learnings to make things of use. In 1820 electrical and mechanical engineering were not yet professions and even the term 'scientist' had not been coined. That year he described his activities by saying: 'Nat[l] Philosophy was my favourite study and particularly electricity'.[J22] He called himself an 'Electrician'[C108] in 1842 when commencing at the Kew Observatory, an 'experimentalist'[J63] in 1846 and, in 1860, a 'schemer'[C375] in the sense of inventor.

Science infused all his activities. He took scientific equipment on his travels so that he could monitor the weather and conduct experiments. Similarly, when his brother Hugh set sail for America, Ronalds armed him with instruments and instructions. Hugh duly sent back his results a few years later:

Dear Frank,

You will receive with this a copy of my account of the Thermometer which I think you requested me to keep before I left England. — I fear you will find it very imperfect — but it is as complete as my opportunities would allow of — the time of making the observation was generally about one hour after sunrise, sometimes a little earlier or later. — I also send you an account of the Thermometer kept by my wife during her Journey from Philadelphia to Lexington — the observation generally made in the afternoon — unfortunately the Tube of your Barometer was broken on its journey so that I could not keep an account of its variations.[C50]

Ronalds' scientific interests were broad, as was common at the time, and extended beyond his principal arena of physics to chemistry, earth sciences and occasionally biology. He wanted to study everything. He commented frequently on local geography and geology, and at one of the Greek islands off the Turkish coast wrote: 'here I will venture a theory upon the formation of <u>pudding stone</u>. I am not often guilty of such things you know.'[J21] He took the opportunity in Sicily to study the marine life:

> I spent almost a whole night once in looking at them *[Phosphoric jellyfish]* and try-
> ing to discern their habits &c in the sea…Two conditions seem…necessary to the
> production of the phosphorescence viz the contact of Atmospheric air and Motion of
> their parts. I should like to have ascertained whether the phosphorescence would
> occur or continue after having been excited in common air, in Nitrogen and other
> gasses in order principally to know whether Oxygen is essential to it. But having no
> opportunity of getting apparatus here, what could I do?[J18]

Revd Joseph Priestley* had discovered oxygen in 1774.

As is generally the case, his scientific curiosity was first aroused by a teacher:

> At Dr Phillips's seminary, Walthamstow, I acquired a little taste for physics,
> (he having a good collection of instruments &c) & a smattering of geometry but
> on removal to Mr Cogan's, at Cheshunt, found that he hated Euclid *[ancient
> Greek mathematician]* as much as he loved Homer *[ancient Greek poet]*; so my
> principal scientific studies here were devoted to the service of a mouse-trap
> company of which David Martineau was the Treasurer & to which my Brother
> Edmund held the office of Tamer. We succeeded very well until no mice remained
> to be caught.[C375]

As well as Phillips, Ronalds would have been influenced by his Uncle Hugh the horticulturalist, who demonstrated that it was possible to pursue interesting science and earn money from it.[A197] Cousin Dr Henry published scientific papers in the medical field and the latter's brother Silvanus was a chemist at the Apothecaries' Company. Ronalds wrote when Silvanus died at age 24: 'I cannot think of our Chemical & other disputes together without great regret that they can never be renewed.'[C25] Uncle Charles Field, brothers Alfred and Hugh, and brother-in-law Peter also had scientific talents, while nephew Dr Edmund became a chemistry academic. Science was a topic of significant interest and discussion across the family. Nonetheless Ronalds would have been self-taught in most aspects of science as well as in the design and manufacture he pursued.

The fundamental motivations for his innumerable scientific endeavours were twofold: the personal satisfaction of proving something to himself, of solving an intellectual challenge, and the desire to create something of benefit. As he wrote regarding his telegraph:

> "Well but joking apart" said Mr Carr (a sensible old Gentleman) before the experts
> were completed, "do you really wish to convince me that you can send me a message

from London to Petersburgh *[sic]* in Ten minutes" "No" was the reply "My only wish at present is to convince myself."[C375]

Electrical science was of particular interest to him because so little was understood about it, it was difficult to study and he could see that it would be of immense value to mankind. Such altruism came naturally and was very satisfying for him, and the ideas were often excellent, but he was not good at promoting their adoption. In reality, many of his inventions simply made his own or his family and friends' activities better, easier, safer or cheaper. He also at times acted in the role of consultant.

Ronalds was sufficiently wealthy and careful with his money that he was not compelled to rely on his science for income, although numerous colleagues of his did, who were employed in government or private enterprises. He obtained just one patent and certainly had no wish to be one of the entrepreneurs of the industrial revolution who commercialised ideas to generate wealth. For most of his scientific life, he followed his own whims in determining priorities: flexibility in his endeavours was extremely important to him. In this sense he was a 'gentleman scientist'. He did sell some of his inventions, like his drawing instruments, but profit was never a primary driver. Consulting advice would have been given free of charge, and many of the devices he made for others were gifts. He was most comfortable working alone or with a colleague although, in the Kew Observatory years, he found himself part of a team at the centre of British science.

Nearly all of his available funds were devoted to science. He was able to establish a well-equipped laboratory at each of the family homes; this was where he conducted experiments and made his devices, and he regarded it as a critical facilitator of good science. He purchased many scientific instruments, along with materials and tools to build more. Even large machines, like his renowned temperature-compensated photo-barograph (Sec. 15.4), were manufactured at Chiswick rather than at the Kew Observatory because the facilities at home were better than those that could be afforded at the observatory.[R18] If he did not build the equipment himself, he would at least stipulate how it should be manufactured and on occasion designed the jigs and other apparatus necessary for fabrication.

Ronalds brought a range of qualities to innovation: he was visionary, creative, learned, practical, prudent and a perfectionist. It is an unusual and powerful mix — the articulation of farsighted concepts is generally associated with confidence more than detailed care and attention.

He valued the input of mentors. Even in later years when he had considerable experience, he would always seek feedback on a favoured idea from trusted friends and family. Once deciding to pursue it in earnest, he shifted easily from the big picture to detail, but only while the object remained important to him and he was able to focus on it. Otherwise he would omit to tidy up the loose ends in the rush to move on to another exciting idea. There are many examples of partially developed concepts and unfinished documents; he was also inclined to forget to put in names, dates and numerical values and did not notice incorrect spelling.

In designing and building any chosen device, however, he combined what the 1897 *Dictionary of National Biography* called 'extraordinary practical ingenuity'[M6] with perfectionism. More than one scientist talked of the 'scrupulous'[A133,B12] accuracy of his work. It could be guaranteed that his machines would be ergonomic, operate smoothly and reliably, and be as precise as humanly possible. He happily trialled modification after modification to achieve the best outcome, independent of cost and even if a client was impatient for results. The apparatus consumed his attention until it was right.

In common with other members of his family, Ronalds loved learning and he knew and understood the literature intimately. His renowned electrical library and catalogue at the Institution of Engineering and Technology is in essence an outcome of what he did as a matter of course. As he wrote in his draft book on turning:

> We hold with the great *[Antoine]* Lavoisier that before any person carries his inventions or a supposed discovery into execution he will always do well (if he can) to discover what has before been done which nearly or even remotely bears upon that which he is desirous to effect. Immense saving of time money & vexation is often obtained by this search [J54]

He had no hesitation in moving into a new, unknown area, and simply started by reading what had come before.

He needed good language skills to comprehend the international literature and when collecting books in foreign countries. He had elocution lessons in French and Italian in adulthood and wrote letters and published a long scientific paper in French. He learnt enough German to read technical works although he benefited in particular from his friend Alexander Blair's detailed assistance in this language (see Chapter 11).

On the other hand, he did not have good recall. He compensated by writing copious notes on his reading that summarised and analysed the work — in his words 'to assist my own wretched memory'.[J17] On several occasions he synthesised his knowledge from the literature with that gained through experiment and experience in drafting a textbook on a topic, but none were completed and published.

Breakthroughs came through his strong grasp of science fundamentals, and his intuition, verified by clever experimentation. He did not attempt mathematical treatments and probably had little or no grounding in algebra or calculus. His geometry was excellent but based on accurate drawing and an ability to visualise concepts in three dimensions. Like many engineers, he liked to hone and then communicate concepts in sketches and detailed drawings, of which many excellent examples survive — his inventions benefited from this strong mix of science, art and design talents. He also made intricate models to help explain the action of new devices to people less conversant in technical drawing.

Concepts were worked up quickly. He spent only about six months designing, constructing, testing, improving and demonstrating his telegraph and its various peripherals and yet it is the invention with which he is always associated. Once he was satisfied, he moved on to something new — there were more ideas than hours in the day. He seldom took the time to document and publicise his work because, as his nephew John Carter recorded many years later, 'He had no desire for wealth or notoriety'.[1] When he did communicate what he had learnt widely, it was in the hope that it would be of assistance to others, more than for personal gain. The exception was his years at the Kew Observatory where he realised the value of detailed annual reports in attracting funding and stakeholder support.

Ronalds' technical papers are not a joy to read. Early publications were too brief to relay their message clearly. Having little self-confidence, he assumed the reader would know much more than he did and dreaded writing anything that might contradict their views. His later papers are the opposite, and contain painful detail of every screw and nut. His goal here was to ensure that others could build his instruments. He wrote in his important booklet *Descriptions de quelques Instruments Météorologiques et Magnétiques* (1855): 'I hope the instrument-makers will not think that I have entered into too minute a level of detail'.[R24,2] The papers suffered from having little overarching context on the purpose of the equipment, its method of operation and results — that was assumed to be understood by anyone who wanted to acquire one.

These technical articles are just one example of his great generosity with his ideas. He always wanted to share concepts that he thought would bring benefit and devoted a great deal of time to developing devices for colleagues. The reward for him was that they valued what he made, and used and obtained advantage from it. We see this trait throughout his story, even when others were in competition. While George Airy* created a contest for primacy between his Greenwich Observatory and Ronalds' meteorological observatory at Kew, Ronalds continued to offer and provide new instrumentation so that Greenwich would have the best.

Ronalds' memory deteriorated noticeably with age and he became more absent-minded about little details as the number of urgent technical issues he was trying to resolve increased. After he left a scientific conference in Edinburgh in 1850, his new Kew Observatory assistant John Welsh advised him: 'I have rescued

---

[1] *Spectator* 20/3/1915 16–17.
[2] Translations are the author's.

your little telescope from oblivion! It had been dropped by you somewhere about the Parliament House — the finder handed it to the Police who gave it up to me on making the proper application.'[C305] He could also lose track of time. His personal communication was not incisive or inspiring and he was inclined to be repetitive about ideas of importance to him. He would have come across as quite vague and disorganised in later years, particularly under pressure.

It is clear that Ronalds' personal make-up influenced his science and his reputation in various ways. His biographer at the Society of Telegraph Engineers wrote in 1880 that he 'was a man of extremely sensitive temperament and of very retiring manners, and this may in a great measure account for his inventions never having been brought prominently before the public.'[M5] He abhorred any form of self-promotion or proactive influence of others and relied on his accomplishments being judged on their merits alone. His modesty ensured that he was understated if not equivocal in explaining the value of his work and his contribution to it — if he publicised it at all. He was also very generous in the credit he gave to others.

His attitude left open the possibility of associates deflecting credit due to him and sadly this seemed to happen a number of times. He detested such behaviour but he did not have the mental strength and confidence to contend it when it occurred. Instead he avoided conflict and let his feelings fester, aided only by the knowledge that he himself had acted with integrity. He gave as one of his reasons for printing a paper that summarised his work at the Kew Observatory:

> a minor object…is the correction of an erroneous report or impression, still obtaining a certain degree of currency, to the effect that I have been employed at Kew under the direction of other persons, and not in carrying into effect plans, inventions, and suggestions of my own…this is done much rather for the sake of guarding against the slightest imputation of *vile* charlatanerie *[fraudulence]* than for the purpose of defending my right to contrivances, &c. which…are not such that any person… would dream of greatly priding himself upon.[R18]

This was written in 1848, five years after the initial actions that promulgated the view that Charles Wheatstone* was responsible for the new equipment at Kew (Chapters 13–14). It had taken him many years longer to respond to the earlier apparent 'borrowing' of parts of his telegraph by Wheatstone and William Cooke without acknowledgement (Chapter 7), and to George Singer* taking the kudos for his mode of electrical insulation (Sec. 6.4). There are hints of similar behaviour directed towards Ronalds by Henry Collen (Sec. 15.2), possibly Michael Faraday* (Sec. 7.3) and Charles Holtzapffel* (Sec. 12.7), but Airy was certainly the worst offender, as described in Chapters 13–15.

Inventors and scientists depend on ideas — it is often all they have — and Ronalds was extremely sensitive to the possibility that his concepts might be perceived not to be original. He was not seeking renown, but simply for those who used his inventions to know that the concepts he called his own were not in fact due to others. He took increasing precautions in later years to prevent such occurrences, including putting a lock on the door of one of the rooms at Kew: 'I wished sometimes primitive trials to be conducted in private for everybody knows how often it happens that disagreeable disputes arise about priority of invention &c'.[J77] He became almost paranoid. By the 1860s, he had matured to the extent of allowing his family to put him forward for knighthood and some of his close scientific colleagues to publicise his key achievements.

Ronalds was also very vulnerable to criticism of his intellectual output, which he took to heart as personal denigration, particularly when he was young. Lacking confidence, it was not easy for him to put forward his work and, being a perfectionist, he only did so when he thought it worthy of attention. His reaction to rebuke was again to withdraw to cope with his feelings privately rather than try to influence others to the contrary. Within the family setting, he demonstrated such behaviour after Peter pilloried his travel journal (Sec. 8.7). The Admiralty's effective ridiculing of his electric telegraph generated a similar response and contributed to his decision not to publicise subsequent innovations (Chapters 6 and 7) — he did not submit another article to a scientific journal for 30 years. It is little wonder that the scientific community was unaware of the breadth of his achievements.

The scientific community did not know it, but Ronalds was highly prolific for over 50 years. His activities may be treated for convenience as having been conducted in phases although, in reality, he was more spontaneous than that. Very little is known about his initial efforts during his schooldays and apprenticeship, but his electrical science and engineering in the 1810s probably produced his first important inventions. These are addressed in two chapters: the first touches on various theoretical concepts and practical devices he developed, after introducing the state of knowledge in electrical science at the time (Chapter 6), and the second focuses on the electric telegraph (Chapter 7). Next follows a period of about 17 years of largely mechanical invention, discussed in Chapters 10 and 12, with the former being devoted to his drawing instruments. The intermediate Chapter 11 describes an important application of these instruments. He then moved to the Kew Observatory, where he focused on meteorological devices. Again his technical achievements are separated into two chapters: Chapter 14 primarily on his atmospheric electrical instrumentation and results, and Chapter 15 describing his photo-recording machines. These are preceded by an introductory chapter outlining his role in developing the observatory. Some final inventions are included in the last chapter, where there is also a section on the Ronalds Library.

Overall, Ronalds worked on 180 different ideas (listed in Table 5.1) of which sufficient is known to enable description in these pages. Arguably the most noteworthy are the telegraph (Chapter 7); the photo-barograph (Sec. 15.4); the hinged tripod stand (Sec. 10.4); the lathe slide rest (Sec. 12.7); the electrical library and its card catalogue (Sec. 16.3); the battery-operated clock (Sec. 6.3); perspective tracing instruments (Chapter 10); electric insulation (Sec. 6.4); electrometers and electrographs (Sec. 6.4 and Chapters 14–15); improved electrostatic generators (Sec. 6.5); the Fire Finder (Sec. 12.14); the airborne meteorological instrument platform (Sec. 14.3); the propelling rudder (Sec. 16.1); and the precise drawings of the Carnac megaliths (Chapter 11). His most important theoretical insights concern retardation of electric signals (Sec. 7.3), the delineation of voltage and current in the battery and in the atmospheric electrical circuit (Secs. 6.2 and 14.4), the mechanism by which the battery worked (Sec. 6.2), and the window of vision in fishing (Sec. 12.11). He also published close to 30 papers and books, which are listed separately in the Bibliography.

Table 5.1:   Approximate Chronology of Ronalds' Documented Inventions and Scientific Studies.

No.	Year	Invention/Study	Reference	Surviving example	Chapter
School Years:					
1	1800	Mousetrap	C375		5
Electrical Science and Engineering:					
2	1810	Insulation of electrical equipment against moisture	R28		6.4
3	1810	Atmospheric electricity observations — Highbury	B12		6.4
4	1811	Electrocution of rats	C375		6.4
5	1812	Atmospheric electricity observations — Hammersmith	C375		6.4
6	1813	Atmospheric electricity recording — electrometers & clock — model 1	B12		6.4
7	1814	Dry pile — use for meteorological observation	R1	London (2)	6.2
8	1814	Dry pile — mechanism of generating electricity	R2		6.2
9	1814	Dry pile — delineation of intensity and quantity	R2		6.2
10	1814	Atmospheric electricity recording — pendulum and indices	R7		6.4
11	1815	Battery-operated clock — model 1	R3		6.3
12	1815	Battery-operated clock — model 2	R5		6.3
13	1815	Dry pile — use to light Volta's gas lamp	R5		6.5
14	1815	Atmospheric electricity recording — rotating resin plate — model 1	R7		6.4
15	1815	Atmospheric electricity observing rod — model 1	R6		6.4
16	1815	Apparatus for making electrometer straws	R7		14.2
17	1815	Apparatus for making electrometer leaves	R7		14.2
18	1815	Electroplating	J85		16.3

(Continued)

Table 5.1: *(Continued)*

No.	Year	Invention/Study	Reference	Surviving example	Chapter
19	1816	Electric telegraph system	R7		7.1
20		dials indicating characters		Edinburgh	
21		message grid		London	
22		expansion joints		Hammersmith	
23		testing posts			
24	1816	Gas pistol	R7		7.1
25	1816	Automated influence electrostatic generator	R7		6.5
26	1816	Electric signal retardation due to induction	R7		7.3
27	1816	Reliable friction electrostatic generator	R7		6.5
28	1816	Battery-operated clock — model 3	R7		6.3
29	1817	Library card cataloguing system	A122		16.6
30	1818	Battery-operated clock — model 4	R7		6.3
31	1818	Atmospheric electricity observing rod — model 2	R7		6.4
32	1818	Air thermometer for electricity	A56		14.2
33	1818	Update of Priestley's *History of Electricity*	J16		6.6
Travel Activities:					
34	1819	Atmospheric electricity observations — Vesuvius	R8		6.4
35	1819	Atmospheric electricity observations — Sirocco	R7		6.4
36	1820	Explanation of mirage	J20		12.11
37	1820	Preservation of Egyptian papyrus manuscripts	J27		8.3
38	1821	Lithography	Fig. 4.4	IET	10.7
39	1822	Analysis of Egyptian pyramid casing	C47		8.7
40	1823	Sulphur extraction and import	J31		9

Mechanical Devices:

No.	Year	Description	Code	Location	Ref.
41	1824	Perspective tracing instrument: from real life — Type 1	R10	London, Edinburgh, Whipple	10.1
42	1824	Perspective tracing instrument: from real life — Type 2	R10		10.1
43	1824	Perspective tracing instrument: from real life — Type 3	R10		10.1
44	1824	Perspective tracing instrument: from real life — Type 4	R10	London	10.1
45	1824	Perspective tracing instrument: from real life — Type 5	R10		10.1
46	1824	Perspective tracing instrument: from plans & elevations	R10		10.2
47	1824	Window-mounted sundial	A39		12.3
48	1824	Table leg level adjustment	J14		12.18
49	1825	Hinged tripod (or hexapod) stand — model 1 — with table	R10		10.1
50	1825	Volute compasses	J14		10.5
51	1825	Letter stamp	J14		12.8
52	1825	Temperature-controlled ventilator	J14		12.18
53	1826	Copperplate engraving and printing	R10		10.7
54	1826	Engraving table	J41		10.7
55	1826	Upside-down window blind	J14		12.18
56	1827	Tracing glass	J14		12.15
57	1827	Watch alarm	J14		12.17
58	1827	Automatic oil feeder to machinery	J14		12.9
59	1827	Trochiametrograph	J14		12.9
60	1827	Tensioned bow to straighten rods	J14		12.16
61	1827	Stable candlestick	J14		12.18
62	1827	Level adjustment for theodolite	J14		12.9

(Continued)

Table 5.1:   (Continued)

No.	Year	Invention/Study	Reference	Surviving example	Chapter
63	1827	Device to draw circles and ellipses in perspective	R10		10.5
64	1828	Geometric pen	R10		10.5
65	1828	Paddle wheel energy storage to limit fuel use	J14		12.12
66	1828	Horse-drawn carriage universal joint	A37		12.2
67	1828	Hinged tripod stand — model 2 — triangular metal head	R10		10.4
68	1829	Rigging or tackle block	A38		12.4
69	1829	Headrest for drawing portraits	C89		10.3
70	1829	Plumbing for multi-storey house	J14		12.18
71	1829	Hay-making by steam or hot air	J14		12.12
72	1829	Artificial grinner — a children's toy	J14		12.18
73	1829	Tools for anamorphosis sketching	J40		10.5
74	1829	Bisecting beam compass	J14		10.5
75	1829	Compact drawing instrument set	J14		10.5
76	1829	Doubly-reflecting pocket surveying instrument	J39		12.9
77	1829	Curvilinear turning attachment for lathe slide rest	B50	London	12.7
78	1830	Turner's pen	C63		12.7
79	1830	Epicycloidal cutting apparatus	J54		12.7
80	1830	Teleographic (or telegraphic) clock	C375		12.6
81	1833	Hinged tripod stand — model 3 — base ties	C239		10.4
82	1833	Equatorial mount for large telescope — model 1	Fig. 12.18		12.10
83	1833	Equatorial mount for large telescope — model 2	Fig. 12.20		12.10
84	1834	Archaeological recording at Carnac and elsewhere	R11		11

85	1835	Window of vision in fishing	B42		12.11
86	1835	Letter printer and typewriter	Fig. 12.13		12.8
87	1835	Steam-driven rotary motion	Fig. 12.26		12.12
88	1837	Turner's Manual	J54	IET	12.7
89	1837	Steam propulsion on canals — model 1 — cogged rail	J55		12.12
90	1837	Steam propulsion on canals — model 2 — smooth rail	J55		12.12
91	1840	Water jet propeller	J14		12.12
92	1840	Equipment to draw panoramas	C92		10.8
93	1840	Equipment to draw stage sets and large perspectives	C92		10.8
94	1840	Risk strategy: preservation of life and property from fire	J56		12.14
95	1840	Fire escape ladder	C99		12.14
96	1840	Self-actuating fire alarm	C94		12.14
97	1840	Fire-locating telescope — horizontal planar map	J57		12.14
98	1840	Fire-locating telescope — panoramic map	J57		12.14
99	1841	Large portable telescope stand	C96		12.10
100	1842	Hinged tripod stand — model 4 — solid triangular legs	R21	London	10.4
Kew Observatory:					
101	1842	Rain and vapour gauge — model 1 — on land	R12	London	14.3
102	1842	Atmospheric electricity observing rod — model 3	R12	London	14.1
103	1843	Improved Volta electrometers	R12	London (4), Oxford?	14.2
104	1843	Improved Bennet gold-leaf electroscope — model 1	R12	London (3)	14.2
105	1843	Improved Henley electrometer	R12	London	14.2
106	1843	Spark measurer	R12	London?	14.2

(Continued)

Table 5.1:   *(Continued)*

No.	Year	Invention/Study	Reference	Surviving example	Chapter
107	1843	Charge distinguisher — Leyden jar with gold leaves	R12	London	14.2
108	1843	Atmospheric electricity recording — rotating resin plate — model 2	R12		15
109	1843	Pluvio-electrometer	R12		14.2
110	1843	Air–earth current observations	R12		14.4
111	1843	Balance anemometer — model 1 — wood	R12	London	14.3
112	1843	Wind vane	R12		14.3
113	1844	Spring anemometer	R12		14.3
114	1844	Atmospheric electricity recording — resin plate — photography	R12	Exeter	15.2
115	1844	Registering or night electrometers	R12	London	14.2
116	1844	Portable electrometers	R21		14.2
117	1844	Modified Coulomb torsion balance — model 1	J62		14.2
118	1844	Storm (or observer's) clock	R20		14.3
119	1845	Bennet gold-leaf electroscope — model 2 — with desiccant	R21		14.2
120	1845	Balance anemometer — model 2 — aligning to wind	J62		14.3
121	1845	Photo-electrograph — model 1 — straws	R15		15.3
122	1845	Photo-electrograph — model 2 — straws & Leyden jars	R15		15.3
123	1845	Thermograph — model 1	R15		15.5
124	1845	Barograph (syphon type) — model 1	R15		15.4
125	1845	Electro/baro/magnetograms — Kew	R15	Cambridge, Exeter	15
126	1846	Declination magnetograph — model 1	R15	London	15.6
127	1847	Modified Argand lamp	R21	London	15.1
128	1847	Temperature-compensated barograph — model 2	R21	London	15.4

129	1847	Airborne instrument platform	R16		14.3
130	1847	Global atmospheric electricity variability	R17		14.4
131	1848	Atmospheric electric potential dataset 1843–48	A77	Exeter	14.4
132	1849	Instrument screen	R21		14.3
133	1849	Horizontal force magnetograph — model 1	R19		15.7
134	1849	Modified Rumford polyflame lamp	R19		15.1
135	1849	Daguerreotype polishing board	R20		15.2
136	1849	Daguerreotype polishing buff	R20		15.2
137	1849	Daguerreotype burning off and fixing stand	R20		15.2
138	1849	Daguerreotype coating box	R20		15.2
139	1849	Ordinate board — linear	R20		15.2
140	1849	Ordinate board — angular	R24		15.3
141	1849	Magnet arc amplitudes apparatus	R21		15.7
142	1849	Gelatine tracing board	J75		15.2
143	1850	Daguerreotype tracings using gelatine sheet	R20	IET	15.2
144	1850	Daguerreotype prints from gelatine sheet	R21	Exeter, IET	15.2
145	1850	Drawing dividers — larger intervals	R20		15.2
146	1850	Drawing dividers — small subdivisions	R20		15.2
147	1850	Outdoor atmospheric electricity apparatus	R21	London	14.1
148	1850	Round-house atmospheric electricity apparatus	R21		14.1
149	1850	Modified Coulomb torsion balance — model 2	J85		14.2
150	1850	Photo-electrograph — model 3	C328		15.3
151	1850	Horizontal force photo-magnetograph — model 2	R21	London	15.7

(Continued)

Table 5.1:   *(Continued)*

No.	Year	Invention/Study	Reference	Surviving example	Chapter
152	1850	Vertical force magnetograph — model 1	R21		15.8
153	1851	Vertical force magnetograph — model 2	R21		15.8
154	1851	Declination magnetograph — model 2	R21		15.6
155	1851	Formal trial of three magnetographs	R21		15.11
156	1851	Photo-electrograph — model 4 — frequency observations	R21		14.4
157	1851	Improved Regnault hygrometer and aspirator	R21		14.3
158	1851	Improved Regnault hygrometer — remote operation	R24		14.3
159	1851	Equipment to make and verify thermometers	J84		13.3
160	1852	Vertical force magnetograph — model 3 — spring supports	C327		15.8
161	1852	Modified Osler self-recording anemometer	Fig. 14.21		14.4
162	1853	Coal gas production and lamp	J85		15.1
163	1853	Temperature-compensated barograph — model 3	J85	Oxford	15.4
164	1854	Thermo/hygrograph — model 2	J85		15.5
165	1854	Rain and vapour gauge — model 2 — on water	R24		14.3
166	1855	Electro/baro/thermograms — Oxford	C350	Oxford, IET	15
Retirement:					
167	1855	Pamphlet holder	J85		16.3
168	1855	Electric candelabra	A94		16.1
169	1857	Combined horizontal & vertical force magnetograph	C368		15.8
170	1858	Consistent pulsing light for lighthouses	C372		16.1
171	1858	Tinting to reduce blackness in photos	J85		16.1

172	1859	Cheap forked telescope supports — model 1	J85		16.1
173	1859	Propelling rudder — model 1 — gondolas	C381		16.1
174	1859	Metal gunpowder magazine	J85		16.1
175	1859	Sundial meridian alarm	J85		16.1
176	1859	Pocket watch box	J85		16.1
177	1860	Electrometer to observe electrical effects of pollination	J85		16.1
178	1861	Forked telescope supports — model 2 — horizontal panning	J85		16.1
179	1862	Propelling rudder — model 2 — steamships	C381		16.1
180		Ronalds Library	R28	IET	16.3

London = Science Museum, South Kensington
Edinburgh = National Museums of Scotland
Hammersmith = Kelmscott House Museum
Oxford = Museum of the History of Science
Whipple = Whipple Museum of the History of Science, Cambridge
Cambridge = Royal Greenwich Observatory Archive, Cambridge University Library
Exeter = National Meteorological Library and Archive
IET = Ronalds Archive, Institution of Engineering and Technology, London.

Chapter 6

ELECTRICAL SCIENCE AND ENGINEERING 1810–19

Ronalds' early work in (primarily) static electricity, both theoretical and practical, against a backdrop of the state of knowledge at the time. His sudden halt after the double rebuffs of his electric clock and telegraph and the death of his mentor Jean-André de Luc.

Although the first evidence of Ronalds' activity in electrical science and engineering dates from around 1810, much of his important work was conducted in a very short period just after he retired from the cheesemonger business in 1814, when he was aged 26–28. He later availed himself of unique circumstances on his Grand Tour to perform additional isolated experiments.

There is a remarkable body of work surviving from this short period, ranging from important theoretical insights to pioneering electrical engineering applications. He made fundamental advances in the understanding of electrical science, including an early quantitative delineation of what we know today as electromotive force and current, the initial description of the effects of induction on the degradation of electric signals in long cables and the mode of operation of the dry pile (an early form of battery). His farsighted experimental investigations included the earliest measurements of atmospheric electricity variations at an erupting volcano. The first electric clock; two effective forms of insulation; the electrograph (to record electrical phenomena); reliable and semi-automatic electricity generators; and of course the telegraph — which is discussed in the next chapter — are some of the useful devices he invented and improved during this brief time. It is unimaginable what might have been achieved had his work continued. As it was, he lost confidence and left the field after the two inventions of which he was especially proud had been dismissed out of hand.

He had exceptional gifts as an experimentalist, inventor and designer. As described in the previous chapter, his strong theoretical understanding was combined with practicality and great originality across the mechanical and electrical sciences and beyond. In scientific activities he was thorough, accurate and patient, although these traits did not always extend to his day-to-day life. The most distinguishing feature of his work, however, was his desire right from the start to deploy electricity to practical ends. Through much of his lifetime, electricity remained primarily a scientific curiosity — there was more than enough to be done just to understand its properties — with few ideas of possible engineering use. Thomas Edison did not patent his incandescent lamp until 1880. Ronalds can therefore be considered to be the first electrical engineer. As research engineer Rollo Appleyard OBE wrote of him in *Pioneers of Electrical Communication* (1930): 'Electricity had become a science, but not yet an engineering science. He bridged the gap that for so long separated tentative efforts from trials…designed to work efficiently and to endure.'[M7]

Ronalds wisely involved more experienced scientists in his early electrical work. He had two primary mentors. The first was George Singer*, who was abreast of the many breakthroughs occurring at the time in electrical science. Being only two years older, he perhaps became jealous of Ronalds' rapid advancement or alternatively lacked awareness of the effect his remarks could have on a person with little confidence and experience. Either way, Ronalds was quickly introduced to the often competitive and catty world of science as their relationship soured in a very public way. He was able to turn at this time to Jean-André de Luc*, who offered sound advice on the personal side of scientific endeavour. De Luc was the opposite of Singer, being an infirm 88-year-old who was set in his ways scientifically.

Various sources are available to assist in uncovering Ronalds' electrical endeavours. The first are his scientific articles published at the time. We also have his reflections on his work at later dates. His book on the electric telegraph written in 1823 includes other hitherto unpublished devices as well as his thoughts on various electrical phenomena. In 1860 he wrote a detailed autobiographical letter to his brother-in-law Samuel Carter, at the request of William Walker who was preparing a memoir on him (Sec. 16.4). The journals from his 1818–20 Grand Tour also include several references to his science. Finally, various comments were made on his work in the literature at the time.

To help set the scene for Ronalds' activities and outcomes, we start this chapter with a very brief history of electrical science up to the time he commenced in the field. His interests spanned across all the topics mentioned — and more.

6.1 Early History of Electrical Science

Today we use words like electric charge, current, potential difference, electromagnetism, fields and so on routinely, but these concepts were not obvious to the pioneers of electricity and were hard-won through careful experimentation and thought. Simply put and in modern parlance, electric charge is the fundamental driver of electrical phenomena and can be either positive or negative. Like charges, that is, two positive or two negative charges, repel each other, while opposite charges attract. Current is the rate at which charge moves past a point, while potential difference, also referred to as voltage or electromotive force, is a useful way of describing what causes charges to move in the first place. Charges move freely through some materials (for example, metals) known as conductors, and with difficulty through others (like glass, wood or resin), which are called insulators.

The electricity existing in nature was the subject matter of the earliest investigations by electrical scientists. It was known in ancient times that some fish could deliver shocks. John Walsh and Lazaro Spallanzani FRS determined these to be electrical in nature in the second half of the 18th century. The suspicion that lightning was electrical was verified through Benjamin Franklin's proposal in 1750 to put a conducting rod high in the air during a storm. The Ronalds Library (described in Sec. 16.3) contains original papers pertaining to both of these discoveries together with engraved portraits of Spallanzani and Franklin.

Electric charge was soon detected in the air in other weather conditions as well. Father Giovanni Battista Beccaria FRS (1716–81) explored the electrical variability of the atmosphere by stringing a 43 m long wire insulated at its ends between a chimney and a tree on a hill outside Mondovi in northern Italy. He monitored the charge in his bedroom using a device called an Electroscope which was linked to the wire. Conclusions from his observations over 15 years included that the air was positively charged in fine weather and followed a cycle through the day. Various weather patterns altered the measured atmospheric electricity. We now understand these effects to be part of a complex global atmospheric electrical circuit, on which research continues today.

Scientists also began to create electricity in order to study it. It had long been known that friction generates static electricity — a familiar example is brushing hair on a cold, dry day, after which the hairstyle may be a little 'flyaway' and lint will be attracted to the brush. William Gilbert (1544–1603), physician to Queen Elizabeth I, investigated this type of electric charge using an electroscope he developed; Ronalds collected three of his works for his library. Revd Abraham Bennet FRS invented a more sensitive gold-leaf electroscope in 1786 that comprised two fine slips of gold suspended together inside a glass cylinder from a conducting cap. Charge applied to the cap spreads to the leaves and they diverge through repulsion. If the cylinder has interior earthing strips, the leaves collapse again after striking them and no charge accumulates on the glass. Further detail and drawings of this instrument and several others mentioned in this chapter are provided in Chapter 14, which discusses improvements made to them by Ronalds at the Kew Observatory.

Friction machines for making electricity became more sophisticated over time. Singer sold the device in Fig. 6.1, which comprises two insulated metal tubes at an adjustable distance from a central glass cylinder. A cushion resting on the glass is linked electrically to one of the tubes. When the glass cylinder is rotated on its axis, the rubbing gives a negative charge to the connected tube and the opposite metal tube is configured to collect the positive charge from the glass. Apparatus is linked to the tubes to perform electrical studies.

Static electricity could also be generated by an Influence Machine. Its basis is the phenomenon of electrostatic induction: when a charged object is placed near a conductor, an opposite charge of corresponding magnitude is induced on the conductor if it is grounded. The forerunner was the Electrophorus popularised by Alessandro Volta in 1775. It comprises in simplest form two discs, one made from an insulator and the other a conducting material with an insulating handle. The insulator disc is first rubbed so that a charge accumulates. When the

Fig. 6.1:   Ronalds' friction electrostatic generator; built by Singer (*c.*1810). Science Museum 1876–0087.

conducting disc is placed on it, charge separation occurs in the metal. Touching the upper surface of the metal drains the charge on this side and the conducting plate is left with a charge opposite to that in the insulator, which can then be used for electrical purposes.

Bennet developed an electricity doubler using the principle of induction to progressively amplify a small charge so that it could be observed in his electroscope. His concept involved three plates and had a mode of operation that may be exemplified as follows. Assuming Plate *A* is positively charged, it can be used to induce a negative charge in Plate *B*, which can in turn induce a positive charge in Plate *C*. If Plates *A* and *C* are then combined, a larger negative charge may be induced in Plate *B*. By continuing the process, the charge almost doubles each cycle. William Nicholson made an improvement to the device in 1788, attaching one of the plates to a crank handle so that it could be passed close to each of the two fixed plates in turn.

The static electricity generated by these machines could be stored in a Leyden jar — the first capacitor. Invented in 1745, it comprises a glass jar partially lined with a conductive material that is connected electrically to its cap. Charge given to the cap flows to the inner lining and the opposite charge accumulates on the earthed exterior metal coating through induction. The charges remain for a considerable period after the electricity source is disconnected.

An early use of electric charge was to light flammable gas. Volta had discovered methane in 1776 and shortly afterwards he invented the gas pistol to analyse gas properties through explosion and also a gas lamp lit by his electrophorus.

Volta's description of the first battery in 1800 was a major turning point in electrical science. Called the voltaic pile, it was made up of a series of zinc and copper discs interleaved with cardboard soaked with brine. A potential difference (measured today in volts) was created across the two ends. If the terminals were joined in a circuit, an electric current would flow for a significantly longer duration than the rapid discharge of static electricity that had been seen to date in electrical science. The new behaviour was termed 'galvanism' after Volta's colleague Luigi Galvani to distinguish it from the familiar 'electricity'.

Of course there was not a ready supply of electrical appliances at hand to demonstrate the utility of galvanism. Nicholson reported one of the ways in which Volta completed the circuit: 'Two blunt probes were inserted in the ears, and the shock passed through the head, after which the communication was kept up. A particular sound, like crackling or boiling, was heard; but the author did not think it prudent to make this experiment repeatedly.'[A5] Many early studies with the voltaic pile focused on the effects of making an electrical circuit by inserting the terminals into a liquid. Sir Humphry Davy FRS and other chemists explored this new field of electrolysis, decomposing fluids and isolating new chemical elements.

To help understand how the battery worked, and particularly the role of the brine-soaked discs in comparison with the electrically dissimilar metals, the dry pile or electric column was developed within a few years of Volta's invention. De Luc was one of its creators, with improvements by Singer and later Giuseppe Zamboni.[A179] Singer's design comprised alternate thin layers of zinc, silver and cartridge paper (neither moistened nor artificially dried). The dry pile was found to have a potential difference but, unlike the 'wet' pile, could not be used for electrolysis. Its potential was also maintained in use whereas a voltaic pile ceased to work after a while. It was all a complete mystery. Controversy over the mode of operation of the two batteries raged for many years; it was nearly a century before the electron was discovered.

Electrical science could be a dangerous activity. Prof Georg Richmann was one of the early electrical scientists to die in the course of his work — he electrocuted himself in St Petersburg shortly after Franklin had suggested his experiment with lightning. Rotation of Singer's electric machine generated 'a remarkable smell'[C10] together with spectacular sparks that were painful if contact was made. Discharge of a Leyden jar could and did on occasion kill — Ronalds made mention of its 'intolerable pain'.[R7] It was natural that many early experiments to understand the effects of electricity were deployed on animals.

6.2 Delineation of the Properties of the Dry Pile

Ronalds' first scientific papers, together with the first invention he documented, all concerned the newly developed dry pile. He would have been particularly attracted to this device for two reasons. Having a sustained source of potential difference, it was a convenient new electrical machine. More importantly, it offered promise as an instrument for studying meteorology and atmospheric electricity — interlinked fields that were very close to his heart — because its electrical properties seemed to vary with the weather. He made piles himself but also purchased Singer's piles for his experiments. His five papers on the topic were published in 1814–15 in the *Philosophical Magazine* edited by Alexander Tilloch, and the first was reprinted nearly 30 years later in the *Annals of Electricity, Magnetism, and Chemistry* (1842).[R1–R5]

Ronalds completed this first paper just 15 weeks after 'retiring' from *Ronalds & Coventry Cheesemongers* and sent it to Singer for review. Singer kindly sent the paper off to Tilloch with a cover note:

Sirs, — I have the pleasure to inclose *[sic]* to you an account of some experiments on the action of the electric column, communicated to me by an electrician of great promise, whose scrupulous attention to the essentials of accurate experimental inquiry it has frequently afforded me pleasure to observe.

It will be seen that these experiments…tend to confirm the opinion I have advanced in the "Elements of Electricity," page 478[R1]

The aim of Ronalds' work was to understand how the pile's electric charge responded to changing temperature and moisture. When a room is heated, the humidity usually reduces. He decoupled this effect and used desiccants and other artifices to adjust the two parameters independently for the study. He performed the experiments both on columns sealed in glass and on those open to the air. The charge was measured using a gold-leaf electrometer.

He ascertained that humidity could affect the column's electricity in counteracting ways: through absorption by the paper segments in the pile itself and through conductivity caused by moisture on the glass container. Unfortunately he did not express these findings clearly —— his early papers were very concise relative to others written at the time. In addition to being inexperienced, he may also have been reluctant to spell out observations that did not align with others' opinions.

De Luc was one of the first to notice the paper and he did not concur with it. Although they were acquainted at this time, de Luc submitted his polite but detailed rebuttal to the magazine, commencing with: 'it will be easy for me to show that this disagreement between us results from a mistake on his part'.[A11] Ronalds would have been mortified. In fact they did have similar views but de Luc had not picked this up from the paper. De Luc was also sensitive about 'his' electric column: the Royal Society in a letter signed by Davy had declined to publish his papers, which happened to disagree in some respects with Davy's work.[A179]

Ronalds' second paper was penned in December 1814. He started by stressing that he 'had no intention to dispute with a philosopher of so great and universally acknowledged acuteness, the truth of a fact which his own elegant experiments have full established.'[R2] He went on to give an early hint of how the pile worked — one which we know today to be correct. In doing so, he summarised the state of knowledge existing at that time regarding the mechanism of generating electricity:

> Mr. De Luc's valuable experiments and observations lead to the conclusion, that the presence of water, or of some conducting fluid, in the substances which have been hitherto interposed between the metals, is necessary to the accumulation of electricity. Whether this effect is occasioned by the presence of water only, because a conducting fluid is essential? which I take to be his opinion; Whether water acts merely as a conductor, differing in some unknown electric relation from the metals? which I imagine to be that of Mr. Singer; or, Whether any kind of decomposition is necessary? are questions not yet determined; but I have no doubt the researches of those intelligent gentlemen will contribute very materially to elucidate the subject.[R2]

It had already been observed by de Luc, Davy and others that current generated by a voltaic pile was associated with oxidation of the zinc. In today's language, the dissolving zinc leaves free electrons in the metal. Positive hydrogen ions in the water combine with electrons at the copper electrode and the net effect is to produce an electric current. Details of this process, however, remained opaque for decades.

Ronalds' comments concerned the electricity generated by the dry pile. De Luc (together with Volta) believed that the paper simply performed the role of connecting the effect produced by successive pairs of dissimilar metals. Singer's view was that moisture enhanced the communication between the metals (what we would now call the contact potential) in some ill-defined way.[B12] The third option Ronalds mentioned — that decomposition was involved — suggested that he already believed the dry pile to behave in the same manner as a wet pile, except that the oxidation was imperceptible in the former case because there was so little moisture and the pile was not used to draw a steady current.

He repeated this view soon afterwards in a letter to de Luc:

It has appeared to me Sir that a decomposition always <u>accompanies</u> Electromotion[C9]

De Luc disagreed, replying:

I still believe that no <u>change of State</u> is necessary for producing electromotion on the Electric-column…the electromotion of the Column, is only owing to the contact of Zinc and copper, the binary groups being separated by some <u>conducting non-metallic</u> substance[C10]

A few years later, Ronalds was even more explicit in his understanding that:

the dry electric column (as it is called) is after all nothing more than a modification of the humid pile of Volta…With all possible deference to the most respectable authorities, I cannot avoid forming this conclusion.[R7]

Nevertheless controversy over the 'chemical' versus 'contact' theories continued for decades.

Another aspect of his second paper was that he also delineated the electrical effects he saw in the dry pile. The degree of divergence of the gold-leaf electrometer was regarded as the 'intensity'. He also noted the frequency of the divergence/collapse cycle as the leaves struck the glass bottle which, he wrote: 'I conceive to be a measure of the quantity of electricity arriving at a certain intensity.'[R2] By these definitions, 'intensity' and 'quantity' are akin to today's electromotive force and current. He noted that quantity and intensity both increased with temperature. The corresponding relationship under changing humidity was more complex and thus the overall relationship between the two defined parameters was often non-linear for a dry pile. He had first made such observations on intensity and quantity with atmospheric electricity around 1810.[R6]

Ronalds' was a pioneering effort from several perspectives. He was not only separating the two physical parameters and their measurement but defining the relationship between them. Prior to this, and indeed for decades afterwards, words were used descriptively and inconsistently rather than with clear physical meaning because explanations of observations were hampered by the lack of knowledge, standard terminology and units of measurement we take for granted today.[A49]

Ronalds did not invent the descriptors quantity and intensity — they had already been employed in electrical science for some years.[A3,A164] In 1801, very soon after Volta announced his pile, Nicholson compared its galvanism with 'electricity' and suggested that the two were essentially the same phenomenon, but distinguished by the wet pile having low intensity and high quantity.[A6] Nicholson had therefore already aligned intensity and quantity with our electromotive force and current, and Ronalds built on that to show how they could be distinguished, measured and related using a common electrometer. At that time, there was still little real understanding of what gold-leaf electrometer readings signified, as exemplified by the explanations in Singer's[A12] and de Luc's[A8] papers.

The very slow subsequent evolution in thinking in the subject underlines the novelty of Ronalds' insight. A new means of measuring current — the galvanometer — appeared in 1820 with Hans Oersted's discovery that there is a magnetic field around an electric circuit. In 1827, 13 years after Ronalds' work, Georg Ohm found a proportional relationship between galvanometer reading and thermocouple temperature for a circuit — now known as Ohm's Law. It was described as heresy in Germany at the time.

Ronalds took the opportunity to raise the subject again when he returned to atmospheric electricity studies at the Kew Observatory. Electric current was now beginning to be deployed in practice, which made the concept of quantity and frequency even more important for him. He wrote in 1844: 'It *[frequency]* seems to form a sort of link between natural high-tensioned (frictional) electricity, and galvanic, or Voltaic or Œrstedic electricity (electro-magnetism).'[R12] As late as 1852, however, Professor George Airy* said in a scientific meeting that 'the connection between electricity and galvanism was not as yet clearly established'[A80] and therefore they should continue to be labelled distinctly.

Latimer Clark's* paper entitled 'On Electrical Quantity and Intensity' presented at the Royal Institution in 1861 advised:

> The modifications of the strength of the electric current in dynamic electricity, and in the amount of charge in static electricity, are at present usually defined by the terms Quantity and Intensity…both terms are equally applicable to electricity at rest or in motion…their joint effects have been ordinarily confounded together and attributed to one cause…but…there exists an absolute necessity for their clear separation before any numerical reasoning can be founded on them.[A106]

In part at Clark's instigation, the British Association for the Advancement of Science formed a committee that year to start the long job of developing and defining the physical bases of these separate parameters and then naming our now standard units of measurement (ampere and volt). The papers by Nicholson, Ohm and Clark are in the Ronalds Library.

Ronalds summarised this phase of his work in 1860 as follows:

In 1814 happily became acquainted with my kind old friend, M Deluc & made, on his dry column, experiments which have some relation to the still vexed questions of chemical action & contact…If Mr Walker is the learned & experienced Electrician,[1] I should feel obliged by his perusal of that note, because I think that those exper^{ts} were the first in which the distinction between "quantity" & "intensity" was observed. (These terms have been ever since much employed). He would perhaps find me in error.[C375]

It is interesting that Ronalds is so positive about this claim when he was usually very modest — and it was founded on his exhaustive study of the literature. He does seem to have been the first person to have given separate definitions of these two critical physical phenomena — based on the readings of a familiar measuring apparatus (the gold-leaf electrometer) — as well as their interrelationship in the particular application of the dry pile. His explanation of the words intensity and quantity align with what decades later came to be adopted consistently by others. His contribution was therefore remarkably farsighted, particularly given the very slow advance in understanding that followed it. It is a shame that the work appears to have been overlooked: this author has found very little previous acknowledgement of it in the literature of the time or since.

6.3 The First Battery-Operated Clock

Back in 1814, Ronalds soon tired of staring at the electrometer leaves and counting their fluctuations to ascertain the 'quantity' of electricity generated under different conditions. As he described: 'idleness (as prolific a mother of invention as necessity, I believe) instigated the contrivance of a little apparatus'.[R7] The device followed the work of several scientists who had used the oppositely charged terminals of dry piles to drive the motion of a small bead suspended on a silk thread through attraction and repulsion. Singer's configuration was based on that developed by Benjamin Forster where the bead alternately struck two charged bells as a chime.

When Ronalds started trials on such an arrangement he was 'surprised at the constancy and regularity of its action'.[R7] This gave him an idea. Rather than try to use the apparatus to observe meteorological variability, he set about doing the opposite — reducing the effects of the weather by making the configuration

---

[1] Ronalds guessed incorrectly. He thought the person who had requested his biographical information through Samuel was Charles Walker*.

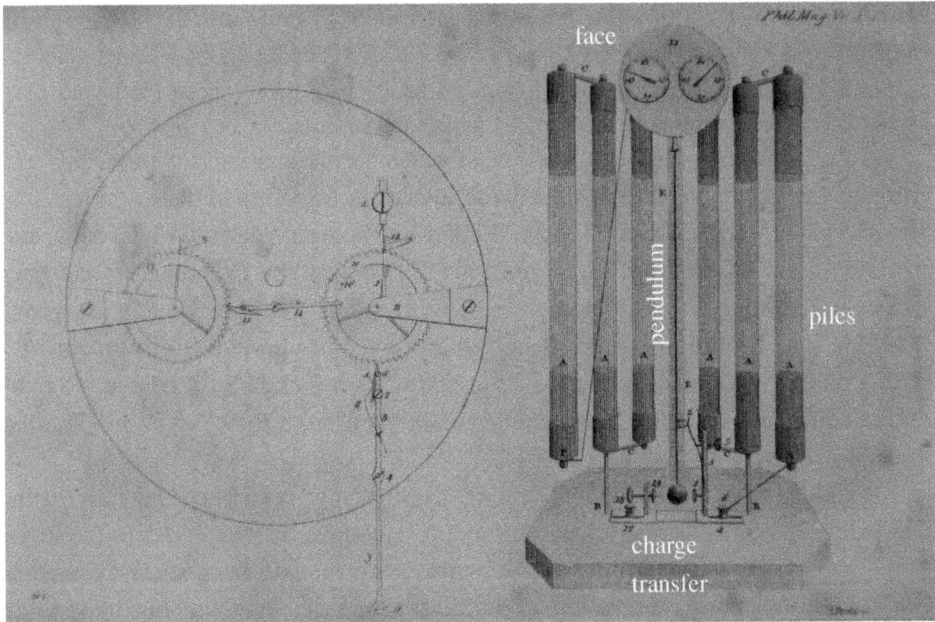

Fig. 6.2:   Battery-operated clock — model 1. Ref. R3.

insensitive to the amount of electricity in the dry piles. He first added stiffness and weight to the moving element to give it the steady periodicity of a pendulum. The role of the electrical attraction was now primarily to boost the pendulum's natural motion to overcome frictional decay. Contact was prevented so that delivery of the pulse did not jar the smooth oscillation — the piles discharged the electrified pendulum through a small air gap that was readily adjustable in width to fine-tune the motion (Fig. 6.2). The arrangement also allowed any increase in electricity to be counteracted by larger amplitude of vibration and additional charge draw-off through the reduced air gap. Finally he applied a mechanism to the apparatus that comprised two toothed wheels, one of which rotated 60 times faster than the other through the pendulum's action. Ronalds had created the battery-operated clock.

   He described the apparatus in his third paper dated 9 March 1815. He wrote modestly:

> If any of the readers of your useful Magazine, by improving upon the method I have
> stated of regulating the power of the column, or by substituting a better, were to
> render it subservient to the measurement of time, it would give me great pleasure.[R3]

He acknowledged the prior work of de Luc and Singer, and noted that he preferred the experimental set up of the former to the latter as the starting point for his own application. He also registered his 'obligation to Mr Gorham, a very ingenious watch-maker at Kensington, from whom I received great assistance'[R3] as well as an improvement suggested by Lord John Henniker*. He was already attracting the interest of other scientists. Henniker wrote to him at the time in his elderly hand:

> Lord Henniker presents his Compliments to Mr Ronalds and requests the favour of his Acceptance of some Glass Rods — which are Strait [sic] when grown[?] — for the use of his Electric Clocks or any other which may be applicable to them.[C5]

Ronalds had used glass stems to provide insulating supports to the dry piles.

The paper was summarised in the 'Report of Chemistry, Natural Philosophy, &c.' in June's issue of the *Monthly Magazine*.[A14] This was the first time he had been associated publicly with a completely new invention and he was naturally proud of it. He joked to an associate: 'I hope I am not more prejudiced in favour of it than a mother may be supposed to be in favour of her child let it be ever so puny and insignificant'.[C14] John Murray later made mention of the clock in his *Treatise on Atmospherical Electricity* (1830).[B33]

Ronalds soon heard about similar work being performed in Italy using Zamboni's dry pile. Zamboni had recently presented one of his devices to the Royal Society in London. Ronalds immediately made contact with the Society's President Sir Joseph Banks, who had kept it for a time at his house along with one of de Luc's piles.[C6] Zamboni's instrument was mentioned in the Royal Society's January 1815 report as comprising a metal needle that oscillated between two piles, but that 'No apparatus to measure time has yet been connected with the simple motion'.[A15] One of Zamboni's associates in Germany advised that he had made a clock very shortly after Ronalds' paper was submitted; this group's early work however does not appear to mention attempts to compensate for meteorological variability.[A17]

Both de Luc and Singer were unhappy with Ronalds' paper, for different reasons. De Luc's concern was the paper's title: 'On Electro-galvanic Agency employed as a Moving Power; with a Description of a Galvanic Clock'. De Luc wished the power source to be acknowledged as 'his' electric column. By now he and Ronalds were in regular correspondence, so he communicated his concerns privately. His letter continued, with its not quite perfect English:

> Your Clock-work is very ingeniously contrived, and your plates describes it very clearly: but you were very fortunate to find near you an cleaver [sic] watch-maker, to execute so well your idea…I am too infirm to contribute to forwarding those experiments which formerly were my delight. But I have told you before, that you are to be looked up for it, and wishing you good health for employing your great talents.[C8]

Nonetheless he was sceptical that the device would actually work:

> though I don't foresee the application of it in the usefull arts, I do not despair that it may in future, and through your indefatigable attempts, answer to that purpose.[C10]

Ronalds was still a novice in scientific publishing as he confessed in his reply of 9 May 1815 to de Luc:

It was sent imprudently without a title as were my former little communications
which I much regret as he *[Tilloch]* has supplied one in this instance which I certainly consider improper [C9]

He continued:

I am sorry that I published it so early for fresh and perhaps preferable methods of
regulating it have since occurred to me [C9]

He described his new ideas to de Luc and included a sketch of the proposed
arrangement. They formed the basis of his fifth paper published some months
later.

Singer chose to go into print with his views. He published his own paper on
the topic of mechanical devices powered by dry piles and, using a now familiar
pattern of damning with faint praise, wrote:

In October last, my friend Mr. Lightfoot...first suggested the employment of an
inflexible pendulum as a means of converting the reciprocating motion...into a
source of rotatory movement; and the correctness of this idea was soon afterwards
practically verified by my pupil Mr. F. Ronalds...

The rotatory motion obtained by this indirect means, is however rather curious than
useful...and has I fear very little chance of becoming at all useful as a time-keeper[A12]

This time Ronalds was annoyed. He mentioned it in a subsequent letter to de
Luc, who replied:

How unphilosophical, as you say yourself, is that meanness of disputing for priority
on some experiment!...

it *[Singer's comment]* certainly implies that you have received a first idea from
Mr. Lightfoot, but don't*[?]* mind it; it is better to overlook these hints...

I have observed in the course of my study of experimental philosophy, that all its
branches cannot be carried on by any single man; for they are so many branches,
some of which takes the whole attention and labour of an individual. Thus it is that
the science is forwarded, provided no jealousy or rivallity *[sic]* put a barr *[sic]* in to
the wheel.[C11]

De Luc's advice came too late. In Ronalds' fourth paper he quoted from
Singer's remarks, giving as his purpose:

to protest against the ambiguous construction of the paragraph, by which, if he does
not mean it, he *appears* desirous of having it understood, that I either directly or
indirectly derived the idea from the suggestion of Mr. Lightfoot, and thus appears to
insinuate plagiarism.[R4]

He also managed to slip in a mention of himself as being 'whom he *[Singer]*
chooses to designate by the appellation of his pupil'.[R4] Singer retorted in the next
issue of Tilloch's Magazine. After advising:

I am not so fond of controversy as to trespass on your pages, or the patience of your
readers, by a lengthened discussion on the contrivance of an electrical toy [A13]

he took up over a page of the journal in a personal attack on Ronalds: some men should 'endeavour to estimate the value of their individual labours with more diffidence' and 'fret and fume…if another wear a sprig of laurel more verdant', and so on.[A13]

At this point, Ronalds was probably wishing he was back in the merchant business. He turned once more to de Luc:

> In the last number of Mr Tillochs Magazine, Mr Singer calls the clock work an electrical Toy altho his Electrical chime surely better *[?]* deserves that name…I feel discouraged from writing any more on the subject. Lest there should exist a shadow of appearance of vanity or quixotism, Mr Singer first induced me to make my name publick *[sic]* in Mr Tillochs Magazine and it was with his advice that I gave the first account of the clock and perhaps his conduct will be the cause of my not troubling the Publick again…but excuse this egotism…
>
> My principal object was to plead not guilty to his apparent charges of Plagiarism & falsehood which whether they concern an electrical Toy or an attempt to accomplish an acknowledged highly desirable and usefull purpose are equally odious vices.[C12]

We will see further examples of his sensitivity to plagiarism in later chapters. Interestingly, Singer seemed to enjoy this style of communication: the previous year's *Philosophical Magazine* contained similar language directed towards Ezekiel Walker.

De Luc encouraged Ronalds to publish the work he had described in his earlier letter. His new mode of weather compensation (Fig. 6.3) employed a vertical rod

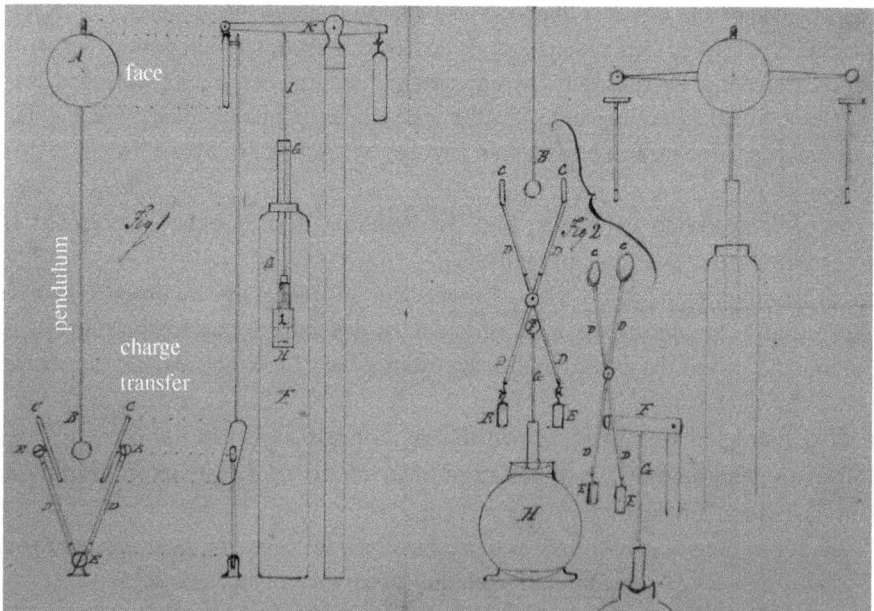

Fig. 6.3:   Battery-operated clock — model 2 (on left) plus other options (1815). IET 1.6.56.

that rested on the surface of the mercury in a large thermometer and was linked to a beam carrying the clock and pendulum. As the mercury rose, literally, the beam was elevated and the pendulum swung further before being charged by brass plates at an adjustable angle. The piles were sealed in a box of cement. He summed up: 'the instrument actually keeps tolerably good time'.[R5] He also included a comment on Singer's letter in his paper:

> As Mr. Singer's account of the invention of this instrument seemed very inconsistent with mine, I wished it to be clearly understood, for the information of his friend in the note to which he has referred, that I derived no kind of suggestion from any person whatever respecting the employment of an inflexible pendulum. That Mr. Singer should honour me with the title of his pupil, I certainly did not at all expect.[R5]

It was 32 years before Ronalds published again in the *Philosophical Magazine*. De Luc and Singer both passed away in 1817, de Luc at age 90 and Singer just one-third of that age. Ronalds kept de Luc's correspondence for the rest of his life, including an unfinished letter he was writing about his telegraph when he received news of de Luc's death. His 1823 description of his various endeavours with the dry pile was in part a tribute to his friend:

> I had, in the year 1814, the good fortune to make his acquaintance; and prompted partly by the opportunity which it gave me of enjoying a little of the society and instructive correspondence of so warm hearted, justly celebrated, and truly amiable a man, and partly by the novelty of the subject, I undertook a few experiments[R7]

He could not find similar words for Singer, although he did acknowledge his various contributions and in particular the 'six excellent columns'[R7] procured from him for his first clock.

Dr Thomas Thomson FRS in his 'Account of the Improvements in Physical Science during the Year 1816' noted:

> *Zamboni's Column.* —— A great number of papers have been published on this new electrical instrument…It is needless to notice the clocks that have been constructed by means of this column as a moving power both in this country and in Germany; because it is obvious that the great irregularity in the motion of these pendulums must render such clocks of no real utility.[A18]

Ronalds started work again on weather mitigation at that time. A reason he gave was that: 'the watchmakers tell me, that their main springs would, if not compensated by the action of their pendulums, exert their force as irregularly as I tell them that my electric pendulum acts'.[R7]

He was never short of an idea. In his 1823 book he outlined several of the many configurations he trialled. He bought some of Zamboni's piles in 1816, but found them unsatisfactory. He reverted to preventing hard contact between the pendulum bob and the charged plates, this time by setting up taut cross wires that cushioned the ends of the swing. Results survive from November 1817 that indicate the clock's gain or loss over several days to have been just tens of seconds.[J15] Other notes survive about model 'F R N° 12'.[J15]

Several pages later in the 1823 book he wrote:

> To state the many fruitless experiments which I have made, with a view to obtain a compensation for the effect of heat, &c. upon the columns, would be to expect almost as much patience from the reader as the experiments themselves required; but having now before me a pendulum, vibrating seconds as *regularly at least* as any common clock...I shall describe it as shortly as possible.[R7]

In this last configuration (Fig. 6.4) the piles were again cemented into a base box and mechanical effects of delivering electric charge to the pendulum were softened by suspending the contact points from fine flexible wire.

His goal in continuing work on the clock was simply to demonstrate that it was possible to use dry piles to keep time in a way that was largely independent of weather effects. He was not interested in any role in commercialising his work. As early as November 1815 he somewhat passively encouraged contacts in Paris to take on his second model: 'if you were to give the subject attention I think a usefull instrument might be produced'.[C14] There is no evidence that his ideas were adopted directly in the clock-making industry.

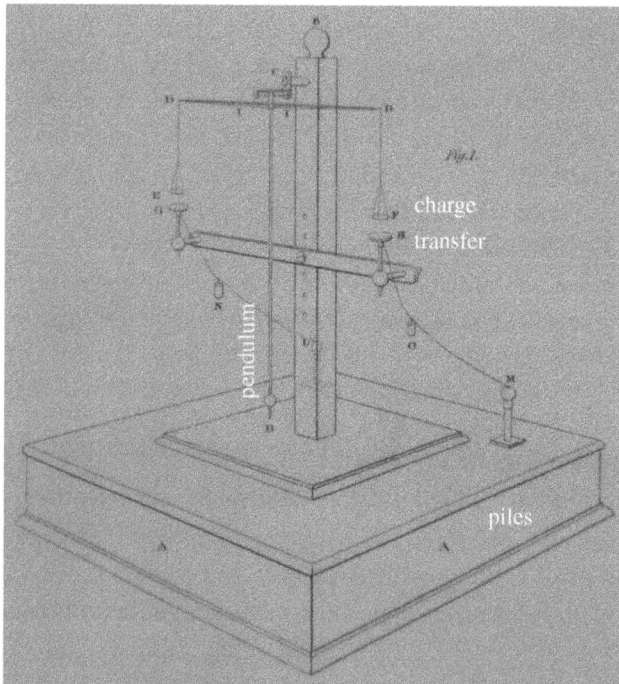

Fig. 6.4:   Battery-operated clock — model 4 (1818). Ref. R7.

Ronalds' 1860 summary of his endeavours reads:

> They *[experiments with dry piles]* led to ~~the~~ my invention in <u>1815</u>…of some wheel-work set in motion by 6 electric columns, made by Singer & which columns are now at the Kew Observatory…In 1816 I set in motion a time piece compensated for the effect of changes of temperature upon elect^c columns. In attempts to measure time by the elect^c column I was however forestalled by foreign electricians [C375]

In giving primacy to European scientists, he was referring to later events — and being unduly modest. He visited Zamboni's laboratory when he was in Verona in October 1820. It happened to be a vacation and the professor and his colleagues were absent, but he was able to inspect the manufacture of the dry piles. He had already acknowledged in his fourth and fifth papers that these piles were superior to those made in England in the electric potential generated. He also believed European clockmakers to be more advanced. He paid homage again in another visit to Zamboni's laboratory 38 years later: 'I there saw several of his vibrating apparatus, and his electrical Clock…*[which]* has been in constant motion…15 years'.[J85] He purchased Zamboni's portrait and sixteen of his publications for his library while there. Several clocks originally powered by Zamboni piles survive in northern Italy.[A177]

Around 1840, scientists turned their attention to clocks driven by electric current, with Alexander Bain being awarded the first patent. His early clocks were analogous to Ronalds' except that the pulse applied to the pendulum bob at the end of its swing was generated by a wet pile.[B146] Charles Wheatstone's* electro-magnetic clock, developed at the same time, had as a forerunner another of Ronalds' inventions; it is mentioned further in the next chapter.

Ronalds always retained his interest in dry piles. He advised Dr Henry Noad FRS, Professor of chemistry at St George's Hospital, in 1853:

> I have some Electric Columns of M. De Luc's kind in glass tubes at Kew, which were made for me by Singer in 1815…Some of them are still very efficient (as you can perceive by trying them with a gold leaf electrometer). They have never been opened or in any way disturbed.[C346]

Two of these dry piles were transferred to the Science Museum in South Kensington when the Kew Observatory closed in 1980. One is 33 cm long and the other 29 cm (Fig. 6.5). It is not known whether they still hold an electric potential difference. The Clarendon Laboratory at the University of Oxford has a chime similar to Singer's made in 1840 that continues to work with the original piles.[A168] It seems likely that Ronalds' invention could not only have approached modern battery-operated clocks in terms of short-term reliability, but well exceeded them in long-term reliability. Dry piles enjoyed a resurgence during World War II as a portable voltage source,[A168] and enthusiasts continue to make them today.

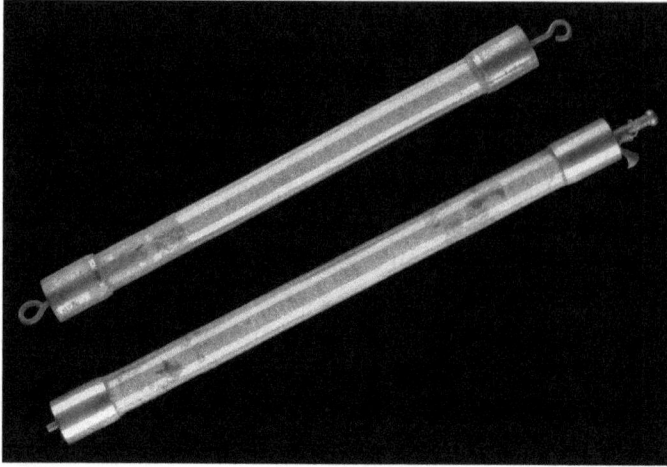

Fig. 6.5:   Ronalds' dry piles. Science Museum 1980–1817 and 1980–1818.

Ronalds' work was rediscovered in the 1970s by Charles Aked, Council member of the Antiquarian Horology Society and first Chairman of its Electrical Horology Group. He notes in his paper 'The First Electric Clock' that 'Ronalds' clock was the very first to be driven by electricity, and his light has been hidden under a bushel for far too long. For nearly one hundred and sixty years Francis Ronalds' work has remained completely unrecognised'.[A162] Ronalds would be surprised by the acknowledgment after all those years. In his lifetime the accolade of inventing the electric clock went to either Bain or Wheatstone.

6.4 Observation and Recording of Atmospheric Electricity

We now again turn back the clock (Ronalds' of course) a few years. The first experiments he performed that found their way into the scientific literature were not about the dry pile at all, but explored atmospheric electricity. By 1810 he had set to work on the topic in his new laboratory in the cockloft at Highbury Terrace, taking advantage of the house's elevated position. He had just received his inheritance, having come of age, and had disposable income for electrical equipment — and seemingly was no longer spending every moment of the day at the office.

The first point of significant interest in this work is the form of insulation provided to the conductor. He developed two improved methods of insulation, which made it possible to measure the very small charges extracted from the atmosphere before they dissipated. He described his first set-up in his 1860 autobiographical letter:

through a round hole (or window) in the south wall, I introduced one end of a long wire, extended down the fields towards Holloway (not then built upon) & insulated it in a new manner, called Singer's (originally mine) on high poles. With this, with pith-ball-Electrometers, Bells, &c, I made a few unimportant observations, of Beccaria's kind...M^r *[Andrew]* Cross*[e]* (a patron of Singer) made experiments at about the same time, & in the same manner on a very extensive scale, but I desisted, because the neighbours were occasionally affrighted by very loud detonations and said that they should be killed by "the Lightning which I brought into the place". (In fact 2 or 3 of my neighbours were killed; but these were only unprincipled rats, experimented upon, dwellers in the Hay-Loft, devourers of my poney's *[sic]* corn.)[C375]

The wire would have traversed the fields the family owned beside their house (Fig. 3.11).

Singer included a write-up of Ronalds' work in his 1814 book *Elements of Electricity and Electro-chemistry*, but from a rather different angle. He started:

In every arrangement of this kind *[for observing atmospheric electricity]*, the principal difficulty is the preservation of the insulation...I have been successful in an attempt to obviate this inconvenience to a very considerable extent, by a new arrangement...

The application of this principle to the perfection of the gold leaf electrometer was the first trial of its excellence; and the result was the most satisfactory demonstration of its utility.[B12]

The method of insulation comprised protecting the charged wire in a narrow glass tube coated inside and out with sealing wax to limit circulation of moist air.

Singer announced 'his' invention in early 1811 in Tilloch's Magazine:

Mr. George Singer has recently discovered a new system of arrangement for the insulators employed in electrical apparatus, by which their insulation is preserved, without the necessity of wiping, through all the vicissitudes of atmospherical change.[A9]

The Ronalds Library Catalogue lists the article with the addendum: 'Note. — This method was derived from my suggestion. — F. R.'[R28] The difference of opinion was possibly the first source of friction between the two men. If Ronalds was correct — and his track record through life is of being generous in giving credit to others — a possible scenario is that his advice to his more experienced mentor was given with considerable timidity and qualification. When the idea was then turned into reality, Singer felt it was his own.

In Singer's gold-leaf electrometer, the wire linking an outer brass piece with the leaves was insulated in this new way through the cap of the bottle. It was described in the 1842 *Encyclopædia Britannica*,[A63] which noted that it met with Faraday's strong approval. Dr Paul Mottelay wrote as late as 1922 that the improved instrument 'is to be found illustrated and described in nearly all works upon natural philosophy';[B137] the design continued to be used well into the 20th century.[A164] Ronalds used it in his own electrometer configurations developed at the Kew Observatory (Sec. 14.2), denoting it to be 'the mode of insulation called "Singer's"'.[R12]

Concerning the application of the insulation to atmospheric electricity, Singer explained:

> An exploring wire for atmospherical electricity has been insulated nearly agreeable to this plan by a very assiduous and promising electrician, F. Ronalds, Esq…The apparatus was erected in a field near Highbury Terrace, Islington, and continued in constant activity for several months: the insulation was tolerably well preserved, but not uniformly so; this he attributes, in part, to the hasty and probably imperfect construction of the apparatus, and partly to the insufficiency of the most perfect insulators, when the stratum of air between the wire and the ground is so moist as to become a conductor of electricity.[B12]

Ronalds himself provided a little further detail on his atmospheric electricity experiments in subsequent documents:

> At the distance of *[space]* feet from my Observatory, and in the house at Highbury, a bell rung by electricity of dew could be distinctly heard[C108]

He had set up a device to communicate the occurrence of increased electric signs so that he could make detailed observations.

> I have by means of a long wire…when a copious dew was falling, collected pungent sparks when the height of the wire did not exceed five feet about the surface of the earth in any part of it.[R1]

No wonder the neighbours were worried — and his mother Jane must have been very understanding. In 1823, he wrote:

> Mr. Singer, in his "Elements of Electricity," p. 283, seems to have misunderstood me in regard to the imperfection on his mode of insulation, adopted in some experiments on atmospheric electricity. It was the imperfect insulation of the spaces of air comprised between the mouths of the tubes and the wire, distances of half an inch, and not that comprised between the wire and the earth, a distance of *fifteen* or *sixteen* feet, to which I *partly* attributed the defect. But my chief enemies were the *spiders*, who frequently obliged me to walk a quarter of a mile, through long wet grass, to destroy the webs which they had spun within the tubes, for they thus established a semiconducting medium between them and the wire. Another source of inconsistency and error arose from the moisture requiring as much time *to get out* of the series of tubes as it had taken *to get in*.[R7]

In 1860:

> Becoming however dissatisfied with the mode of Insulation hitherto used… I devised, & M[r] Brande kindly printed (& figured) in his Quarterly Journal…the mode which I used at Kew afterwards.[C375]

The new approach was published in 1817, the year that Singer died, as Ronalds' sixth paper. A modified version was described in his 1823 book. The insulation was incorporated into an improved device for collecting and observing atmospheric electricity that was significantly more sophisticated than the earlier

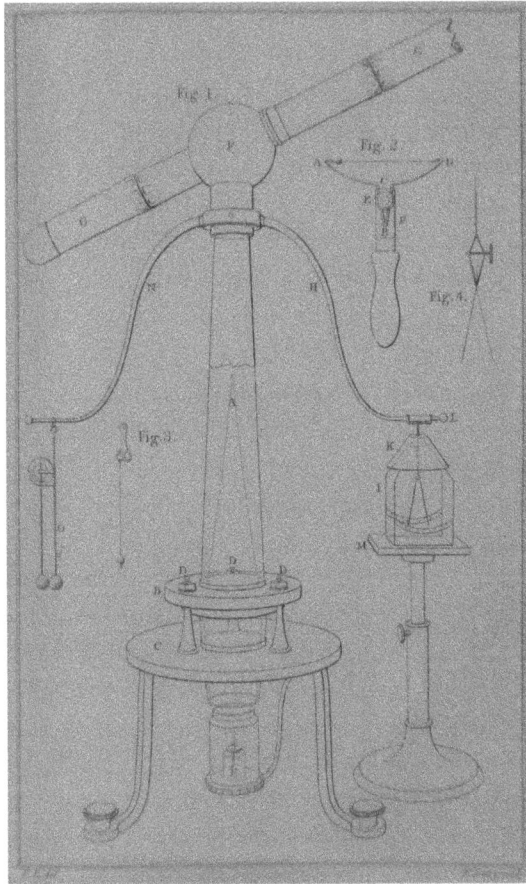

Fig. 6.6: Atmospheric electricity measuring apparatus (*c*.1818). Ref. R7.

configurations illustrated in Singer's book. To prevent condensation Ronalds used a spirit lamp to heat a partially hollow glass column that supported the end of the gilt bamboo conducting rod (Fig. 6.6). Multiple electrometers of differing sensitivity were deployed to measure the electricity; the connections were configured to enable their rapid change-out as weather conditions varied. He would have made the elegant stands using his lathe, as well as the instruments.

Ronalds' revised mode of insulation was also written up in the journal *Annales de Chimie et de Physique*.[A19] The editors noted that an alternative was to place the lower part of the rod in a glass receiver containing a desiccant such as calcium chloride or concentrated sulphuric acid. Ronalds later explained that he had already tried hygroscopic substances in this application and gave several reasons why he found the approach less successful than the warmed glass.[R7] He did use desiccants to keep electrometers dry however, as described in Sec. 14.2.

Ronalds was one of the first to realise that the electrometers did not measure the absolute charge in the air at the height of the wire. The observation was the potential difference between the higher stratum and the lower electrometer bottle, or the earth's surface if the electrometer was grounded. He mentioned this effect in his first paper and described it in detail in his notes and in his 1823 book, acknowledging there the work of Berlin physicist Paul Erman. Erman had postulated in 1803 that the positive atmospheric electricity observed in fair weather was in fact due to the surface of the earth having a negative charge. A 1925 publication notes that Erman's ideas were 'received with incredulity' and the paper was thus 'seldom referred to in the literature of Atmospheric Electricity'.[B138] Some years later Ronalds described his own views rhetorically: 'Do not we and the plants dwell in a double or treble electrophorous *[sic]* of Volta's semiconducting kind?'[C108]

While early electrometers were remarkable devices, for atmospheric electricity studies they required extended manual observation, which was not Ronalds' forte. He soon set about automating the recording so that data could be gathered from unattended instruments — a concept for which he would achieve international reknown many years later at the Kew Observatory. He gave a justification for his interest in such endeavours when describing his pilgrimage in 1818 to the site of Beccaria's work: '*[I]* rather envied him *some* of the pleasures of his *serene* occupation; but who, that could have *calculated* the tedium of the task, would ever have undertaken them?'[R7] Supplementing his visit, he collected 28 of Beccaria's papers and his portrait for his library.

He was able to put his ideas into action when the family moved to Hammersmith in 1813. His new atmospheric electricity apparatus included 'a series of registering electrometers (described in Singers Elements of Elect[y] &c…). His notice of which, and a slight mention of the Highbury wire, in Brewsters Encyclopædia *[Ref. A42]*, gave me a little encouragement.'[C375] Singer described Ronalds' apparatus as comprising a rotating dial that allowed electrical contact to be made with each of series of electrometers at different times. Electrometers were designed that would retain their readings to enable the daily variation to be seen at one observation. The concept was reused at Kew in the period 1845–48 to enable 24-hours-per-day measurement (Sec. 14.2).

Ronalds also trialled alternative systems at this time. One was similar in concept to his initial electric clock. He had first developed the clock mechanism in 1814 'for the purpose of registering the number of vibrations which occurred in my absence'[R7] — the 'strikings'[R2] of the charged pendulum on the discharging points were counted using the pawl-and-ratchet and the dial. The frequency or 'quantity' of electricity under different atmospheric conditions could thus be determined. Adding bells to this type of device would also provide an audible signal for the observer, as mentioned above.

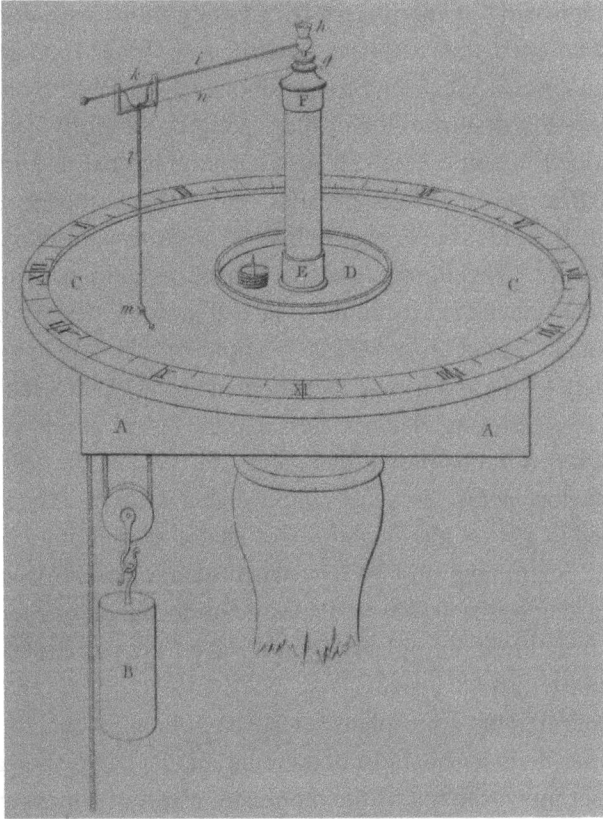

Fig. 6.7: Atmospheric electricity recording apparatus with rotating resin plate — model 1 (1815). Ref. R7.

The basis of his next self-registering apparatus (Fig. 6.7) was a horizontal tin clock face of 20 cm diameter, coated with resin and attached to a timepiece. As the circular plate rotated, a fine thread gradually wrapped around the spindle and a small gold bead at its end described a spiral on the resin. Electricity transferred to the bead, via a connection to the spindle, heated and softened the resin according to the sign and intensity of the charge. The varying track became visible when dusted with 'common *dry* hair-powder'.[R7] The plate could be attached to the mechanism of either the minute or hour hand, or other arrangements, depending on the atmospheric electricity phenomenon to be recorded, for example the daily cycle versus a storm.[R7]

The approach was not altogether new. Georg Christoph Lichtenberg FRS (1742–99) had found that resins showed a pattern of the electric discharge they had received when they were sprinkled with powder. The approach was honed by Bennet, who described in his *New Experiments on Electricity* (1789)[B6] a method of taking an impression of the pattern by pressing moist paper or linen on to it. Bennet also mentioned that Marsilio Landriani (1751–1815) had contemplated transferring atmospheric electricity to a resinous plate moved by a clock.

Ronalds' was the first machine of the type to be carefully designed and built and then described in some detail. It was still highlighted and illustrated in the *Encyclopædia Britannica*[A62] four decades later. The complete section on the Electrograph in Ronalds' book was published verbatim in Tal Shaffner's 1859 *Telegraph Manual.*[T7] The device had been used as evidence in a law suit in America as it was considered by some to be a step in the development of electric telegraphy. Ronalds used a modified version of his apparatus at the Kew Observatory in 1843–44 and it was also adopted at the Greenwich Observatory.

The series of registering electrometers and the electrograph were not the only devices he developed in this period that were still 'state of the art' when he set up the Kew Observatory nearly 30 years later. He also employed the heated insulator he had described in 1817. Other examples of its use were in Italy in 1819. That he took the fragile instrument and its associated electrometers on his Grand Tour illustrates his strong desire to observe specific atmospheric phenomena that were quite different from those in England. Data survive from two experiments — at Mt Vesuvius and in Sicily.

Ronalds measured the atmospheric electricity of the volcano in June and early July 1819. He was there in the midst of an eruption that had just created a new lava flow path. He set up on the top of the mountain, about 450 m away from the craters, with a vertical rod 4.3 m high connected by a 6 m long wire to his electrometer. Temperature, pressure, humidity and magnetic effects were also monitored. The results showed the air to always have net positive electricity. The magnitude of the charge varied frequently depending on the occurrence of explosions or whether the wind was blowing the fumes of the craters or aqueous fumeroles towards him. The fumes from the two-week-old crater were positive while those from the older crater seemed to be predominantly negative. The observations were included in his 1823 book, published in Brande's* Journal[R8] and later summarised in Murray's *Treatise on Atmospherical Electricity.*[B33]

Ronalds was in all likelihood the first person (silly enough) to measure atmospheric electricity at an erupting volcano. As recently as 1967, the authors of a scientific paper entitled 'Water and the Generation of Volcanic Electricity' noted that 'the admirable suggestion that one should make electrical measurements near an erupting volcano to determine whether the clouds carried a positive or negative charge…is still pertinent today.'[A156] In recent decades, electrostatic field monitoring at volcanoes has become an area of active research. Measurements using remote sensing techniques are used to detect eruptions and ash fallout as well as to understand effects of volcanic activity on the global atmospheric electrical

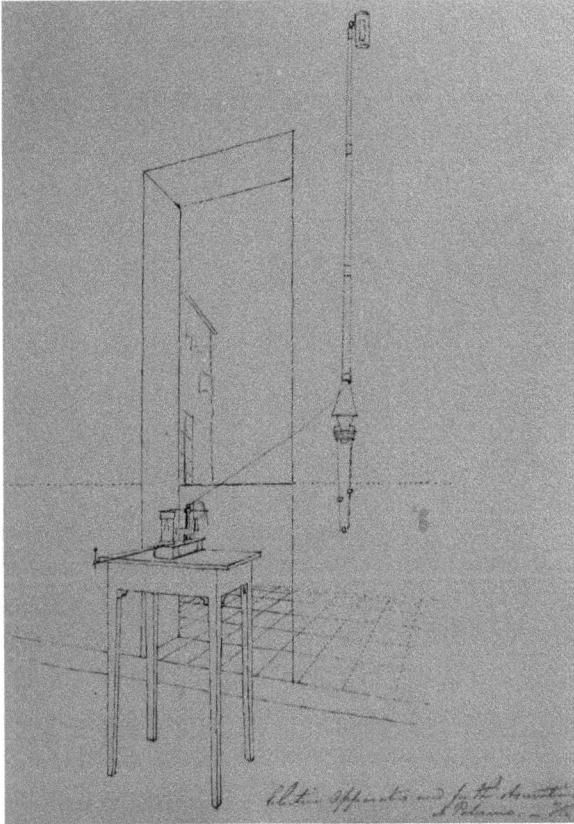

Fig. 6.8: Atmospheric electricity measuring set-up at Palermo (1819); includes collecting lamp at tip of rod (see Sec. 14.1) and several electrometers. IET 1.6.60.

circuit. At a more fundamental level, the work informs theories as to the development of life on earth.[A185]

Ronalds conducted a second set of experiments in July 1819 at Palermo during a dry Sirocco wind. On several days he positioned his apparatus (Fig. 6.8) on nearby Mt Pellegrino or on the roof of his hotel and measured the changing atmospheric electricity regularly over a nearly 24-hour period. The charge was always positive but first rose and then fell in magnitude during the day and night. He was most intrigued how this was 'diametrically opposed'[R7] to his and others' observations in serene weather in northern Europe, where the atmospheric electricity reduced near the middle of the day.

His results were probably the first documentation of atmospheric electricity at a location so close to where the sirocco develops in the Sahara Desert. More importantly, they would have been the first to exhibit an essentially single daily cycle, as seen in Fig. 6.9. The double cycle that had been observed to date (and continued to be for another 150 years) is now understood to have been influenced

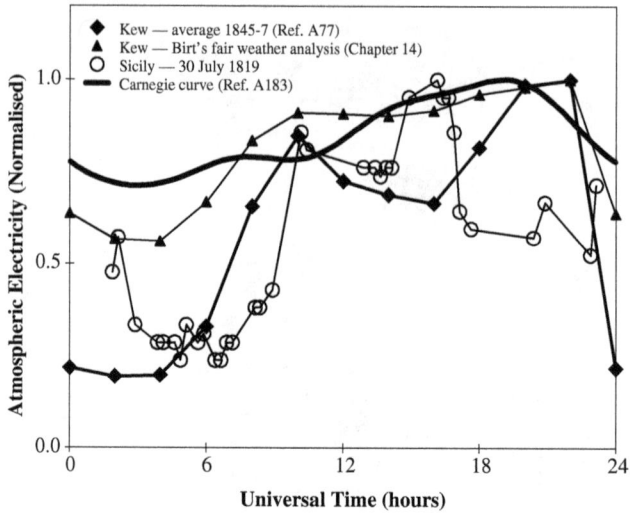

Fig. 6.9:   Qualitative daily cycles of atmospheric electricity potential observed by Ronalds in Sicily and at the Kew Observatory, compared with the characteristic Carnegie curve.

by smoke pollution. It was discovered in the early 20th century through observations over the oceans that the characteristic diurnal variation of electric potential in clean air is single-peaked; it is called the Carnegie curve.[A183]

The sirocco has long been believed to induce effects in humans. In a letter to his sister Emily, Ronalds explained:

> Perhaps it was during the time of a Sirocco wind that I wrote the disgraceful letter which has given your kind heart so much pain, when, whatever is badly written is put down to <u>its</u> account in Sicily & Naples. Thus if a man or woman writes a stupid book or <u>journal or letter</u> they say, Oh it was written in Time of Sirocco.[C43]

The greater positive ionisation of the air at these times is now also understood and some people use negative air ionisers for health reasons.

Ronalds presented his results in his 1823 book on the electric telegraph. Both the volcano and sirocco measurements were repeated in the *Encyclopædia Britannica*.[A62] In total, both the 1842 and 1855 editions of the *Encyclopædia Britannica* covered nine aspects of Ronalds' early electrical work. He partially summed them up:

> In about 1817(q), A warmed Electrical machine, a resinous-plate Electrograph, a pendulum Doubler, moved by a Clock, & capable of charging the Telegraph, each described in the Encyclopædia Britannica...together with many undescribed affairs of a like kind were devised & executed.[C375]

The Encyclopædia additionally explained his atmospheric electricity conductor, improved electrometers, telegraph and gas pistol (modified from that invented by

Volta). The 'Electricity' entry under which all these appeared was written by Sir David Brewster* and incorporated 11 of Ronalds' illustrations.

Ronalds also used the opportunity of his 1823 book to describe various other unusual meteorological phenomena he had observed on his travels. His comments on the regular reappearance of thunderstorms, which interested him as an atmospheric electrical phenomenon, were picked up by the 1823 *Edinburgh Philosophical Journal*,[A24] also edited by Brewster.

Another example where Ronalds' trip embellished his science was his investigation into the electric charge delivered by torpedo rays to stun prey and for defence. With an assistant, he dissected a specimen 'to fish for new information about their electric organs':[J18]

> The only thing I did was to confirm my discoveries made long ago viz that my knowledge of anatomy as well as of almost every Branch of Literature, Science & the Arts is far too limited to allow of my <u>looking at what is to be seen</u> in such a journey as I am taking.[J18]

6.5 Improved Electrical Machines

As yet another aspect of his early electrical activities, Ronalds offered improvements to several types of electrostatic generator then in use. The friction machine was temperamental, particularly in humid conditions. Family friend Mary Nichols* documented, for example, that when she and her sisters attended an evening lecture on 'mixed Philosophy', 'the experiments on Electricity failed owing to the wet weather'.[A173] Ronalds found that by applying the heat of small spirit lamps under the two metal tubes (Fig. 6.10) his machine was '*always* effective'.[R7] He also made design suggestions to aid portability of the machine. In addressing such a prevalent problem, the first innovation was picked up widely; it was included in the 'Scientific Intelligence' section of the *Edinburgh Philosophical Journal* as well as the *Journal of Arts and Sciences*, the *Glasgow Mechanics' Magazine*, the *Royal Cornwall Gazette*, James Ferguson's *Essays and Treatises* and various other publications of the day.[A23,A25,B21] Noad also included the hint in his 1859 *Manual of Electricity*.[B77] It appears that the application of heat came to be adopted routinely — Professor William Thomson FRS, later Lord Kelvin, made mention of it in his 1874 Presidential Address to the Society of Telegraph Engineers.[A119] Kelvin is introduced more fully in Chapter 14.

Next, Ronalds developed an influence machine that he called the Pendulum Doubler. 'Since everybody, who may have had so much patience and indulgence as to read thus far, may not happen to have seen the *revolving* doubler of Nicholson or Bennett *[sic]* described',[R7] he outlined its operation before describing the changes he had made. He automated the charging process by converting the pendulum bob

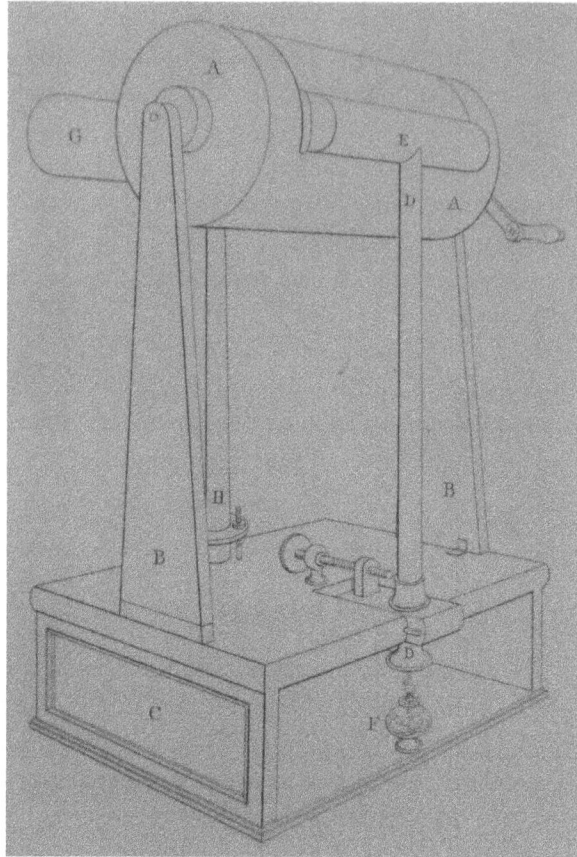

Fig. 6.10:   Reliable friction electrostatic generator (*c*.1816). Ref. R7.

of a clock to be the third plate, as seen in Fig. 6.11. Various wires were set up to earth the plates and provide electrical contact between them at just the right points in the cycle. The accumulating static electricity was stored in a Leyden jar. He developed the device to demonstrate that he could keep the electricity in his telegraph wire continually topped up to overcome any small losses through imperfect insulation.

The idea of deploying a pendulum had first been suggested by Bennet and was applied by his friend Dr Erasmus Darwin FRS (Charles' grandfather) in 1800 to provide electricity to a pot plant.[B9] Largely removing manual effort from the generation process was highly novel and Ronalds' elegant manifestation was reprinted in the 1823 *Edinburgh Philosophical Journal*. The article started: 'In a work just published by Mr Ronalds…we find, among many other ingenious contrivances, one of a Pendulum Doubler, which we consider highly worthy of notice.'[A22] The device was described in an article on influence machines in the *Journal of the*

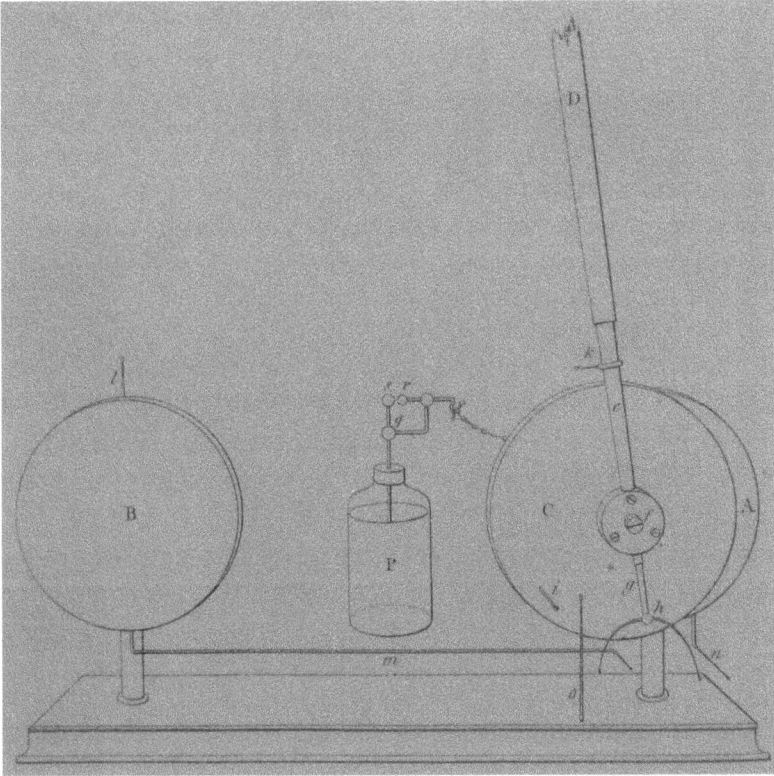

Fig. 6.11:   Automated influence electrostatic generator with 10 cm diameter plates (1816). Ref. R7.

Society of Telegraph Engineers in 1888.[A129] Two years later, and more than 70 years after its invention, John Gray included the machine with an illustration in his book *Electrical Influence Machines*, noting that 'This apparatus might be very convenient where a constant discharge of sparks for a considerable length of time was required.'[B107]

Finally, Ronalds recommended that a Leyden jar charged by a dry pile was a useful substitute for the often unreliable electrophorus to light Volta's gas lamp or the newly developed town gas.

6.6 The End of Electrical Invention

Ronalds' electrical inventions slowed dramatically in 1817 after the double rebuffs of his clock and his telegraph, and the death of de Luc. He turned his hand to writing an update to Priestley's* *History and Present State of Electricity* (1767). Considerable material survives in the Ronalds Archive, some of which has small edits in another hand, but the book itself was never published. He set out various propositions and illustrated them with some of the innumerable experiments he performed to help understand the phenomena. He was quite plucky:

> Least I should try nothing in this kind of research myself, I first rendered a piece of citron arsenic of 4 Ounces, which was then at hand electrical by friction, so that it attracted and repelled minute Bodies. I frequently took off the electricity excited in this manner with the naked hand. Then still holding the same piece of arsenic in the hand I rendered myself electrical on a stool of Pitch by electricity constantly flowing to it from an electrical machine more than a quarter of an hour so that my hair stood on end, & those standing round drew the strongest sparks out of me: but I did not observe the smallest change in my Body.[J16]

It can be surmised that his mother was not at home at the time.

The books and papers he acquired and catalogued in this period were the kernel of the extensive library he went on to develop. He continued during his Grand Tour, noting of this time: 'In 1818–19–20. Worked in Italy on a Bibliography of Electricity & Magnetism; collecting Books & notices of Books &c.'[C375] Unfortunately, many of these early books probably never reached London. In the long, emotional letter to his sister Charlotte in April 1819, mentioned in Sec. 4.6, he wrote:

> I can't proceed with the purpose I have had in vain these 2 years viz that of continuing D[r] Priestley's History of Electricity without them. I can't surely bear the idea of having thrown away two years of my life on it even supposing I had all the materials ready. Think then what must be my feelings on loosing so great a part of them…With half the vexation I have already experienced about this History of Electricity I might have received a hundred times more satisfaction in the end by employing myself in a more usefull means perhaps[C24]

Ronalds continued to follow the scientific literature during his travels, even though he had ceased active research himself. He made a request to his brothers Charles and Nathaniel in early 1819:

> Be so good also to look over the table of contents of the Journals of the Royal institution and Tilloch's Magazine which have come out since I left London and copy down any thing you think most likely to interest me.[C21]

After returning to England, however, Ronalds focused his creative energy elsewhere. His preface to his 1823 summary of his earlier electrical activities reads:

> It was the original intention of the author, on taking leave of a science, which once afforded a favourite source of amusement, to prepare for the press many other portfolio scraps, as little deserving of publicity as these, perhaps: but…he is *already* compelled to bid a cordial adieu to Electricity[R7]

Ronalds had arranged to travel again to France and Italy to meet people regarding a new business opportunity. He left London three weeks after writing those words and did not return to his electrical science and engineering work.

Chapter 7

WHO INVENTED THE ELECTRIC TELEGRAPH?

The evolution of the electric telegraph through the 19th century, with emphasis on Ronalds' early contributions, his ongoing interactions and feelings through the period, as well as the behaviours and views of other developers, scientists and commentators.

Mr Barrow presents his Comp^s to Mr Ronalds, and acquaints him with reference to his Note of the 3rd Inst, that Telegraphs of any kind are now wholly unnecessary; and that no other than the one now in use will be adopted.

Adm^y Office

5 Augt am [C17]

Sir Francis Ronalds is hardly a household name but, if anything is known of him, it is the story of the blunt rejection of his electric telegraph in August 1816. 'Mr Barrow' (later Sir John) was the Second Secretary of the Admiralty, while 'the one now in use' was an optical communication system — the semaphore. The story gained such notoriety in Ronalds' time that Sir John's son felt the need to write to the Editor of the *Pall Mall Gazette* to explain that his father acted as 'the organ of the Board, and not on his own responsibility'. 'Well might Mr. BARROW'[1] quipped the *Express*. The rebuttal had delayed the incorporation of both electricity and rapid communication into the industrial revolution by up to 20 years. A large monument to Sir John Barrow FRS survives at his birthplace of Ulverston.

---

[1] *Express* 22/11/1866 2a–b; IET 1.9.2.

7.1 Ronalds' Telegraph

Ronalds' vision for the telegraph was truly profound. First he foresaw the age of electricity:

> electricity, may actually be employed for a more practically useful purpose than the gratification of the philosopher's inquisitive research...it may be compelled to travel as many hundred miles beneath our feet as the subterranean ghost which nightly haunts our metropolis *[reticulated gas street-lighting had been introduced in the 1810s]*...and...be productive of, at the least, as much public and private benefit.
>
> ...give me *materiel* enough, and I will electrify the world *[he acknowledged here that he was borrowing Archimedes' phraseology].*[R7]

He then homed in on a specific application:

> why should not our kings hold councils at Brighton *[a favourite residence of George IV]* with their ministers in London? Why should not our government govern at Portsmouth *[Royal Navy base]* almost as promptly as in Downing Street? Why should our defaulters escape by default of our foggy climate?...why...add to the torments of absence those dilatory tormentors, pens, ink, paper, and posts? Let us have *electrical conversazione offices*, communicating with each other all over the kingdom[R7]

The flowery language reflects his recent immersion in classical Greek and Roman cultures during his Grand Tour. His words are also remarkably prophetic. He is not only describing what would be the first deployment of electricity at scale, but going very much further to imagine the world we have today — two centuries later — with global communication by 'electrical conversations' rather than pen and paper. He was 28 years of age when he completed the telegraph and had stopped running the family cheesemonger business just two years earlier.

He mentioned in his autographical letter that it was when he was experimenting with atmospheric electricity at Highbury using a long wire 'that the idea of transmiting *[sic]* intelligence, by <u>dis</u>charging an insulated wire at given intervals, first occurred to me'.[C375] That was around 1810. A few years later:

> in the summer of 1816, I *amused* myself by wasting, I fear, a great deal of time, and no small expenditure, in trying to prove, by experiments on a much more extensive scale than had hitherto been adopted, the validity of a project of this kind. I believe I succeeded to the entire satisfaction of several very eminent scientific friends[R7]

Ronalds combined his stunning foresight with considerable practical nous. He took over his mother's long back garden at Hammersmith for the job. Initially he set up a thin iron wire nearly 13 km long by stringing it back and forth between insulated supports on two large wooden frames (Fig. 7.1). He then installed a buried cable in a trench 160 m long and 1.2 m deep under the gravel walkway down

Fig. 7.1: Thirteen kilometre long aerial telegraph arrangement; Ronalds' original drawing. IET 1.6.50.

the west side of the property. Its single copper wire was connected at its ends to equipment in his laboratory above the stable and in the tool shed near the rear of the garden.[J82]

It was natural for him to choose static electricity rather than current to power his telegraph: 'Some German and American savans first projected Galvanic or Voltaic telegraphs…But the other form of the fluid appeared to me to afford the most accurate and practicable means of conveying intelligence'.[R7] It was 16 years since Volta had invented his battery, but they remained unreliable. Even two decades later they proved problematic for William Fothergill Cooke, who wrote: 'When I first directed my attention to the Electric Telegraph, the best battery would only remain in use for a few hours, and became rapidly weaker.'[B72]

The lines were charged initially with a friction electrostatic generator (labelled *D* in Fig. 7.2) at each end; these were purchased from George Singer*. 'Wishing to convince *some* of those, with whom seeing only is believing, that it was possible to keep my telegraphic wire constantly electrified with a very small source of electricity',[R7] he also developed an electrostatic influence machine, described in Sec. 6.5, so that charge could be provided with minimal manual intervention. For full-scale operation he suggested substituting a small steam engine for the clockwork mechanism that brought the three discs into proximity and contact with each other cyclically.

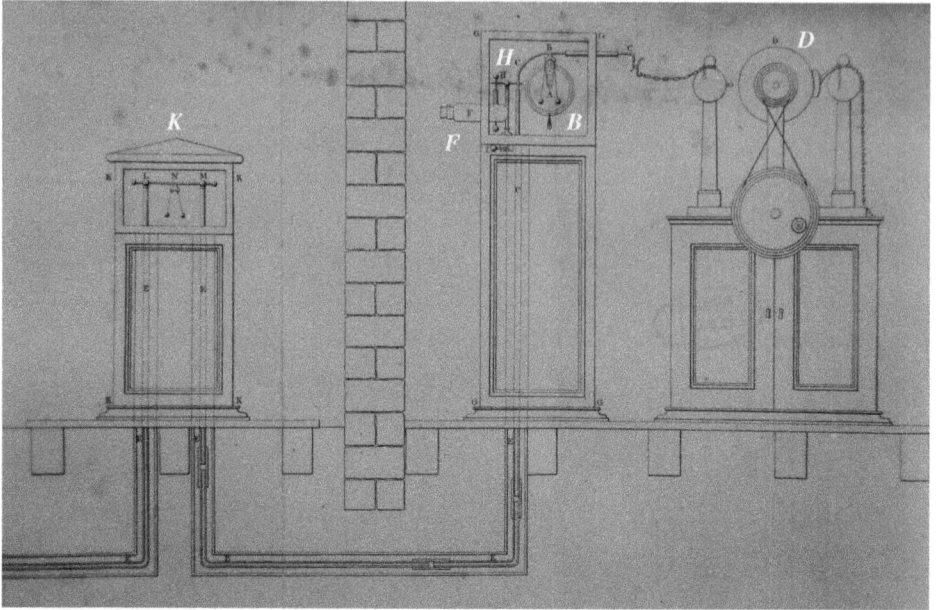

Fig. 7.2:   Arrangement of subterranean telegraph. Ref. R7.

B Revolving alphanumeric dial, *D* Electric generator, *F* Gas pistol, *H* Switch connecting generator to pistol, *K* Testing post.

Ronalds decided upon synchronised dials as the means of data transfer. He placed a pair of concentric circular brass plates, shown in Fig. 7.3, at each end of the wire. The rear plate turned on the seconds arbour of a clock mechanism and was marked into 20 subdivisions, each having a letter, digit and a message helpful to the communication process. The front plate was stationary and had a slit to show just one of the divisions. When the wire was earthed momentarily, a diverged pith-ball electrometer hanging in front of the plates dropped, and the information then visible in the sliver of the plate was noted.[C388] He developed a protocol for checking at the beginning of the communication that the two clocks were indeed synchronous. First, the sender overcharged the wire to explode the gas pistol *F* and alert the recipient to a forthcoming message. He or she then diverged the pith-balls electrically a little more than usual when the message PREPARE was visible on their dial. The receiver adjusted their cover plate if necessary to ensure they would see the same sign at that second and, when complete, sent back the signal READY. Other useful messages included on the dial may be seen in Fig. 7.3.

As an alternative to spelling out words, he devised a Message Grid to speed up the communication process. It comprised a book with ten numbered leaves, each having ten columns and rows. Discharging the wire sequentially when each of three numbers were in view would take the recipient to the correct page, column

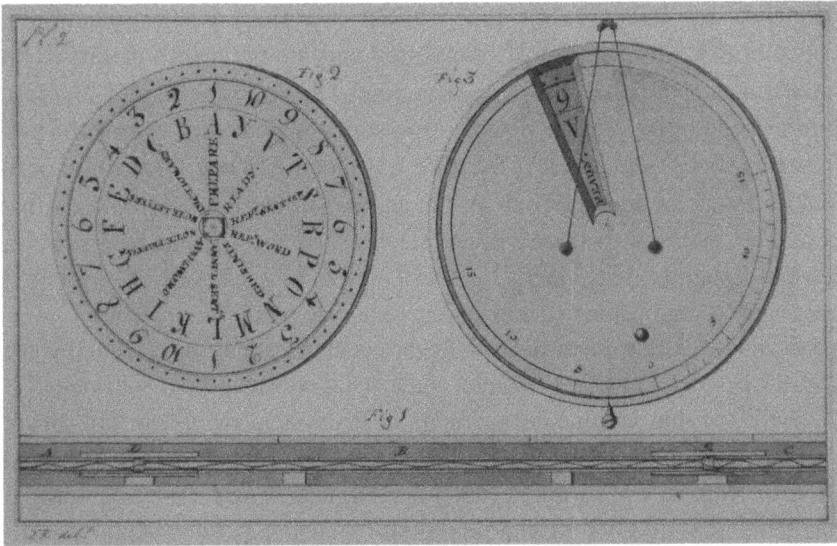

Fig. 7.3:   Revolving dial and insulated cable; Ronalds' original drawing. IET 1.6.46.

and row of the table to read the message in that box. He noted that the process took less than a minute.

Particular attention was paid to the design details to ensure a workable system. The most obvious issue, insulation of the buried wire, was not of concern to Ronalds after his experience with atmospheric electricity and the deployment of coated glass tubes (see Sec. 6.4). He first laid a 5 cm square wooden trough in the trench, lined inside and out with pitch. The wire itself ran through glass tubes resting on small blocks in the trough. The ends of each tube were coated in soft wax and housed in glass sleeves to prevent moisture ingress while allowing thermal expansion or contraction in different weather conditions. The detail has been described as the first such expansion joint.[T20] A wooden lid was screwed onto the trough while the pitch was hot and the trench was then refilled.

He also incorporated a Testing Post *K* (Fig. 7.2) to enable the location of any failure in the line to be pinpointed. Here the wire rose out of the ground, protected by an outer box, and the connection could be broken on either side of an electrometer to determine which section was at fault. He noted that he had no cause to use the post in six months of experimentation with his telegraph.

Even the warning pistol he developed was later written up with illustrations in the *Encyclopædia Britannica*.[A62] It was activated by a spark passing through the switchable connection *H* (Fig. 7.2) to enter the barrel through a brass ball and rod projecting from the top. On ignition the flammable gas in the vessel discharged a cork at the far end.

Ronalds also addressed the practicalities of rolling out the system at scale. A particular risk was damage. He suggested burying the cables in cast iron pipes with redundancy provided through extra lines. Testing posts could be located at post offices and other suitable places. A person would keep an eye on the pith-ball electrometer there to ensure that the wire was electrified and, if a failure occurred, it would be sourced and repaired. Watch stations might be required for some sections of line, such as river crossings. He finished the risk management section somewhat flippantly by saying:

> let us have…kings that love their subjects enough to prevent civil wars…
> and should they still succeed in breaking the communication…hang them if you can catch them, damn them if you cannot, and mend it immediately in both cases.[R7]

We will see that various components of his system reappeared in subsequent telegraphs. Benjamin Silliman, Professor at Yale, noted in his *Principles of Physics, or Natural Philosophy* (1869) that Ronalds 'used a movable disc, carrying the letters, the type of all dial telegraphs'.[B89] Thomas Dixon Lockwood, engineer and patent attorney for American Bell Telephone Company, wrote of the set-up in 1896:

> This telegraph was by far the most ingenious and simple that had yet been proposed, and is besides noteworthy as being a foreshadowing of the synchronous printing systems…and of many selective or individual signaling systems which have within the last twenty years been so frequently devised for telephone calls.[T18]

Concerning the testing posts, telegraph engineer and author John Fahie noted that 'here, again, the telegraph engineers of the present day *[1884]* have followed out his ideas almost to the letter.'[T15] Indeed telephone junction boxes are seen today at regular intervals along sidewalks and still have the same purpose.

Ronalds nonetheless distinguishes himself from other characters in this story in maintaining until he died that he was *not* the inventor of the telegraph — those who followed him were the opposite. He knew that numerous scientists had contributed to the subject before him. The opening sentence of his 1823 book *Descriptions of An Electrical Telegraph* reads:

> Dr. Watson and this friends, Lord C. Cavendish, Mr. Martin Foulks, Dr. Bevis, Mr. Graham, Dr. Birch, Mr. P. Duval, Mr. Trembly, Mr. Ellicot, Mr. Robins, Mr. Short, and some other gentlemen, so long ago as the year 1748, proved, that electrical shocks might be conducted through long circuits with *immeasurable* velocity.[R7]

Interestingly, not everyone believed it was even possible. Singer wrote of Dr William Watson's work in his 1814 textbook on electricity:

> There is some doubt as to the accuracy of these experiments; they were made at a very early period, and have not, I believe, been repeated…Metals, although the most perfect conductors we have, oppose some resistance to the motion of electricity, and a charge will even prefer a short passage through air to a circuit of 20 or 30 feet through thin wire. It is therefore rather uncertain that the charge of a small phial has ever passed through an interval of four miles.[B12]

Ronalds collected nine of Watson's works for his library.

Some other scientists who had contemplated sending electric signals through wires were unknown to Ronalds at the time. In particular, he was excited to discover in Como many years later that his hero Volta had formulated the idea. A 'C M' had proposed another permutation in the 1753 *Scots Magazine*.[B187] It was to comprise 26 lines (one for every letter of the alphabet) suspended in air. A small ball would be hung from the ends of each wire, which would be electrified briefly in turn to signal letters at the other end. This story was also rediscovered in the 1860s.

Ronalds welcomed anyone who wished to see and participate in his experiments. Decades later, when there seemed almost to be a need to verify that the trial had taken place, several eyewitnesses could be found who were just young boys at the time. In an 1889 article we learn:

> Both for building the frames for his eight-mile circuit of wire, and for the troughs for the underground line, Mr. Ronalds employed the services of the late Mr. Eyles of Hammersmith…His son, Mr. Silas Eyles, who was a boy at the time, has stated that he distinctly remembers seeing the telegraph in the garden, which his father assisted to build.[T16]

The article also quotes a letter from Ronalds' first cousin Thomas Field Gibson. He had the benefit of now being an elder statesman after his role as one of Prince Albert's Royal Commissioners for the Great Exhibition of 1851:

> How well I remember when a school boy 55 years ago, seeing the clock apparatus in your little upper room over the stable connected with another at the bottom of the garden, of the meaning of which I had but a very hazy apprehension — also the lines of wire stretched from frame to frame across the grass plot, and a day which I daresay you have forgotten when some old Non Londinensis came to inspect and be enlightened.[C403]

Alfred's youngest son Alexander documented his father's reminiscences: 'he *[Alfred]* assisted his eldest brother (Sir Francis Ronalds F.R.S.) in his school holidays to perfect his — my uncle's — invention of putting electricity to words'.[C418] Alfred and Thomas were both around 13 years of age.

Ronalds mentioned the visits of several scientists. Dr Abraham Rees FRS was a nonconformist minister in London and a trustee of Dr Williams's Library; his most important work was his 45-volume *New Cyclopædia or Universal Dictionary*. Other guests included Lord Henniker*, William Brande* and Professor Karl Knorre,[C227] a friend of Uncle Charles Field from what is now Mykolaiv on the Black Sea.

Two other names that are very important in this story — Wheatstone and Cooke — also participated in the demonstrations. Ronalds later noted in his diary:

> At about the time of my having completed my "Electric Telegraph" Mr Cooke, then a Surgeon at Brentford, well acquainted with our family, and Father of Mr W. F. Cooke, called on us at the Upper Mall, Hammersmith: on which occasion I explained to him minutely, and proved the <u>perfect practicability for general</u> use of my

Telegraph, making him the recipient of many messages, by it...and receiving messages in return from him.

NB Mr W. F. Cooke was a very young boy at this time.[J82]

The young William Cooke would have been between 10 and 12 years of age. Alfred Frost's biographical memoir of Ronalds, written in 1880 with the assistance of the family, advises that 'Wheatstone, then a boy of about 15, was present at many of the principal experiments at Hammersmith.'[M5] Charles Wheatstone* had by then purchased Volta's book and was conducting the electrical experiments in it with his younger brother.[B183] Ronalds' efforts were the stimulus for him to later begin investigations into telegraphy.[C400]

Once scientific colleagues had approved his set up, it was obvious to Ronalds that he should offer it to the Admiralty — it was barely a year since the Battle of Waterloo had brought an end to the long wars with France. He wrote to Robert Dundas FRS, Viscount Melville, First Lord of the Admiralty, on 11 July 1816:

> Mr Ronalds presents his respectful compliments to Lord Melville and takes the liberty of soliciting his Lordship's attention to a mode of conveying Telegraphic Intelligence with great rapidity, accuracy, & certainty in all states of the Atmosphere, either at night or in the Day and at small expense...Having been at some pains to ascertain the <u>practicability</u> of the scheme it appears to M^r R and to a few Gentlemen by whom it has been examined to possess several important advantages over any species of Telegraph hitherto invented & he would be much gratified by an opportunity of demonstrating those advantages to Lord Melville[C15]

There was a reply from Melville's Private Secretary Robert Hay two weeks later:

> I am desired by Lord Melville to acknowledge the receipt of your letter of the 11th Instant. — His Lordship has left Town for some Weeks, but he has requested me to see you on the subject of your discovery, if you desire it.[C16]

Ronalds excitedly wrote back in the affirmative. Barrow's rejection letter came by return post.

It transpires that his timing could hardly have been worse. His competition — the semaphore — had been honed by the enemy in the 1790s with lines constructed across France to assist communication in the war. The first example of a simpler set up designed in England was opened by the Admiralty literally the week before he sent his letter. It comprised a series of tall poles, each having two pivoting arms that a person moved using winches to relay messages in code. Ronalds could not compete with its inventor Admiral Sir Home Popham — Popham's system was even awarded a Gold Medal by the Society of Arts.[B187] After the initial trial, the Admiralty built a permanent link between London and Portsmouth. Parliamentary returns show that the aim was for it to be useable seven hours per day in the seven summery months and five hours per day in 'winter', but it was actually inoperable on 30 per cent of days.[M5] It was also expensive, costing over £2,000 per annum to run.

Ronalds was well aware of the alternative of patenting his concept. His brother-in-law Peter and the latter's cousin John Martineau had taken out a patent the previous year for their new way of decolourising sugar during the refining process.[P1] Seventeen months later, his brother Hugh patented a tanning process. It did not make sense for himself as there was very little chance of a civilian market emerging within a patent's 14-year lifetime. This is summarised nicely in a major review of the telegraph written in 1854:

> Since the year 1821 the principles of action of the working telegraphs of the present day were known to scientific men, and the question naturally arises, how was it that it still took so many years to make the telegraph a working fact? The answer is that the combination of circumstances necessary to bring it to perfection had not arisen…
> No one imagined that it would ever become a necessary social engine, or that it would pay 'seven per cent.' to a public Company. The patronage of the Government could alone have been looked to by any of the proposers of the new method of telegraphy[A86]

Ronalds wrote, not quite convincingly, in 1823 of Barrow's letter:

> I felt very little disappointment, and not a shadow of resentment on the occasion, because every one knows, that telegraphs have long been great bores at the Admiralty. Should they *again* become *necessary*, however, perhaps electricity and electricians may be indulged by his Lordship and Mr. Barrow with an opportunity of *proving* what they are capable of in this way.[R7]

In fact he would have felt very foolish having expended so much money and time in a rather public way on something that was so roundly dismissed. He expressed this feeling 54 years later to his friend Latimer Clark* when his pioneering work was at last recognised formally by Government: 'I believe it is almost needless to tell you that the honour conferred is much less valued by me than the testimony it helps to afford to the validity of my early labours in the matter'.[C404] Barrow's letter was the end of the telegraph for Ronalds. There was no point in trialling improvements when there was no market and, with his sensitive disposition, he lost confidence and interest to pursue related electrical engineering activities.

7.2 Roll-out of the Electric Telegraph

After his contribution, things went quiet on the telegraph front in England for some years. Cooke recorded that he first became aware of the concept in 1836 on seeing a demonstration of an apparatus recently conceived in Germany; he was: 'Struck with the vast importance of an instantaneous mode of communication, to the railways then extending themselves over Great Britain'.[B66] Cooke's insight was brilliant. The railway era had just started: various short, isolated lines had been built and the first trunk railway was to begin operation the next year. Something was needed to control train operations. He immediately started to develop tele-graphic equipment. Within months, he had met Wheatstone, agreed a partnership with him, and lodged their first joint patent for 'improvements'[B44] to the electric telegraph. Years later, he said in a letter published in the *Reader*:

> Had it been my fortune to have known of Mr. Ronalds' labours, when I returned from Heidelberg in 1836, I cannot have the slightest hesitation in saying that I should have sought his advice, even before that of the great Faraday, and also his co-operation as a partner in my patent.[2]

At the time Cooke's decision was an obvious one. Wheatstone was now a renowned professor in electrical science who had just been elected to the Royal Society. Ronalds, on the other hand, had dropped the subject 20 years earlier and stayed under the radar ever since.

In the years since Ronalds' work there had been a number of critically impor-tant discoveries in electrical science, and Wheatstone and Cooke took advantage of them. John Daniell* created in 1836 what became known as the Daniell cell. It generated a much longer-lasting and more reliable current than Volta's pile and enabled this form of electricity to power practical technology for the first time. Faraday* had discovered that a current could also be created by relative motion of a coiled wire and a magnetic field, but it would be several decades before electric generators on this principle would supplant the battery.

Hans Oersted FRS had observed the converse in 1820 — that electric current generates a magnetic field; a magnetic needle thus became a very convenient way of displaying electric pulses. Soon after Oersted's work, it was found that the effect was amplified by winding the wire around the needle; it then registered even along a very long wire. The electromagnet was developed within a few years, where iron was magnetised only when current flowed through the wire coiled around it. In this way, mechanical work could be done at a distance by applying a current. With the aid of the new relay switch, the electric signal could also be repeated at intermediate locations along the wire to increase its transmission dis-tance. Finally, in 1827 Ohm published his relationship between electromotive force, resistance and current, which enabled circuits to be designed mathemati-cally. All the elements were now available to put together a new generation of electric telegraphs.

Cooke and Wheatstone's first paying job was a 1.5 km line for the London & Birmingham Railway in 1837. They chose an apparatus with five magnetic nee-dles developed by Wheatstone and described in their patent. Two needles deflected at any time to point sequentially to letters of the alphabet arranged around the board on which they were mounted. Each needle required its own wire and the five

---

[2]*Reader* 8/12/1866; IET 1.9.2.

wires were buried. It was a clever arrangement but not practical for longer distances, and the railway halted the trial after a few months. Their second client was the Great Western Railway in 1839. This 20 km long system deployed four-needle instruments but with a separate return wire to allow a circuit with a single-needle, and a spare — a total of six wires. This was also little used.

Britain's Heritage of Science (1917) by Schuster and Shipley records of the time:

> When the electromagnetic effects of currents had been discovered…The main difficulty was to reduce the number of wires, which were at first thought to be necessary for indicating the twenty-five letters; in this respect Ronalds had been ahead of his successors.[B133]

In 1840 Wheatstone and Cooke patented their dial telegraphs to address the problem. Cooke made contact with Ronalds at this point via their mutual friend George Cooper who was a surgeon in Brentford:

> I have been trying for some time to borrow an account of Mr Ronalds' Electric Telegh published by himself many years ago — meeting you yesterday reminded me that you were acquainted with Mr R, and I shall esteem it a favor if you will procure the loan of it for me as soon as you can.[C100]

Cooper immediately sent the letter on to Ronalds[C101] who duly posted a copy of his book to Cooke. Wheatstone also applied the dial concept to a telegraph that printed type and an electric clock.

John Finlaison wrote shortly afterwards of Wheatstone and Cooke's apparatus 'that the signal disc…is identically the same as that of Mr. Ronalds".[B49] Finlaison was the patron of Alexander Bain; Bain and Wheatstone (his former colleague) were in dispute over which of them had first developed the electric clock and the printing telegraph. Wheatstone himself acknowledged Ronalds' prior work on the dial (but not the electric clock) in 1856. Writing in the third person, he noted:

> Mr. Wheatstone…endeavoured, first, to ascertain whether it was possible to construct an efficient telegraph with a single circuit…With this object in view, one of his earliest ideas was to adopt the principle of Mr. Ronalds' Chronometric Telegraph; but by substituting a magnetic needle for Mr. Ronalds' electrometer, and by using its double motion to point to letters…seen through apertures placed before the dial[B67]

Ronalds was again being mentioned because of a conflict — this time between Wheatstone and Cooke. He would not have enjoyed being used as a pawn in this way. Nevertheless, he swallowed hard and made a pointed remark in reply to another letter from Cooke in 1867:

> You are good enough to say it would have been a fortunate thing if in 1837 you had known of my previous labours…Mr Wheatstone, whom I had long known, was well acquainted with what I had done. (It would seem indeed that Two figures *[of the dial]* in the second plate of that pamphlet were closely imitated in your Specification of 1840)[C392]

The telegraph along the Great Western Railway was upgraded and extended to Slough in 1843. It now deployed overhead wires and simpler two-needle machines. Elements of this system were developed by both Cooke and Wheatstone and it became the common choice for more than a decade. A certain number of needle deflections to the left and/or right could signify either a letter, or a phrase printed in a chart. The former was similar in concept to a semaphore or Morse code, while the latter strongly resembled Ronalds' message grid. Queen Victoria's household quickly began to use this telegraph. The extended line facilitated royal communication just as Ronalds had foretold, except that Brighton was not a preferred residence for the new monarch. *The Times* noted in 1844 'that a message can be expressed by the aid of the telegraph from Windsor Castle to Buckingham Palace in 25 minutes'.[3]

Perversely, Cooke and Wheatstone's big break was provided by the Admiralty. In 1845 it commissioned the partnership's first long-distance telegraph between London and Portsmouth, alongside the London & South Western Railway. Its success signalled the demise of Popham's semaphore. That year business outgrew the partnership and the Electric Telegraph Company was formed, purchasing Wheatstone and Cooke's joint patents for a huge sum approaching £150,000. Cooke remained a director until the company was nationalised in the late 1860s to become part of the Post Office. British Telecommunications (BT) is a direct descendent of the Electric Telegraph Company, having split off again in the early 1980s.

The new company was able to expand rapidly in the boom years of railway construction. By the end of 1847, it not only had the skeleton of a national network of telegraph lines but had built underground connections from the train termini to an Electric Telegraph Office in the city. The subterranean lines were housed in cast iron pipes coated inside and out with pitch, incorporated redundancy and had testing posts every 500 metres.[A86] Outside the city centre, linemen patrolled the overhead wires where required. All are reminiscent of Ronalds' recommendations. The network extended over 3,000 km by the early 1850s. Public use of the telegraph increased rapidly, driven in particular by the demand for timely news and information. A newspaper called the *Daily Telegraph* began circulation in 1855 (and still exists today).

7.3 Signal Retardation due to Induction

As the lines became longer, and particularly as they started to go underwater, a critically important phenomenon emerged, one in fact that Ronalds had

---

[3] *The Times* 23/7/1844 7a.

highlighted much earlier. It concerned the rate of propagation of electricity along cables in different situations. We start the story with a contribution by Wheatstone. In describing his entry into the field of electric telegraphy, he had noted that 'the experiments of Ronalds and others had failed to produce any impression on the scientific world; this want of confidence resulted from the imperfect knowledge we possessed of the velocity and other properties of electricity'.[B67] He was able to demonstrate in 1834 through a clever experiment that the 'velocity'[A48] of an electric spark along a long copper wire was not infinite. In his words, 'the results were witnessed by the most eminent philosophers of Europe and America'.[B67] Faraday postulated four years later that the passage of electricity was retarded due to induction.[A54] Their thinking was a little naive at this stage — in fact the 'velocity' is never infinite, even in a vacuum.

Ronalds retained his diary entry for 30 April 1853 on a related topic:

Attended Mr Faraday's Lecture.[4]

Conversed with him (afterwards) about Induction &c. Mentioned the long continued, or, so called perpetual charge of the Electrophorous, notwithstanding the metallic contact of both surfaces of the resinous plate with each other, and with the Earth...

I then introduced the subject of the insulation of my Telegraphic-Wire and said that a great mistake had been made by Mr Carpmeal,[5] in his last Lecture, & by many others, in saying that there was want of Insulation. I stated that so far from this having been the case, the wire contained in the Glass tube retained its electricity too long, sometimes, even when brought into conducting communication with the Earth. Mr Faraday asked if he should mention this in his next lecture, and I thanked him for, & agreed to the suggestion.[J85]

Five months after that lecture, on 4 October, Faraday and Professor George Airy* witnessed experiments performed by Clark on extremely long telegraph cables coated with gutta percha (a natural latex insulation). They showed that it took an appreciable time (of the order of a second) for a 'wave of electricity'[A87] to travel the length of the line, and the cables also retained electric charge for some time. These effects occurred only when the insulated cable was immersed in water or buried, but not in air.[A87] Similar results had been seen by Dr Werner von Siemens (founder of today's global electrical and telecommunications giant Siemens) in Berlin in 1850.[A88] Faraday explained the observations in a lecture at the Royal Institution in early 1854, noting them to be 'illustrations of the truth of my ancient views of the nature of insulation, induction and conduction'.[A88] The paper made no mention of the much earlier work done by Ronalds, who missed the lecture anyway as he was on the Continent.

---

[4]Faraday was in the midst of a six-lecture series on static electricity at the Royal Institution.
[5]William Carpmael, a leading patent lawyer and engineer, was similarly giving a six-lecture series on the electric telegraph.

The lessons were learnt yet again a few years later in connection with the expensive failures of early submarine cables. By 1861, Clark had realised that they weren't the first to witness the phenomenon and he wrote to Ronalds in Italy:

> I will forward you an interesting report on an investigation connected with Submarine Cables which I have lately undertaken —
>
> It is most interesting to me to see the phenomenon of "retardation of current" which has so greatly occupied the attention of Electricians of late years in connection with long Submarine Cables, distinctly foretold & described in your work published in 1823![C377]

Ronalds had concluded back in 1816 that retardation was the biggest risk of all to a buried electric telegraph line and he addressed it first in his risk management section:

> That objection, which has seemed to most of those with whom I have conversed on the subject the least obvious, appears to me the most important, therefore I begin with *it*; viz. the probability, that the electrical compensation *[an early word for induction]*, which would take place in a wire enclosed in glass tubes of many miles in length (the wire acting, as it were, like the interior coating to a battery *[Leyden jar]*) *might* amount to the *retention* of a charge, or at least might destroy the *suddenness* of a discharge…even although the wire were brought into contact with the earth.[R7]

Ronalds saw the wire surrounded by moist soil with an intermediate insulator as being akin to a long Leyden jar that generated induction and retained the charge (see Sec. 6.1); Faraday used the same analogy in his paper nearly 40 years later. Ronalds went on to describe methods of minimising its occurrence in extremely long insulated wires of the type used in his experiments. Many years later he admitted to Clark: 'my difficulty of discharging my wire was greater than that of preserving a charge'.[C388] Wheatstone appears to have overlooked this behaviour in viewing Ronalds' experiments as he did not attempt to investigate the influence of the properties of the wire and its surrounds on electricity transmission.

The story is told in the introduction to the Catalogue of Clark's Library[B128] and was underlined in his Presidential Address to the STE in 1875.[T11] German physicist Karl Eduard Zetzsche argued that Clark had overstated Ronalds' understanding. In his 1877 book *Geschichte der Elektrischen Telegraphie*,[T12] he queried whether retardation would have been discernible in Ronalds' set up and doubted that he had really envisaged what was later understood by the concept of induction in cables. Other authors of this later period followed Clark in crediting Ronalds' early insights.[B108,T13,T15,T19]

Faraday's thinking formed part of what is now known as Faraday's Law of Induction. It would be the 1880s before the effects of resistance, capacitance and inductance on the propagation of an electromagnetic wave along a cable were quantified in the 'telegrapher's equations'. Wheatstone wasn't quite right that

Ronalds' work 'had failed to produce any impression on the scientific world', but the speed of transmission of his pioneering concepts was certainly very slow! Speaking privately to Faraday, for example, rather than raising the point during the lecture, would have been Ronalds' natural way, but it resulted in his ideas receiving little notice.[A120]

7.4 Recognition of Ronalds' Contribution

Expansion of the telegraph had quickly inspired people to document the history of its development. Ronalds' demonstration was noted in nearly all of these accounts, both in Britain and abroad — like it or not, he was beginning to lose his anonymity. Publicity also generated occasional demand for the 1823 book. Clark had written to him in 1861:

> I have long wished to have the opportunity of asking you to allow the copies of your most interesting work on the Electric telegraph, which I understand are still in existence, to be published.
>
> Your work is extremely rare and except by quotations few know of its existence. For a long time I advertised for it and offered a Guinea for a copy but in vain *[it was originally offered for sale at six shillings]* — at last I met with it[C377]

We will see this story repeated regarding another of Ronalds' publications, *Sketches at Carnac*, in a later chapter; he consistently underplayed his work.

Ronalds' telegraph was mentioned in the 1824 *Encyclopædia Britannica* and, by the 1842 edition, was the subject of a feature article, complete with diagrams.[T1] This was repeated almost verbatim in Alfred Vail's book *The American Electro Magnetic Telegraph*.[T2] Vail was employed by Morse, about whom more anon. Sir John Rennie* was perhaps the first engineering leader to give Ronalds the honour of inventorship of the telegraph, in his 1846 Presidential Address to the Institution of Civil Engineers.[A69]

Writing in *Traité de Télégraphie Electrique* (1849), Abbot François Moigno* underlined the disadvantages of the configuration:

> Mr. Ronalds would have completely solved the problem of telegraphy, if he had not met two insurmountable obstacles; the difficulty of establishing the essential synchronism between the two clocks, and the inability to clearly identify what would succeed static electricity.[T4]

The Great Exhibition of 1851 was well timed to celebrate international development of the telegraph. The *Reports of the Juries* noted: 'As might have been expected,…the Exhibition is rich with a large number of very ingenious contrivance, applicable to every stage of electric telegraphic communication.'[B61] Those receiving medals included Bain and Siemens in addition to the Electrical Telegraph Company. In putting the recent innovations into historical and technological context, Ronalds' pioneering effort in 'making a telegraph that did some actual work' was highlighted; he would have appreciated the judges using this phrase. His clockwork dials and mode of insulation were also described.

Recognition of Ronalds' telegraph continues today — various representative articles from different periods are listed in the Bibliography. The trend in recent decades has been to portray his work largely as a one-off, primitive experiment using outmoded static electricity.

7.5 Attitudes of the Inventors

Perusing the literature of the period, it is apparent that numerous descriptions of the Wheatstone and Cooke telegraphs were labelled as being Wheatstone's alone. As early as 1841, the *Mechanics' Magazine* noted:

> Public opinion, however, having...chosen to assign all the merit to one of these gentlemen only, to the prejudice of the other, two mutual friends, Sir Marc Isambard Brunel, and Professor Daniell, were very judiciously called in to adjust the respective pretensions of the parties[A59]

After some deliberation, the Arbiters crafted a carefully-worded statement:

> Whilst Mr. Cooke is entitled to stand alone, as the gentleman to whom this country is indebted for having practically introduced and carried out the Electric Telegraph as a useful undertaking...and Professor Wheatstone is acknowledged as the scientific man, whose profound and successful researches had already prepared the public to receive it... it is to the united labours of two gentlemen so well qualified for mutual assistance, that we must attribute the rapid progress which this important invention has made during the five years since they have been associated.[B66]

The statement was signed off by the partners but did not stop their bickering. Thirteen years after the arbitration, Cooke published a 48 page pamphlet entitled *The Electric Telegraph: was it invented by Professor Wheatstone?*[B66] Wheatstone penned *A Reply to Mr Cooke's Pamphlet*[B67] in early 1856, which ran to 74 pages. Cooke's follow-up two months later, *A Reply to Mr Wheatstone's Answer*,[B72] took 152 pages. Cooke conveniently republished all three documents together with a detailed preface and additional materials in 1857 — a total of more than 550 pages over two volumes.[B73] They are all in the Ronalds Library. Their purpose was to determine which of them deserved the credit for the electric telegraph. It did not occur to them that anyone else might share the honour.

Cooke argued in his first pamphlet that he was the primary inventor because his was the first telegraph to contain the three elements essential for practical use: the alarm, detector and reciprocal communicator. The alarm alerted a person at the far end to a forthcoming message, while the detector determined if the wire was transmitting electricity. By a reciprocal communicator Cooke was referring to his united system with both sender and receiver seeing the same signals in a mutual communication between two distant places.

It could be contended that Ronalds' system also had these three attributes, which is what Wheatstone did in his reply, albeit to belittle Cooke rather than to honour Ronalds. Concerning the reciprocal communicator:

> Mr. Cooke's "principle" was no novelty…having been developed completely and effectively by Mr. Ronalds [B68]

Cooke disagreed with Wheatstone's assessment and argued in his reply:

> Mr. Ronalds' telegraph was not a combined system of apparatus, constructed so as to work reciprocally, but three separate and independent elements of communication, viz., the two clocks and the line of insulated wire, arranged into a kind of partnership, for the purpose of producing a combination of effect by a coincidence of action. [B72]

Cooke's was certainly a more sophisticated 'combined system' than Ronalds' following from its different form of electricity, but his early dials still worked independently.

Wheatstone argued in turn that the accolade was his because their first patent in particular 'was entirely and exclusively Mr. Wheatstone's invention'. [B67] He acknowledged however that he had 'applied…principles discovered or developed by a succession of eminent men of science'. [B67]

In the meantime, another claimant for the title had emerged. Professor Samuel F. B. Morse wrote in 1837:

> There is to be a contest, it seems, for priority of invention of this Electric Telegraph between England, France, Germany and this country. I claim for myself and consequently for America priority over all other countries in the invention of a mode of communicating intelligence by electricity. [B185]

Morse was actually an artist by profession; he was commissioned in 1825 to paint General Lafayette's portrait during his and Fanny Wright's tour of America (Sec. 4.7). He filed a caveat for a telegraph in 1837 and a patent in 1840. By 1844 it was in action along the Washington to Baltimore railroad, funded by Congress. The system devised by Morse and his scientific advisors with its forerunner of the famous dot and dash code had several big advantages over Cooke and Wheatstone's multiple needle arrangement. First, it was simpler, in part because it required only a single wire (it used an earth return). Second, it provided a permanent record of the message in marks on a tape. Most important was the later realisation that the clicks of the device could be interpreted by ear after a little practice, which allowed very rapid messaging. Morse's telegraph became the standard across continental Europe as well as America.

As early as 1848, Ronalds' cousin Henry Jnr in rural Canada wrote home:

> I must tell you that the march of improvement is going on with rapid strides as the electric Telegraph is established from Montreal to London [*Ontario*] about 65 miles

from here, so that we receive the news of a packets arrival within two days of its touching the shores of America.[C203]

America's transcontinental telegraph line was completed in 1861. By chance, a relative of Ronalds' brother Hugh sent the first message from Washington Territory: Governor William Pickering had married the sister of Hugh's wife Kate in Albion, Illinois. Pickering's friend, President Abraham Lincoln, replied: 'Your patriotic despatch of yesterday received, and will be published.'[B148]

Other forms of telegraph were also developed and patented. Ronalds' early work was on occasion used to defend new patents as he had previously described various aspects of the operation included in Cooke and Wheatstone's and Morse's specifications.

We have relatively few glimpses into Ronalds' frame of mind about the telegraph through this long period. The first hint that he was following the development of the technology is his suggestion in 1840 to deploy the telegraph in alerting the authorities to a fire; it is described in Sec. 12.13. Others had similar ideas but the concept was not taken up in London for another 40 years.

By the late 1840s, Ronalds had realised that telegraph infrastructure could exhibit unexpected behaviour during certain meteorological and geomagnetic events. The occurrences were of great interest to him in his role at Kew Observatory, as discussed in Sec. 14.4, and he was in contact with several railway engineers. A number of his Kew colleagues became aware of his telegraph work at this time — he would have enjoyed sharing his book with them and explaining what he had done all those years ago.[C311] A Greenwich Observatory associate wrote: 'the perusal of it has given me much entertainment, and the undoubted right of considering you the inventor of the Electric Telegraph'.[C211]

Fortunately Ronalds was in Europe during Cooke and Wheatstone's pamphlet war, but colleagues kept him informed, for better or worse. Charles Weld, assistant secretary and librarian at the Royal Society, wrote to him in Paris in early 1855:

> Mr Cooke is at length making considerable stir respecting the Electric Telegraph and its introduction by him into England prior to Wheatstone — but I am glad to find your name honourably and very properly mentioned in the history. — M^r C mainly attacks M^r W — other parties talk of you.[C359]

Ronalds let it bother him. He made a diary note late that year that started: 'Perused (again) a part of Mr Cooke's Pamphlet "The Electric Telegraph" &c'[J82] and went on to contend a point in it. Someone must have sent the booklet over to him.

By 1857, he was at ease with his place in the history. He wrote in his diary in Switzerland:

> At Vevey I walked up to the Church of St Martin & took my seat on the same bench under the same tree which protected my (not then balled [sic]) pate from the same August Sun Thirty seven years since & gazed at the same peaks...Little did I, in 1817 dream that my electro-telegraphic wires would scale such peaks [*Switzerland*

had quickly developed a national network], would stretch themselves between Ireland & America would make short work of all terrestrial distances & difficulties in so short a period as Forty years. —

The Abbe Moigno (in his Hist. d Telegie electque) calls me "one of the Fathers of the Electric Telegraph". I am contented with this Title. I think however that I was a tolerably effective Father.[185]

His perspective in 1860 is recorded in his autobiographical letter requested by his brother-in-law Samuel Carter:

In 1816, amid many scoffs & jeers & a few imputations of insanity, I worked hard at the electric telegraph…

Much has been said about my difficulty of insulation, by those…who did not seem to be aware of the facility of insulating conductors circumstanced as my buried wire was (i e) serving a like office to that of the interior coating of the Leyden Jar…The buried wire actually retained a part of the charge too long sometimes & even a small renovation of signs (due to certain effects of humidity I believe) sometimes occurred spontaneously. This was a slight but real objection, about equivalent, perhaps, to that arising from the retention of Magnetism in the electro-magnetic Telegraph.

I am fully aware that what I have now said is open to discussion (but this is not the place for it) and am very far from advocating the cause of static in lieu of magnetic electricity for telegraphic purposes (sufficient not perfect insulation could be attained) & a provision was contemplated "for keeping up a sufficient & constant supply of electricity"…

but I will say that if the electric telegraph of 1816 had been fairly examined, an effective instrument might have been in the hands of the government & that after Dr Œrsted's…experiments an improved telegraph might have been in the said hands; also that messuages [sic] might have been conveyed thereby as cheaply in England &c as they are in America "a'most."

…You must excuse all this egotism, my dear Carter. I cannot avoid it…

In 1823. Had a little better success in the Descriptions of an Electc Telegh &c so often alluded to; but it was written in great haste ~~and in love~~ [C375]

In arguing that the development of telegraphy had been delayed significantly by the Admiralty's decision, Ronalds was building on the observation that technological improvements had accelerated as markets appeared and experience was gained.

7.6 Knighthood and Commemoration

Globalisation of the telegraph had started some years prior to Ronalds' comments. The 1866 completion of the mammoth line between the UK and USA across the Atlantic precipitated a string of knighthoods: to Directors of the company, to a part-owner of the ship that laid the cable, its captain, the person responsible for laying the cable and the cable-maker. These were not the first honours for the electric telegraph. That went to Clark's later business partner Charles Tilston Bright, who was Chief Engineer for the first Atlantic cable crossing in 1858. He was knighted within days of its completion, which was fortunate for him as the line failed irrevocably within a few weeks.

In amongst the celebration banquets, people began to raise the question once again of who had invented the telegraph and shouldn't they be honoured too? Supporters of Cooke and Wheatstone called for their recognition. Ronalds had returned to England from his travels and his name was also put forward by Samuel Carter — it was now exactly half a century since his demonstration had been conducted. The newspapers were captivated:

> HONOUR DESERVED BUT NOT OBTAINED — The old story of the failure of justice is repeated with respect to the invention of the electric telegraph. If any name may be set at the head of the noble list in this new roll of honour, it is that of Wheatstone. To Wheatstone we owe the fact that the whole civilised world is now brought within an instant of time.[6]

Promoting Cooke:

> Who really invented the electric telegraph? The *Times* of late has been advocating that honour for Professor Wheatstone, but the advertising columns of the same paper contain a formal award made in 1841 between Mr. W. F. Cooke and Professor C. Wheatstone, signed by Sir I. Brunel and Professor Daniel, which denies to Wheatstone either the sole invention or the leading share in it…So the question may be regarded as settled.[7]

Samuel's letter to the *Daily News* included:

> Neither Mr. Wheatstone, then, nor Mr. Cooke, is entitled to the honour of the first invention of the electric telegraph, great as no doubt their merit is in its application and improvement. "Suggestions" as to the feasibility of transmitting communications by means of electricity had been previously made, and experiments tried…but to Mr. Ronalds belongs the honour of having been the first to invent and publish to the world a really efficient electric telegraph…In a matter of so much scientific interest and public concern the claims of all parties who have aided in this great discovery should be fairly considered; and as a friend and relative of Mr. Ronalds…I venture to solicit a space for this letter in your valuable journal.[8]

Even before Samuel's intervention, Ronalds had been noticed in several newspapers and the tale of the Admiralty's rebuff was retold many times. Support from newspaper editors and the public continued to grow:

> WHO INVENTED THE ELECTRIC TELEGRAPH?
> …It seldom happens that the author of a great discovery, after failing to attract attention to his application of science, lives to see his own invention universally adopted. Mr. Ronalds appears to be the least pushing of original inventors, and it is just that

---

[6] *London Magnet* 15/10/1866 4e; paraphrasing *The Times* 10/10/1866 8e-f.
[7] *Guardian* 31/10/1866 20c; Ref. B87.
[8] *Daily News* 3/11/1866; IET 1.9.2.

in his later years he should have the satisfaction of knowing that he is appreciated by his countrymen.[9]

Another contribution:

> The honour of the invention of the electric telegraph has been claimed on all hands, and by almost every country; but...there is no doubt, — and the fact is admitted by Professor Wheatstone and Mr Cooke...— that the first to make the practical step of laying down an electric wire of considerable length, connected with dials, and capable of transmitting intelligible messages, was Mr Ronalds[10]

Ronalds' instinct would have been to cringe from this publicity, but he had also been reliving his own actions and those of others for 50 years and his unassuming nature had matured and hardened. With his claim mounting, the *Reader* published several comments by Cooke in late 1866 on the topic of 'the hunt for telegraphic inventors, quite a fashionable sport for the moment':[11]

> The merit of the invention must then consist, in a very great degree at least, in the *practical realization* of that which had been before an idea or an experiment. To the merit, such as it may be, of this practical realization, I have maintained, from first to last, one consistent claim...I now expect a final confirmation of the same unpretending claim from the justice of the Public.[12]

He added, more modestly:

> "There is a tide in the affairs of men." Mr. Ronalds was before the tide...We were all practical men, but it was my good fortune to take the tide at its turn.[2]

Nonetheless Cooke's argument upset Ronalds. He believed above all that he had addressed and solved the problems attendant on making a working system, although certainly it was not as effective as those possible with newer technology.

Cooke had sent some of the articles to Ronalds. He struggled in his reply and it was not completed for over three weeks; he was also very ill at the time and not sure how much longer he would live. In it he wrote:

> Pray accept my thanks for the kind expressions in your letters, to myself, & to the editor of the "Reader"...You will I trust excuse me for not here entering upon the topics, alluded to, in the latter further than to protest against the assumption that my "Telegraph was not adapted for practical use". This was not the opinion of competent scientific gentlemen who saw it in action...
>
> You will also, I am sure pardon the strong natural feeling, which I experience on a subject which formerly cost no small labour & anxiety[C392]

This is the same letter in which Ronalds made his pointed comment about Cooke and Wheatstone's dial telegraph looking much like his.

Cooke had intended his letter to be a friendly one. He started by writing:

---

[9] *Saturday Review* 17/11/1866; IET 1.9.2.
[10] *Brighton Herald* 8/2/1868; IET 1.9.2.
[11] *Reader* 22/12/1866; IET 1.9.2.
[12] *Reader* 1(?)/12/1866; IET 1.9.2.

> Many years ago, when you were a young man, and I was a boy, my Father, afterwards D[r] Cooke, of Durham, lived at Brentford —
>
> I think I recollect your living at M[rs] Nairn's Cottage in the Butts —
>
> If I am right, it is a singular fact that two men so much associated with the Electrical Telegraph, should have been residents in the same small town [C389]

To which Ronalds responded:

> I duly received your letter of the 11th ~~instant~~ ultimo, & the number of the "Reader" and ~~beg~~ have to apologize for the delay in acknowledging them. In reply I ~~must~~ will first beg leave to say that it must have been your acquaintance with my Cousin, D[r] Henry Ronalds, to which you refer, but that I had the honour of knowing your respected Father & well remember an obliging visit from him, when…I explained & exhibited to him the Electric Telegraph in ~~action~~ operation [C392]

Cooke had thick skin and was not perturbed: we will see that he wrote much the same thing again to Ronalds three years later.

Wheatstone received his knighthood in early 1868. The Cooke machine moved into overdrive:

> WE read some months ago that the honour of knighthood had been conferred upon Professor Wheatstone, "in consideration of his great scientific attainments, and of his valuable inventions". The former qualification has never been doubted, but the mention of the latter has called our attention to the controversy…as to the amount of credit due to him on account of the one great invention with which his name has always been connected. [13]

Cooke's brother Revd Thomas Cooke published the *Authorship of the Practical Electric Telegraph of Great Britain*[B87] in 1868, which ran to 131 pages. The next year came the 43 page *Invention of the Electric Telegraph: The Charge against Sir Charles Wheatstone of Tampering with the* Press.[B88] Ronalds collected them both for his library. Cooke received his honour shortly afterwards 'for great and special services in connexion with the practical introduction of the electric telegraph'.[B122]

The 'inventor' accolade was still open. Ronalds was knighted a few months later (at age 82) 'in consideration of his having been the original inventor of the electric telegraph'.[B122] The first thing he did was to write to *The Times*:

> We are requested by Sir Francis Ronalds to state, in reference to a paragraph which appeared in our impression of the 4th inst. (of which he was not previously cognizant) that although he invented and employed a perfectly efficient electric telegraph in 1816, and fully described it in 1823, he disclaims the appellation of "original inventor of the electrical telegraph," many schemes of the kind having preceded his. [14]

Not everyone listened to Ronalds. The *Encyclopædia Britannica*, in its 'CHRONOLOGICAL TABLE Of the principal events of political and military

---

[13] *Atlas* 13/3/1868 11b.

[14] *The Times* 7/4/1870 7f; IET 1.9.2

history, with notices of…the most important inventions and discoveries, from the earliest times to the close of the year 1875' includes in the year 1816: 'Electric telegraph invented by Ronalds.'[A130]

Two early letters of congratulations received by Ronalds were from brothers William and Thomas Cooke. Sir William's started:

My dear Sir Francis,

Death has been heavy among my dearest friends & relations during the last fortnight, or I should have written sooner to congratulate you on Her Majesty's acknowledg[t] of your telegraphic labours. It is singular that Brentford should have furnished two of the men who were most practical in their original views; I might say, the only two men who upon the year 1837 realized in their minds the Electric Telegraph, as a future fact.

I have before said, that you were before the time — had you taken the subject up again in 1835 — it would have been all your own, I fully believe — or had we jointly taken it up in 1836 — we should have divided the honours — At all events I am sincerely rejoiced that you are at last one of the number who have rec[d] national recognition.[C405]

There is no corresponding note in the archive from Wheatstone, although it may have been lost. Clark's letter survives, to which Ronalds replied:

Your kind congratulations are peculiarly esteemed; for I have every reason to believe that your expressions of approbation &c. of my humble labours in Electro-Telegraphy have been influential in Her Majesty's gracious act in my behalf…

Thanking you Dear Sir with all my heart for your assistance sympathy & good wishes[C404]

News could still only reach the Antipodes by boat. The ship carrying the 'April Mails from England'[15] arrived in Adelaide on 31 May 1870. The story quickly spread around the country 'by electric telegraph' that 'Mr. Francis Ronalds, the original inventor of the electric telegraph, has been knighted.'[15] In New Zealand, the *Taranaki Herald* ran an article that had first appeared in the *Pall Mall Gazette*: 'The name of Mr. Francis Ronalds was lately recorded as having received the honor of knighthood. Who is Sir Francis Ronalds? was a question more likely to be asked than answered. Sir F. Ronalds is neither more nor less than the originator of our telegraph system.' The story ended by noting that 'Sir Francis Ronalds is uncle to Mr. F. Ronalds of this town *[Edmund's son]*.'[16]

The most remarkable example of the family's pride in Ronalds' achievements is Alfred's memorial in Ballarat. Although Alfred was long dead in 1870, and is today more widely known than his older brother, his gravestone reads:

---

[15] *Argus Supplement* 6/6/1870 1a-b.
[16] *Taranaki Herald* 17/9/1870 3e; National Library of New Zealand.

Fig. 7.4:   Alfred Ronalds' gravestone in Ballarat, Australia. Ronalds family papers.

Sacred to the Memory of ALFRED RONALDS, BORN 1807 *[sic]*, DIED 1860. BROTHER OF SIR F. RONALDS, F.R.S. THE FIRST INVENTOR OF ELECTRIC TELEGRAPH IN 1816.

His son Alexander had the memorial made after he had achieved some success in business (Fig. 7.4).[B165]

In just a few years' time these far-flung colonies would be linked into the global telegraph network. Australia was connected to London via Singapore and India in 1872, and the line between Australia and New Zealand was completed in 1876.

News of Ronalds' knighthood stimulated demand for the original book on the electric telegraph. Copies soon ran low and it was reprinted in October 1871. The journal *Nature* noted that 'SIR FRANCIS RONALDS has done well in republishing this'.[T9] Both editions of the book are retained in numerous libraries around the world — it is certainly his most successful publication. Inscribed copies he presented to family and friends are also sold on occasion by booksellers.[B134]

The telegraph itself was still buried in the garden at Hammersmith. A later occupant Captain Henry Hill was probably the first person to look for it and he was successful 'after several months' search'[B94] in 1862. Hill's neighbour James Atkinson Peacock was a friend of Ronalds and described it to him:

Several yards of copper wire were found and…a glass tube or the greater part of one with the copper wire in it…and one of the joints with a short tube... the copper wire seemed to be in perfect order. The wooden trough and pitch…had become consolidated with the earth which was as hard as, and formed an opening like, that of a drain tile[C407]

The specimen was soon on the international circuit. Hill presented a portion to Brighton Museum in the grounds of the former residence of George IV. As Peacock wrote to Ronalds:

> Mr Hill's gift to this public museum is an excellent means of making public the fact that an electric Telegraph was perfected and in actual operation as early as 1816 by F. Ronalds Esq^e and in confirmation of any statement of this fact by any person he will only have to refer for the proof of it to a portion of the actual Telegraph in this public Museum where it may be seen by any one sceptical on the subject.[C393]

Apparently Ronalds' work was open to doubt in 1868.

Hill then loaned a sample to the Special Loan Collection of Scientific Apparatus organised in 1876 in South Kensington (Chapter 10). The adjacent item was 'the original wooden model of the dial'[B94] of the telegraph. This was provided by Clark who had received it from Samuel Carter. By 1880, the two components were combined and in Clark's possession. It made its way to the first International Electrical Exhibition held in Paris in 1881 where it formed part of an historical display that also featured apparatus by Volta, Oersted, Faraday and Wheatstone. This became an annual event and the sample travelled to the 1883 Exhibition in Vienna via the 1882 show at the Crystal Palace.[A125,T15] Clark later gave the sample to the General Post Office Museum,[C415] which passed it to the Science Museum in 1894. It has been on display much of the time since and is now in the Information Age gallery.

Morton Stephenson, a subsequent owner of the Hammersmith house, unearthed another section of the wire in 1930. Part of it is at the museum there and the rest was donated to the Science Museum by his wife[17] and is now housed in the National Museums of Scotland. Further fragments have been found by more recent occupants.

The other surviving relic of the telegraph is one of its friction electrostatic generators (Fig. 6.1). Ronalds had taken it to the Kew Observatory in 1843, from where it was donated to the Special Loan Collection. It is on display in the Science Museum.

A professional body for the industry called the Society of Telegraph Engineers (STE) was established in 1871, as discussed in Sec. 16.3. Here was a forum where Ronalds' early contribution could be, and was, acknowledged very frequently. Early Society President (and Clark's brother-in-law) Sir William Preece FRS, for example, noted to the membership in 1887:

> in the year 1823 Mr. Ronalds wrote a paper on underground telegraphs that would do credit to any member of this Society if written in the year 1887. It is perfectly astonishing how that man's instinct saw the various troubles that were likely to be

---

[17] *The Times* 6/2/1930 19c.

met with in the construction of long underground lines…it is a pamphlet that is well worth studying by everybody here.[A128]

With such ongoing recognition, Ronalds' renown in the electrical engineering community was assured for many years.

7.7 Conclusion

Returning finally to the question of this chapter, there is of course no simple answer. Ronalds was right to acknowledge that ideas from numerous people contributed to the telegraph. Many different talents are also required through the long discovery, development, demonstration, deployment, improvement and take-up phases of a new technology, and these very seldom reside in a single person. Cooke, Wheatstone and Morse chose to interpret the terminology 'inventor' in different ways to emphasise the particular stages of the innovation process that suited their purposes; they seemed to crave the primary credit for a device that had completely changed the world. Ronalds desired something a little different — acknowledgement for his part in a long chain, which he believed after decades of reflection to be quite significant. He received more than that, although he had to live a long time to get it.

His documented demonstration came to be regarded as a seminal input to the development of telegraphy. As Fahie wrote in 1884:

> Ronalds will always occupy a high position in the history of the telegraph, not only on account of the excellence and completeness of his invention, but also for the ardour with which he pursued his experiments, and endeavoured to bring them to the notice of his countrymen. Had he worked in the days of railways and joint-stock enterprise, there can be no doubt that his energy and skill would have triumphed over every obstacle, and he would have stood forth as the practical introducer of electric telegraphs; but he was a generation too soon, and the world was not yet ready for him.
>
> His little *brochure* of 1823 is the first work ever published on the subject of electric telegraphy, and is so marvellously complete, that it might almost serve as a text-book for students at the present day[T15]

Ronalds was the first to articulate the scope of the telegraph to transform communication. Needing to use static electricity, his ability to 'render this wild and wayward form of electric force subservient' to his purposes and 'effectually controlling it'[B61] was highlighted at the Great Exhibition. Both Cooke and Wheatstone acknowledged his influence in their own work — Wheatstone noted that Ronalds' configuration incorporated all the elements essential for practical use and also cited Ronalds' contributions in the design of his own dial telegraph and in his experiments on the 'velocity' of electricity. Cooke wished Ronalds had been his partner. The testing posts Ronalds conceived were another essential component of the roll-out of the telegraph, underlining the realistic nature of his design. Lastly Clark, one of the most important early telegraph engineers, was the first to draw particular attention to Ronalds' advanced understanding of the effects of induction in telegraphy. Through these diverse contributions towards such a pivotal technology, 1816 remains the year of Ronalds' life that is best known. He might even have become a household name if his timing had been a little different.

Chapter 8

THE GRAND TOUR

Ronalds' travels in Europe and the Eastern Mediterranean; his objectives, modes of transport, lifestyle, challenges, who he met there and what he experienced.

There was a change in tempo after 1 September 1818, when Ronalds embarked on his extensive overseas tour. His route passed across France to Italy, then down through Sicily and Malta to Egypt and the Holy Land. Turning for home, he travelled via Cyprus, Turkey, Greece and the Balkan coast to Italy and then through Switzerland, Germany and Belgium (Fig. 8.1). He reached London again just a few days before Christmas in 1820, warning his family: 'Invite me to the Christmas party wherever it is as I cannot invite myself, I'm serious.'[C43]

Undertaking the trip addressed a number of priorities for him. Travel and adventure was in his family's blood and his knowledgeable and creative mind was eager to experience new things. He had already had a taste of the Continent himself (Table 8.1). He heard more about it from family members, including Uncle Hugh Ronalds who travelled in search of unusual plants for his nursery, his brother-in-law Peter Martineau who loved Italy, and Edmund's father-in-law Dr James Anderson. His cousin Betsey Ronalds had toured Belgium, France and Switzerland two years previously with her aunt and other companions.[J89]

With peace in Europe after the Battle of Waterloo, the 'Grand Tour' was again a popular pastime for well-to-do English gentlemen. He even referred to the Grand Tour in jest to Charlotte:

I shall be very learned you know after making the Tour of Europe…The grand tour, how very grand it will sound. I must teach you to ask me when the Prince so & so comes to England & whether the Duke so & so is musical and when I heard from

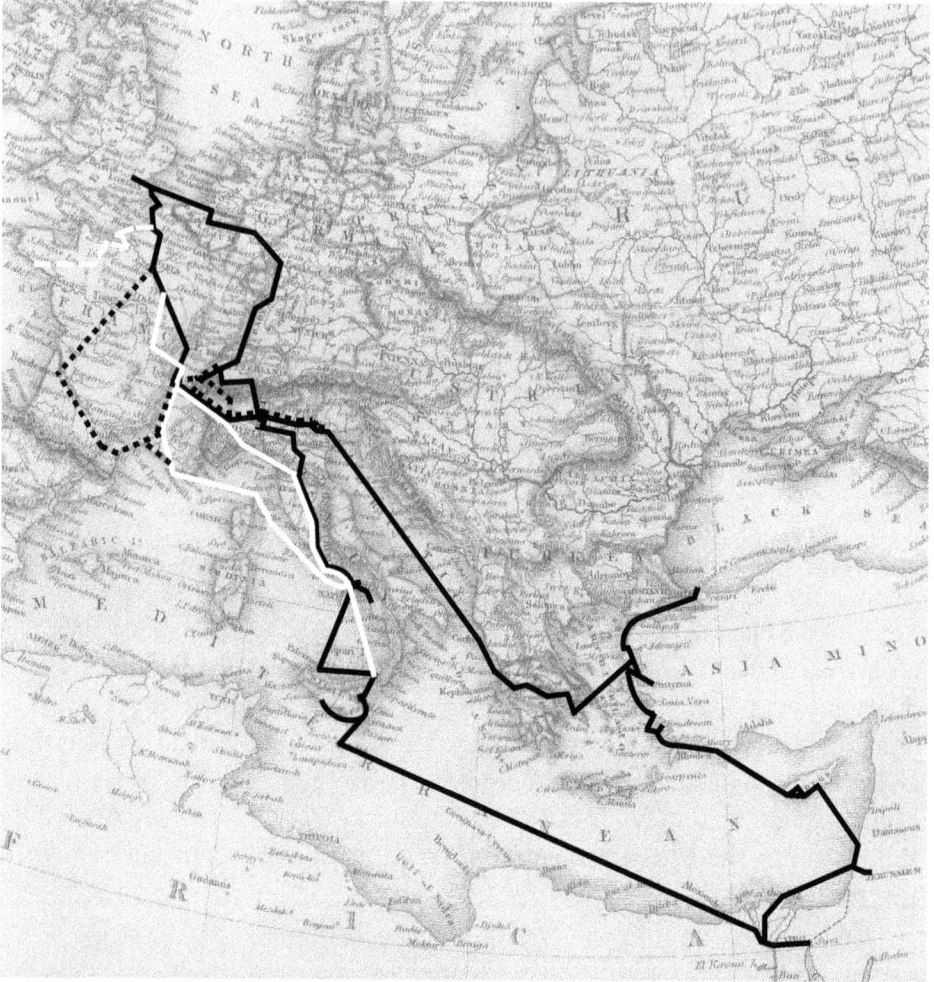

Fig. 8.1:   Routes of Ronalds' various travels. Modified from David Rumsey Map Collection, available at http://www.davidrumsey.com.

Black Solid line: 1818–20 Grand Tour, White Solid: 1823–24 Sulphur Import, White Dashed: 1834 Carnac, Black Dotted: 1853–62 Retirement.

my friend the General in the Court &c & whether I have thanked the Duke for his presents of minerals &c &c. But I must give you some special cautions for every body travels nowadays you know.[C23]

He also had more specific objectives. One was to advance his scientific endeavours. On his travels he collected scientific books and mineral samples, visited scientists and universities, and conducted experiments; this was perhaps the most premeditated aspect of his trip. Another was to look after his health. He already suffered various complaints including rheumatism and digestion problems

Table 8.1: Ronalds' International Travels.

Dates	Duration	Destinations	Companion	Purpose
Aug 1814	?	France, Switzerland	Hugh Ronalds	Electrical science, vacation
Sep 1818–Dec 1820	2 years 4 months	Europe, Near East	Alone	Grand Tour
Jul 1823–Feb 1824	7 months	Italy, France	Alone	Sulphur importation
Oct–Nov 1825	6 weeks	Paris	Maria Ronalds	Bereavement
Sep–Nov 1834	11 weeks	Carnac	Dr Alex Blair	Archaeology
Dec 1853–end 1862	9 years	Italy, France, Switzerland	Alone	Library, assisting observatories

which he hoped better weather and a more active lifestyle would ameliorate. Nonetheless he did not live as healthily as he might:

> Bisogna distrarsi quite belli giornata — you must take recreation this fine weather — says my landlady. I fear I have been <u>poring</u> rather too much last week over travels to and accounts of Vesuvius & Etna for I look yellow. I must recollect I came here more on account of my health than to study and you may as well bear this in mind also least you should expect more studied remarks than you are likely to get.[C21]

He had a further goal concerning his role as the senior male of the family:

> I will acknowledge that <u>one</u> of my motives for <u>leaving</u> home was to give a little space for other members of my dear family to put in a word or two or as many words as they liked & I sincerely hope that my absence has allowed the influence of other opinions to operate in <u>family</u> concerns. I seemed to egotize too much.[C43]

Ronalds would also have been spurred on to leave the family home for a while by his younger brother Hugh's decision to migrate to America earlier that year. It has additionally been surmised that he wished to overcome the rejection of his telegraph by the Admiralty in 1816. He makes no allusion at all to the telegraph in his journal and letters home, although he broaches many other heartfelt subjects — it seems he put it right out of his mind.

8.1 Style of Travelling

The trip involved various modes of transport (Tables 8.2–8.3). Sailing was essential of course and fortunately Ronalds had good sea legs:

> we had scarcely peeped out of the harbour when a violent <u>in shore</u> gale obliged us to stand out to sea a long way and it increased so much that the Captain & the whole crew were frequently obliged to go upon their knees & say prayers by the dozen before we could reach our Port in safety[J19]

> I have been rolling about on the Sea at a great rate *[sailing round the Sicilian coast and then to Malta]* having made three attempts to get here and driven back twice… but the sea air agrees with me very much[C28]

On land, sailing down a canal or river where it was available was 'less fatiguing'[C19] than using the relatively poor roads. He was the first Englishman to pass up the new Mahmoudiyah Canal, connecting Alexandria with the Nile; the canal proved to play a significant role in revitalising Alexandria's economy. In returning down the Nile from Cairo:

> My companions de voyage were about 80 turkish soldiers & their arab peasants or farmers with their wives. I had by this time learnt to get rid of my comfortable English prejudices a little[J20]

> I have constantly adhered to the recommendation of Dr Anderson in my transits from one place to another viz to chuse public rather than private modes of conveyance where I can.[J20]

With the popular alternative of travelling with friends in a hired vessel or vehicle:

I should have met with fewer difficulties perhaps but then I should have talked English, and I should not have indulged my Vanity in overcoming them.[C19]

Public transport was also cheaper, which was always a consideration for him. The types of boat he utilised included merchant vessels, courier boats, brigs, schooners, skiffs, and local vessels like the *Speronaro* in Sicily and the *Cangia* and *Germ* in Egypt.

Moving from town to town by road, he usually travelled in a stagecoach, known as a *Postwagen* in Germany and which he called by the French name of *Diligence* (Fig. 8.2). On the trip from Paris to Chalons:

My fellow passengers were a fat Italian Abbé, a lace maker of Lyons, a sensible man enough, a Merchant of Provence and two other Messieurs...I made numerous blunders in speaking, some of which made us all laugh heartily, for instance I called to the Conductor once for my Culottes instead of my Capote *[a long cloak or coat, usually with a hood]*...Then they all wanted to learn some English expressions, the Abbé was highly delighted when he could understand distinctly and pronounce tolerably the words God damn.[C19]

Later, crossing the Alps into Italy:

The road was very good all the way and exhibits indelible marks of Napoleon, ask who did this or that great work the answer is generally Napoleon. I arrived here yesterday in company with the fat Abbé, a Pole, a Neapolitan, a Maltese, a Piedmontese & a frenchman and his servant. So we had a variety of nations in the Diligence and almost as great a variety of animals to draw it up Mont Cenis for there were 7 Horses, 2 Oxen, 2 Cows, & 2 Women who belonged to the Cows & oxen.[C19]

In more remote areas he travelled by mule, ass, horse, camel or on foot.

The pace of travel was naturally slower than it often is today. It took him five days to go from Paris to Lyon and another five on to Turin. The trip from Corfu to Venice up the Adriatic Sea in a Trabaccolo (cargo boat) lasted 15 days. He described this trip as his worst, with the boat being buffeted by gales for nine days.

He lingered in places he liked. His longest stay was six months in Naples; after two months there, he wrote:

I intend now to enjoy Naples, every thing I have hitherto seen and done here has been seen and done with a view rather to fulfil a sort of obligation to see all the Lions[1] and Kill the time. I intend to paint, or read, or sail, or walk, or ride, or in short to make myself at home.[J18]

He took lessons in painting and Italian elocution. Two months later:

my life has been quite as monotonous as if I were at home. I get up not very early in the Morning, breakfast, paint clouds (if there are any), go to mass (sometimes, particularly on Sundays, by way of distinguishing the end of the week from the middle

---

[1]'Lions' — tourist attractions — originally referring to those kept at the Tower of London. Ronalds' cousin Betsey used the same phrase 'to see the lions' in her travel journal.

Table 8.2:   1818–20 Grand Tour — Approximate Itinerary.

Year	Country	From	To	Days	Location	Science Interests	Companions	Travel to Next Location	Days
1818	France		01/09		London			By coach via Canterbury	0
	France	01/09	02/09	1	Dover			By boat across English Channel	0
	France	02/09	03/09	1	Calais			By diligence	2
	France	05/09	21/09	16	Paris		Dunnage	By diligence & boat via Chalon	5
	France	26/09	28/09	2	Lyon		Abbé	By diligence via Col du Mont Cenis	5
	Italy	03/10	08/10	5	Turin	Superga, site of Beccaria's experiments		By diligence via Vercelli	2
	Italy	10/10	22/10	13	Milan	Building with extraordinary echo	Coins, Apothecary	(Day trip to Monza) / By boat down new canal via Certosa	1
	Italy	23/10	27/10	4	Pavia	University where Volta based	Coins, Apothecary	Left 5 a.m. by Veterino via Cremona, Bozzolo, Mantua, Modena	4
	Italy	31/10	05/11	5	Bologna		Coins, Apothecary	Left 4 a.m.	2
	Italy	07/11	16/11	9	Florence	Collected books on electricity		Via Terni	6
	Italy	22/11	04/01	43	Rome		3 Irishmen	By veterino along 'Via Appia' via Pontine Marshes, Mola di Gaeta	5
1819	Italy	09/01	07/07	179	Naples	Vesuvius (3 trips): electrometer measurements / Mineral samples	Irishmen, Apothecary, Wade, General, Smith	(Trips to Pozzuoli, Paestum, Caserta) / By merchant vessel	3
	Sicily	10/07	04/08	25	Palermo	Mt Pellegrino: measuring heights with barometer / Atmospheric electricity measurements	General, Henniker	(Trip to Bagheria) / By speronaro via Capo Peloro	2
	Sicily	06/08	15/09	40	Messina	Collected volcanic mineral samples / Procured minerals from Lipari Islands / Phosphorus polyps	General, Henniker	(Trip to Charybdis and Scylla by skiff) / By speronaro and mule via Taormina and Aci	3
	Sicily	18/09	05/11	48	Catania	Measurements on Etna / Minerals and geology / Dissected torpedo ray fish for electric organs	General, Gioeni	(4 day trip to Etna by mule via Nicolosi) / By speronaro	1
	Sicily	06/11	11/11	5	Syracuse			(Trip to Epipoli) / By schooner during gale	3
	Sicily	14/11	16/11	2	Girgenti			By schooner during storm	4
	Sicily	20/11	27/11	7	Syracuse			By brig	3
	M	30/11	06/01	37	Valletta	Bought books	Allingham	By Turkish brig *Besiktasi*, sailed close to S coast of Crete	14
1820	Egypt	20/01	24/01	4	Alexandria		Allingham	By cangia via newly opened Mahmoudiyah Canal & Nile	7
	Egypt	31/01	08/02	8	Cairo	Geography of Nile Delta		(2 day visit to pyramids at Giza. Trip to Suez by camel) / By germ via Damiata branch of Nile	5
	Egypt	13/02	27/02	14	Damietta	Mirages		By Turkish boat	5
	Holy Land	03/03	04/03	1	Jaffa			By mule via Ramla	3
	Holy Land	07/03	04/04	28	Jerusalem		Connor	(3 day trip to Bethlehem on foot; 3 day trip to Dead Sea by ass) / By horse via Ramla	2
	Holy Land	06/04	07/04	1	Jaffa			By mule via Caesarea	2
	Holy Land	09/04	14/04	5	Acre			By boat	3

Country	From	To	Days	Location	Science Interests	Companions	Travel to Next Location	Days
Cyprus	17/04	21/04	4	Larnaca			By boat	1
Cyprus	22/04	03/05	11	Limassol	Pudding stone formation	Davidson, Boggie, Coats	(5 day trip to Paphos & Mount Olympus by mule) By boat via Kastelorizo, Rhodes, Bodrum, Cos, Calimno, Didyma	28
Turkey	31/05	07/06	7	Smyrna		Davidson, Boggie, Coats	By schooner via Mytilene, Assos	5
Turkey	12/06	17/06	5	Constantinople		Davidson, Boggie, Coats	(Day trip to Black Sea by schooner & rowboat via Buyukdere) By schooner via Dardanelles	2
Turkey	19/06	21/06	2	Canakkale		Davidson, Boggie, Coats	(Trip to Trojan Plain by horse) By schooner via Lesbos, then horse from Piraeus	3
Greece	24/06	30/06	6	Athens		Davidson, Boggie, Coats	By boat via Eleusis then by mule	1
Greece	01/07	02/07	1	Corinth		Davidson, Boggie, Coats	By boat	4
Greece	06/07	08/07	2	Patras		Boggie	By boat via Araxum, Carnia, Sta Maura	7
Greece	15/07	07/08	23	Corfu	Electric storms, water spout	Boggie	By trabaccolo	15
Italy	22/08	04/10	43	Venice			By courier boat up Brenta river	1
Italy	05/10	06/10	1	Padua	Sparks emitted from hair being combed	Professor	By diligence via Vicenza	1
Italy	07/10	13/10	6	Verona	University where Zamboni based	Belzoni	By diligence via Brescia	1
Italy	14/10	23/10	9	Milan		Davidson, Coats	By veterino via Macone?, Arona	1
Italy	24/10	25/10	1	Domo d'Ossola	Snow accumulates first on south side		By diligence via Simplon, Brig, Sion	1
Switzerland	26/10	27/10	1	St Maurice	Alps minerals including quartz		By diligence via Vevey	1
Switzerland	28/10	29/10	1	Lausanne			By coach via Coppet	1
Switzerland	30/10	08/11	9	Geneva			By diligence via Lausanne	1
Switzerland	09/11	12/11	3	Berne			By diligence?	1
Switzerland	13/11	17/11	4	Bale			By diligence?	1
F	18/11	30/11	12	Strasbourg	Ordered German books on electricity		By diligence	1
Germany	01/12	06/12	5	Carlsruhe			By postwagen	1
Germany	07/12	10/12	3	Frankfurt			By diligence	1
Germany	11/12	11/12	0	Limburg			By postwagen	1
Germany	12/12	14/12	2	Cologne			By postwagen	1
Germany	15/12	16/12	1	Aachen			By diligence?	1
B	17/12	18/12	1	Bruxelles			By boat via Antwerp?	1
				London				

1820

Table 8.3:   Typical Modes of Transport, Journey Times and Costs.

Trip	Mode	Duration (Days)	Cost (Local Currency)	(£.s.d)
Dover → Calais	Boat	1	7 Shillings	0.07.0
Calais → Paris	Diligence	2	48 Francs	1.19.2
Rome → Naples	Veterino	5	4.8 Ducats (share)	0.17.5
Valletta → Alexandria	Brig	14	20 Spanish Dollars	9.10.1

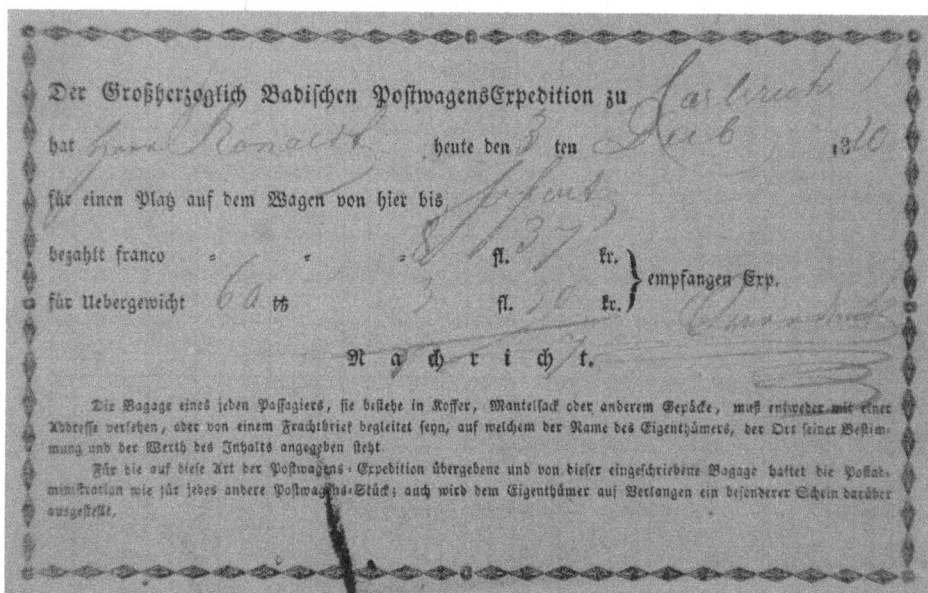

Fig. 8.2:   Stagecoach ticket from Karlsruhe to Frankfurt (6/12/1820). IET 1.1.54.

and for other little affairs which it is not necessary to detail *[a friend, perhaps?]*), read or lounge at Booksellers, dress, Walk in the Villa Reale, or make a little mineralogical excursion, or an excursion of observation amongst the Natives, dine at a Trattoria…go to S[t] Carlo's or the Fondo (where I fall in love about twice a week with *[Girolama]* Dardanelli *[an opera singer]*) or to the Accademie dei Nobili (where the newly arrived Squirrels get me by the Button to know what is to be seen at Naples), go home, go to bed, go to sleep, dream of England and of you. But I beg your pardon for all this egotism[118]

In Rome, where he spent six weeks:

I get up at <u>about</u> 9, then breakfast, then take a tour with a piece of the Guide in my pocket to look at some Ruins or pictures…then I go to M[r] Vasi's[2] library where I

---

[2]Giuseppe Vasi (1710–82), Italian engraver and architect who published a series of guides.

search for information about anything I have seen which appears to deserve my pains (I am wretchedly ignorant of Roman history, don't <u>crow</u>) then I read the papers a little, then I take a stroll in the Corso or on the Monte della Trinita, then I dine…then I call at the grand Caffé to see a little billiards or lounge on the sofas to digest my dinner, then I put my hands in my pockets and walk home sometimes with an italian acquaintance who nicks me out of some tea, sometimes without, then I drink tea, then I read or settle any other matter I may have upon my hands such as writing to you, then I gāpe 3 or 4 times and go to bed (sound the a in gāpe long).[C21]

Ronalds enjoyed a variety of sleeping arrangements. Where possible he stayed in a good quality inn. In Milan:

My room is at least 25 feet long and 20 wide, it is well furnished with muslin Curtains and White bed and the Pillow cases trimmed with lace and very clean, and I pay only 1 lire Milanese per night (about seven pence english), but the staircase is not of the best sort, I am obliged to be very cautious in going up and down it.[C19]

And in Valletta, Malta: 'I am entombed every night in a Sea of feathers.'[J19]

Outside the major towns, a monastery or the local consul's residence were preferred sleeping places. Heading to Jerusalem:

After we had passed the gate about a mile we bade him *[the Consul]* a cordial adieu for we had been lodged in his house and lived with him all the time of our stay at Jaffa and were treated with the greatest hospitality and kindness (by the by I had the <u>honour</u> to sleep upon the couch which Buonaparte used when there).[J20]

And at Jerusalem (Fig. 8.3):

Fig. 8.3:   Ronalds' sketch of Jerusalem's walls (1820). IET 1.1.9.

We were soon safely lodged in the chambers of the latin Convent which have been occupied by all the frank travellers during more than *[blank]* years as is shewn by their having <u>immortalized</u> upon the doors until there was scarcely space left for our names.[J20]

On leaving, Ronalds gave the convent a $7 donation.

A local house was another possibility:

I returned to lie at the village *[near the Pyramids]* in a regular arab cottage, to sleep was impossible for the flies actually dyed my shirt red at the part about the collar.[J20]

Alternatively, he slept out-of-doors. In present-day Israel:

I was requested by the muleteers to alight in a solitary place near a pond so crowded with frogs & toads that the noise was absolutely stunning; here I learnt we were to take up our lodging for the night amidst the long damp grass...Some turks who travelled*[?]* with me gave a good example of their usual politeness in offering me their beds (for mine had been sent by sea) but of course I could not accept them, & the cold & noise of frogs & crickets effectually prevented sleep.[J21]

At sea, he preferred to sleep on deck rather than in the cabin, even in bad weather. Sailing to Sicily:

Our Berth was the best in the vessel viz the ship's boat on the Deck, where we spread our hammocks on the oars and slept in the open Air and with no other covering than our Shirts & trousers[J18]

The voyage from Egypt to the Holy Land was less romantic:

She was a turkish crazy old non-descript vessel freighted with rice (the Pasha's*[³]*), Lentils, 20 or 30 mamelukes *[military caste in Muslim societies]*, 2 or 3 arab women, 3 Saints (i.e. a sort of wandering priests generally pretending to be mad) and a very large assortment of animalerle *[to do with animals]* vermin. She was the best vessel I could get. Every body & every thing was soon disposed in some sort of order. I after much contention got myself, my servant & my luggage stored in the ship's boat, for the cabin was notwithstanding all my efforts to overcome my "<u>prejudices</u>" absolutely unbearable.[J20]

Six weeks into his travels, Ronalds wrote: 'the irregularity of living and the being obliged to eat dishes which don't suit my digestive powers have rather thrown me back.'[C19] His eating and drinking habits continued to evolve with the diversity of his destinations and as he grew hardier. For example, in Milan:

O' for a nice basin of ArrowRoot or a Mutton Chop with the gravy in it and a hot mealy potatoe *[sic]* [C20]

then in Naples:

The wine I drink here at 8 pence a bottle is excellent, they call it Pagliarella, it has something the flavour of verry *[sic]* old port and the colour of tourney port...The Lacryma Christi *[Tear of Christ, a wine produced on the slopes of Mt Vesuvius]* so much boasted of is by no means capital [C22]

[³]Leader of Egypt, although the country was still nominally under Ottoman rule.

Fig. 8.4: Pigeon houses on the Nile (1820) — a source of food for Ronalds. IET 1.1.9.

Farther afield, sailing up the Nile River delta (Fig. 8.4):

> I killed *[shot]* myself an excellent dinner of some beautiful little birds called in Italian Gallo el Paese *[country rooster]* very much like our wood cocks.[J20]

In the Cyprus interior:

> each person was attended by a servant behind his chair. We ate a la turque (i.e. we pulled a young roasted Kid to pieces with our fingers) and we retired out of doors into the Horse yard to take our coffee & Pipes. This yard also served for the Kitchen.[J21]

His luggage grew as he travelled. It started off weighing less than 18 kg and reached 27 kg in Germany, even though he had posted home books, rock samples and gifts along the way. He carried a variety of apparatus including a thermometer, barometer, hygrometer and electrometers, a telescope, looking glass, compass, perspective compasses, drawing and painting equipment including a stool, knife, rope, hammer, and bedding. He also had a wide range of maps, travellers' guides and historical texts, as well as the 'sweet scented Bible'[C18] given to him by his mother, and Shakespeare, so that he could 'take a stride back to England for an hour or two'.[C20] He carried a pistol and regretted not having a gun. He also carried his seal.

For everyday sightseeing he wore trousers, shirt and waistcoat, cravat, long tailcoat, gloves and a tall hat, similar in shape to today's top hat. He had an overcoat for cooler weather and an umbrella to shield the sun or rain. He also needed evening wear for dinner, the opera, etc., breeches and stockings for formal functions and costumes for masquerade balls. On the road, Ronalds had his clothes

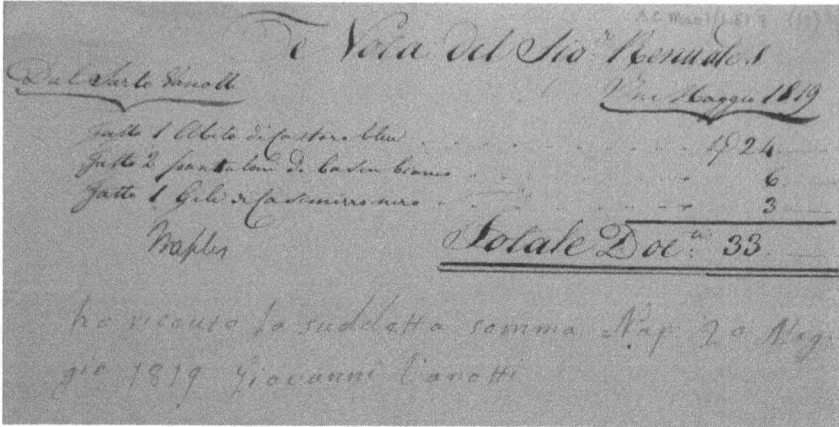

Fig. 8.5: Tailor's receipt for blue dress coat, pantaloons and waistcoat in Naples (May 1819). IET 1.1.51.

washed regularly, and garments were mended and new ones tailored as required (Fig. 8.5). His last letter home included a similar request for his brother Charles:

> My Dear Vicar. I smell very strong. My cassock *[ankle-length robe]* is grown musty. The smell of the German pipes in the port Houses is very disagreeable to me & will be to every body else by the time I reach home, therefore order me a new Coat, blue dress (if still in vogue), a waistcoat (white) & a pair of dress pantaloons or trousers or breeches…The warmest coat but moderate.[C43]

8.2 Travelling Companions

Many people embarked on the Grand Tour with a party of friends but others, like Ronalds, set off alone. This would have been quite a novelty for a man who had grown up in a large family with servants. Being fluent in French and Italian, he became friendly with people of various nationalities: 'Coffee house acquaintance is certainly of the most convenient tho' not of the most confidential kind for a traveller.'[C19] In Egypt, however, Ronalds required the services of someone who could assist him in his interactions with the local people:

> I have forgotten to introduce you…to my companions de voyage…my servant Ferdinando *[Beccari]* whom I hired at Alexandria, a stout healthy Livornese who speaks Arab and Italian, a good cook but not a very finished valet[J20]

Ferdinando proved to be invaluable:

> To ward fleas *[primary carriers of the Plague]* & other insects *[including malaria carriers]* I make Ferdinando sew me up every night i.e. he sews up all the openings of my shirt and sews the pantaloons to it also by this means, and with a veil of Gauze I succeed tolerably well.[J20]

Ferdinando was also very protective of Ronalds when he faced any risk. When it seemed necessary to return to Egypt from Cyprus to collect money while the Plague was raging, Ferdinando undertook the task 'for he has not the slightest fear of the infection having lived many years at Alexandria in the midst of the Plague'.[J21]

Ronalds gave some of his 'companions de voyage' a nickname. Several of these have already been mentioned, including 'the Abbé', who 'quitted me at Turin for like many Italians as well as Englishmen he seemed to possess no curiosity concerning the Curiosities of his own Country'.[J17] On the Diligence from Turin to Milan, he met 'Coins' and 'the Apothecary' and the three travelled on together to Pavia, Bologna and Florence. 'Coins' was numismatist Giulio di San Quintino, who was 'most labouriously [sic] engaged in a work intended to illustrate the histories of some ancient Italian families by the Coins which they had struck'.[J17] The Apothecary (Mr Graham) 'came into Italy to learn Italian & Taste':[J17]

> The Apothecary is hammering away at Italian with my books. He has got the Padrone of the House to teach him to read & I am obliged to put up with his marching into my room now and then with a Turkey-cock strut and hear him gobble out a few mouthfuls of wretched stuff which the Sign[r] Padrone, who is also a Cook, has crammed into him.[C19]

People travelling at that time were often well known, or became so, and touring was still sufficiently novel that they published their experiences. Many of Ronalds' acquaintances can be identified in this way. He met 'M[r] Rennie (the son of the architect)'[J21] and his travelling companions on a brig at Piraeus, Athens; Sir John Rennie* later described his journey in his autobiography.[B92] Rennie's sailing friends included John Hodgetts Foley, subsequently a British MP, and clergymen (later Dr) George Waddington and Barnard Hanbury who together published *Journal of a Visit to some parts of Ethiopia*[B18] in 1822. Ronalds already knew Waddington from Sicily, and at Piraeus 'received with great pleasure his congratulations on having compleated the same tour nearly which he was then beginning'.[J21]

Ronalds ran into the famous chemist Sir Humphry Davy near Rome: 'who had come there to shoot. I saw him have 6 fine shots running but (he discharges a Eudiometer[4] better than a fowling piece) missed all.'[C21] He spent time with Volta's colleagues at the University of Pavia but 'was much disappointed in not having an opportunity of conversing with this Princeps in re electrica *[prince of electrical science]* himself who was at his seat at Como'.[J17] He was however able 'to see the collection of original apparatus with which Volta made his very important Discoveries'.[J17] Ronalds made contact in Sicily with physician and naturalist Dr Mario Gemmellaro (who made regular observations on Etna), and offered him his barometer. In Naples he became acquainted with the Spanish geologist Carlos de

---

[4]Laboratory device for measuring changes in volume of a gas mixture.

Gimbernat and undertook overnight experiments with him on Mt Vesuvius. Gimbernat had developed an apparatus by which an artificial spring was formed by condensing volcanic vapours, which were aqueous:

> Mr Gimbernat succeeded in his object of establishing another fountain at an acid fumarole in which I hope I was of some little use to him, &…I passed a few hours with my electrometer very satisfactorily [J18]

Another person he befriended in Naples was Matthew Wade, who was:

> Governor of the Castle dell'Uovo, which is almost opposite my lodgings about a quarter of a mile distant in the Sea. He is a jolly old Cock…Baldpate, long pigtail, red face, bottle nose, stick, nap gout guts, a regular Uncle Toby. I used to go there to paint Vesuvius, now I go to get a little of his old Port…He tells me all about Lord Nelson who very frequently supped upon fish with him and Sir W. Hamilton and Lady Hamilton. He says Lady Hamilton always drank a Bottle of his port after supper & sometimes two and grew fat upon it [J18]

Several of Wade's letters to Lady Hamilton survive.

Ronalds' circle in Naples also included 'the General', Joseph-Antoine de Gourbillon*. Their time together included ascents of both Mounts Vesuvius and Etna. De Gourbillon described the journey in his book *Voyage Critique à l'Etna en 1819*, published in English in 1820 as *Travels in Sicily and to Mount Etna in 1819*.[B14] He imparted significant drama into the story, where he was able, in comparison with Ronalds' matter-of-fact accounts.

It is interesting to compare some of their passages concerning the expedition to Mt Etna. They first attempted the climb to the volcano's summit. Ronalds wrote:

> At about 3 in the morning we arose with the intention of gaining the summit…but the passage over the lava before arriving at the foot of the cone was…extremely difficult and even dangerous…Indeed the General who is short sighted was actually obliged to return [J18]

He described the view at the summit as a 'grand spectacle': 'If I possessed but one little spark *[of poetic fire]* surely it would kindle whilst*[?]* I describe the scene which surrounded me, but no, I have not one little spark.'[J18] De Gourbillon noted in his book: 'I was fortunate in the resolution I had taken…to proceed no further at that time' because 'after incredible fatigue' there was 'no view'.[B14]

On entering the Southern Crater, de Gourbillon recorded:

> The descent became more and more rapid; the friendly shelves of lava began to diminish…and…were at length succeeded by loose and crumbling *scoriæ*…A frightful silence succeeded. A large block gave way, and slided *[sic]* from beneath my feet; and, instead of lodging further below, I saw it fall perpendicular, and heard its sound as falling in the abyss — an abyss into which another step would have taken me with it. At the cry which my terror forced from me, my companion stopped and saved himself…I found myself with my face on the ground [B14]

Fig. 8.6:   Mt Etna's craters (1/10/1819); IET 1.1.9.

De Gourbillon believed they were the first people to descend the crater. Ronalds wrote simply:

> We got as low as the base of the cone of the eruption of *[space]* without the least difficulty[J18]

Turning their attention to the new crater and cone (Fig. 8.6) which de Gourbillon explains had formed only four months earlier, he wrote:

> At the first glance which I cast into this abyss, I uttered a cry of more than surprise. Admiration and terror were the causes of my agitation...Here all was in action...The volcano murmurs; and from the spot where the liquid lava is boiling to the edge of the crater where I am, the immense extent of the gulph *[sic]* offers a scene the most awful and grand.[B14]

Ronalds summed up:

> I consider my journey to Etna a verry barren one[J18]

Although it had been a more challenging expedition than Mt Vesuvius, it was of less scientific interest to him.

Ronalds met up with his friend Sir Frederick Henniker* in Sicily, and they and de Gourbillon sailed down the north east coast of the island together. 'On one fine evening...as we lay at full length in our little speronaro, the romantic scenes inspiring romantic thoughts, we shook hands upon a compact to meet in Egypt.'[J18] This was not to be: the General changed his mind and Ronalds was delayed in Sicily by financial and passport issues, as explained below.

Henniker soon moved on. Ronalds recorded at the Ear of Dionysius at Syracuse (Fig. 8.7):

> I ascended by the help of 6 or 7 men (who are ropemakers living and working in another larger cavern hard by) and a Pulley, and Ropes, and here I found with great

Fig. 8.7:   Ear of Dionysius in Syracuse — climbed by Ronalds and Henniker (1819). IET 1.1.9.

pleasure a card of Sir Fred[k] Henniker left by him there for me no doubt (I shall return the call on the top of the 2nd Pyramid of Gizeh if I have the good fortune to reach the top…)[J19]

Henniker then wrote to Ronalds from Malta:

I sail Oct[r] 2nd for Alexandria — reckon upon ten days passage, & 4 or 5 days more to Cairo, where I propose staying at least a fortnight, in hope that you will ere that have join'd me there…my plan upon leaving Cairo will probably be to proceed up ye Nile as far as ye 2nd Cataract but of this we will talk at Cairo ye Grand — in ye interim I have only to express my best remembrances to ye General & for your mutual welfare.[C26]

Henniker sent subsequent messages through Peter Lee, the British Consul in Alexandria. On 27 April 1820:

I avail of this opportunity of…informing you that Sir Fred[k] Henniker left Suez in an English Ship for Tor, where he was to land and after visiting Mount Sinai &c would proceed to Jaffa & Jerusalem, and intended reaching Smyrna the latter End of June — & that you may have an opportunity of joining him, if such be your wish[C30]

and a month later: 'Sir Fred[k] Henniker…was at Jaffa the beginning of this month.'[C33] They continued to miss each other however. Henniker had left Cairo before Ronalds could get there and returned after Ronalds had left. He followed Ronalds' route homeward but was always a couple of months behind him.

Ronalds was by then travelling with Walter Stevenson Davidson and his friends Dr Thomas Coats and William(?) Boggie from Bombay. The four of them sailed for nine weeks up the Turkish coast to Istanbul, then across to Athens and the Peloponnese. Davidson and Coats met up with Ronalds again in Milan on their way back to England. Davidson had imported merino sheep into Australia with his friend John Macarthur, the renowned wool industry pioneer. He was also a merchant and banker, and before his travels in the Eastern Mediterranean was dealing in opium and other commodities in Canton. Coats was a surgeon in Bombay who also had literary interests and Boggie was probably a former employee of the East India Company.

In Constantinople, Ronalds and his three sailing friends dined twice with Sir Robert Liston, British Ambassador to the Ottoman Empire, and his *Chargé d'Affaires* Bartholomew Frere: 'Lady L is as fine an old scotch Lady as you ever meet with.'[J21] They were shown around Athens by Giovanni Battista Lusieri, whom Ronalds later called 'the best sketcher in the world':[C375] 'We found him a very obliging old intelligent gentleman and in due time I got him to show us his celebrated drawings.'[J21] As well as his watercolours of landscapes and buildings, Lusieri was also known for removing the Elgin marbles from the Parthenon while in the employ of Lord Elgin; these were bought by the British Government in 1816 for display in the British Museum. Ronalds did not agree with this action: 'We preserved a profound silence on the subject'.[J21] Half a century later, Sarah Flower née Martineau expressed sentiments similar to her uncle's during her visit to Athens: 'if ever the Greeks restore any portion of these buildings, England ought to send back these marbles to be put in their right places.'[B151]

Ronalds and Co also dined in Athens with Royal Navy officers James Mangles FRS and Charles Leonard Irby, who published in 1823 their *Travels in Egypt and Nubia, Syria, and Asia Minor, during the Years 1817 & 1818.*[B24] They had accompanied Giovanni Belzoni up the Nile. Ronalds himself accompanied Belzoni from Verona to Milan, and he 'was good enough to show me his work which is not yet published and to give me a great deal of very interesting information about Upper Egypt'.[J22] Belzoni had been a strongman in a travelling circus before he and his wife went to Egypt. There he was commissioned by British Consul-General Henry Salt (whom Ronalds also met in Egypt) to transport the seven tonne bust of Ramesses II to England. The monolith sailed from Egypt in 1817 and remains on display at the British Museum. Ronalds had already mentioned Belzoni twice in his journal: when entering the Pyramid of Chephren, the entrance of which Belzoni had discovered in 1818, and when viewing the Egyptian statues he had given to his hometown of Padua. Belzoni's book *Narrative of the Operations and Recent Discoveries within the Pyramids, Temples, Tombs and Excavations, in Egypt and Nubia*[B15] was published in late 1820. Ronalds interacted again with Belzoni in England in 1821–22, with their mutual friend Henniker, and paid homage to his memory when he was in Padua in 1859 (Sec.16.1).

The time at which Ronalds was travelling was important also in the deciphering of Egyptian hieroglyphics, in which there was rivalry between English and

French scholars. He encountered people who assisted both groups in making breakthroughs. At Didyma Turkey he met Jean-Nicolas Huyot, an esteemed French architect. Huyot had sent inscriptions to Jean-François Champollion that helped the latter finally decipher the script. In Jerusalem, Ronalds spent time with 'Mr Grey a gentleman of very great litterary & antiquarian attainments'.[J20] He developed a method to enable Revd George Francis Grey to preserve papyri 'which he had himself taken from off the Breasts of Mummies'.[J27] Grey had a few months earlier travelled with Henniker up the Nile from Cairo to Thebes. Grey later gave these papyrus manuscripts to Dr Thomas Young FRS — physicist, physician, Egyptologist and famous to engineers for Young's Modulus of Elasticity — who noted they were 'in excellent preservation'.[B25] One proved particularly important in the deciphering quest.

In Sicily, Ronalds came across Charles Allingham:

> I learned that an english gentleman was lying very ill *[with shaking fits associated with malaria fever]* on board a brig which had been driven into Syracuse by the same storm which I had encountered…Therefore I immediately determined to quit the schooner and take my passage to Malta in the same vessel with him, which determination was an extremely fortunate one for I derived very great pleasure in being of some little utility to him.[J19]

After his recovery, Allingham and Ronalds went sight-seeing together around Malta. Allingham stayed on the island for the rest of his life and became a respected portrait painter.

Not unexpectedly, all of these companions on Ronalds' travels were male. His sisters worried that he could not also enjoy female company. In Paris, he wrote:

> The women are certainly by far the better sex but they required too much french from me & I required too much english from them.[J17]

When Charlotte enquired about 'the young ladies'[C24] in Naples, he explained:

> Why they look a little like the summer flies in winter, it's not very easy to get a very good introduction [C24]

To Emily:

> You pity my want of Ladies' society. Thankée; don't do so any longer but pity me for having too much of it, I am lodged in the same Hotel with a French lady & a half who contains as much society as 50 english women at least.[C22]

He related an amusing story about her:

> Being in the midst of a thorough wash, bounce came open the door of my room which communicated with the French Lady's and in bounded Madam apparently somewhat flurried — but as it was a regular Scene I think I can tell it best in Scena [J18]

– which he goes on to do.[5] It transpired that she was hiding from a former lover.

---

[5]The story is recounted at http://www.sirfrancisronalds.co.uk/naples2.html

Socialising with women was more difficult in the Turkish States:

> The turks were surly and when I attempted to converse with the Arab women through Ferdinando they were immediately severely reprimanded and ordered away from me.[J20]

Arriving at the British community of Smyrna in modern Izmir, Turkey:

> The Bazaar is fine, the harbour is fine & the convenience of living is good but there is nothing half so fine or so good in Smyrna as the young ladies, at least so it appeared to us after having been so long accustomed to the arabs & turks and none at all.[J21]

8.3 Cultural Interests

In France, and especially Italy, Ronalds often spent the evening at the opera or ballet, but he did not always enjoy the experience. As he admitted at the ancient theatre at Taormina:

> The Romans must have yawned at their theatres as much as we do I think for they always seem to have chosen spots for building them where something might be seen besides what was going forward in them, here they had pitched upon the most advantageous point for a View of Etna[J18]

In Paris:

> I am just returned from the french opera. Could scarcely make out a word of it…The Ballet. Verry pretty dancing, verry fine, marveilleuse.[C18]

In Milan:

> [*Violante*] Camporese is the prima Donna, her powerful bursts electrify every body except the card players.'[J17]

> The object of the Ballet master seems to be to present every moment some new picture by putting his puppets through every species of attitude & variety in grouping. Unity of action is compleatly caricature, for instance two or three dozen of astonishers (all in one precise attitude), a competent number of astonished, a given quantum of terror, a corresponding portion of terrified &c &c &c compose one of the changes, & when the box is rattled again & the bustle & confusion has subsided (which takes about half a minute) another picture is presented. An Italian Ballo always puts me in mind of a Kaleidoscope.[J17]

Turin:

> The Opera here is not superb either in Respect to the Building or the Singing and nothing can possibly be worse than the dancing.[C19]

Naples:

> You're going to have in London the first Violin player of Italy viz [*Niccolò*] Paganini but I hope he's not going from Naples just yet, his execution is astonishing but not superior to his taste & expression…He played some Airs with variations on the fourth string only of his Violin which drove the audience mad the other night. It was the first time I ever saw a mad audience.[C23]

He was in Naples for the funeral of the exiled Spanish King Charles IV:

His Majesty is now lying in State here. An immense square platform about 8 feet high is erected...and on this stands a sort of bin in a sloping position which serves his majesty to recline on in full uniform and <u>looking very well</u>, his household and officers standing round as if he were giving an audience...Tomorrow he will proceed in grand procession to take possession of his subterranean palace...

Hey dey, what a Row. The drums are beating and preparations making for the Funeral. I will endeavour to shew you what passes — The streets are already crowded...Now the fine black eyes begin to twinkle and smiles and bows to cross each other in all directions from the Balconies, and knowing glances and old acquaintance like nods from the house tops. Blue, Red, Yellow, Green, no one colour seems more fashionable than another. I think they are very fond of gaudy dress but there <u>are some</u> smart looking girls <u>too</u>. I must shave...Here they come. Here they come. Here — no, a false alarm. Diable, says the French lady. Baugh, says the Prince, the prince Gallitzin[?]. Damn it, says I...Now we begin to talk knowingly. Now we being to gaaape a little. Now we crack jokes. Now the jokes are all cracked so we gaaape a little more and the wit grows flat. Now we tuck in a few Oysters... Now, Now, Now they <u>are</u> coming, <u>really</u>...Some tell me that the King will be placed upon a Seat in the church and Dinners &c served to him during 2 or 3 days as usual, of which not partaking or replying to certain interrogations put to him by the proper officer, this latter will declare his firm conviction to be that his majesty is actually dead and consequently must be buried. Others say that he <u>will</u> reply to some of the questions that he will decline politely to eat and will declare his intention of not returning to the Pallace. I must ascertain which is the <u>correct</u> account.[118]

Ronalds studied how the locals fished off Messina in Sicily:

What's that fellow perched on the top of the mast in the boat for like an old crow <u>looking out</u> on a tree? = He is one of the paternity of Fishermen who take the Sword Fish. He is watching the approach of a fish by the appearance of a little streak which it makes by its track in the Water. Now he spies one and is giving notice to his companions in little boats below...rowing with all their might in the directions which he points out to them...You may also perceive that a large plank is laid across the little boat projecting about a yard and a half beyond the sides and that the oars which are very long have their fulcrum at the ends of that plank. This contrivance is to facilitate (Ah the fish has gone down, they have lost him), this contrivance (here comes another, Tally ho my Boys, it's as good sport as coursing is it not), this contrivance is to facilitate the (see how he dodges them), to facilitate the sudden turning of the boat. Bless us, how he dodges them, but they are round in a moment. Now they come up with him, the harpoon and line fastened to it is all ready — Whiz — the fish is struck, away he goes, he shews them some play...Now he begins to grow <u>sick,</u> here he comes. They will have him along side the boat in a minute. = Poor fellow. = Aye, poor fellow, tagliata di spada *[swordfish steak]* Sig[r] Marinaro.[118]

He also tried a Turkish bath in Alexandria:

I was invested with a turban of coarse linen and a cloth round the waist. Another man then came and led me into a smaller octagon Hall paved in Mosaic and containing also a handsome fountain. This was very hot I think equal to 110…my conductor muttered a prayer and desired me to lie down on my back upon a marble slab, which I had no sooner done than he soused me all over with a large bowl of scalding water… This made me roar a little but he seemed perfectly regardless of that and repeated the dose 2 or 3 times more seeming to enjoy the fun. Then the merciless rascal put on a glove of seal skin and proceeded to scrub my skin off almost…This part of the operation resembled very much that of scalding & <u>scraping a pig</u>. Then he soused me again and then shampooed me i.e. he pulled all my joints untill they cracked as loud as a drayman's whip and pressed down my shoulders almost to dislocation I think. He would now have taken some horse hair and soap and given me a second working with them had I not resisted for as it was I am sure I had undergone a much more thorough wash than ever good Mother Cogan *[schoolmaster Eliezer Cogan's wife]* gave me and was as happy to get out of his clutches as ever a lap dog from a washing tub. All the rest was delightfull *[sic]*. He returned me into the custody of the first performer in the dressing place, my servant brought my linen & some perfume and I reclined upon the cushions for about an hour drinking the very best Moka coffee and smoking excellent tobacco & feeling a <u>little</u> initiated. I took a looking glass and arranged my mustachios like all the other grave personages [J20]

Ronalds enjoyed observing people, both as communities and individually. Even in the private setting of his journal, he shows considerable balance in his comments:

I am quite ready to acknowledge that a residence of a month or two in a place does not furnish a sufficient opportunity of observing accurately the character of a set of people and therefore speak with the greatest diffidence on the subject [J19]

Concerning the French:

To say that you <u>get more</u> by a frenchman's vanity than his good-will may be too severe but is not this the case with <u>almost</u> all sorts of casual acquaintance. [J17]

Moving on to the Italians:

In a word I like Naples but I <u>hate</u> the People, nobody hates them more cordially & I don't think they have <u>many</u> admirers. [C23]

The Egyptians:

The arabs altho' obliged to <u>knock under</u> to the *[Turkish]* soldiers kept themselves as much as possible aloof from them and in all the intercourse that did take place shewed a degree of <u>pride</u> which I did not expect from people whom we are almost inclined to call savages…If we will treat them as savages we must expect that they will be rather savage towards us I think. I can clearly discover great affection and hospitality towards <u>each other</u> and is not this the proper test of character? [J20]

He often made special mention of the female population. In Paris:

I must pay <u>some</u> homage to French <u>female</u> vivacity. [J17]

And in Syracuse:

Of the ladies I can only say how they <u>look</u> for I have spoken to but one. They look fairer than some other Sicilians, rather prettyer *[sic]* than some others, rather coyer than some others, capital eyes, good teeth, good whigs *[sic]*, not <u>verry, verry</u> good figures. [J19]

The single woman mentioned here must also be the subject of another rather cryptic comment:

> I met here *[at the amphitheatre in Syracuse]* with a rather curious adventure in ~~the sentimental~~ Sterne's line but…I shall not tell it. I put down this hint merely that I may not forget it <u>myself</u>.[J19]

Laurence Sterne's novel *A Sentimental Journey Through France and Italy* (1768)[B4] describes the hero's amorous adventures, including awkward encounters with a chamber maid in his hotel room involving a kiss to her cheek. For Ronalds, the emphasis of his interaction would likely have been humorous more than romantic.

Being interested in politics and societal attitudes, he also commented on local institutions and customs in places he visited and generally found them inferior to those in England. Passports appeared to cause particular difficulty and were required to be issued or signed when travelling across provincial borders within a country as well as internationally (Fig. 8.8). In Sicily:

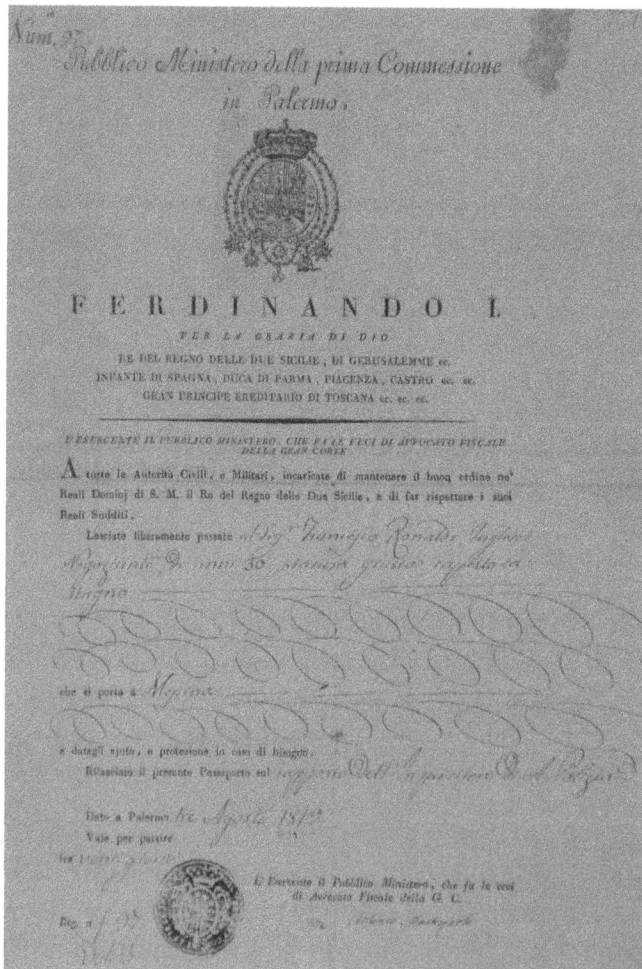

Fig. 8.8:   Ronalds' passport issued in Palermo to travel to Messina (3/8/1819). IET 1.1.22.

All my diligence to get through Syracuse proved abortive by the scandalous conduct of the Prince *[blank]*, who ought to have signed my passport instead of indulging in a sound sleep at the proper hour of business...By his negligence I lost three weeks, I lost the society of Sir Fred^k in Malta & Egypt & my money for the passage, so you may guess that I made noise enough to awaken his Excellency but you cannot imagine with what perfect sang froid *[coolness]* first my entreaties and afterwards my abuse were received.[J19]

Paris:

None of you except Peter *[Martineau]* can imagine how almost every place is warped. One can't do one's own private business in any degree of comfort without paying some holder of a place, and if the occupier of a place commits an act ever so dirty he escapes without animadversion, and no modesty is displayed in the scramble for a good place.[C19]

Leaving Jerusalem:

Happy Europe! Thrice happy little England! Tomorrow for the first time during 19 long months I bend my way homeward again to bask in the sunshine of pretty & honest English faces, again to abuse my rulers' heads without endangering my own, "to speak daggers but use none",[⁶] to talk of Pilauf *[local rice dish]* but eat none.[J20]

8.4 Evolution of the Journey

The scope and duration of Ronalds' tour was not planned in advance; he enjoyed the flexibility of being able to change his course. Initially, for example, there was 'no conception that a Journey in Egypt, Syria, Greece &c could ever have come into my head'.[J22] A few months into his travels, he wrote:

I hope*[?]* wandering about the face of the earth will not become an habitual medicine like opium.[C21]

When he agreed to accompany de Gourbillon in August 1819, he wrote:

we think of going to Malta then to Tunis & Carthage & god knows where else. When I can write & have anything to say I will.[C25]

This was soon revised:

from the last information we got it is doubtful whether we shall go to Tunis for they say the Plague is there and the quarantine is 60 days.[C25]

On the day he and de Gourbillon parted company in Catania (Fig. 8.9), Ronalds made enquiries about taking a passage to Leghorn or Genoa. After making the decision to go on to Egypt to try to catch up with Henniker, he wrote in November 1819:

At any rate I am determined to get home by June if possible.[C27]

In December 1820 he mused:

Accident pure accident has determined my long stay abroad. I have floated about on the world a little like a feather in a duck pond in whatever direction the wind blew — There's a great pleasure in this altho' some pains[C43]

---

⁶After Shakespeare's *Hamlet*.

Fig. 8.9:   Ancient baths at Catania (1819). IET 1.1.9.

Throughout his travels, he missed his family and wanted them to share his experiences. Describing his journey from Rome, he wrote in his journal in Naples:

But I am impatient to bring you to Naples, to me, to this Paradise inhabited by Devils [J18]

In a note to Edmund about wild-fowl shooting:

I wished very much for a Harlequin's wand the other day in passing through the Pontine Marshes to conjure you and Dido *[the family dog]* and two good doubles over, good god what sport! [C21]

This desire was mastered when he developed a writing technique of 'conjuring' Charlotte to join him at a place he enjoyed, as described in Sec. 4.7:

Come my little Whig — Hocus Pocus — come make haste…I want you to take a walk with me on the Marino. It is both more convenient and pleasant to me to show you about than to set down Raw Memo's…I am very glad to have discovered myself in possession of this conjuring art. [J18]

Ronalds also showed his homesickness when describing his first Christmas away, in Rome:

Having determined most magnanimously to see every thing that was to be seen on Xmas eve I sallied forth dressed out in stockings at about 6 o clock with 3 Gentlemen to the Quirinal *[Palace]* where we heard some fine music <u>now</u> and <u>then</u> during a space of about 5 hours, then we went to the Church of St Luigi dei Francesi where we spent about 2 hours and heard some still finer music <u>now</u> and <u>then</u>, & lastly we went to the Church of Sta Maria Maggiore where being locked up in the Chapel Borghese and seated upon a beautiful cold red granite Sarcophagus until ½ past 6 in the morning we <u>now</u> and <u>then</u> heard some of the very finest music and the ceremony was certainly fine, but I could not avoid fancying that one of the <u>Tunes</u> went very much like some song about "How different is thy fate from mine" and envying your comfortable circle round Uncle William's *[William Field Jnr's]* Xmas Port on the little table before the good rousing fire.[C21]

Replying to queries from home about his plans, he wrote as early as February 1819:

These <u>parts</u> are so interesting that no one circumstanced as I am would think of returning without killing all the lions that come in his way or even going a little out of his way to encounter them. But I must acknowledge that my valour has some what diminished…I feel no very great relish for the Combat…I undertake it <u>now</u> rather with a view to avoid the shame of a retreat than for the <u>glory</u> of the Conquest[C22]

He later wrote:

Whilst we were there sprawling upon our uncomfortable bed of sand…we could not help thinking of our homes, our arm chairs, our Books, our friends and asking why we gave ourselves so much trouble to come away from them and to stay away from them so long. These reflections struck me more forcibly than my friend for he has no mother or other relations whom he cares so much about as I do, but I excused and consoled myself with the idea that I shall now soon turn to northward again and…I should reproach myself for not having completed my project[C25]

Eventually the novelty of travel had worn off completely, compounded by ongoing financial and bureaucratic challenges, poor weather and homesickness:

I can't bear the vagabond life I am leading any longer & will cut and run as fast as I can.[C31]

I have nothing to wish for but to join you all once more. I am tigerish sulky &… every thing disgusts me in this bearish country *[Germany]* [C42]

8.5 Money Matters

Ronalds managed his outlays carefully and kept a detailed ledger of income and expenditure during his travels (Table 8.4). He told Charlotte: 'How much have I spent since I came out? No very great sum. I think not more than if I had been at home. Your purse is not yet worn out, it only wants a stitch or two.'[C27]

There were limited ways in which a long tour could be financed in those times and he ran into difficulties despite his care. Before he left home he had organised sufficient funding for the scale of travels he originally envisaged. He received £50 from the family business account, which he carried mainly as gold with about £5 'cash in purse'.[J26] Further funding could be obtained *en route* via a network of correspondents affiliated with a bank in London (Fig. 8.10). These regional houses were

Table 8.4:   Ronalds' Initial Expenses in Egypt.

1820		Dol[s]	Piast[s]	Paras
Jan[y] 20	Passage from Malta in a Brig	20		
	Cabin boy & Servant	3		
	Custom House expenses & Janissary		5	
	Ass to Pompey's Pillar		2	
	Looking Glass		3	
21	Coffee for 2		2	
	Snuff		1	
22	Washing		8	
23	Tobacco, Pipes, Coffee		9	
	Bath		2	5
	Ass		2	
24	Bill at Alexandria 4 days	5	6	
	Provisions &c &c for voyage to Cairo p canal (a cheat)	8	4	
	Waiter[?]		4	
	2 Asses, 1 Camel to Canal		3	
	Passage from Alexandria to Cairo	2	5	
25	Bread at Foueh about 5 lbs		1	
	Honey, Coffee, Candles at Foueh		4	5
	Oranges 5			50

All further expenses in this region are recorded in Italian, aligning with Ronalds' interactions with Ferdinando.
Source: IET 1.1.40.

important focal points for travellers: they could also facilitate introductions into local society, provide advice and manage mail. He carried bank letters for several places he intended to visit. In Milan: 'I presented my letter of Introduction to day and secured money…but no civility',[C20] but in Naples 'I find the House… (DeWelz & C°) much more disposed to civility…for they have returned my call.'[C21]

When he first contemplated venturing further afield, he wrote to Edmund requesting another letter of credit. It took 12 weeks to arrive but luckily he was still in Naples. He then tried using a bill of exchange in Palermo to speed up the process, but again did not receive this money until three months later.

By now he was in debt and accepted financial support from de Gourbillon. He confided to his mother:

I am already out of Elbows…However don't be alarmed for the Agent of Mess[rs] DeWelz promises to pay my bill at the Hotel and General de Gourbillon…is as

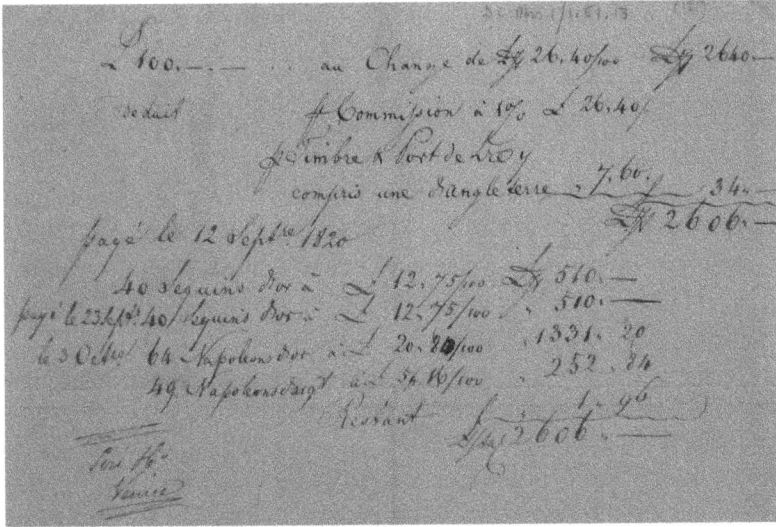

Fig. 8.10: Receipt for £100 from letter of credit on Siri & Co, Venice (Sep 1820). IET 1.1.51.

pressing in his offers of assistance as a man can possibly be. The principal obstacle I have to overcome is my pride therefore and the idea that I can never have an opportunity of <u>returning a favour of this kind</u> perhaps renders it a very perplexing case. I am now in doubt whether I shall wait here untill my money arrives or whether I shall go on with the General to Messina in a day or two, for he declares he won't go on without me & yet I know he wants to go on. Did you ever hear of such an awkward quandary. It's the first time I ever <u>felt</u> myself in such a one. It's not at all <u>funny</u> I assure you. When I get up in the morning I say "what shall I do"? When I have dined I say again what the devil shall I do? And when I go to bed I say Good god what shall I do? & I think at least of that part of the Lord's prayer Give us day by day our daily bread. Then I dream of Bankers and Gold and letters of Credit and bills & so forth all night sometimes, but it's very odd that I frequently quite forget my condition and am as happy as ever…Now the die is cast, I have accepted of some money from the General, I am no longer independent [C25]

As to his *[Charles']* question when do I mean to come home, I can only say that the longer I stay away the more uncertain it becomes when I shall return. And as I have hinted, having by my want of <u>management</u> involved myself in some measure in an obligation not to quit my friend, it must depend a little upon his intentions and Views. [C25]

Travelling with an acquaintance in this type of relationship and for a significant period can be challenging. It was exacerbated by them having different interests and by Ronalds' strong sense of loyalty to stay with the person who had supported him. He hinted at one of the sources of irritation:

As we approached Catania we passed between walls so high as to exclude entirely the view of anything beyond them…but these are constructed with materials which compensated me in some measure for their height namely the Lavas of the surrounding country, & they exhibit all the varieties of composition, fractures &c &c. In fact they constitute a <u>museum</u> of Lavas…The General paid me off handsomely for all my quizations…To stand gaping at a blank Wall with a Poet at my elbow already exasperated by my former tardy progress. Think on it.[J18]

The General, in complete contrast, described himself as 'being the child of feeling rather than of reflection'.[B14] A present-day writer on Mt Etna, Antonio Patanè, suggests he had an 'egocentric nature, difficult and grumpy'.[A187] De Gourbillon provided his own perspective of his relationship with Ronalds in the introduction of his book:

Of all voluntary compacts, the bargains for travelling together in company are the most hastily concluded, and the most speedily repented of. To travel alone, is a very dismal thing. But to have for a companion one whose humour, opinions, and manner of seeing and feeling, are in direct opposition to one's own, is one of the greatest of human miseries: it is better, a thousand times, to renounce the most inviting journey, or to hang oneself the moment it is commenced.[B14]

De Gourbillon and Ronalds went their separate ways after a total of eight months.

Ronalds' monetary challenges increased greatly when he moved on to the Eastern Mediterranean, due to the unreliable mail combined with his uncertain itinerary and more rapid pace of travel. He warned Charlotte at this time:

Never be anxious about me untill you hear bad news. I beg this as a favour and if I thought you would not at least do all in your power to grant it me I should be extreemly uncomfortable for I am now going to places where the uncertainty of the post is greatly increased or even the possibility of writing at all is somewhat doubtful.[C25]

A letter of credit needed to be able to be used in multiple places, and eventual receipt was dependent on the agent in each place following instructions to send letters on to Ronalds' subsequent destination:

I requested Edmund to send me a letter of Credit (no more of your plaguey Bills) on Malta, & Cairo & Alexandria. If it is not sent let it be also on Smyrna…and send a letter directed to me to the same house on which is drawn at Valletta to be forwarded [C27]

I believe Hammersley's *[bank in Pall Mall, London]* correspondence extends as far [J23]

Larnaca was later added to the correspondence trail.

This request was made in October 1819. Edmund, who was busy at home running the family business (which was severely damaged by fire during this period) and with a new wife and son, was perhaps less responsive to Ronalds' increasing needs than the latter would have liked. Ronalds' requests also tended to be a little vague and variable. He had still not heard from home in May 1820 when he reached Cyprus. Help came along at that moment in the form of Davidson, Coats and Boggie, 'three of the most obliging people I ever encountered'.[C35] Davidson lent Ronalds £20 and invited him to join them on their boat while Ferdinando sailed for Egypt where the letter of credit was believed to be. Ronalds added in his letter to the agents Briggs & Lee that Ferdinando carried:

I have further to request that you would be so good as to render assistance to my servant in procuring a lodging should it be necessary & to pay his passage and other little expenses at Alexandria should he have to wait there. I gave him 9 dollars.[C32]

Ronalds ended up receiving £100 in June in Smyrna through the kindness of brothers Peter and John Lee. John, a merchant in Smyrna, advanced him the money on a bill to Edmund's business, at Peter's request and guarantee in Alexandria:

> I cannot be sufficiently thankfull *[sic]* to Mess[rs] Davidson for his kindness in bringing me here *[to Smyrna]* from Cyprus nor of M[r] Lee for furnishing me with the means of getting away again.[C35]

Lee was able to explain the problem: overlapping requests for letters of credit had resulted in this critical one being annulled because: 'Your brother had heard from you that you had received the £100 you had expected from Naples.'[C34] The Lees also arranged a final payment to Ferdinando from Ronalds.

8.6 Travel Risks

It can be seen that travel entailed some risks in those times. Ronalds mentions several sources of risk; however, with his understated writing style he gives little impression of experiencing real danger. In addition to running short of money, other hazards included transport itself (particularly by sea), and his choice of adventures, such as exploring live volcanos. A fourth was illness, including from the plague and malaria. He wrote in Smyrna: 'It is dreadfully hot here and the plague has today begun to make its appearance, we fly its attacks tonight.'[C35] He was required to spend a total of 35 days at Lazarettos (maritime quarantine stations) off Corfu and in Venice on his way home. He tried to make the best of it:

> In the Lazaretto at Venice one of my principal evening amusements was that of making the Guardiano *[guardian]* (a sly old white headed cock) sing La Biondina in Gondoleta, which we all joined in loud Chorus[J24]

'Banditté'[C24] posed a further risk. One of Ronalds' companions was killed in Naples. He wrote home simply:

> There have not been so many robberies as lies, and the principal one was owing to an ostentatious display of rings, Clothes, silver, shaving & dressing apparatus etc. The gentleman who was killed dined with us frequently and was a very accomplished dandy but a very reasonable & even learned man too.[C24]

Ronalds was threatened by bandits several times in Egypt and the Holy Land. One of his more dangerous excursions was to the Dead Sea:

> I started at day break one morning Solo (for nobody would accompany me *[not even Ferdinando!]*) dressed in the shabbyest *[sic]* style which my wardrobe would permit and carried nothing about me but a pocket full of Paras. I mounted a sorry ass, slept at night on the bare ground and arrived on the shore where I bathed the next day

without molestation. On my return the day after I met some arabs who gave me some
pilauf & took all my paras from me & so got back safely.[J20]

Henniker, undertaking a very similar journey a month later with a guard and servant, had his face and ear severely slashed and was stripped naked in an attack. He
lost significant blood and took a month to recover in Jerusalem.[B23]

Ronalds wrote of another trip in the Holy Land:

I was treated with great rudeness and incivility all the way by the turks (the only
instance of the kind I had met with) and a scuffle once ensued about the right of
boiling the pot at the fire with Ferdinando which had like to have terminated in mischief for knives were out in a moment.[J20]

He noted many years later in his autobiographical letter that he 'Visited Egypt,
(where I got some teeth knocked out)'[C375] but gives no hint in his journal as to
where and how that occurred. Finally, on his way home sailing up the Aegean:

we passed Skyros & prepared our arms for Pirates. What degree of danger there was
I cannot determine for the people in these parts allways use great magnifying powers
in looking at such things.[J21]

About this time, he wrote to his family:

And so may god keep you all untill I come home to brag of my wonderful adventures
& achievements in the East over some of your proper old port round your english
fire sides — I am very well and intend to continue so a dio vola *[as God flies]* as the
Italians say. Tell Peter I have not even yet got rid of my absurd english prejudices,
that particular prejudice for a pretty English face sticks by me most of all.[C29]

Ronalds certainly ended his long adventure a stronger person than when he set
out. Not only did he succeed in his original goals, but his self-confidence increased
and his experiences and observations were the inspiration for a variety of his later
scientific undertakings.

8.7 Back at Home

When Henniker had also returned home, the two interacted frequently, sharing
travel stories and information. There was a flurry of letters in mid-1822 when
Henniker (denoted H in the following short excerpts) was working on his book
Notes during a visit to Egypt, Nubia, The Oasis, Mount Sinai, and Jerusalem.[B23]
Unfortunately only a few of Ronalds' (R) contributions to the exchange have
survived.

H: should you pass this way I should like to speak concerning a drawing of the
Papyrus plant Title page & many &c &c…if you admire music there will be
some very fine in a neighbouring Church tomorrow *[Sunday]* — Must be there
at eleven if you like to call — say you will come & breakfast with

Yours truly [C46]

R: I am very much obliged by your invitation for tomorrow to breakfast and as
much disappointed at being obliged to escort some ladies here *[Ronalds'*

mother and sisters] to a Chapel *[probably Essex Street Unitarian Chapel].* But I will take my chance of finding you at home on Tuesday morn^g before 2 by which time I hope to have almost completed the destruction of the 2nd Pyramid.[C45]

H: Have you any drawing of the Papyrus plant — or do you know of any — is the enclosed papyrus. — Come on Tuesday as you say and let me in the interim know what will be requisite for the safe opening of my papyrus scroll, and we will make a job of it.[C46]

H: Your sketch of the Anapus *[river in Syracuse where papyrus grows]* is quite the thing…but more of this on Tuesday [C46]

H: I have now to ask do you know of any drawing of a candgy *[Egyptian sailboat]* for my Title page…I am quite nervous today.[C46]

R: I am extreemly glad to hear that the book is so near its best & wait with no small anxiety for a copy from Murray on the 1st December *[1822].*[C48]

Henniker had asked Ronalds to review the book. Ronalds also analysed a piece of the casing from the Second Pyramid that Henniker had sent home from Egypt. He determined that it was limestone with a coating that 'the action of the blow pipe'[C47] showed was not resinous:

Yet I think it still extreemly probable that some substance has been applied to the surface which has caused the <u>scaly</u> effect so easily perceived by the use of a Knife.[C47]

The angle which the polished surface of the casing makes with the edge is 126°.[C47]

Henniker included the results in his book. This smooth casing on the top third of the pyramid (Fig. 8.11) made its ascent very challenging and Henniker is believed to be first European to have reached the top.

Ronalds' own travel journal was intended for his personal use and for the amusement of his immediate family: 'I don't wish all the nonsense I have written… to be handed about amongst those who do not know how to indulge me in my egotism and trifling.'[C19] Some family members, particularly his mother, Emily and Charlotte, greatly enjoyed early instalments he sent home, but they attracted criticism from his brother-in-law Peter. Calling Ronalds a 'grumping old Batchelor',[C23] Peter encouraged him to express his emotions rather than just his understated,

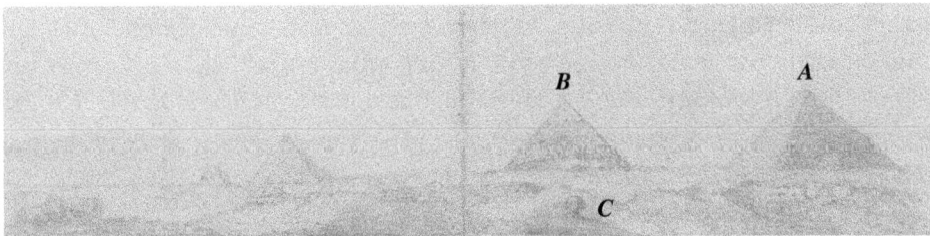

Fig. 8.11: Pyramids in Giza (1820). IET 1.1.9.

A Great Pyramid (climbed by Ronalds), *B* Second Pyramid with smooth upper casing (climbed by Henniker), *C* Great Sphinx.

scientifically styled observations. Ronalds turned to Charlotte: 'I appeal to you or any reasonable creature, What would have been the consequence of my "unbending myself" in <u>my</u> letters, of giving way to my imagination, of talking of sighs & serenades'[C23] and getting rid 'of my absurd English prejudices';[C29] 'he wants me to…experience or pretend to experience a sort of mawkish enthusiasm for antiquities and an affected admiration for the Ladies and the Music.'[J18] The criticism hurt him deeply and Peter's phrases recur through the remainder of the journal. Ten months later he wrote:

> Why should I care for Peter (said I one morning as I was wiping my Razer upon a piece of my poor disgraced Journal). I have been despairing myself of amusement because *[?]* <u>that fellow</u> don't like it & I have suffered my vanity to be wounded by his complaints of my "<u>grumpiness</u>" &c[J19]

Several of his siblings encouraged Ronalds to continue writing and later to publish his work. He mentioned to Emily just before he arrived home:

> As to publication which Charlotte hints at I have no such intention, there are too many Squirrel travels in my trunk already — I'll sell you the Copyright for a Kiss[C42]

Brother Hugh's perspective was more serious:

> I expected before this to have heard of the publication of your travels on the continent and in that case I hoped to receive a copy — it seems to me the duty of a Batchelor out of business to obtain information and put it into such shape that those who are more deeply engaged in the affairs of life may get at it in the shortest possible way — by this means the Traveller makes himself as usefull a member of Society as any other person taking his due share in the common division of Labor — but if he confines his knowledge to his own immediate circle it seems to me that he must rank amongst the unprofitable members — who consume the produce of a Country without returning an equivalent.[C50]

Ronalds summarised this period in his autobiographical letter:

> In 1821 Spent much time upon drawings from the foreign sketches & on the Journal which I had kept endeavouring <u>vainly</u> and in vain to render them & it a little amusing. (Sir Fred Henniker succeeded in producing a very pretty Book.)[C375]

Ronalds refers to his 'MS. Journal of a Tour in the Levant &c'[R11] in his 1836 publication with Blair about Carnac and the quote differs from the words in the original journal. Unfortunately the travel book was not completed and very little of the more polished material survives. In this respect Ronalds remained one of Hugh's 'unprofitable members' of society.[7] He did however publish his weather and atmospheric electricity observations during the tour, as discussed in Sec. 6.4, and other learnings soon prompted a return trip. We travel again to Italy in the next chapter.

---

[7] Ronalds' original travel journal and sketches are now available at http://www.sirfrancisronalds.co.uk.

Chapter 9

A SULPHUR BUSINESS OPPORTUNITY?

Ronalds' secretive follow-up trip to Sicily and Naples, probably with the farsighted view of establishing a business to import sulphur into the UK as a feedstock for the industrial revolution.

There is an air of mystery about Ronalds' 1823–24 trip to southern Italy. He and his family left detailed evidence of when, where and how he went, and what he did there, but virtually nothing about why he did it. His autobiographical letter says simply: 'went again abroad with a very unscientific object in view'[C375] and even this was later scribbled out. It is known that he travelled via Paris and Marseilles to Messina in Sicily where he spent three and a half weeks. He then spent more than four months in Naples before heading for home through Italy and France to Calais (Table 9.1). His journey is included in Fig. 8.1 in the previous chapter.

The travelling itself was mixed. Ronalds would have enjoyed the leg across the English Channel, which had just started to be served by steamboats, and he also noted with interest the increasing use of steamers in other locations. The passage on a brig from Marseilles to Messina was less pleasant. Three days into the journey he noted: 'Child of one of the passengers died at 5 AM. Put it in a box and buried it at Monte Christo *[a small island off the east coast of Corsica]* at 3 PM.'[J31] Anchoring off Pozzuoli near Naples three days later they were 'told that the child's death would occasion a longer quarantine than usual'[J31] before the vessel could continue on to Sicily.

The five-day trip from Messina back to Naples was worse, notwithstanding a warm recommendation from an acquaintance: 'The Bearer is Padrone…of a beautiful Speronaro that is <u>ready</u> to start'.[C51] A speronaro is a long, open boat rowed

Table 9.1: Ronalds' 1823–24 Trip to Italy — Itinerary (Incomplete).

	From	To	Days	Location	Travel to next location	Days
1823		05/07		London	By steamboat and coach via Gravesend	0
	05/07	06/07	1	Dover	By steamboat	0
	06/07	06/07	0	Calais	By diligence via Boulogne, Abbeville, Amiens, Clermont	2
	08/07	11/07	3	Paris	By diligence via Fontainebleau, Nevers, Moulins	3
	14/07	15/07	1	Lyon	By small boat via Valence	2
	17/07	19/07	2	Avignon	By diligence	1
	20/07	29/07	9	Marseilles	By brig *La Providenza* via Montecristo	6
	04/08	11/08	7	Pozzuoli	By brig *La Providenza*	2
	13/08	06/09	24	Messina	By speronaro	5
	11/09	18/01	129	Naples	By private carriage via Mola di Gaeta, Terracina	1
1824	19/01	21/01	2	Rome	By diligence via Viterbo	2
	23/01	26/01	3	Florence	By diligence via Lucca, Chiavari	2
	28/01			Genoa		
	30/01			Turin		
	03/02			Chambery		
	08/02			Paris		
	13/02			Calais		

with oars but also having sails. Even Ronalds, a good sailor and always under-stated, noted: 'Very bad passage. Boat heavily laden & embarrassed with Passengers. Hard*[?]* gales & foul. Nearly swamped off Capri. Praying.'[J31]

He also suffered various diplomatic difficulties, and showed less patience and greater confidence in handling the bureaucrats with the experience of his earlier travels. In Paris:

Applied for Passport for Lyons. Refused…

Vexatious detention of passport, was informed that the Prefect had detained it. Applied to him & to the Ambassador.[J31]

On the third day of the quarantine at Pozzuoli:

Protested against the confinement on board (Heat 87) & demanded a Guardiano & a Chamber. Refused. Threatened to represent the case to the Government of Naples. Was told that even if the Mal aria *[literally, evil air]* fever which prevails at Baia should kill any of the Passengers no redress could be had…and no assistance procured in case of illness but on the contrary the quarantine would be more rigorous.[J31]

Day four:

Repeated threats &c & found the deputies more reasonable…refused positively to make any present to them. Violent discussions. &c &c[J31]

Day five:

> Applied to them again & demanded an explicit answer before witnesses. Received
> our release or (Pratica) *[dossier]*.[131]

Ronalds' trip notes are written as a business diary. He provides details of who
he interacted with each day and evening, with brief summaries of what was (and
was not) discussed, as well as letters sent and received. In Marseilles he met three
people while awaiting his passage to Sicily. His notes from each of these conversations comprised:

> Little Business in Marseilles.[131]
>
> Houses & Living cheap at Marseilles.[131]
>
> Business of Marseilles. In[t] *[Interest]* low.[131]

He also kept the detailed notes one of these people had provided to him in French
on current trade at Marseilles in a range of commodities — wine, spices, metals,
minerals (including sulphur in three different forms) — with current prices and
brief comments on the state of their market: 'dear', 'cheap', 'reasonable', 'good
price',[130] etc.

On the day he arrived at Messina, he wrote:

> Called on M[r] A Furse who invited me pressingly to live at his house. Refused & went
> to Lion d'Oro. He called at ab[t] 4 and conversed about the object of my visit to
> Messina &c...First impressions favourable to the idea of establishing at Messina.[131]

In his conversation with Furse three days later:

> He wished to know further particulars...gave him to understand that I wished to gain
> more information relative to the Business of Messina, Society &c &c before I could
> propose any thing.[131]

The morning after arriving at Naples, Ronalds called on Mr F. Degen, to
whom he had been given a letter of introduction by Abraham Furse:

> M[r] Degen called and conversed about affairs &c (see my letter to M[r] Furse dated
> 13th). He shewed me M[r] Jn[o] Furse's letter[131]

Three days later:

> M[r] Degen called and shewed me a letter which he was about to send by this day's
> post to Mess[rs] Heath's[131]

A fortnight later:

> M[r] Degen called. He said he would draw up a sort of proposition and expressed
> himself very desirous of having something fixed upon soon.[131]

And the next day:

> Wrote to M[r] A Furse inclosing proposal &c &c[131]

Nearly two weeks on:

> M[r] D said...that he had no doubt that some very good business might be done at
> Messina.[131]

Ronalds and Degen then waited two months for Abraham Furse to come up to
Naples. Just six days after that, Ronalds departed for Rome in Degen's carriage on
his way home.

Ronalds would have raised his business idea initially with *Heath Furse & Co* in Old Jewry, London, through John Furse. In addition to their merchant operations, the principals were bankers and at a later time had heavy industry interests; they might also have been Ronalds' maternal relatives. The company was a forwarding agent for mail as well, with its international footprint including Messina and Leghorn, and the affiliation with *Degen & Co* in Naples. The presence of British companies like *Heath Furse & Co* in the region had been encouraged inadvertently by Napoleon — King Ferdinand had been protected in Sicily by the British during the war. The Kingdom of the Two Sicilies he formed in 1816 had its capital at Naples.

It seems plausible that the opportunity Ronalds discussed with *Heath Furse & Co* concerned the volcanic mineralogy around Messina and Naples. He had developed a strong interest in the local geology during his Grand Tour, which he continued to pursue back in London. After visiting the Solfatara volcanic crater near Naples in January 1819 he had written:

> It is very curious to observe the process of evaporation for procuring Alum carried on by the natural heat of the earth, what a famous plan it would be for Hugh's tanning scheme or for Peter's Sugar Pans.[J18]

> My ride…today has furnished much to think about but nothing to write about, for you don't want to know about Alum & Sulphur & I don't choose to describe what has been well enough and often enough described by others.[J18]

What he had observed that day was the release of hot sulphurous fumes from the magma through an overlying aquifer. Oxidising occurs to form a deposit of yellow elemental sulphur on the rocks and the sulphur also reacts with the rock to form Alum. Alum was already used in leather manufacturing. Sulphur itself, however, was soon to become a critically important element in the industrial revolution as its uses multiplied.

It was at this time, January 1819, that Ronalds first wrote that he was contemplating travelling to Sicily. He studied the mineralogy at Vesuvius and later Etna at length, collecting samples and procuring more from the Lipari Islands on the volcanic archipelago between them. In Messina:

> I have been mineralizing in the neighbourhood on a Donkey or rather <u>with</u> a donkey to carry my Stones (which has been my principal amusement during the greater part of the time I have spent here).[J18]

Sending the minerals home he wrote:

> be so good as to get somebody to enquire for them, to pay the expenses on them, and get them safely lodged amongst my other rubbish.[C27]

> if you can…see the box opened so that my stones in an interior box…may not be injured…you will oblige me very much.[C28]

And then:

> It gives me great pleasure to hear that my little treasures from Sicily & Malta are arrived[C38]

He studied these further after his return to London; some of his notes on the crystals and other minerals in the lava and the classification of his rock samples survive in the Ronalds Archive. He wrote to Henniker*:

> In the arrangem^t of this mere skeleton of a collection of Volcanic minerals I have followed almost entirely *[Déodat de]* Dolomieu *[French geologist after whom the Dolomites are named]* whose Classification I have availed myself in making the enclosed outline…This has really been an additional pleasure for me whilst making the little collection for myself to divide my spoil with you…I am only sorry that I have been obliged to leave many drawers*[?]* in the box still empty. If I can get specimens for myself I hope you will allow me to divide them also with you. I have written*[?]* to our common very good friend Mr Gioeni at Catania for some.[C49]

Cavaliere Gioeni oversaw the museum housing his brother Giuseppe's renowned geological collection. Ronalds made contact with him again when he returned to Messina in 1823.

He also looked into financial matters in preparation for his return trip. He copied sections out of Cohen's 1822 *Compendium of Finance*[B20] on the national budget of the Kingdom of Naples and also on Neapolitan 5 per cent stock, including payment of dividends and exchange rates.

Advising him in what he called the '~~proposed~~ contemplated establishment'[J31] were George Tatlock and a Dr Smith. The latter was possibly Thomas Southwood Smith. He was the same age as Ronalds, a Unitarian preacher, physician to the London Fever Hospital, and had a strong interest in public health; he and others held the belief that diseases were spread by bad air. Smith's advice to Ronalds might have concerned the properties of the sulphur in the volcanic rock. Sulphur in various forms had long been valued therapeutically but, when combusted, produces toxic sulphur dioxide gas which harms people, animals and vegetation.

Ronalds' first cousin George Tatlock was a silk broker in London. Sadly he died on 21 November 1823. Ronalds would have heard this news prior to his final discussions with Furse and Degen in Naples but not in time to notify Furse before he set sail there.

If Ronalds had spotted the sulphur opportunity, his timing was impeccable. Sulphur (known then as brimstone) was mined near Naples in Roman times and Sicily was producing by the 18th century — it was mined there using picks and then purified by igniting heaps of the rock and collecting the molten sulphur. Much of the 'refined' sulphur output was shipped to Marseilles, which was then a leading chemical manufacturing centre. The British Government changed the rules in 1822, however, by announcing that its salt production tax would be repealed the next year. UK sulphur imports began to grow rapidly from the mid-1820s as a result (Fig. 9.1). In these early years, most of the sulphur was used to produce oil of vitriol, now called sulphuric acid, which became the fundamental feedstock of

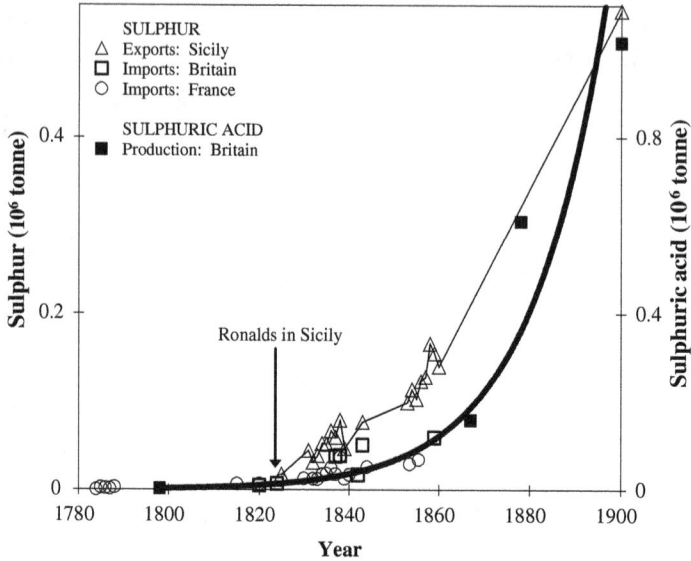

Fig 9.1:   Growth of the Sulphur Industry. Refs. A61, B51, B69, B99, B144, B147, B156, B191.

the chemical industry. Sulphuric acid, together with common salt, were principal ingredients in making soda through the new Leblanc process. Soda, in turn, was employed in glass-works, soap-making and in bleach and dyes for textiles. UK soap production alone more than doubled in the quarter-century 1825–50.[B147] A very important later use of sulphuric acid was in making synthetic fertilizers. In 1843, the renowned German chemist Justus von Liebig (under whom Ronalds' nephew Dr Edmund completed his PhD) wrote: 'we may judge, with great accuracy, of the commercial prosperity of a country from the amount of sulphuric acid it consumes'.[B48] Interestingly, perhaps the best mid-19th century description of the workings of the then well-developed sulphur industry is in *Chemical Technology*, written by Dr Edmund and his co-authors. For many years Sicily provided nearly all of the world's elemental sulphur needs.

It is not altogether surprising for Ronalds to have identified the looming possibilities in sulphur. Chemistry and earth sciences were strong interests for him, he had broad and deep understanding of technology and business affairs of the day, he was a visionary, had spent significant time in Sicily and Naples, and had experience as a continental merchant. He had perhaps developed a new way of producing or processing sulphur, or his interest might have been in particular applications, such as leather-making or textiles (in which his family had businesses). What he probably lacked, however, was the drive to negotiate a deal to closure, and this would have been exacerbated by his cousin's untimely death.

British merchants did indeed capitalise on the Sicilian sulphur boom. One of the early players in the industry was James Rose, whom Ronalds had met at a 'profuse english dinner'[J31] hosted by Abraham Furse a week after his arrival in Messina. *Heath Furse & Co* were London agents to Benjamin Ingham, the most successful of the British merchants in Sicily. Ingham had entered the sulphur business by 1830 and in 1841 had the majority shareholding in a sulphuric acid plant in Palermo.

Ronalds did not focus solely on business while he was abroad, but also took time to indulge his wider interests. He had cause to note *en route* to Marseilles: 'Good Roquemaure 8^d p bottle delivered at Marseilles. Fine wine season at L'Hermitage.'[J31] He met many British merchants and other business people socially and appeared from surviving receipts (Fig. 9.2) to entertain guests himself. He also

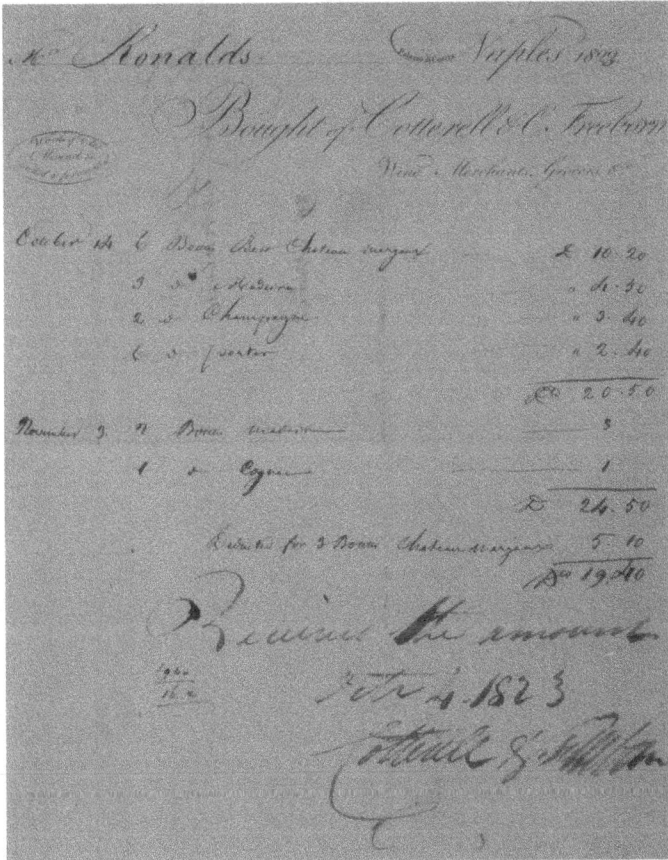

Fig. 9.2:   Receipt for 20 bottles of alcohol, Naples (Oct 1823). IET 1.1.71.

enjoyed the opera, concerts and balls (Fig. 9.3), particularly in his beloved Naples. Each Friday evening there was a party at the home of the British Envoy to the Two Sicilies, William Richard Hamilton.

Maintaining an element of intrigue: 'Count *[blank]* called. Politics. Discovered him to be a Spy.'[J31] Ronalds also met 'Baron de Belem'[J31] and later lent him 100 and then 20 Ducats (a total of nearly £20). A month later the 'Baron' departed

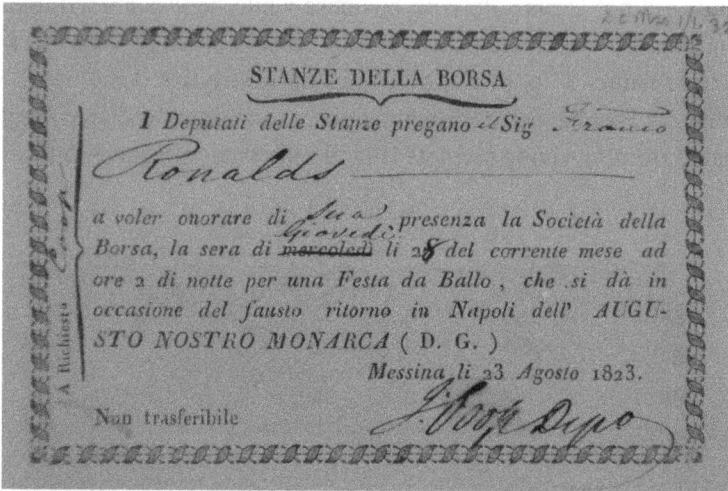

STANZE DELLA BORSA

I Deputati delle Stanze pregano *il Sig* _____

Ronalds _____

a voler onorare di *sua* presenza la Società della Borsa, la sera di mercoledì li 28 del corrente mese ad ore 2 di notte per una Festa da Ballo, che si dà in occasione del fausto ritorno in Napoli dell' AUGUSTO NOSTRO MONARCA (D. G.)

Messina, li 23 Agosto 1823.

Non trasferibile

Fig. 9.3:   Ticket to a celebration in Messina on the return of the monarch (Aug 1823). IET 1.1.32.

Table 9.2:   Estimate of Annual Expenses of House Keeping.

		in Messina	in London
House rent & taxes	£	25	70
Board 2 Persons at 1 pezzo p day		80	180
Dinner parties 5 at 3£ each		15	30
Conversazioni		10	20
Servants 1 Man £25 Woman 10		35	70
Washing (1/- p day)		20	30
Fire & lights		10	30
Dress		70	80
Theatre Box 50 nights at 1½ pezzi		15	40
Carriage		100	300
Sundries		20	40
		400	890

Source: IET 1.1.37.

suddenly, without a passport, and also owing money to others. Ronalds had by this time drawn on *Heath & Co* for £50, with Degen providing him the money in local currency as needed. Other bankers he met included Carl Mayer von Rothschild who founded the Naples arm of that banking giant. Ronalds compared the cost of living in Messina and London (Table 9.2), and later with Naples. Naples was 20 per cent more and London over twice as expensive as Sicily. He took advantage of the prices to order several sets of new clothes.

Concerning his science interests, Ronalds called on Giuseppe Saverio Poli, a physicist, biologist and natural historian whose collections were the foundation of the Zoological Museum of Naples. Ronalds obtained four of his publications for his library and proudly presented in return his brand new book on the electric telegraph. He had earlier given a copy to the *Académie des sciences* in Paris.

Ronalds' surviving papers give no hint as to the outcome of his discussions in Naples with Furse and Degen and have no further mention of Sicily after his return to London in 1824. Although he enjoyed his trip, it almost certainly came to nothing for him personally as a business venture.

Chapter 10

PERSPECTIVE TRACING INSTRUMENTS

Motivations for Ronalds' work on drawing instruments, descriptions of his devices, their manufacture and the successful business he created. Reproduction of the family's drawings, and the ongoing use of his tripod.

This next chapter of Ronalds' life is noteworthy in that it concerns ideas that he himself turned into a business. His tracing instruments also foreshadowed the modern era of precise pictorial recording and, as an aside, generated one of his longest-living but least recognised legacies — the portable tripod stand. He devoted significant time to his drawing instrument business over nearly two decades.

The original idea, like several others, had come from his travels, which offered many stimuli to his fertile mind. When he first returned home from his Grand Tour, he funnelled his energy into documenting his trip. He reworked his journal and sketches and assisted his friend Sir Frederick Henniker* to produce his travel volume. He also classified his rock samples and catalogued book purchases. Next, he wrote up his earlier work on the telegraph and other electrical inventions as well as his recent experiments at Vesuvius and Palermo. He then developed a business plan for an opportunity he had espied in Italy, and visited Sicily and Naples again with the aim of bringing it to fruition. In 1824, he turned his attention to a machine to aid perspective drawing. He later explained that he had: 'Prosecuted a sketching tour in Sicily with Sir Fred^k Henniker; when we both became aware of the great advantage derivable from the use of mechanical sketching instruments'.[C375] Ronalds is being characteristically generous here in sharing credit for his idea with his friend. He and Henniker were only together for a very short time in Sicily, although they did discuss the merits of accurate drawing.

Ronalds (and perhaps Henniker) had made use of existing instruments on his travels. He wrote home early in his trip of his ride across the mountain pass from France to Italy:

> I am very sorry I had no time to sketch during the passage of Mont Cenis, one ought to remain in those super-terranean Regions several days at least to have but a faint idea of the beauties. I found the Camera Lucida and Perspective Compasses of use in purchasing by the loan of them the civility of Monsr Messum[?]. They are scarcely known here abouts.[C19]

He soon acquired his own compasses, which he used in Egypt in 1820:

> At day break I sallied forth again and got on the Top of the Great Pyramid with considerable labour but not so much as I expected from travellers' accounts. The air was tolerably clear and I could see the Pyramids of Dashur and Memphis such enough to sketch them in their exact relative positions by means of a pair of perspective compasses [J20]

Devices to aid perspective drawing had been in existence for a long time. One type was the perspective compasses, which were developed in the 17th century. They resembled modern drawing compasses, having two extendible arms with end points. An eyepiece on the central hinge facilitated sighting along the arms to distant objects and their apparent separation or size could then be marked on the drawing. Christoph Scheiner invented the pantograph at about the same time, which was a mechanical linkage based on the parallelogram. When an image was traced, a pen making a corresponding motion would reproduce it at larger or smaller scale. The camera lucida (latin for light chamber) was patented by William Wollaston* in 1807. With this instrument, the artist was able to look down at the drawing paper, rather than at the scene, but trace an image of the subject. The image is generated by light reflected to the eye through a prism.

In another alternative — the camera obscura (dark chamber) — light from an external scene passes through a small hole in a dark box or room and projects an inverted image onto a surface. The concept had been known since ancient times. A lens was later added to assist in focusing and more portable machines became available in the 18th century, which travellers employed to aid drawing by tracing the image. Pinhole cameras of this type were used to take the first photographs by capturing the image on a light-sensitive material. Exposure times were measured in hours, until Louis Daguerre and Henry Fox Talbot separately developed chemical means to accelerate the process in 1839–40. By 1851, the *Reports by the Juries* of the Great Exhibition noted that it was just a few years since the Daguerreotype 'required that a person should sit without moving for twenty-five minutes in a glaring sunshine'.[B61] Further improvements in convenience of use followed progressively and, in 1900, photography was opened to the mass market when George Eastman introduced the Kodak Brownie.

The drawing instrument that Ronalds developed followed other well-used mechanical concepts. He was aware of the earlier examples — in describing his own device he wrote:

> A work, printed in Paris in 1646, by Father Francis Niceron,...entitled "Thaumaturgus opticus," contains the description of a perspectograph, which more nearly resembles the present invention than any other which seems to be known.[R10]

He might have seen the book in the British Museum's Library or perhaps during his time in Paris in 1823. While there was certainly similarity between his and Niceron's instruments, there were also differences brought by time and progress. Oxford Emeritus Professor Martin Kemp, a leading expert on visualisation, notes in his 1992 book *The Science of Art*:

> Perhaps the best example of how the delights and precision of scientific instruments in the later eras could breathe new life into old ideas is provided by the new generation of perspectographs which appeared in the first half of the nineteenth century. The most prominent of these was the machine patented in 1825 by Sir Francis Ronalds[B175]

Ronalds had several goals for his instrument. In addition to being well made, he wanted it to be accurate and convenient to use. He had foreseen a problem common with devices of this type, 'which have been invented by unartist-like mechanics, and rejected by unmechanical artists'.[R10] The instrument he developed was light to carry, quick to set up, easy to master and in use gave freedom of movement similar to that of rapid sketching. As a result it saved time over traditional methods of determining perspective and was a pleasure to use. He also felt very strongly about accuracy: 'The ignorant and unfeeling are *now*, thank God, almost the only apologists for the employment of bad perspective on *any* occasion. A distorted limb is not more hideous than a distorted mountain.'[R10] He achieved all this by bringing together two of his talents and interests — science and art. In particular, his machine demonstrates his strong understanding of geometry and theories of perspective, his practical knowhow, experience in sketching, an appreciation of art history that had been enhanced by his travels and, most of all, his superb mechanical design skills.

He was quick to acknowledge however that his instrument did not serve all purposes. Concerning the need to study perspective:

> Machines applied *habitually* to the purpose of sketching from nature...would, so employed, unquestionably oppose to the proper use and cultivation of that most beautiful and convenient little camera-obscura and perspectograph, the human eye, which, *when* well trained, never has been, and never will be, rivalled...by mortal machinery.[R10]

He also welcomed alternative media. In a prophetic passage written at this time concerning what he called a 'Chemical Camera',[J14] he mused:

> Place any substance or infusion upon the paper or other surface receiving the image which by the action of light would <u>stain</u> the paper…illuminate the object as strongly as possible by means of lenses in any number or any size…If this were practicable we should obtain far better mechanical <u>painters</u> than human ones of light & shade.[J14]

The words were prompted by studies he was conducting to improve his painting skills and in particular on how the appearance of colour, light and shade vary with parameters such as distance, viewing angle, illumination and atmospheric conditions.

Ronalds did not invent just one tracing instrument, but a whole range. He could not resist altering and improving the design to suit different purposes. The instruments fit in two classes, which he called 'kinds' — those used to draw from real life and those to draw perspectives from plans and elevations. Each had several variants and 'optional extras'.

10.1 Perspective Tracing Instrument — First Kind: Drawing from Real Life

Ronalds' perspective tracing instruments of the first kind are configured as tee squares in two planes with a shared transverse member; the latter is supported in rollers pinned to a table. An angled handle with a ball and socket connection to a slider on the horizontal tee plays the role of the drawing implement. Holding it '*precisely* as if it were a pencil',[R10] the hand can move freely around the paper through the action of the sliders and rollers and an attached pencil draws the resulting curve. A small bead (labelled *M* in Fig. 10.1) moves corresponding distances in the vertical plane. Its vertical movement is enabled by the thin thread it is suspended upon being attached to the handle and kept taut by a weight inside the vertical tube. When the operator views the bead through an eyepiece clamped to the table and traces the object behind by moving the handle, the object is drawn on the page.

The flagship of the range was the 'pocket form'[R10] where the metal bars pack up into a velvet-lined wooden or red leather case, 30 cm long and 5 cm wide (Fig. 10.2). It could produce sketches of up to 50 cm by 22 cm in size. At least three examples survive in British museums, two of which were sourced from Thomas Henry Court (1868–1951), a dealer and collector of scientific instruments. That at the Science Museum in South Kensington was acquired in 1894, while Robert Whipple — founder of the Whipple Museum of the History of

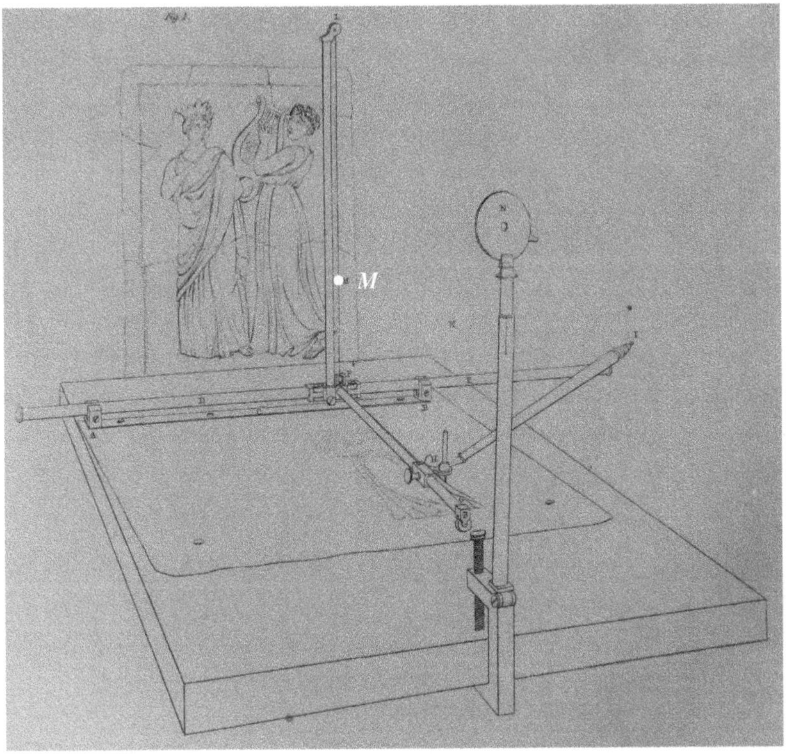

Fig. 10.1: First type of tracing instrument of first kind — pocket form. Ref. R10.

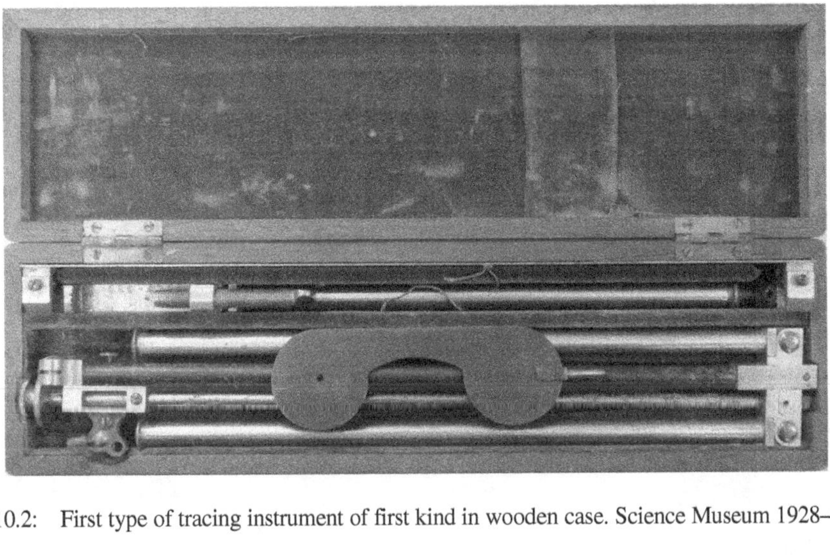

Fig. 10.2: First type of tracing instrument of first kind in wooden case. Science Museum 1928–0939.

Science in Cambridge — bought his version from Court's estate. The instrument at the National Museum of Scotland in Edinburgh was donated by Robert Orr in 1931. He had first offered it to the Science Museum: 'They refused it as a gift, but if it is any use to you I shall be glad to be rid of it.'[C420] The devices still come up for sale on occasion at auction houses.

The second type of the first kind of instrument was housed in a mahogany case that also served as a drawing board (Fig. 10.3). When closed, the box measured 30 cm by 13 cm by 3 cm thick. It was less portable but a little more convenient to use than the first type.

The third type (Fig. 10.4) was configured with elongated members to enable larger sketches. In particular, the vertical bar is attached well beyond the cross bar to give the longer distance from the eye to the vertical plane of the bead necessary to maintain a natural angle of vision over the whole drawing. The eyepiece is also

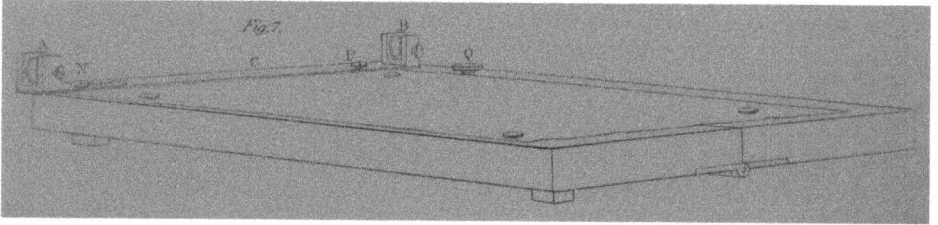

Fig. 10.3:   Second type of first kind of tracing instrument — drawing board. Ref. R10.

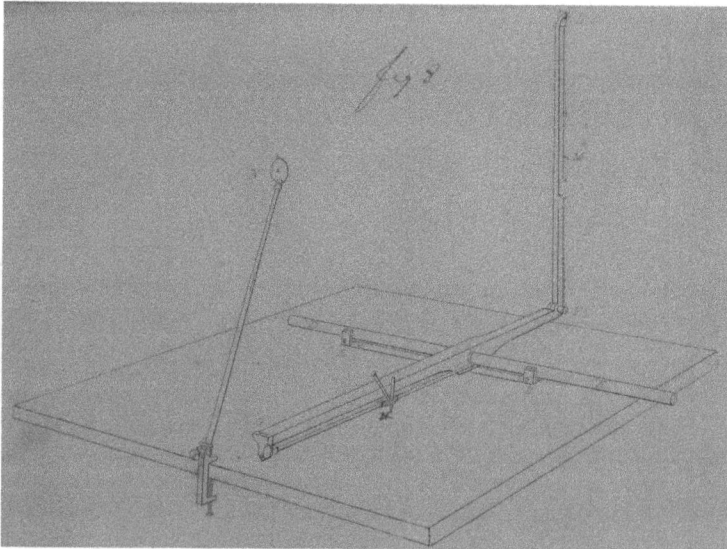

Fig. 10.4:   Third type of first kind of tracing instrument — large drawings. Ref. R10.

Fig. 10.5:   Fourth type of first kind of tracing instrument — with tripod table. Ref. R10.

angled forward to allow the artist to reach further over the table. The instrument came in various sizes with cases up to a metre long.

   In the fourth type of the first kind of instrument (Fig. 10.5), some of the metal elements were attached permanently to a mahogany drawing board that was supported on a tripod (or perhaps more correctly hexapod) stand. The table measured $52 \, cm \times 47 \, cm \times 5 \, cm$ when folded, making this variant more bulky than the others. The tripod legs could be carried in a fishing rod bag. Ronalds assured potential purchasers that it took just three minutes to fit the legs to the underside of the table and then complete positioning of the drawing apparatus. An example of this instrument is held by the Science Museum, although the drawing table does not survive. It belonged to Ipswich-based genealogist Frederick William Campbell and was later collected by Sir Henry Wellcome (1853–1936) who founded the Wellcome Trust.

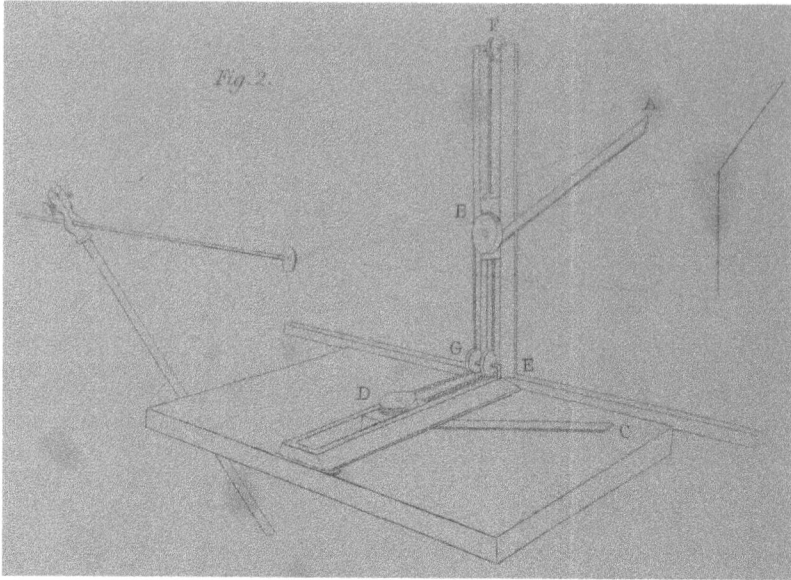

Fig. 10.6:   Fifth type of first kind of tracing instrument — for drawing numerous straight lines; also probably the basis of another device to draw ellipses. Ref. R10.

Ronalds developed yet another variant of the machine, a fifth type, to facilitate drawings with many straight lines (Fig. 10.6). Evidence also survives of other options. The original circular eyepiece could be replaced with one shaped a little like spectacles but with only one hole so that the artist did not need to keep an eye closed (Fig. 10.2). Eyepieces that allowed holes of smaller diameter to be slid into place were also available, which he recommended for detailed work like portraits.

10.2   Perspective Tracing Instrument — Second Kind: Drawing from Plans and Elevations

The second kind of perspective tracing instrument was quite different from the first in purpose and form. It was designed 'to facilitate that sometimes intricate and always tedious operation'[R10] of making accurate perspective drawings from plans and elevations (or key dimensions) even before the object was built. Ronalds would have developed the machine to illustrate his inventions, but it also served the growing need as the industrial revolution progressed for engineers and architects to communicate through technical drawing.

The drawing board and paper are vertical in this instrument and the ground plan is pinned onto a horizontal table at any desired orientation. A bar (*C* in Fig. 10.7), kept vertical by three polished steel pieces on its round base, is able to move freely around the table so that a point on its end (clearly visible through

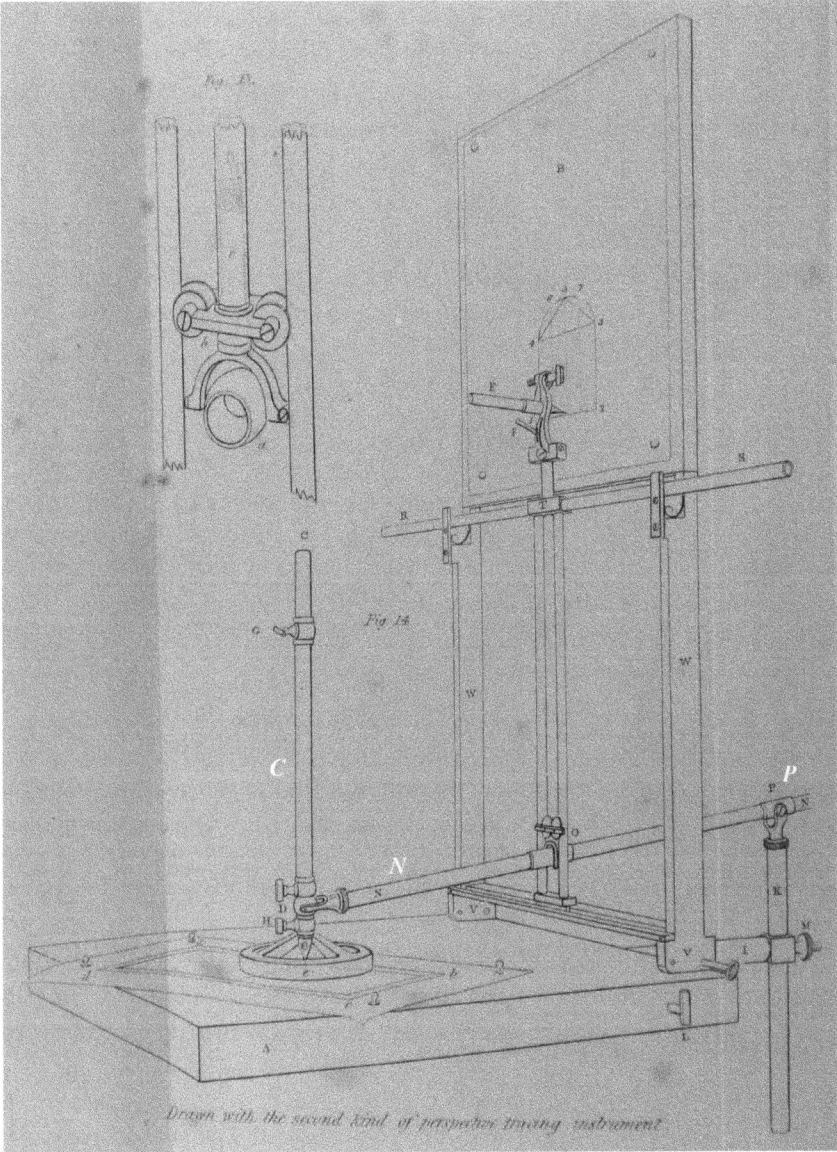

Fig. 10.7: Tracing instrument of second kind. Ref. R10.

apertures in the base) can trace any shape. This movement induces similar but different movements at the pencil, which is pressed onto the vertical board by a spring when a small lever is lowered. The relationship between the two sets of movements is dictated by the chosen orientation of a bar N between the vertical member C and an adjustable rear sliding support P. The point P is effectively the position of the eye and N is a light ray from the imaginary three-dimensional object on the table. Moving bar C along a face of the object in plan therefore produces a perspective view of this line on the vertical drawing board. Raising the connection point of *N* up *C* draws a perpendicular from the ground plan, while clamping the connection at this upper point and tracing around the plan once more generates the perspective of the outline at a height above the ground. The apparatus can thus draw rectilinear shapes in axonometric projection very rapidly. Ronalds provided detailed instructions regarding the ready incorporation of more complex architectural features such as pediments.

Again the instrument came in several sizes. The smallest variant suited quarto or octavo paper and closed up to dimensions of 71 cm × 36 cm × 6 cm. The largest model was fixed to a table and had a treadle to control contact of the pencil on the drawing board by foot.

10.3 Headrest for Drawing Portraits

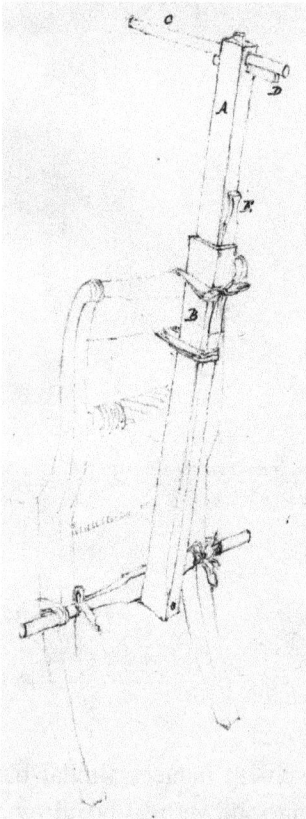

Fig. 10.8:   Headrest for drawing portraits (*c.*1829). IET 1.6.29.

Another option in the tracing equipment suite was a rest to steady the sitter's head while drawing a portrait (Fig. 10.8). Ronalds described its use to Lord Stanhope*:

> the Subject of the Sketch sits with his back easily and naturally resting against the back of the chair as usual, & <u>after</u> any unconstrained position of the head has been also assumed, the squared piece is drawn upward & the cylindrical piece pushed forward, untill the end of the latter touches the head. The little wedges are then used to fix both pieces in the proper places. Whilst the sketch is in progress the Subject should contrive to feel the end of the stick with his head yet also endeavour to forget the rather ridiculous circumstances of his (pilloried) situation otherwise an expression of the kind seen in the Sketch which Mr Cary will shew may become too apparent…The Pillory is quite at your lordship's service for the purpose of putting any subject into for experiment…(or as a pattern for a carpenter). I have no use for it at present having pilloried several of my best friends here.[C89]

Gaston Tissandier's 1876 book *Photography* described a somewhat similar headrest as 'an instrument which no portrait studio should be without'.[B96] A sketch of a rather alarmed sitter survives (Fig. 10.9) which might be the one referred to by Ronalds.

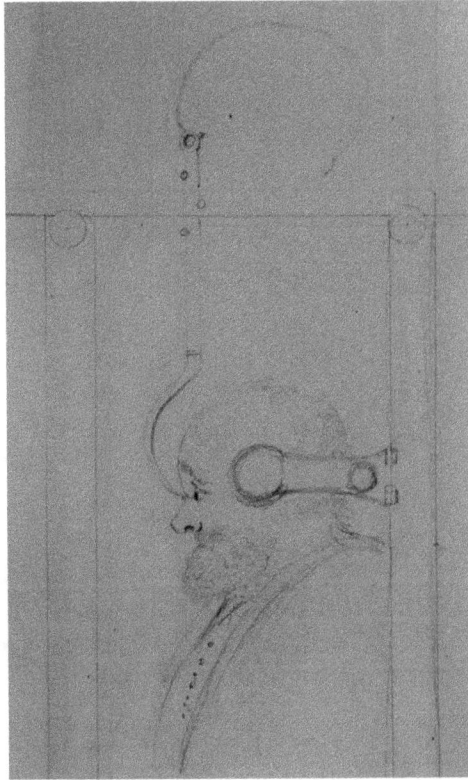

Fig. 10.9:   Extreme equipment for drawing accurate portraits. IET 1.6.5.

10.4 Portable Hinged Tripod (or Hexapod) Stand

Ronalds was always striving to make his pastimes more convenient and enjoyable. To this end he contrived a further accessory for his first kind of tracing instrument — a light but stable tripod to support his drawing board when travelling. It turned out to be one of his most enduring inventions.

His first model was developed by early 1825 and constituted part of the fourth type of tracing instrument. Its versatility was increased greatly by adding a triangular head made of '*free* gun-metal'[R10] (Fig. 10.10). A roller was positioned along each of its sides. There were two angled wooden legs per roller that met at their base and they formed in aggregate a triangle of three triangular frames. The legs ended in a sharp point to give purchase in the ground. The whole weighed one or two kilograms depending on the length of the legs and whether intermediate joints were required. The drawing board was fixed onto the tripod head with a central bolt and nut, and when not in use the triangular head could be stored

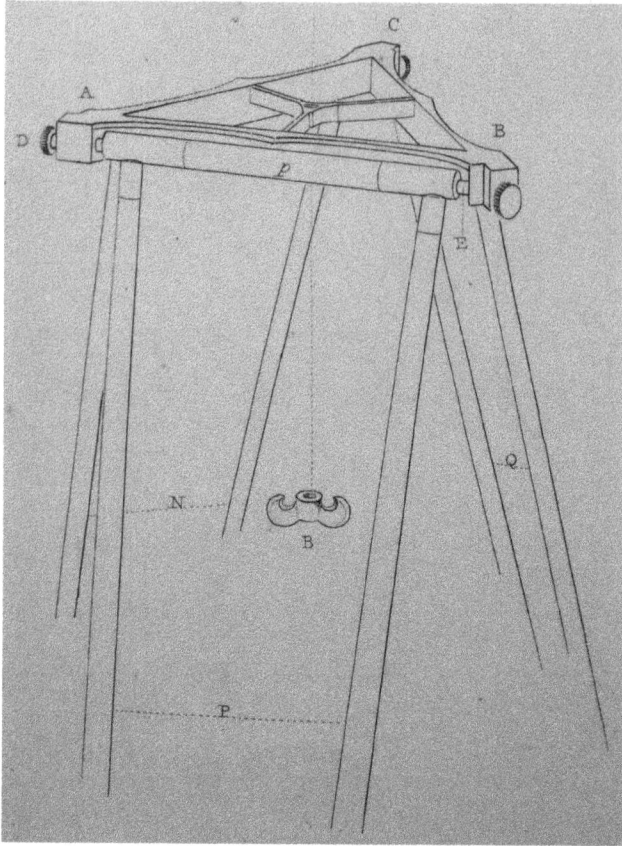

Fig. 10.10:   Portable hinged tripod (or hexapod) stand. Ref. R10.

inside the closed board. Details of the head could be adjusted readily to support other optical instruments.

Further variants followed. Ronalds added the option of ties that rested in the forks of the leg pairs and pinned together to form another triangle, thus linking the legs at their base to reduce the effects of accidental jolting on smooth, hard surfaces. It is seen in Fig. 14.5. As he advised his later colleague Edward Sabine*: 'Before using this affair I sometimes contrived to kick down the whole concern and spoil my sketch.'[C239] In another model, the hinged triangular legs were solid rather than framed to provide even greater strength and sturdiness (Fig. 14.4).

If the overall concept appears a little commonplace, that's because it is. He wrote as early as 1828 that it 'has been much approved as constituting also a remarkably solid and convenient staff for telescopes, theodolites, levels, and all other instruments which require great steadiness, combined with portability and lightness'.[R10] He did not claim originality for the initial idea. He was advised that a Mr Leicester had outlined a similar concept for a table. Ronalds developed his thoughts independently and was almost certainly the first to hone the design of a stand-alone hexapod stand, and then organise its manufacture and offer it for sale. Today the configuration remains the basis of wide-frame tripods used in surveying and for other purposes. Ronalds' frame is a little broader at the top than most because in its original form it supported a board rather than an instrument. At least one of his tripods survives in the Science Museum.

10.5 Curve-drawing and other Instruments

Ronalds makes mention of further attachments he developed to assist in drawing curves:

> An apparatus is sometimes added to these machines which is employed to rule a great variety of right lines, and to represent perspectively (and with very little trouble of adjustment) circles and *ellipses described about their perspective centres* — which saves much time in finding points.[R10]

A little more detail is provided in a letter to his friend Edward Clive Bayley:

> Now the conditions necessary for describing upon a stationary surface an epicycloid are that a point shall describe a curve about another point with certain proportional velocities whilst that other point describes a circle...The conditions required for describing an ellipsis *[now called ellipse]* are that a point shall describe a curve about another point whilst that other point moves in a right line...It is upon this latter plan...that I some time since constructed an apparatus for describing elliptic arches & applied it to my second kind of perspective tracing Instrument...which I have found very efficient & usefull.[C85]

Ronalds labelled the first device he mentioned to Bayley as the 'improved geometric pen, for describing epicycloids'.[R10] The name 'geometric pen' had been used for some years to describe instruments where circular motion is employed to draw complex curves. The wheels are toothed to allow regular relative rotation without slippage while a pen attached to one of the rolling wheels describes the curve. A simple and familiar example is the drawing toy 'Spirograph' developed in the 1960s.

George Adams' book *Geometrical and Graphical Essays*,[B7] first published in 1791, credits the invention of the geometric pen to John Baptist Suardi. The device pictured in the book comprises a central wheel about which roll two additional wheels along a radius. There were several challenges in reducing the concept to practice. All wheels must be well off the paper to avoid clashing with the pen, and the motion needs to be induced in some way. In Adams' instrument, the pen aided in these roles: it provided support to the outer wheels and was guided by the hand to produce the motion, both of which could potentially adjust its trajectory. Overall, it was a rather rickety apparatus that could get out of adjustment easily.

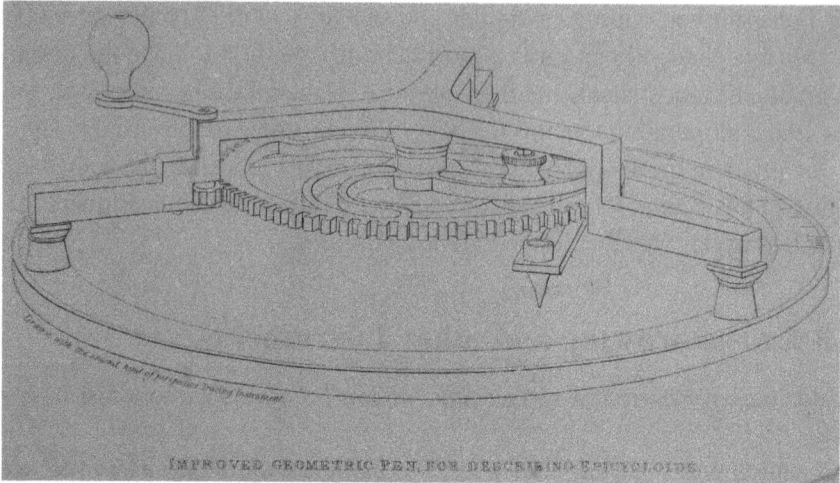

Fig. 10.11:   Geometric Pen for drawing epicycloids. Ref. R10.

Ronalds was not impressed with Adams' manifestation of the device — he hated poor design. He wrote to a colleague: 'On considering the figure & description in Adams's Book maturely one cannot avoid the impressions that it (<u>so</u> constructed) must be very unsteady and liable to error.'[C63] His solution was to provide a surrounding wheel, toothed on both sides, which supported the small wheels along a spoke and also drove the apparatus via a pinion (Fig. 10.11). Turning the handle rotated the outer wheel around a fixed central wheel, which then drove any number of inner wheels. The large wheel rested in a sturdy triangular frame and the whole device was held in place by three pointed supports on the base of an outermost ring. As an aside, he was more impressed with Adams' book *An Essay on Electricity*, of which he had six different editions in his library.

Ronalds' geometric pen could draw an almost infinite array of curves, not just epicycloids. Means of modification included adjusting the length of the arm holding the pencil, changing the number of inner wheels, or their diameter and number of teeth, or swapping them around, or attaching the pencil holder to another wheel. The outermost ring was graduated to facilitate precise positioning of the pencil when drawing a sequence of similar curves in close proximity.

He used the same concept in designing a turner's pen to engrave curved patterns on wood. Drawings of this apparatus that help to clarify its mode of operation are given in Sec. 12.7 in the discussion of his turning inventions.

The device for creating ellipses that he described to Bayley was probably based on his fifth type of the first kind of tracing instrument for drawing straight lines (Fig. 10.6). Another incomplete sketch survives of a circular frame that carries a toothed wheel meshing with a cogged rail (Fig. 10.12) — this might be an early version of the design. He has annotated it as 'by Hozapfell [*sic*] jun^r',

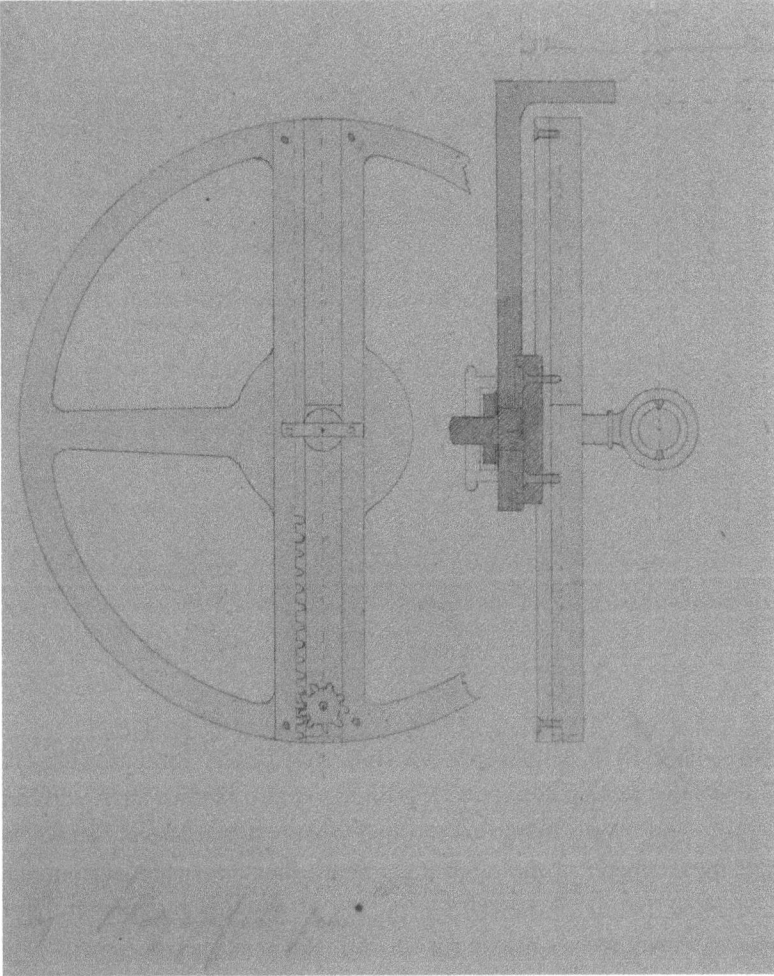

Fig. 10.12: Holtzapffel's sketch of a device, possibly to draw ellipses. IET 1.7.22.

suggesting that he and toolmaker Charles Holtzapffel* were working together on such instruments. Holtzapffel was also the probable recipient of his comments on the geometric pen in Adams' book. Various curves produced by the devices have also been retained in the Ronalds Archive (Fig. 10.13).

Ronalds conceived other, simpler, drawing instruments in this period as well. In 1825 he documented a modification of the compasses to enable the drawing of a volute (spiral) rather than a circle. The compasses turned about a grooved cone spiked into the paper from which a string unwound. The selected taper of the cone determined the tightness of the spiral. The legs were also separated by a spring, or an 'arc with rack work if you chuse'.[J14] An instrument of this type was patented by a Mr H. Johnson and brought into manufacture more than 30 years after Ronalds' concept.

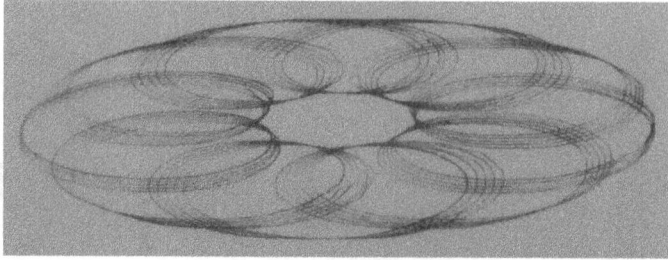

Fig. 10.13:   Epicycloidal curves drawn in perspective. IET 1.7.94.

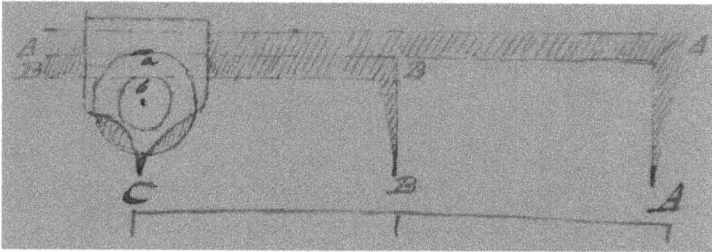

Fig. 10.14:   Instrument to bisect a line. IET 1.6.136.

In November 1829, he described a Bisecting Beam Compass (Fig. 10.14) to divide a line into two equal halves. It comprised two beams, each with a point at its end. They were engaged by two toothed wheels that rotated together, with one being half the diameter of the other. One beam thus advanced twice as far as the other. The same month he sketched a 'pocket'[J41] drawing instrument that through its compactness would be suitable for travelling. The basis was a sector with a pair of hinged rulers, but additions enabled it to also be deployed as a parallel ruler, set square, protractor or compasses.[J14] Finally, he outlined simple methods for creating distorted images by anamorphosis, noting that 'the effect would perhaps be rather amusing'.[J40]

10.6 The Business: Using, Documenting, Manufacturing and Selling the Drawing Instruments

As with most of his inventions, Ronalds developed the various sketching devices for his own use, and he continued to employ them until his last years. His original motivation as we have seen was to be able to draw scenes accurately and relatively quickly when he was travelling. Soon he started to use the instruments in making technical drawings of his inventions. The most valuable example of his perspective art is his archaeological record of the megaliths at Carnac in France, which is the subject of the next chapter. He also put considerable effort into portraits. Along the way he found that the machines could be a source of entertainment among family and friends — they had broader appeal than his other mechanical, electrical and meteorological inventions, which were often of interest only to fellow scientists (Figs. 10.15–10.17).

He was invited to demonstrate the first kind of machine in action in the library of the Royal Institution at its weekly evening meeting on 28 March 1828.[A32] Family friend, Mary Nichols*, documented a probably typical instance of the instrument's use in a social context. Her diary entry for 12 March 1834, when she was 20, tells us:

> Went to my drawing lesson. Mamma Emma and I called on Mrs. Ronalds. Mr. F. Ronalds shewed us his drawing instrument. Mamma and I had our portraits taken. Mr. Ronalds and the three Misses Martineau *[Ronalds' nieces aged five to twelve]* then went with us to the Horticultural Gardens *[at Chiswick].*[J33]

Ten weeks later the Ronaldses called at Mary's home: 'Mr Ronalds shewed me how to use his Perspective Machine.'[J33] Mary became an accomplished artist in adulthood. Perspective drawings made by other friends and relatives are held in the Ronalds Archive. One is by obstetrician and amateur artist Dr Robert Batty and another, of mushrooms, is by Sarah Fitton who wrote several books on botany. Both of these were drawn at Hastings.

Fig. 10.15:   Ronalds' self-portrait drawn with the first kind of perspective tracing instrument, using a mirror (*c*.1830). IET 1.7.79.

Fig. 10.16:   Bust drawn at Hastings with perspective tracing instrument; Ronalds has noted that the shading was done using the instrument. IET 1.7.37.

Ronalds applied for a patent soon after the instruments were developed. The specification covered the basic forms of both the first and second kind of instruments; none of the accessories were itemised, although an accompanying drawing showed the upper portion of the tripod legs.[P3] The patent was sealed on 23 March 1825 and enrolled two months later. It was the only patent he filed over his career, although he did explore the idea for other inventions in 1828 and 1837. It was also the first instance where he took on a leading role in the commercialisation of an invention. He would have been encouraged on both counts by family and friends who could see the potential popularity of his instruments. His brother Charles appears to have started to provide business and legal support to him at this time and it was he who witnessed Ronalds' signature on the patent.

In 1828, Ronalds printed a book that documented a number of the drawing devices. It also gave the theory behind the development of the machines, directions for assembling and maintaining them, and many practical hints to facilitate their use in different circumstances. The book had the snappy title: *Mechanical*

Fig. 10.17:   Perspective of Temple of Neptune at Paestum drawn from ground plan and elevation using the tracing instrument of the second kind (*c.*1828); Ronalds had visited Paestum in February 1819. IET 1.7.80.

Perspective; or, Description and Uses of an Instrument for Sketching from Nature, Accurately and Conveniently, Every Kind of Object: and of a Machine for Drawing in Perspective Architectural and other Subjects, from Measurement, or from Ground Plans and Elevations. In two Parts.[R10]

Interestingly, the first perspective drawings of the devices were made and coloured by 'I Thompson 1 Wellington Street Strand'.[J35] Ronalds soon gained confidence to publish his own technical drawings created using his instruments, and it is these that are included in *Mechanical Perspective*. Later copies of the book also contain additional perspective drawings of scenes and objects to demonstrate what could be achieved using the instruments (Fig. 10.18).

He donated copies of the book to various institutions at different times (Table 10.1); these included the Royal Society, the National Art Library at the Victoria and Albert Museum, and the Society for the Encouragement of Arts, Manufactures, and Commerce (now known as the RSA), of which he was a member as early as 1815.[A16] It is also held in the Ronalds Library at the IET. The Wheeler Gift Collection held Latimer Clark's* copy but it is now lost. Others were purchased for about four shillings by people who bought his instruments.

Fig. 10.18:   Croydon Palace, adjacent to the family home (1827); drawn and engraved by Ronalds. Ref. R10.

Ronalds provided considerable information on the manufacture and sale of the instruments in a letter to Sabine in late 1848, prompted by him seeing one of the tripods in Sabine's office. In it he noted:

> I had a man at Work (at Croydon) for several years, to make these tripods and the Perspective Instruments &c and contrived machinery to save labour, which it did but it saved no money, the cost took the gold from the gingerbread. Cary had the chief profit on the Instruments and Holzapffel *[sic]* on the Machinery.[C239]

A detailed equipment list survives in Ronalds' hand in which he itemised each step in the production of the first kind of instrument along with the tools needed. Manufacture would have taken place at the premises he leased in Upper Church Street (see Sec. 3.7).

He included a pricelist in *Mechanical Perspective* (Table 10.2) and advised where they could be purchased:

Bate's, 20–21 Poultry
Cary's, 181 Strand
Dollond's, 59 St Paul's Church Yard
Holtzapffel's, 64 Charing Cross
Jones', 62 Charing Cross
Newman's, 122 Regent Street
Reeves and Sons', Cheapside

Table 10.1:   Provenance of Copies of *Mechanical Perspective* in Public Collections.

No.	Current Location		Provenance	Scenic Views Included?
	Institution	City		
1	British Library	London	Legal Deposit?	No
2	Science Museum Library	Swindon	—	No
3	Library of Congress	Washington	Copyright depository 1849	No
4	British Library	London	From 'Patent Office Library'	Yes
5	IET	London	Ronalds Library	Yes
6	National Art Library	London	'With Mr Ronalds' Compliments' 1871	Yes
7	Royal Society	London	'Ex dono Auctoris' 1831	Yes
8	Royal Institute of British Architects	London	From RSA	Yes
9	Whipple Library	Cambridge	Presented by David Dewhirst, Professor of Astronomy	Yes
10	Yale University	New Haven	First acquired by John Darlington, London, 1847	Yes

Table 10.2:   Instrument Price List.

	£	s	d
Perspective Tracing Instrument — First kind: from nature			
Type 1 — pocket form	3	15	0
Type 2 — with mahogany box cum drawing board	3	17	0
Type 3 — extended for larger drawings	3	17	0
Type 4 — with stable mahogany drawing table	5	10	0
Perspective Tracing Instrument — Second kind: from plans and elevations			
Smallest size	10	0	0
Larger sizes	15–20		
Attachments			
Device to draw circles and ellipses in perspective	10	0	0
Tripod stand	2	10	0
Drawing board to fit stand	0	6	0

Source: Ref. R10.

Stockists received a 20 per cent discount and generally took the financial risk of the subsequent sale.[C69] Reeves is still an artists' supplies brand today, while Dollond & Aitchison remained a high street store for prescription spectacles until recent times. Mary Nichols documented a dinner party at her parents' home where the guests included George Dollond FRS as well as Ronalds and his mother.[J33] Ronalds continued to deal with all of these firms with his subsequent inventions.

Advertising took several forms. Ronalds put notices in the *Literary Gazette* over several years while John Cary listed the items in his catalogue.[B38] Holtzapffel offered in 1834 to include the instruments in his catalogue[C69] but for some reason they did not appear.[B39] Holtzapffel did submit the instruments to an exhibition in London, where they could be demonstrated and purchased. Perhaps the best advertisement was the drawings Ronalds created with the devices.

The instruments won both expert and public approval. The 1826 *London Journal of Arts and Sciences* by William Newton was perhaps the first magazine to feature them; his description of the two kinds of instruments ran to three pages with two additional illustrations.[A27] In a summary of an annual display of inventions outlined in Chapter 12, the *British Magazine* headlined the second kind of drawing instrument:

> This most ingenious piece of apparatus especially demands the attention of the visitor...The neatness and accuracy with which the instrument fulfils its design is extraordinary.[A34]

Luke Hebert — civil engineer, mechanical draftsman, patent agent and inventor — devoted much of the description of 'Perspective Instruments' in his

MECHANICAL PERSPECTIVE.

—— Mr. RONALDS' Patent Perspective Tracing Instruments are now on Sale and View, at Mr. Cary's, Optician, No. 181, Strand.

This Perspectograph is wholly devoid of the Inconveniences and Difficulties attendant on the Use of the Optical and Mechanical Instruments which have been hitherto invented, and will enable any person unacquainted with the Laws of Perspective, to Sketch with great facility and accuracy from Nature. It constitutes, in fact, a species of natural Pantograph, which, with a little habit, will furnish the Operator with the means of copying with great delicacy, every Line and Feature of an Object, or Assemblage of Objects, however complicated.

Also his new and remarkably firm, yet light and portable Tripod Staff, for the support of Drawing-Boards, Plane-Tables, Theodolites, Telescopes, &c. &c.

Literary Gazette 23/10/1830 694a

Second edition.
MECHANICAL PERSPECTIVE; or, Descriptions of a Pocket Instrument for Sketching accurately and conveniently from Nature, and of a new remarkably firm and portable Tripod Staff, &c.

By F. RONALD, Esq.

Hunter, St. Paul's Churchyard.

N. B. The Instruments themselves are sold by Mr. Cary, 181 Strand, (where specimens of their action may be seen); by Mr. Dollond, St. Pauls Churchyard; by Messrs. Holzapffel, 64 Charing Cross; Mr. Newman, Regent Street, &c. &c.

Literary Gazette 31/8/1833 559a

1836 *Engineer's and Mechanic's Encyclopædia* to the first kind of machine, and included two of Ronalds' diagrams:

although it [the cameral lucida] must be admitted to be a very portable and beautiful instrument, the acquisition of the art of using it is extremely difficult to all, and to some persons impossible...

Many instruments have been contrived for finding the various perspective points, but the process...is extremely slow; even the most simple figure would require to have many points found in it before its outline can be produced...

Mr. F. Ronalds, of Croydon, has, however, contrived and patented an apparatus, by which *the lines themselves, of whatever form or arrangement they may be, may be drawn directly from the object* with the same facility as tracing them...

We have in this most ingenious instrument a simple and elegant adaptation of the foundation laws of the science of perspective; it may be called a teacher of perspective as well as a perspectograph.[A66]

His article was reprinted in the *Magazine of Science and School of Arts* in 1844.

Volume 2 of the *British Cyclopædia of the Arts and Sciences* (1838) described both kinds of instrument in two different articles and highlighted the value of the first for portraiture:

> We have seen that portraiture commenced with a simple outline, formed by a shadow *[silhouette]* that was afterwards filled in…A nearly similar but still more ingenious apparatus for portraiture has been contrived by Mr. Ronalds…any person possessing that simple apparatus may readily take portraits[A57,A58]

In the pre-photography era, a simple means of creating likenesses would have been very appealing.

His was certainly not the only sketching aid on the market. In addition to the camera lucida and camera obscura, another mechanical drawing instrument developed in France by Charles Gavard was patented in England in 1831. In describing it, Hebert noted that 'the instrument does not differ materially from that invented by Mr. Ronald, and already before the public.'[A41] The device did however include a curved attachment for sketching panoramas. A later example of the apparatus was awarded an Honourable Mention in the *Reports by the Juries* of the Great Exhibition in 1851.[B61]

Demand grew quickly for Ronalds' instruments. One of his first customers was John Taylor FRS, a Unitarian mining engineer and later Treasurer of the British Association for the Advancement of Science, who bought two of the instruments with tripods in June 1825.[C239] Sir Purcell Taylor, who might be a descendant, believed he came into the possession of Ronalds' original instrument in 1889.[M6] The first sales of the more versatile tripod with the triangular head were to Captain Henry Kater FRS (who had interests in surveying and precision measurement) and to the camera lucida inventor Wollaston's son. Both were sold by Cary within weeks of being available. Later, 'Old Troughton had one and was much pleased with it.'[C239] We meet telescope-maker Edward Troughton in Sec. 12.10.

Ronalds' best customer would have been Lord Stanhope. Stanhope bought his first tracing machine not long after they entered the market and another in 1839. He arranged with Ronalds to match it with a tripod with shorter legs. He preferred to draw while sitting and it was more convenient to carry, Stanhope noting: 'I intend to take them always with me when I travel.'[C87] He also acquired a 'pillory' for portraits. A month later he purchased further equipment for a friend.[C91]

Stanhope made many complimentary remarks about the instruments and was also enamoured with Ronalds' art:

> I was very much struck with the admirable perspective of the Prints annexed to your work, & with the Profile which you had drawn, & which he *[Cary]* assured

me is a perfect likeness…the instrument is so perfect in its principle & in its performance [C90]

If any illustration were requisite to shew the extreme utility & importance of this most ingenious & admirable Instrument, I might refer to the experience of a Gentleman who was at Naples when I was there, & who is not an Artist, but fond of scientific pursuits. He had been enabled by this Instrument to make a Copy which he shewed to me, & which was drawn with the utmost accuracy & neatness, of a Print representing the Pope going in Procession, & containing a great number of figures, &, as he told me, to copy even engraved Stones, & a hand writing with such perfect exactness that the writer supposed it to be his own. This Instrument supersedes altogether both the Camera Obscura, & the Camera Lucida, the latter of which I could never succeed in using [C88]

Through sheer luck, several pages of Ronalds' makeshift diary survive from this period — he kept the notes he had made on the back. The jottings are illuminating in regard to the extent of the instrument business and his personal involvement. In May 1833 he sent off instrument number 150 and tripod number 80, while instrument 179 and stand 84 were made.[J45] He dealt with most of his outlets — Cary, Holtzapffel, Newman, Dollond and Reeves. He later estimated that in the decade 1830–40 he sold 140 of his improved tripods.[C239] The drawing instruments also continued to be sold into the 1840s, and so perhaps in addition around 300 of these were made. Even in the 1850s, visitors to Kew Observatory often studied the instruments to understand the means by which Ronalds made such accurate perspective drawings of his complex meteorological and magnetic apparatus.

It appears that demand extended to the Continent: his pamphlet was translated into French[J37] and a sketch of one of his instruments has a French label. The French market had been informed as early as 1826 through the *Journal Hebdomadaire des Arts et Métiers*:

We have previously noticed several devices for mechanical drawing either of manmade items or from nature; but we believe what we describe today to be the most ingenious and the simplest of all those we have offered for perspective as well as for elevations, geometric plans, etc.; at least this is how it is judged in England.[A28]

For many years Ronalds took personal interest in each sale. He loaned equipment to help potential customers decide if they wished to buy and on occasion delivered purchases himself. He was very happy to make changes to any item to better suit the user. A record survives of the sale of instrument number 159, tripod number 78 and a panelled drawing board to Mr Simpson. Ronalds' letter accompanying the package extended to advice on how to become accustomed to the instrument before making any sketches: 'take the handle in the fingers, rest the hand upon the table as usual in writing and then try to write freely; this is easily accomplished and creates the necessary confidence.'[C68] At least one model surviving in the Science Museum retains this complete freedom of movement after nearly two centuries.

One of Ronalds' potential customers was John Isaac Hawkins, who had patented a mechanical pencil in 1822 and also the upright piano. Hawkins borrowed Ronalds' geometric pen in 1839: 'I should have sent home your Epicycloidal Pen long ago but I contemplated affixing an everpointed pencil to it. I have attempted it & failed, there not being room for it to pass under the wheels.'[C83] He ended the letter: 'with best respects to your excellent Sister *[Emily]*'.

Once he had found himself in the tracing instrument business, Ronalds would have adopted instinctively this hands-on approach to quality and customer satisfaction. He would also have enjoyed making ongoing improvements to his machines. On the other hand, he would soon have tired of promoting his inventions and providing sales assistance and yearned the freedom to turn his attention elsewhere. He hints at both of these feelings in a letter to Stanhope in late 1839:

> I have had great satisfaction in selecting, finishing & testing the best ins^t in my possession for you & sending it to Mr Cary's...
>
> the tedium of providing Mr Cary (& other shops) with these Ins^{ts} is no longer relieved by contriving little mechanisms for the Construction &c [C92]

Fortunately, a new challenge arrived at this time in an idea from Stanhope, as described below. From then on, other activities increasingly took his attention and the drawing instrument business slipped.

The devices continued to sell. As Ronalds later noted to Sabine regarding the tripod:

> How many were sold after the latter time *[1840]* I know not, but I found that it became partially imitated & sold by *[probably Thomas]* Robinson, long since, & more recently by others, and I have very lately seen one, at Mr Newman's, which came from Paris & was called the French tripod and every little particular of mine was imitated (à la Chinoise *[in the Chinese way]*).[C239]

As late as 1855, several of the tripods were exhibited at the follow-up to the Great Exhibition — the *Exposition Universelle* in Paris. Ronalds' atmospheric electricity apparatus rested on one, as discussed in Sec. 14.1. A detailed wooden model of another form of his atmospheric electricity apparatus complete with supporting tripod was also on show. Nearby was a display by Charles Piazzi Smyth FRS, Scotland's Astronomer Royal and Professor of Astronomy at the University of Edinburgh. Smyth made use of the tripods to support his portable telescopes. He published a short book on his exhibits, in which he wrote:

> ground...may be rough or smooth, hard or soft... the most effective form of support whereon to place the telescope...is without question an expanding tripod, with six

legs at top, joined to make three at the bottom; similar to that exhibited…and with it very great firmness is procured… it was arrived at…by Mr Ronalds…and great indeed is the advantage over the ordinary theodolite tripod. I had long used Mr Ronald's stands, wherever firmness combined with portability was required, but without knowing, until very recently, to whom observers were indebted for the improvement.[B68]

Smyth sent a draft of his words to Ronalds for his approval, which Ronalds toned down a little to give the version published. The incident illustrates how the tripod was by then employed as a standard accessary in astronomy and similar pastimes without thought of its origin — which is what Ronalds would have wanted. He appreciated widespread use of his inventions and cared little about personal recognition or reward for his efforts. His letter to Sabine quoted earlier was written simply because this close colleague also had no previous inkling that Ronalds had developed the tripod he used daily.

10.7 Reproduction of Drawings

Ronalds combined his artistic efforts with an interest in the reproduction of his work. He and his family (certainly his brother Alfred and cousin Betsey) made use of two media: lithography and copperplate engraving/etching.

Lithography was developed in Germany in the late 1790s by Alois Senefelder. It is a chemical process based on the mutual repellence of oil and water. The procedure begins with a limestone that absorbs both of these substances. A waxy crayon is used to draw the image onto the limestone surface, after which the plate is treated with a solution of gum arabic and nitric acid which adheres to the non-oily parts. When oily ink is applied, it is attracted only to the greasy drawing. Paper is then pressed onto the inked stone face to produce an impression. The print is of course a mirror image of the drawing, as it is with engraving, and skill is therefore required in executing both the subject itself and any text onto the plate back-to-front. The development of lithographic printing in England owes much to Charles Hullmandel, who spent time with Senefelder during his Continental travels. By 1819 he had a lithographic press in London and in early 1824 he published *The Art of Drawing on Stone*,[B27] which became an essential manual of the craft.

Engraving and etching are much older printing technologies than lithography. In both techniques, cavities, which later hold the printing ink, are scored into a metal plate. This ink is then transferred to damp paper by applying pressure. For etching, the plate is first covered with a protective material called etching ground. After thin segments of the ground have been scratched out to form the image, the plate is dipped in acid, which bites into the exposed areas to form the grooves. In engraving, the grooves are gouged directly into the plate using a sharp implement.

The earliest example of the family's involvement in the printing process is Fig. 4.4 in Chapter 4. It is the sole surviving specimen from Ronalds' efforts in 1821 to publish a book on his Grand Tour (Sec. 8.7) and endures only because he later used it as scrap paper. The print is annotated: 'F.R. del. Printed by C. Hullmandel'. He had therefore quickly embraced the new craft of lithography — he had perhaps seen it in Germany. He was also one of Hullmandel's early collaborators. Ronalds would have enjoyed honing his skill in transferring his sketches onto stone and would also have relished interacting with Hullmandel on the science of producing a good print. It is likely that both Alfred and Betsey experimented together with him in the new craft — Betsey later worked with Hullmandel in illustrating her father's book on apples.[B35,A198]

Before long, Ronalds required numerous copies of drawings of his perspective tracing instruments and their results for inclusion in his instructions for use.[R9] He noted in *Mechanical Perspective*:

> The figures and other sketches which accompany these "Descriptions" may be considered as tolerably accurate specimens of the action of both the Perspective Tracing Instruments; since they have been etched and engraved for the most part from black-lead pencil lines which were transferred by pressure from the paper of the original sketches to the etching grounds of the plates[R10]

He had already worked with renowned engraver Wilson Lowry FRS in creating the plates for *Descriptions of An Electrical Telegraph*. Unlike that book, the plates in *Mechanical Perspective* are not annotated as to artist and engraver. It was Ronalds' nature for him to acquire this additional skill himself. He recorded around this time that he contrived an engraving table,[J41] which was likely an improvement on that developed in the late 1700s by Abbé Joseph Longhi. As well as making and etching his perspective drawings (Figs. 10.19 and 10.20), he probably also printed at least some of them using a press he built or bought. There is some evidence that Alfred participated in this exciting new venture — Fig. 12.3 was drawn by Alfred at the family home with the aid of Ronalds' tracing instrument.

Ronalds became so interested in this mode of reproduction that he researched a book on engraving and printing. The introductory chapter survives, entitled 'Origin & Progress of Engraving in Copper & Wood'.[J42] After outlining early developments in France, Germany and Italy in printing playing cards and book illustrations, he defined three stages in the evolution of printing using metal plates. The first was exemplified by Maso Finiguerra (1426–64). He was a silversmith who made niello pieces, where the engraved lines are filled with black niello inlay to accentuate the pattern. Ronalds quoted the art historian Giorgio Vasari (1511–74) who claimed that Finiguerra took impressions on paper of some of his engraved

Fig. 10.19:   Ramsgate Harbour (1829); drawn and engraved by Ronalds. Ref. R10.

work using a roller. In the second stage of evolution, after an image was engraved onto a plate in reverse, it was fixed onto a board and paper pressed forcibly upon it. Blue tints were most commonly used. The final stage was signalled by the development of the press, the separation of the craft of printing from goldsmiths, and engravers beginning to sign their work. This is another example of Ronalds' wide interests that was not published.

Alfred went on to publish *The Fly-fisher's Entomology* in 1836. He noted in his preface, written in the third person, that 'he has been induced to paint both the natural and artificial fly from nature, to etch them with his own hand, and to colour, or superintend the colouring of each particular impression.'[B42] It is again not certain whether Alfred also printed the plates, and the illustrations are unattributed. Ronalds assisted Alfred with his book, later recording that he

Fig. 10.20:   Dover Castle (1829); drawn and engraved by Ronalds. Ref. R10.

'furnished…part of, & edited, a little work for my brother'.[C375] They would also have shared their drawing and printing equipment.

In 1848, Alfred migrated to Australia. He immediately established himself as a professional lithographer, engraver and copperplate printer, 'HAVING brought with him from England the complete apparatus for carrying on the above business'.[1] He left Ronalds in charge of producing the fourth edition of his book. Two letters from Ronalds' Kew Observatory colleague Charles Younghusband in March 1850 hint at his task:

> I see by the publication of the book on flyfishing that you have got rid of another piece of business which is always a relief[C281]

> I received this morning the book on Flyfishing for which I beg you to accept my very best thanks — I had not time to do more than see the plates were very beautifully executed.[C283]

It seems that Ronalds' responsibilities might have included printing the plates and supervising their hand-colouring.

Ronalds retained his artistic interests throughout his life, although he could spare little time for them once he had taken on his role at Kew. His drawing

---

[1] *Geelong Advertiser* 10/4/1849 4d.

reproduction experience proved valuable at the observatory, however, in creating permanent records of photographic traces from his meteorological instruments; their discussion forms part of Sec. 15.2. Lithography, engraving and printing thus evolved into important capabilities for him. He also encouraged and assisted his more artistic family members Alfred and Betsey, whose work has endured through reproduction to be still well known and highly revered today.

10.8 Large Drawings, Panoramas and Stage Sets

It was on 6 December 1839 that Stanhope wrote to Ronalds with an idea about his tracing machine:

> It has occurred to me that another very useful application of that invaluable Instrument would be what I may call its converse operation, by delineating correctly on a plane surface, such as a Wall, or in a circular room, as for a Panorama…any… image that might be desired. For this purpose it would only be requisite to trace with the pencil the drawing on the Board, & keeping the eye to the eye piece to direct an assistant to mark the different points which appeared at the edge of the Bead onto the wall.[C91]

Not unexpectedly, the idea set Ronalds' mind racing. He sent a detailed reply before Christmas including descriptions and sketches of two possible new instruments. He first thanked Stanhope for his 'most curious, ingenious & I believe perfectly original idea'.[C92] In fact it was not original: the concept had been illustrated by Lodovico Cigoli (1559–1613) in an unpublished treatise called *Prospettiva Practica*[B175] but that was unknown to them. Ronalds continued: 'After proving the advantages of employing this happy invention in the simple manner which you have mentioned I should perhaps be tempted to complicate & spoil the operation in trying to render it still more expeditious.'[C92]

Ronalds' complications were of two types and concerned both the shape and method of projection. Starting 'I wish to beg leave to remark',[C92] he reminded Stanhope that his current instruments were based on planar projection and would be inaccurate for a circular panorama. A panorama could be drawn with the correct cylindrical perspective by adapting a machine described by Egnatio Danti in his 1583 book *Due Regole della Prospettiva Practica*. In those times that device had often been used inexactly to find selected perspective points for drawings on a plane surface. The basic concept was described in *Mechanical Perspective* and involved sighting an object along a bar and at the same time marking its position on a cylindrical surface with a metal stylus or pen aligned along a parallel lower bar (Fig. 10.21). Ronalds called the instrument a 'Panoramagraph'.[J35]

Ronalds' overall scheme was to first trace the scene from nature to produce a small panoramic sketch on the cylinder in pencil. To facilitate the inverse

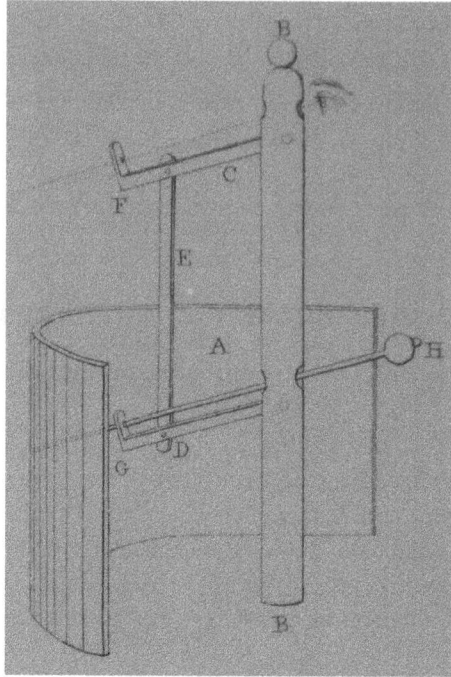

Fig. 10.21:   Schematic of device to draw panoramas. Ref. R10.

process the upper bar would be replaced with a modified form of magic lantern to shine a point of light on the wall. The small drawing would be retraced and simultaneously projected by the light for the artist to mark with his chalk 'as rapidly as he could follow it'.[C92] With this second innovation the concept was thoroughly original.

A magic lantern comprises a concave mirror and lenses to concentrate a light source and project the image of an object onto a wall — it is the precursor of the mid-20th century slide projector. In this particular application, the light would have simply been focused into a spot that could be seen in a darkened room. Ronalds intended to use limelight, an intense illumination created when an oxy-hydrogen flame is applied to quicklime (calcium oxide). The effect had been discovered in the 1820s and was first used as a form of stage lighting at Covent Garden Theatre in 1837. Ronalds had already talked to Cary, whose specialism was microscope building, about a suitable optical apparatus.

Large painted panoramas of remote places and events were a very popular form of public entertainment at that time — Crabb Robinson* always commented on the latest display. The first had been painted in Edinburgh in about 1790, and Robert Burford later presented his work in a large rotunda in London. Ronalds had also conferred with Burford about his ideas and the difficulties of converting a succession of planar drawings into a panorama before writing to Stanhope.

He had even more interest in another potential application he saw in Stanhope's idea — painting theatrical sets. Ronalds paid significant attention to stage scenery. On his Grand Tour it was mentioned several times in connection with his evenings at the opera or ballet; he wrote in Milan:

> The Scala did not fail to attract me regularly every night…The scene painter struck me as being the best performer…is not Scenic illusion a higher art than we generally take it for? If it had more encouragement our theatrical amusements would be greatly heightened…I visited the scene loft twice and was much amused with the manner & facility of producing the effects. I even envie[?] the man…his apparent content & self complacency while he walked about with his long brushes giving a dab here & a splash there, growing a tree or building a house as large as life in about 10 Minutes.[J22]

Ronalds could see that the inverse action of his instruments would render a picture for the audience exactly proportional to the original drawing even across an irregularly shaped stage at varying distances from the eye. As he wrote excitedly to Stanhope:

> even the theatrical construction of solid objects as Thrones, Steps, Balconies, Bridges &c &c might be ascertained with I presume an immense saving of labour and extreme correctness. Scheiner-like *[inventor of the pantograph]*. I already hear in the back of our opera pit the exclamation "Che bella cosa è la prospettiva Stanhopea" *[what a beautiful Stanhope view].*[C92]

Stanhope responded to the concept just after Christmas:

> I am convinced that it will by your ingenuity…become well adopted to its object…I intend to make a trial of copying on a dead wall a Portrait, for the defects in it would be more easily perceived than in a Landscape…I should be very curious to see the result [C93]

Ronalds continued to explore ideas. A scribble survives relating to an adaptation of the first kind of perspective tracing instrument: 'Lens with spot in tortoiseshell-frame in place of the bead. Light in place of the eye'.[J36] Moving the handle over the drawing on the board would thus trace the image on to the wall. On 12 November 1841 he developed a concept that applied the 'system of Inversion to drawing a very large perspective View from a Ground Plan and elevation'.[C92] This time he modified his second kind of perspective tracing instrument so that the lines traced from the drawing could be shone onto the wall via his magic lantern (Fig. 10.22). A further drawing (Fig. 10.23) depicts a high level rail along which runs a bogie holding a pen to aid in marking the picture on the wall.

Ronalds' autobiographical letter gives the following summary of his efforts with the 'inverse' apparatus:

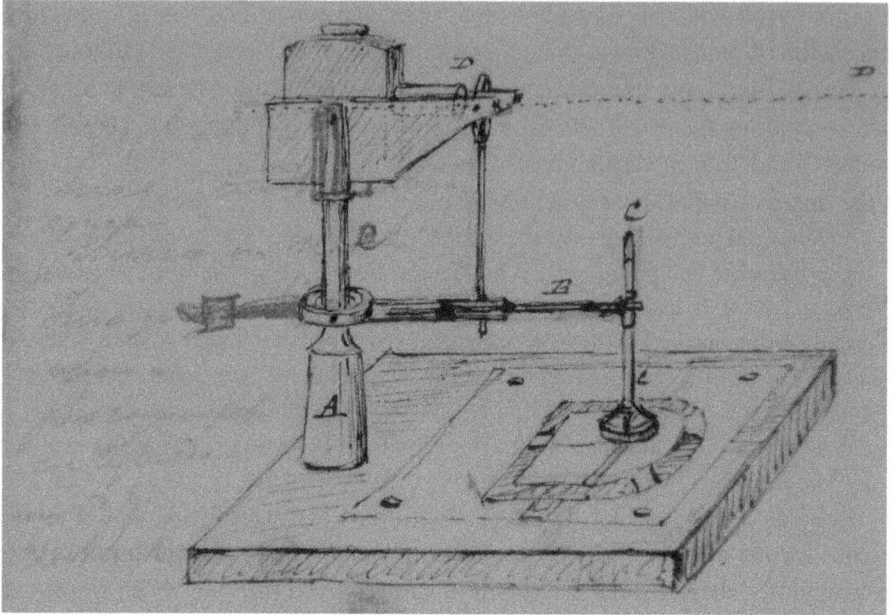

Fig. 10.22: Device to project a perspective drawing onto a wall, traced from ground plan (12/11/1841). IET 1.6.25.

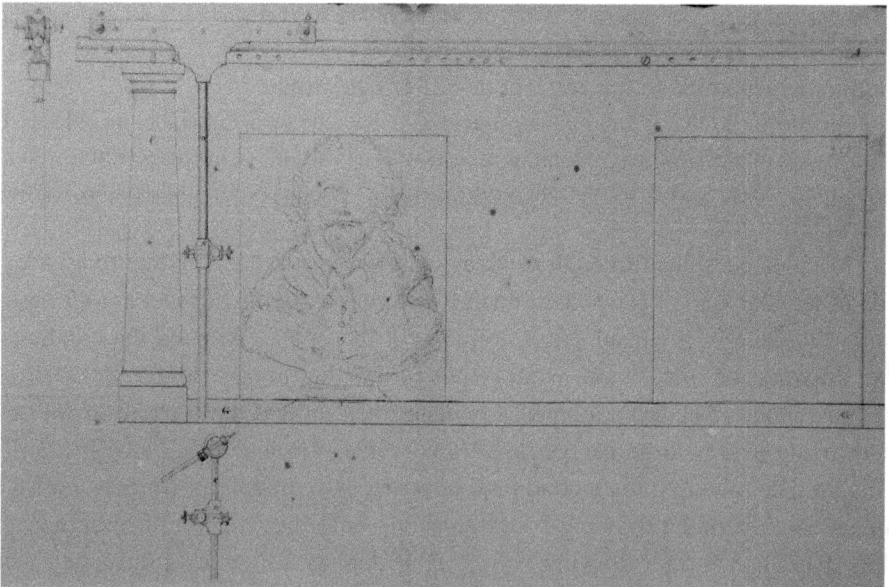

Fig. 10.23:   Mechanism to draw large pictures on walls. IET 1.6.116.

Influenced by suggestions of his Lordship, in the cource of a very kind & flattering correspondence, I constructed an Instrum^t for panoramic sketching & for employment in copying, upon Walls, & upon Theatrical Scenery, greatly magnified views executed, in piccolo *[small]* & dal vero *[real life]*, by its means, or otherwise. Part of it was like a <u>very</u> <u>old</u> Machine (false for <u>plane</u>-perspective) Invented by Ignatius Danti as a <u>plane</u> Perspectograph merely. It was in either the first or the second locality of the Polytechnic Institution. It is now (with much other such lumber) at the Pantechnicon.^C375

The Polytechnic Institution (described in Chapter 12) was an exhibition centre, while the Pantechnicon was built in 1830 in Motcomb Street, Knightsbridge, and comprised display, sales and storage facilities. He stored his possessions at the latter during his 1853–62 European travels. Further detail on what was displayed at the Polytechnic Institution does not survive, nor has evidence been found that the devices were adopted professionally.

Concerning the early tracing instruments, he summarised:

In 1824, et seq. Devised 2 Perspective Instruments One for drawing dal vero the other from plans & elevations, also a Staff for the support of the former...I was long occupied in London, Birmingham, Croydon & Chiswick in contriving mechanisms for, and in superintending the manufacture of these patented affairs. Many of the first kind, patronised by <u>you</u> *[Samuel Carter]*, by Lord Stanhope and by other good friends, were sold (By Cary &c) until photography opposed an obstacle; but the hexipod staff...has been, to this time, very extensively used for supporting many other kinds of instruments requiring great stability.^C375

These instruments serve to illustrate Ronalds' talent and success from several perspectives. First, they show that he could invent devices with popular appeal and go on to commercialise them, albeit not in a way that maximised personal profit. One of these devices — the tripod — remains ubiquitous today. Second, they provide a surviving example of his skills in mechanical design and manufacture. A third feature is his practice of applying his inventions to create additional value — in this case, in illustrating many of his later innovations, in honing his and his family's skills in the reproduction of drawings, and in recording the Neolithic monuments at Carnac. The significance of the Carnac work lies in the extreme precision that was possible with his instruments. In this sense they effectively bridged the gap between previous tools to assist perspective art and practical photography, and created an earlier beginning to the era of accurate visual recording. We turn now to how he created this legacy. In Chapter 15 we will see that he went on to embrace photography almost immediately after its invention for the purposes of recording data automatically, using as a starting point the inverse operation of his drawing instruments.

Chapter 11

ALEXANDER BLAIR
AND THE CARNAC MEGALITHS

Ronalds and Blair's archaeological trip to Brittany, the slow and partial dissemination of their work, and the eventual recognition of the value of Ronalds' drawings in providing almost photographic accuracy of the ruins at such an early time.

Around 1830 Ronalds took on another interest, turning his hand to archaeology. He travelled once again to France in pursuit of his new hobby in 1834, this time with his friend Dr Alexander Blair. Their target was the megalithic monuments near Carnac in Brittany. With Blair's nature being pivotal to how their work was (not) completed, and little being known of him, we start with a somewhat detailed sketch of his life and talents. The remainder of the chapter describes the pair's activities in Brittany and after they returned home, the manner in which their work was gradually discovered and how it was received. We finish by referring to some recent expert remarks on the enduring value of their achievements.

11.1 Dr Alexander Blair (1782–1878) and his Family and Friends

Alexander Blair (seen in Fig. 11.1) was a writer and became a good friend to the Ronalds family. He was Ronalds' only known co-author although, in the end, not a well chosen one: he was a highly original thinker but also disorganised, despondent and a ditherer.

Residing for many years in Birmingham, Blair made a lifelong friend in Samuel Carter, despite their nearly 23-year age difference. Ronalds would have

247

Fig. 11.1: Alexander Blair aged 90; oil painting by Hugh Carter (1873). © The Hunterian, University of Glasgow GLAHA 44150.

met Blair and Samuel when he spent time there in the 1820s concerning the manufacture of his drawing instruments; this was perhaps through Samuel's uncle Revd John Corrie FRS who was long-time President of the Birmingham Philosophical Institution and a Unitarian minister. Some years later, when Samuel married Ronalds' sister Maria in Chiswick, the witnesses were Ronalds, Blair and Corrie. Samuel and Maria named their first child Alexander.

In his younger years, Blair chose to stand in the shadow of another close friend and, to understand Blair, we need to meet John Wilson. He was a larger-than-life character — hugely intelligent, physically large and athletic, melodramatic, domineering, restless and a brilliant orator. A statue of him still stands in Princes Street Gardens, Edinburgh. The American author Edgar Allan Poe wrote of him that 'No man of his age has shown greater versatility of talent, and few, of any age, richer powers of imagination. His literary influence has far exceeded that of any Englishman who ever existed.'[A60] Part of Poe's accolade should be reserved for Blair.

Blair, however, was very different from Wilson. The kindest description of him is that given by Wilson's daughter Mary:

There was almost a timidity of character expressed in his bearing at first sight; but the wonderful intelligence of his countenance...dispelled that impression, and the real meaning was read in perceiving that modesty, not fear, conquered his spirit[B81]

Regarding their relationship, Mary wrote: 'The two men were mutually invaluable to each other.'[B81]

Certainly Wilson depended on Blair intellectually, professionally and emotionally throughout his prolific literary career, even when Blair's dispirited mood prevented effective support. Blair's, and especially Wilson's, many surviving letters show an intense relationship with commensurate ups and downs. A not untypical request for Blair's assistance reads:

Since I wrote to you much misery has been mine. It kills me Blair, & I am sick of this life. I cannot sleep, nor write, nor think, nor act. — Nothing is done...Your presence can alone keep me in life...I enclose five pounds — and hope to see you very soon...I will work like a slave before you. But till you come cannot hold a pen.[C56]

Blair was a perfectionist, always struggling to complete what he started. Frederic Hill, the son of his friend Thomas Wright Hill, illustrated this in his autobiography:

On one occasion, when we were desirous of assisting a friend in obtaining some medical appointment, we applied to Dr. Blair to compose our circular. He good-naturedly undertook the task, but his scholarly precision soon brought him to a standstill. He confessed he could not make up his mind whether at one part of the document he should use the word "that" or the word "which". As despatch was important, one of us was obliged to take the work from the doctor's hands.[B111]

Blair also confided to Wilson at the age of 58: 'I have begun a great many letters to you & ended some — but not sent — not liking[?] — them.'[C103]

Wilson was very demanding of his friend, in part to overcome this trait and stimulate his work. He also encouraged Blair to write freely and informally[C44] and then paraphrased these ideas in his own publications. Wilson's ongoing advice to his friend included: 'write. Do not stop when you think you write ill; but go on. Be plain & perspicuous at all times — and avoid all needless minute details.'[C37] Writing was both Blair's calling and his handicap.

But back to the beginning. Born with a twin brother Richard, Blair had grown up in London; his father, also Alexander, was an industrialist who later lost his fortune, and his mother Mary a society lady and author. He studied languages and philosophy at the University of Glasgow, where his classmates included Wilson.

After University, Blair assisted at his father's soap-works and probably taught at Hill's school. His and Wilson's circle soon also encompassed the Who's Who of the British literary world, including William Wordsworth, Samuel Taylor Coleridge, Sir Walter Scott, Thomas Carlyle, Poet Laureate Robert Southey and Thomas De Quincey.[B16] Blair spent significant time in the Lake District where several of these luminaries resided. Wordsworth also visited Blair in Birmingham, with his wife writing: 'remember me affectionately to Mr Blair'.[B162]

Wilson joined the editorial team of Blackwood's Edinburgh Magazine in 1817. Under his leadership, the magazine gained a strong reputation, with Crabb Robinson* noting it was 'certainly a miscellany of considerable ability & very amusing'.[1] The pressure was soon on Blair. One of Wilson's many requests to him reads:

> I sometime ago released you from the task of the Magazine…If you could…send me down a good large packet in a fortnight or so, you would be serving[?] me truly… One Translation of some strange & striking German thing — or Spanish — an original Essay — and some two or three lively things — or a good Review of some interesting work — or in short any thing…Surely this is not being very unreasonable.[C36]

Blair's sister Mary Margaret Busk also contributed numerous literary pieces to the Magazine.[A153] Whereas Blair and Busk published anonymously, Wilson (with Blair's ongoing help) penned many articles under the alter ego Christopher North, by which he became very well known.

Wilson was appointed Professor of Moral Philosophy at the University of Edinburgh In 1820. He immediately recruited Blair's assistance in preparing the course material through an extensive series of letters over several years:

> I would fain hope that your useful and enabling letters do not interfere too much with your own pursuits, whatever these may be…I wish you to send me two or three letters, if possible, on that division of the passions regarding religion…With respect to metaphysics, do not fear on any subject to write, provided <u>a conclusion is arrived at</u>.[B81]

Blair did find time to publish several books in his own name, including *Graphic Illustrations of Warwickshire*, printed in eight parts in 1823–29 and featuring drawings by esteemed artists.

In late 1827, a Chair in English was advertised at UCL — the first at a University in England. Blair applied for the position, with the University of Glasgow having just awarded him an honorary Doctor of Laws.[B116] Wilson prophesied: 'I fear it would make you unhappy.'[C53] Blair was appointed Professor of English and Rhetoric in 1830. He returned Wilson's earlier compliment by sharing with him his thoughts on his new teaching duties:

> I shall lecture upon a Textbook…My general plan…will be 1st to state shortly the scope & points of the Lecture: then to read it <u>word for word</u>, observing upon it, as

---

[1]Ref. J13 14/2/1819.

I go along…Every Student must have the book, & <u>follow me</u>…I am very much puz-
zled what I had better do about examinations, & exercises. I am half inclined to shun
them this Session.[C65]

With his retiring nature, Blair would not have relished his oratorical duties as
Wilson did. He also queried the University Warden: 'Will the Council think it an
unreasonable request…to be spared the ceremony of a formal Introductory lecture
[attended by Council members, other Professors and the public]'.[C64] His request
was accepted. Blair remained at UCL for six years.

After travelling to Carnac with Ronalds in 1834, Blair struggled increasingly
with depression. His University role would have been a primary trigger, much as
Wilson had foretold. He was able to keep excellent physical health and continued
to enjoy long visits with Wilson and the Carters; Samuel and Maria's son Hugh
Carter later painted his portrait. He lived to the age of 95.

11.2 The Trip to Carnac

Ronalds' and Blair's initial aim in travelling to Carnac was to help document the
standing stones in the area. As they noted in their book *Sketches at Carnac
(Brittany) in 1834; or, Notes concerning the Present State of some Reputed Celtic
Antiquities in that and the Adjoining Communes*:

> The undoubted high antiquity, immense number, and reputed Celtic origin of the
> monuments…have rendered them objects of great interest to antiquaries. Their inti-
> mate relationship with some of our oldest, most celebrated, and much discussed
> remains entitles them, we think, to a greater degree of *general* interest than they
> appear to possess in this country *[Britain]*.[R11]

Their book was written in the hope of encouraging and facilitating further travellers to explore the ancient and extensive remains.

For two intellectuals who loved travelling and learning, a trip to Carnac would
have been an appealing idea. For Blair, the original motivator was probably
aligned with his expertise in languages. A pencil scribble on the inside cover of his
journal reads:

> The Celts! — & who were they?
> — Why the Celts you know were the Aboriginal inhabitants of all Western Europe
> — that is to say, provided only that there were no others before them; — who to this
> day speak by their lineal representations; Welsh in Wales, Breton in Brittany, Gallic
> or Erse in the Highlands of Scotland, & if Gen Wellesley*[?] [Duke of Wellington
> — Irish-born military leader and British prime minster]* will allow them, wild Irish
> in Ireland[J47]

Ronalds on the other hand would have enjoyed the combination of practical and
theoretical scientific endeavour and his tracing instrument was ideal for recording
accurately the state of the monuments.

In preparation for the trip they studied the works of esteemed antiquaries. Three in particular are referred to in their book. The first had recently been printed by Ronalds' friend John Nichols* and he borrowed it a week before their departure.[C71] The author, Revd John Bathurst Deane, had visited Carnac in 1831 with a French expert and again in 1832 with professional land surveyor Murray Vicars. The *Gentleman's Magazine*, also published by Nichols, advised their work to be 'altogether of the highest and most interesting antiquarian character.'[A43] Crabb Robinson agreed:

> Continued this morn[g] read[g]...Rev[d] M[r] Deane. He writes to prove that <u>Carnac</u> in Brittany has one of these huge Temples intended to represent the Serpent... The research is far out of the reach of my ignorance yet the investigation much interests me.[2]

Deane was later one of the founders of the Royal Archaeological Institute of Great Britain and Ireland.

In France Abbé J Mahé, Canon of the Cathedral of Vannes, produced an essay on the monuments in 1825, after which the *Société polymathique du Morbihan* was formed in Vannes. Twenty years earlier Jacques de Cambry had published *Monumens Celtiques, ou Recherches sur le Culte des Pierres*. He was the founder of the Celtic Academy.

On their journey, Ronalds and Blair crossed the channel to either Dieppe or Le Havre in the second week of September 1834 and made their way down to Brittany. Their trip home is better documented. At Châteaugiron they hired horses for three days to visit a 'stupendous Dolmen'[R11] called *la Roche aux Fées* (the Fairies' Stones). They then travelled via Mayenne, Domfront, Condé, Caen and Château d'Harcourt to Honfleur on stagecoaches, where they caught a steamboat to Le Havre and Dieppe. A further stagecoach trip via Abbeville, Montreuil and Boulogne took them to Calais and the ferry. The route is shown in Fig. 8.1. This extended journey suggests that Blair was perhaps a less accomplished seaman than Ronalds and wished to minimise the time on the open water on their return. They probably arrived back in London on 22 November 1834.

In line with their goal of increasing the ease and enjoyment of seeing the megaliths, they provided a recommended travel itinerary in *Sketches at Carnac* based on their experiences. For example:

> From Rennes, a pleasant excursion of a few leagues (occupying one long, or two short days, should be made through Château-girond, Piré, &c., to the great ROCHE AUX FÉES OF ESSÉ *[Fig. 11.2]*...
> The rock F, which is at about the centre of the roof...is six feet thick, and its weight must far exceed that of any *supported* stone at Stonehenge or Carnac.[R11]

---

[2] Ref. J13 14/2/1835.

Fig. 11.2:   (a) Lithograph of La Roche aux Fées Dolmen; original drawn Nov 1834. Ref. R11.

Fig. 11.2:   (b) La Roche aux Fées Dolmen in 2012, situated SE of Rennes via Essé.

Comparison of photo with the sketch gives a sense of the accuracy of Ronalds' perspective tracing instrument.

Their advice to the traveller continued:

Arrived at his inn, in the little Bourg of Carnac...he must be content to find bed-rooms scarce, but not beds, — or fleas. Excellent fish (red mullet, &c.), poultry, flesh occasionally, a little game, with bread of seigle *[rye]* or wheat, eggs, chestnuts, admirable potatoes (8 inch girth by 15), fair Bourdeaux *[sic]* wine, milk, coffee, roasted apples, capital pears and grapes, will constitute his *hard* fare. His purse holds out well. "Money", said our landlady, — the fat, hospitable Veuve *[widow]* Gildas, — "appears once a year; you turn your head and it is gone." Horses may be had at a reasonable hire. The only carriage at Carnac, M. Michel's, serves all the purposes of all the people, not excepting those of the Juge de Paix *[Justice of the Peace]*, and some of the visitor's purposes too, if his bones are not very tender.[R11]

We dare not recommend a sojourn, of even one night, at Loc-Maria-Ker, — the abode, at the time of our visit, of cholera, fever, death, and silent dismay[R11]

They had based themselves at Carnac for a five-week period commencing on 25 September. For the first week they were accompanied by a Mr C.F. (almost certainly Ronalds' uncle Charles Field) before he returned to London via Paris. They travelled each day to explore the surrounding area, noting their detailed observations in a journal. On 10 October, for example:

we avoided the wind, in sitting down to lunch under the lee of Roch Crion on bread boiled chesnuts *[sic]* pears which we brought with us from our inn, & a bowl of new milk, bought at a neighbouring peasant's for 2 sols, w[ch] we corrupted with drops of brandy.[J47]

The journal is written in Blair's hand and style with numerous practical annotations by Ronalds: insertion of missing details such as directions, factual corrections and suggestions for improvement. One of the more facetious comments is to Blair's phrase 'After a dinner of eggs at "la Grande Auberge" of Plouharnel...',[J47] where Ronalds inserted 'It was not'.[J47] Ronalds' major contribution was in making the drawings and plans, as well as providing mineralogical knowledge for the study. He documented his compass bearings and dimensions in a separate book,[J48] which also survives in the Ronalds Archive, along with a number of his original pencil drawings of the ruins.

The tools for their task (Fig. 11.3) included a telescope, compass, 'Sir Howard Douglas's reflecting semicircle'[R11] (described in Sec. 12.9), '3 carefully marked measuring rods',[J47] measuring tape, string and rope, plumb line, lamps, Ronalds' perspective tracing instrument and his tripod stand. They employed Jean-Marie Kermorvin, the son of a local carpenter, for 13 Francs per week to assist them. On 11 October, they:

Proposed borrowing a chair in Keriaval to be used in drawing. Jean Marie doubted chairs existing at Keriaval[J47]

Fig. 11.3:   Lithograph of Obelisk at St Cado — Ronalds positioning tape for Blair to take reading; original drawn Oct 1834. Ref. R11. De Closmadeuc joked of this 'effet de lune *[mooning]*': 'Oh prudish lady, look away! it is shocking!'[A133]

They were, however, successful:

> The <u>chair</u> of yesterday has produced the <u>hare</u> of today…it proved to be a fine heavy
> one: the price 30 Sols. We bought it in compensation for the chair[J47]

The 'chasseur' *[hunter]* who provided it joined them for lunch where they were sketching and:

> tells us indeed that he served forcé *[obligatorily]* of course, under Bonaparte;
> as a Cannoneer: & in that capacity that it was frequently his painful duty to
> point his guns at English ships & English men; but that he spared himself the
> painful part of this duty by taking care so to point his guns as that they should
> never hit.[J47]

Ronalds and Blair described the megaliths in five categories and identified particular examples using the local name or after a nearby village. Arguably the most impressive visually today is the category they named 'Lines',[R11] noting them to comprise a series of nearly parallel rows of granite rocks (Fig. 11.4). Then, as

Fig. 11.4:   (a) Lithograph of Kerlescan Lines; original drawn 18/10/1834. Ref. R11.

Fig. 11.4:   (b) Kerlescan Alignment in 2012: looking east down an avenue about two-thirds of the way from the south edge.

today, these alignments extended over several kilometres, interspersed with 'Inclosures'[R11] *[sic]* (the second category) of various shapes and other gaps. Ronalds later wrote that 'Mystery, Wonder & Magnitude are probably the bases of the sensation which one experiences on a first visit to them.'[J49]

They found no enclosure 'in anything like a complete state'.[R11] For this reason they conducted a detailed survey of one — that at the head of the Kerlescan Lines — to record its appearance before further deterioration occurred (Fig. 11.5). The survey and sketches took six days, during which Blair noted:

> The old peasant, of whom we have spoken as our friend, living upon the corner of our field visited us to day. One of his <u>daughters</u>? was with him, a pretty little girl whom Wordsworth might have fancied.[J47]

Nearly 60 years later, surgeon and archaeologist Dr Gustave de Closmadeuc (1828–1918) wrote that their work here 'has great importance for archaeology'. By then additional standing stones had disappeared at this 'the most remarkable of the enclosures'. Not only were Ronalds' perspective drawings and highly detailed plan of great value, but 'Nothing to date has been written that approaches this description'[A133] of the monument in Blair and Ronalds' book.

The next category they documented included 'Scattered Stones and Obelisks'.[R11] One, the so-called Stone of Sacrifice (Fig. 11.6):

> is deeply, and indeed fearfully, channeled. Cavities... in the upper part...receive and just fit the shoulders of a person lying down on his back; his head meantime falling back over a little ridge...into another cavity, from whence channels, that may be supposed to have conducted the blood of the victim, lead to the lower part of the stone, where they seem to have united with a kind of trough...If the life-blood of a *human* offering placed in the above-described position ever gushed through this mouth, it must have presented a truly horrid spectacle[R11]

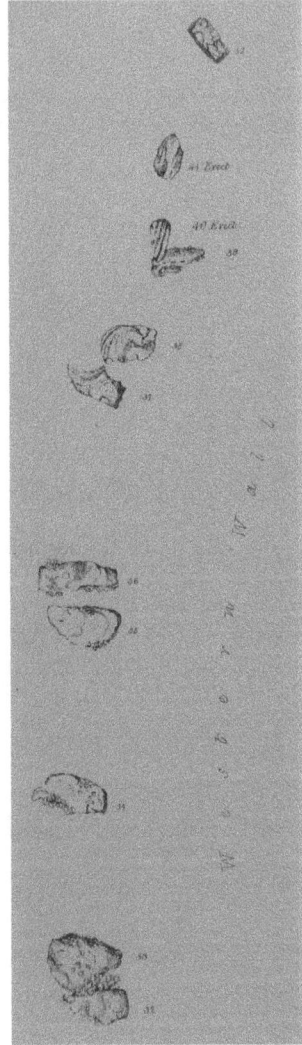

Fig. 11.5: Small section of plan of Kerlescan Enclosure; survey made Oct 1834. Ref. R11.

The Fourth Class was 'Tumuli',[R11] of which St Michel is the largest in the area (Fig. 11.7):

> Whatever may have been the original Celtic destination of this work, the zeal, or power, or both, with which the design was carried into execution, is magnificently attested to by the many thousand tons of stone which have been carried up the natural hill...to form even the present enormous pile. Yet it is said to have been originally much higher...

Fig. 11.6:   Lithograph of Stone of Sacrifice; original drawn 23/10/1834. Ref. R11.

Fig. 11.7: Lithograph of Tumulus St Michel near Carnac with M Michel's charrette *[cart]*; original drawn Oct 1834. Ref. R11.

We could not discover that it had ever been excavated to the base, and were much pressed to undertake the task, work being offered at a cheap rate. If it really has not been opened, the search might very probably be repaid by the discovery of Celts *[axe-like tools]*, coins, &c.[R11]

The first excavations of the tumulus were not commenced until 1862 by René Galles. The burial chamber was found to contain cremated human bones as well as stone chests, jewellery and celts.

The final category of monuments was that of 'Dolmens',[R11] of which there were innumerable (Figs. 11.8 and 11.9). Blair noted:

> The country actually teems dolmens. I exaggerate: but we have said to each other that we never go a new way to an old place without finding one: & that I believe was very nearly or literally true.[J47]

The Dolmens were in various states of preservation. Concerning that at Crucuno, for example: 'A farmer now uses it in summer for beating hemp and flax in, and in winter for storing fuel; therefore, perhaps, its utility may for a time protect its integrity.'[R11]

Ronalds and Blair must have quickly realised the quality of the work they were undertaking: they were astounded how their observations varied from those of the earlier experts. On their first daytrip from Carnac, they noted:

> Visited the parallels of le Maenac, and having, after some difficulty, satisfied ourselves that we had found the points from which M[r] Deane's two Views are taken, were surprised at their extreme inaccuracy. His account of dimensions as far as we proved them, we found little to be trusted.[J47]

> The most astonishing disproportion between the representation & the facts is to be found in M. Cambry's plate...As drawn the field appears to be one of stones from 25 to 30 ft high; they are not of half that height.[J47]

Three days later they acknowledged: 'our disappointment...in short our conviction of the utter untruth, intended or strangely unfortunate with which it *[the Lines]* has been delivered to the world'.[J47]

Blair subsequently made a sarcastic note in the privacy of their journal regarding a Dolmen near Locmariaquer now known as La Table des Marchands:

> It is a very fine one, described with admiration, and I may say drawn to ours by M[r] D. for he appeared to us to have left out half that was before him, & on the other hand to have succeeded in introducing as much that was behind him. We refer to our own two sketches for an exact representation. The back stone...is most remarkable...by the sculptures on its interior face w[ch] M[r] D has delineated in a manner wonderfully unlike the original. We flatter ourselves that we have been a little more successful.[J47]

They were more circumspect in their book. The artwork by Ronalds (1834), Deane (1832) and Dryden (1869) for this Dolmen may be compared in Figs. 11.10–11.12. Sir Henry Dryden (1818–99) is known for his excellent draughtsmanship and the beautiful illustrations of antiquities he made using a camera lucida.

The journal includes descriptions of local customs Blair and Ronalds witnessed, including a wedding celebration. They talked with local inhabitants as part of their research, with many of these conversations also being recorded. The locals

Fig. 11.8:   (a) Lithograph of Dolmen at Crucuno; original drawn 13/10/1834. Ref. R11.

Fig. 11.8:   (b) Dolmen at Crucuno in 2012: situated between Plouharnel and Erdeven away from the sea.

Fig. 11.9: (a) Original sketch of Dolmen at Kerroc'h (Oct 1834) perhaps including the proud house-owner. IET 1.7.58.

Fig. 11.9: (b) Dolmen at Kerroc'h in 2012: situated between Carnac and Plouharnel towards the sea.

Fig. 11.10:   Lithograph of Dolmen near Lochmariaquer; original drawn by Ronalds 15/10/1834.
Ref. R11.

were often eager to show them interesting stones. In addition to pinpointing and documenting the megaliths in their current condition, Ronalds and Blair also wished to probe their possible former state: there were widespread indications of removal of stones for use as building material or to clear fields for tilling, as well as natural decay. They garnered very little evidence, however, to support Deane's view that all the Lines were once connected to form a 'Serpent'[A46] at least 13 km long. They did hear traditions as to how the monuments were formed. The Lines were a Roman army who were turned to stone by Carnac's patron Saint-Corneille — or they might have been placed by monks. Dolmens were houses built by small but very strong men. The conclusion of Blair and Ronalds that these Dolmens were tombs (which was later verified) contrasted with Canon Mahé who believed them to be altars.

Fig. 11.11:   Dolmen near Lochmariaquer: drawn by Deane (1832). Ref. A47.

Although working seven days a week, they achieved significantly less than they hoped for during their trip. They were pressed for time: Blair had looming lecturing commitments at UCL and Ronalds would have wanted to spend time with his brother Hugh and his family who were visiting England from Illinois. It would have been with mixed feelings that they left Carnac on 31 October.

Fig. 11.12:   Dolmen near Lochmariaquer: drawn by Dryden (1869). Guernsey Museum and Art Gallery GMAG 7679.

11.3 Documenting and Disseminating the Findings

After returning home, they prepared their work for review and dissemination. Ronalds appeared to organise their fieldwork while Blair would have written the first draft of the majority of the text. Ronalds took possession of Blair's journal at some stage and, using it and his own notes, drafted the remaining sections. He would have also provided the momentum to finalise the book for printing.

Drawings selected for publication included Ronalds' plan of the monuments, which extended over an area of more than 40 km^2. He was characteristically modest in giving it the title:

> Map of Some Celtic Antiquities in the Neighbourhood of Carnac. From Mr Murray Vicars's Survey in 1832, in the Possession of the Revd John Bathurst Deane: to which are added a few objects surveyed in 1834 by Mr Fra^s Ronalds.[R11]

It is annotated 'C. Hullmandel's Lithography', which suggests that Hullmandel transcribed the drawing onto stone and then printed it (see Sec. 10.7). This was probably due to the large size of the plan — 67 cm × 35 cm. The numerous perspective tracings in the book are about 40 per cent of this size and have no annotation. It seems likely that they were drawn on stone and printed by Ronalds: he was certainly still following the literature on lithography and printing. (He may well also have printed the unattributed lithographs in his Uncle Hugh's book *Pyrus Malus Brentfordiensis* (1831).[B35,A198])

In Ronalds' mind his drawings of the Carnac megaliths were scientific rather than artistic in nature. He noted in the book that 'picturesque effect and embellishment have been little regarded. Correct outline has been sought, and, we confidently hope, obtained'.[R11] Nonetheless, they show his significant ability and growing confidence in drawing and reproduction — de Closmadeuc called him 'the consummate artist'.[A133]

Perhaps the first public airing of the work was described by Crabb Robinson on 6 May 1835:

> I went to the London University lecture on Celtic antiquities by Mr Blair — A more disappointing lecture I never heard. He explained 4 different drawings of Stonehenge *[which would have been made by Ronalds]* but I did not hear a single thought worth recollecting abo^t the age or purpose of these antient *[sic]* monuments. He s^d he had made a journey to Carnac but he told us nothing abo^t their resemblance to or diversity from Stonehenge or Abury *[Avebury]*, of which he also s^d nothing. I chatted with a number of acquaintance there…Home between 11 & 12. read abo^t Carnac in the Archaeology &c &c.[J13]

Robinson was clearly hoping that the lecture would compare monuments in different places and focus on their origins more than their current appearance. His comments would have done little to boost Blair's (and Ronalds') self-confidence.

Ronalds sent a draft of the book to Deane in January 1836, with the wish that he should 'criticize'[C73] it. Deane noted in his reply:

> Upon the whole I think your Notes very important; and if you can prevail on any Englishmen to visit the place, they will be invaluable…You ought to send several copies to France — there they will, I think, sell well.[C73]
>
> I have only made one or two trifling remarks[C73]

His criticisms concerned the extent to which his own work was represented in the text. A typical comment was:

> You have not mentioned in your preliminary remarks what my opinions really are. As I was the first investigator, it might perhaps prevent some Frenchman hereafter from robbing us all, if you were to record this fact: & say that I had decidedly pronounced my Verdict *[that the Lines were formerly a continuum used for Serpent worship]*.[C73]

The authors duly incorporated Deane's comments but in a way that indicated that his theories did not necessarily coincide with their own views.

Fifty copies of the book were printed in 1836 by family friend Richard Taylor, whose firm survives today as the Taylor & Francis Group. Copies were received by Oxford and Cambridge Universities and precursor institutions to the British Library and the National Library of Scotland. Expensive books were sometimes excluded from the legal deposit requirement, so this was perhaps the authors' generosity. At some stage the volume was apparently offered for sale at £1.5s, although no advertisements have been found. They also passed copies to family and friends: one went to electrician Charles Walker*. Deane advised that Sir Richard C. Hoare 'is certainly entitled to the Compliment of a Volume for he is the Father of Antiquaries'.[C73]

This rather lacklustre dissemination suggests that it was no longer a priority for the authors to publicise the Breton monuments through the medium of *Sketches at Carnac*. It was intended to be just their initial work: they described it as 'the unripened fruit of our visit'.[R11] A number of Ronalds' drawings had been omitted and they had configured the book without page or plate numbers so that subsequent efforts could be inserted readily. The plan now, however, was to eclipse it with a second book, more learned and broader in scope. As Blair noted in his research: 'The Monuments are effects: a philosophical spirit asks their causes.'[J52] Crabb Robinson would have been pleased.

Blair was to take the lead on the text. Several draft sections in his handwriting are retained in the Ronalds Archive. He explored the history of settlement and traditions of Celtic tribes in various places through ancient writings, using his expertise in philology and relationships between language and culture. He similarly traced the Germanic peoples; this would enable him to contrast monuments in different parts of Britain, as well as in Scandinavia where numerous antiquities were preserved. One of his anecdotes reads:

> the old Norwegians were so jealous of proceeding in all matters of conflicting interests according to the strict course of law, that they would not go upon a bear-hunt, without previously summoning the bear to appear in his own defence, & upon his non-appearance proclaiming him out-lawed: which sentence of course made it free for every man to take away his life.[J52]

Blair was pedantic, wishing his writing to be perfectly argued. As a result it often became ponderous and convoluted in attempting to address the many ambiguities and anomalies existing at that time; for example: 'to speak first of the Celts, we find ourselves compelled to admit (at least) the ~~eight~~ seven following variations from these demarcations *[where subsets of the 'race' had settled]*'.[J51] Pragmatic Ronalds kept bringing the work back to their central goal of helping to understand the monuments they had visited, described and sketched. In doing so he tried to insert structure and clear conclusions into the writing. He also prepared the drawings and maps for the book, including a map of Celtic and Germanic territories, probably in Caesar's time. Deane provided critical comments on part of Blair's work addressing serpent worship theory.

They had given themselves a very challenging task and progress was slow. It was not until modern times that the Carnac stones were dated to the Neolithic period, well prior to the settlement of Celtic tribes. The function of the alignments and their enclosures remains a mystery even today. Blair was additionally unwell through the period. Wilson wrote to him in 1836:

> I hear not of your long illness till I heard…that you were well again. God be pleased! and may you be long healthy & happy…I am glad that you are going to give up the Professorship.[C74]

Wilson's much-loved wife died the next year. On receiving the news, Blair started a reply to his friend on 24 April 1837 which indicates his frame of mind at this time:

> Such a startling communication forces me to reflect on the interruption, almost cessation of our once frequent correspondence…It forces me to reflect on the cause; which I must ascribe much to myself, who feeling myself perhaps born with some part of some kind to perform upon our mortal stage, desiring to find & execute it, & yet performing none, am burthened, & I believe <u>drawn into myself</u> in a continual undirected, or many-ways-directed, fruitless effort…I walk on, as if I were on my way to leave life, without having understood it. God's will be done…
>
> I have been trying to write a book or part of one w[h] I hardly can do…
>
> I shall have great comfort in hearing from you which I much desire.
>
> > Ever yours affectionately
> >
> > Alex Blair
>
> This letter should have regarded you more & myself less: but write.
>
> May 13. I am now with my brother…I had no conception that this had lain so long by me, unsent.[C75]

In 1840, Ronalds sent an abridged manuscript to the renowned Egyptologist Samuel Sharpe* for his comment and advice. He replied:

> I have been very much pleased with your book…Any thing I could say on it would only pass for compliments, but I would strongly urge upon you the publishing it just as it is, and let the larger mass of letter press *[written material]* be published separately in 800.
>
> Could you come so far as to dine with us on Tuesday next the 5th Aug at 5 oclock?[C98]

The dinner probably took place, but publication did not. Ronalds dreaded causing offence and certainly did not have the effrontery that Wilson showed repeatedly (and Blair probably needed) in adopting his incomplete work for his own purposes. It is a pity: publication would probably have provided a valuable record of the state of various ancient monuments in the UK in the first half of the 19th century, as was delivered for Carnac. Other evidence, discussed in Sec. 12.9, suggests that they might have been studying and documenting monuments for some years before they visited Carnac — it made eminent sense for them to start at home. If so, their portfolio of work must have been considerable.

Blair wrote to Ronalds from Fontainebleau France in August 1842:

That Book, which we began together, continually stands before me, as if it were the thing for which I live…

Whilst I write, I think that I am doing what is interesting & applicable: — when I have written, — I fall into despair:

…*[Samuel]* Carter told me you were half-inclined to ~~publish~~ take off a few ~~plates~~ ~~more~~ more impressions of the Plates, & give the Public what there is done.

I wish you would follow any wish you have about it…

I am vexed beyond measure, at having implicated you in an affair, in which I seem to be quite unable to render my part.[C105]

Wilson's daughter, in comparing the two men, later summed up Blair perfectly: the other, never satisfied, always doing and undoing, has unfortunately given but little to the world; and it is to be feared the grave will close over this remarkable man, leaving no other trace of his rare mind and delightful nature than that which friendship hallows in its breast.[B81]

Interestingly, both of the dates 1836 and 1843 are listed for *Sketches of Carnac* in Allibone's 1908 *Critical Dictionary of English Literature*.[B126] Was demand increasing? The evidence suggests the answer to be 'no' and that no second print run was made. Some were starting to become aware of the work however; Ronalds was advised in 1844: 'D^r *[William]* Buckland *[an eminent Oxford geo-historian]* wishes…to know something about your drawings of Carnac.'[C136]

Within a few months of receiving Blair's letter from France, Ronalds started work in earnest at the Kew Observatory. He later summarised the period in his autobiographical letter:

wasted much time & labour in a journey to Britany *[sic]*, and in contributing drawings plans & descriptive matter to a still unfinished (I believe) work on the Celtic Remains of Carnac [C375]

His new path did not extinguish his interest in prehistoric ruins. In the 1842–43 Christmas–New Year period, he visited various ancient earthworks and stones in the environs of Brighton. He kept neat records, noting places that 'should be accurately drawn and described'.[J59] A similar note survives from Leicester in 1837. As late as 1855 Charles Weld, who was a travel writer and artist as well as being assistant secretary of the Royal Society, wrote to him in Paris: 'Thank you for your kind proffer respecting Celtic remains — If you are within sight of that famous Dolmen near Saumur I should like a sketch of it from your pencil.'[C359]

With Ronalds' creative energy now focused elsewhere, however, any chance of the second book being completed dissipated. Even the authors had realised this before the end of decade, and began to distribute remaining copies of the original *Sketches at Carnac* to interested antiquarians (Table 11.1). Blair's friend Revd

Table 11.1:   Provenance of Copies of *Sketches at Carnac* in Public Collections.

No.	Current location		Details of original owner			
	Institution	City	Name	Given by	Relationship	Year
1	British Library	London	Legal Deposit?			1836?
2	National Library of Scotland	Edinburgh				
3	University of Oxford	Oxford				
4	University of Cambridge	Cambridge				
5	National Art Library	London	Charles Field	Ronalds	Ronalds' uncle	1836?
6	New York Public Library	New York				
7	Royal Society	London	'Ex dono Auctoris'			1844
8	University of St Thomas	St Paul	George Francis	Ronalds	Ronalds' acquaintance	1848
9	Bibliothèque nationale de France	Paris	Charles Lenormant	Ronalds	Ronalds' acquaintance	1848
10	Société polymathique du Morbihan	Vannes	Jane Carter	Blair	Ronalds' niece	1850
11	National Library of Wales	Aberystwyth	James Yates	Blair	Blair's friend	
12	Society of Antiquaries of London	London	Royal Archaeological Inst.	Blair	Via James Yates	1850
13	Metropolitan Museum of Art	New York	Soc. Antiquaries of London	Blair	Via James Yates	1870

Note: There is no copy in the Ronalds Library.

James Yates FRS should be credited for the book's emergence by encouraging them in this regard: he would have been one of the first to realise its true value. A Unitarian minister and scientist, Yates was also one of the founders of the British Association for the Advancement of Science (BAAS). Crabb Robinson described him as 'a man who perplexes with all his learning rather than enlightens'.[3]

Blair was invited to speak at the annual conference of the BAAS in Birmingham in September 1849. Ronalds would also be there, wearing his Kew Observatory hat. It was at this time that a flustered Blair scribbled this undated note to Ronalds:

Mr Banks has asked me for an abstract for the Athenæum by 7 this evening —
Shall I say —
"Dr B described some remarkable monuments visited on the South Coast of Brittany & upon the comparison of these with others in Scandinavia, to the known manners of the ancient inhabitants, claimed — "
I can make nothing of it.
Shall I send him nothing at all…
Come to discuss [C263]

The talk Blair delivered was entitled: 'On some remarkable Primitive Monuments existing at or near Carnac (Britanny [sic]); and on the Discrimination of Races by their local and fixed Monuments.'[S1] It was summarised in the conference proceedings and in the Athenæum. Blair noted that ancient stones in Scandinavia were either the graves of battles — sometimes arranged in the shape of a boat — or courts of law, and both reflected the people's culture. The Carnac megaliths in contrast appeared to serve a religious purpose reminiscent of Druidical society. His conclusion was given as: 'The importance of thus identifying the characters and monuments of nations, was urged in an ethnological view.'[S1] He also noted that their book was 'not yet published'.[S1] The summary prompted a note from Ronalds' friend Dr John Lee*:

I have read in the Athenæum the report of Dr Blair's Paper on the Ruins of Carnac — and I intend to bring it forward before our Syro-Egyptian Society in November [C270]

The story was finally starting to spread.

The next year Blair made a donation to the Royal Archaeological Institute, which had been founded six years earlier:

Dr Blair begs to present to the Archaeological Institute, at the request of their Member, Mr James Yates, a copy of his friend Mr Ronalds' & his Sketches at Carnac.
Dr Blair is sorry to perceive on one or two of the plates a slight stain contacted during the fourteen years, that the books have lain at the printers warehouse. [C284]

Also at Yates' request, Blair gave his last copy to the same Institute in 1870. Yates explained to members that the work was 'the only full and correct representation of these [megalithic] structures.'[A114]

---

[3]Ref. J13 4/6/1839.

Ronalds presented the book to the Royal Society on his election to that august body; it suggests that he believed it to have at least some scientific merit. In 1848 he gave a copy to George Grant Francis, a Welsh antiquary. The same year, he also made a presentation to the *Bibliothèque nationale de France* via Charles Lenormant — he was a curator at the library and a well-known archaeologist and historian. Two years later, in mid-1850, he was asked by Major-General Sir Charles Pasley FRS of the Royal Engineers for a copy, but by then he had run out.[J80]

In time, the antiquarian community began to make use of *Sketches at Carnac*. An early reference to it is in *Rude Stone Monuments in All Countries*, published in 1872 by Dr James Fergusson FRS, for which Ronalds' plans and several of his perspective tracings were redrawn. De Closmadeuc described how he first managed to see the now coveted book in 1891:

> I had long wanted to study this book. Finally, this year, one of my learned friends in England, after much research, was able to find a copy that had belonged to Mathew Moggridge...He posted it to me, allowing me to keep it as long as I wished[A133]

Moggridge was an associate of George Francis.

De Closmadeuc continued:

> The book seemed so curious and instructive that I translated it roughly in full. It must be said now, our national pride must suffer, as nothing so important has yet been published on the alignments in the Carnac area. This book is a masterpiece of accuracy and scientific acumen...many modern writers, who have in their turn described the Carnac monuments, would not have wasted their time, if they had the chance to get their hands on Blair and Ronalds' book.[A133]

His remarks form part of a 34 page article entitled 'Deux Archéologues anglais à Carnac en 1834'. It was published in the *Bulletin de la Société polymathique du Morbihan* in 1891 — unfortunately after both Ronalds and Blair had died. In addition to literal translation and paraphrasing of sections of *Sketches at Carnac*, de Closmadeuc comments in detail on each of its plates. He also provides context to illustrate the importance of the book, which centres primarily on its scrupulous precision — be it of the plans, drawings, dimensions or descriptions — in comparison with what had come before or since:

> Needless to say...the measurements have been taken accurately...
>
> To repeat, the best thing about the book's plates, regardless of the merit of their execution, is the perfect loyalty of the pencil, down to the smallest details.[A133]

Previous antiquaries had romanticised their depictions of ancient ruins. The novelty of accurate recording at such an early time has resulted in a fuller knowledge of the megaliths:

> Blair and Ronalds' book is all the more precious because it was composed at a time when the alignments..., although already mutilated, were however much more complete they are today.[A133]

Regarding the drawing and detailed description of the semicircle of stones at Crucuny, for example, de Closmadeuc notes: 'This monument is now totally destroyed.'[A133] It was restored in the 1920s by Zacharie le Rouzic, perhaps aided by Ronalds' drawing.

With de Closmadeuc's article, the secret was out in France, although the book itself would have been harder than ever to obtain. Still, it is very fortunate that *Sketches at Carnac* saw the light of day at all, with the authors (particularly Blair) feeling so little pride in it. Archaeologists today continue to hold a different view. Oxford Professor Andrew Sherratt, regarded as one of the most influential archaeologists of his generation, wrote in 2002:

> Even more impressive *[than Dryden's later watercolours of Carnac]*...are the achievements of the amateur archaeologist Sir Francis Ronalds...Ronalds' "sketches" and plans have long been admired for their antiquarian and topographic appeal (and their plan of the Carnac alignments...)... the mechanical drawing instrument which he patented, and was used in pioneer archaeological fieldwork in Brittany...fill*[s]* in an important chapter of archaeological recording before the use of photography.[A182]

Similarly, a 2009 French publication on the megaliths from the *Centre national de la recherche scientifique (CNRS)* notes:

> No less remarkable *[than earlier beautiful French lithographs]*, but for their quality and precision, are the drawings made in 1834 at Carnac by two Englishmen...The 24 drawings by Ronalds...have almost photographic accuracy. Both plans published in the book are also, by their precision, great advances over their predecessors.[B193]

Ronalds could never have dreamed that approaching two centuries after his hurried trip to Carnac, he would retain a reputation as a 'pioneer' 'amateur archaeologist'. Far from being 'wasted...time & labour', he and Blair generated unique visual, numerical and descriptive data of a very important set of ancient monuments. Doing so earned his tracing instrument a place in the history of visual recording and also demonstrated his skills as a surveyor. This chapter of his life is tinged with sadness, however, in that professional recognition and use of their introductory work came largely after their deaths. Furthermore, the 'ripened fruit' of the visit was never published and thus the many additional perspective drawings Ronalds must have made of ancient monuments in various places do not survive.

Chapter 12

SCIENCE EXHIBITIONS: A GLIMPSE INTO RONALDS' MECHANICAL INVENTIONS 1824–41

Ronalds' many (but little known) inventions of a mechanical nature, including those for which display in early science exhibitions or scribbles in his Ideas Book are the only surviving evidence.

The tracing instruments in Chapter 10 mark the beginning of a long line of inventions that are categorised here loosely as 'mechanical'. This was more than a discipline change from electricity. Ronalds had also chosen to leave the rather rarefied atmosphere of traditional scientific enquiry, written up in learned publications, for a more practical and potentially anonymous new world. Mechanical engineering was burgeoning in support of the steam engine and all it generated, plus myriad other facets of the manufacturing economy, and numerous inventors were developing new ideas and products.

For Ronalds, this period became effectively a very long lull in his career — not in terms of reduced creativity, but because his interest and ability to introduce his work to the public was at its lowest ebb. There is limited evidence of wide use for many of his concepts, and information on the devices themselves is scant. Of a number there is probably no record. Miscellaneous sketches, notes or letters concerning others survive in the Ronalds Archive. One very useful source is a notebook containing 100 pages of ideas that he jotted down up to December 1829. Some are recognisable as the embryos of later devices but others probably never left the Ideas Book.[J14]

What can be gleaned of his work at this time provides interesting insight into his hobbies, know-how and character. He mastered the lathe to the degree of drafting a book on turning, as well as making lathe accessories for friends and a modification to the slide rest that has been adopted widely up to the present day. With his surveying skills, he similarly improved tools of that trade. Dabbling in steam-driven transport (the big thing of the time), designing the mounts for large telescopes, conceiving an early typewriter cum letter stamp and helping his brother Alfred delineate a critical phenomenon in fly fishing illustrate his versatility. Other efforts, such as a suite of suggestions to help London combat fire, demonstrate remarkable vision as well as his public-spirited nature. Some endeavours are in contrast quite mundane and not necessarily new but simply aimed to entertain the children or assist his mother. Various concepts described in the Ideas Book under the heading 'Domestic Economy'[J14] helped reduce wastage around the home. His efforts sum up to a large portfolio — the reader may wish to peruse just some items.

Finally, the stories that can be pieced together of this phase of his life show that he was a well-regarded engineer sought out to solve challenging problems and to devise new instruments for particular purposes. One example is a letter from a Mr Gifford in 1837 enquiring how best to connect two wheels to allow variable eccentricity while retaining uniform rotation:

> I send you my problem preceding my apology for troubling you, and yet I think you will grant me pardon, when I disclaim any wish to take up much of your valuable time, having a strong impression that you will clear the way at a glance when I have been striking in the mud for days…I know well your <u>eccentric</u> abilities…
>
> If you could describe a curve in the direction of Gloucester Place…it would give me the greatest pleasure, I was not fortunate in finding you at home when I called at your house…
>
> My respectful Comp[ts] to M[rs] Ronalds and the young ladies.[C78]

The wording suggests that Gifford was a customer of Ronalds' turning devices or geometric pen (Sec. 10.5).

The best summary of Ronalds' activities in this period — although there are innumerable omissions — is that he provided in 1860:

> When the Polytechnic Institution *[actually the National Repository]* was first established…I was asked to contribute toward the object of it &…placed 6 or 7 little things there, as a new Fore-bed for Carriages:– a new Instrument for describing the Ellipsis & the Conchoids:– an addition to the Slide-rest used, (much now vide *[see]* Holzapfell) in turning curved surfaces:– a semi-transparent Sun-dial, shewing mean time, & the Perspective Instruments. I cannot remember whether 1 or 2 other articles were there or not. They were all put into the descriptive catalogue. I believe that a model of what I called a Teleographic-Clock was not there. It indicated the time by variations in the height of one of three large black balls visible at great distances <u>&c</u>
>
> When this institution was removed to its present site…sent there a self acting Fire-alarm (now at Kew) which I mention chiefly for the sake of venting a little spleen.

Verily we schemers are not always very honest. A <u>Gentleman</u> had another fire-alarm there, & mine happening to be the best, The ticket on it was removed from the one to the other. A similar kind of manoeuvre was executed at the great Paris exhibition of 1855.[C375]

As he intimates, this part of his career additionally coincided with the introduction of public displays of inventions and manufacture in London. One of the motivations for this new phenomenon was that people were becoming curious about science and engineering as industrialisation spread — they wished to understand more about the changes occurring to their lives and surroundings. Descriptions of these events provide additional information on several of Ronalds' devices. The exhibitions were also the means by which a number of his actual instruments have survived to the present time (Table 12.1). Given their importance to the story, we start this chapter with a very brief sketch of the evolution of scientific displays.

One of the first exhibitions of recent innovations had the full title: 'National Repository for the Exhibition of Specimens of New & Improved Productions of the Artisans and Manufacturers of the United Kingdom'. Its goal was to support trade by stimulating manufacturing improvements and commercialisation opportunities, which it would do through an annual exhibition showcasing the talents of inventors and artisans. King George IV became its Patron. Ronalds family friend Dr George Birkbeck was Chairman of the Committee that selected items for display. Ronalds also knew other Committee members: his uncle Thomas Gibson had responsibility for the important area of textiles and Charles Holtzapffel* covered tool manufacturing. The exhibition was set up in a long gallery at the King's Mews, just north of Charing Cross, and opened its doors in mid-1828. The entry fee was one shilling, but free to exhibitors, who were also able to sell their wares.

The catalogue for the 1828 exhibition is held at the RSA. It showcased the manufacture of both useful and decorative household items ranging from fabric, tableware, furniture and art to early refrigerators and washing machines. As well as full-scale equipment such as looms, it also had models of machinery, structures and anatomy. More unusual items included two patented portable water closets, a portable warm bath, guns in the shape of a walking stick or whip, and 'Belts for securing Lunatics'.[B32]

The concept, being new, attracted numerous critics.[A29] *La Belle Assemblée* opined: 'Seriously, this pompous effort is calculated only to disappoint natives, and to make us appear contemptible in the eyes of foreigners.'[A30] An article in the *British Magazine* was more positive, and ended by 'earnestly recommending our readers to join the throngs who daily visit this most interesting exhibition, where they may find, not merely much to interest, but much to instruct;…seeing and examining for one's self is the best method of acquiring knowledge'.[A34]

Table 12.1:   History of Display of Ronalds' Inventions.

No.	Object	Ronalds Alive				Ronalds Deceased	
		National Repository	Polytechnic Institution	Great Exhibition	Exposition Universelle	Special Loan Collection	Science Museum
Electrical Science and Engineering:							
1	Dry piles	—	—	—	—	—	1980–today
2	Singer's friction electrostatic generator for telegraph	—	—	—	—	1876	1876–today
3	Wooden model of telegraph dial	—	—	—	—	1876	1894–today
4	Buried insulated telegraph wire	—	—	—	—	1876	1894–today
Mechanical Devices:							
5–6	Perspective tracing instrument: from real life	1828–1830	1838–?	—	—	—	1930–1942
7	Perspective tracing instrument: from plans & elevations	1830	—	—	—	—	1894–today
8	Device to draw circles and ellipses in perspective	1830	—	—	—	—	1981–today
9	Horse-drawn carriage universal joint	1830	—	—	—	—	—
10	Hinged tripod stand — model 2 — triangular metal head	1831	—	—	1855	—	1867–1921
11	Hinged tripod stand — model 4 — solid triangular legs	—	—	—	1855	1876	1876–today
12	Window-mounted sundial	1831	—	—	—	—	—
13	Rigging or tackle block	1831	—	—	—	—	—
14	Curvilinear turning attachment for lathe slide rest	1834?	1838–?	1851	—	—	1920–today
15	Self-actuating fire alarm	—	1840–?	—	—	—	—
16	Equipment for large drawings, panoramas and stage sets	—	1840–?	—	—	—	—

Atmospheric Electricity and Meteorology:

No.	Item			1855	1876	Range
17	Insulated atmospheric electricity observing rod	—	—	1855	1876	1876–today
18–21	Improved Volta electrometers	—	—	1855	1876	1876–today; 1957–today; 1980–today
22–24	Improved gold-leaf electrometers	—	—	1855	1876	1876–today; 1980–today
25	Improved Henley electrometer	—	—	1855	1876	1876–today
26	Spark measurer	—	—	1855	1876	1876–?
27	Charge distinguisher — Leyden jar	—	—	—	—	1957–today
28	Registering or night electrometer	—	—	—	1876	1876–today
29	Round-house electrical observatory model	—	—	1855	—	—
30	Rain and vapour gauge	—	—	1855	1876	1876–today
31	Balance anemometer	—	—	—	1876	1876–today
32	Improved Regnault hygrometer and aspirator	—	—	1855	—	—
33	Saussure's hygrometer	—	—	—	1876	1876–today

Photo-registration:

No.	Item			1855	1876	Range
34	Temperature-compensated barograph	—	—	1855	1876	1876–today
35	Declination magnetograph	—	—	—	1876	1876–today
36	Horizontal force magnetograph	—	—	1855	1876	1876–today
37	Ordinate board — linear	—	—	1855	—	—
38	Barograms & thermograms — Oxford	—	—	1855	—	—
39	Barograms & magnetograms — Kew	—	—	1855	—	—
40	Daguerreotype copies using gelatine sheet — Kew	—	—	1855	—	—
41	Improved thermo/hygrograph (after Ronalds) — 1867	—	—	—	1876	1926–today
42	Improved barograph (after Ronalds) — 1867	—	—	—	1876	1926–today
43	Improved magnetograph (after Ronalds & Brooke)	—	—	—	—	1915–today

Dates denote years displayed or housed.

The gallery was demolished in 1832 to make way for the new National Gallery and Trafalgar Square. The National Repository continued briefly in Leicester Square but was never a great success. Evidence suggests that seven or eight of Ronalds' mechanical devices were exhibited there, as described below.

A more successful attempt, the Royal Polytechnic Institution, opened in 1838 in a purpose-designed building with entrances in Regent Street and Cavendish Square. Its great hall featured a large model canal. The Institution's purpose differed from the National Repository in wishing to entertain as well as educate the public. There was significant emphasis on demonstration, with steam-driven machinery in operation, experiments and lectures, and the opportunity to descend in a diving bell. As early as 1841, customers could have their photographic portrait taken in Richard Beard's studio. An article in *Punch* gives some sense of the vibrancy: 'People of weak nerves should venture very cautiously in the Polytechnic Institution. For, at first entrance, there is such a whirlwind of machinery in full action — wonderful things going up, and coming down, and turning round all at once'.[A65] It appears that at least four of Ronalds' mechanical inventions were included in the exhibits. In the early 1880s, the building was reborn as a technical education institute and the model for other Polytechnics around the UK. It was renamed the University of Westminster in 1992.

The concept of display went global with the 'Great Exhibition of the Works of Industry of all Nations', held in 1851 at the Crystal Palace in Hyde Park. It was spearheaded by members of the RSA under the Presidency of Queen Victoria's husband, Prince Albert. The Prince's organising principles for the event were exhibition, competition and encouragement. Displaying the wonders of industry and manufacturing from around the modern world would celebrate free trade and support ideas-sharing in a time of peace. Ronalds' first cousin Thomas Field Gibson was on the Royal Commission — he had followed his father into silk manufacturing and was one of a handful of Commissioners who were not titled and/or a parliamentarian.

Ronalds played a role in developing several technologies that were showcased at the Great Exhibition. Photography was highlighted in the *Reports by the Juries* as 'the most remarkable discovery of modern times',[B61] while the electric telegraph was 'now the science of the age'.[B61] His contributions in these two arenas and others were acknowledged, as discussed in other chapters. Ronalds himself was occupied with the Kew Observatory in 1851 and did not submit any items to the display.

The Great Exhibition quickly captured the public's attention, although his cousin Betsey painted a lightly different perspective in August that year: 'the Chrystal *[sic]* Palace is beginning to flag a little & am not sorry for I really believe that People will never settle down to work again.'[C326] Its popular and financial success spawned activity in several directions, including the first steps towards a permanent science museum and a pattern of World Fairs that continues today. Ronalds' work was represented at subsequent *Expositions Universelles* in Paris in 1855 and 1867: 16 of his items were on display in 1855 and his telegraph was commemorated at the latter event. He was not the only member of the family to have contributed to these major exhibitions. Alfred's daughter Maria Shanklin was awarded a Gold Medal for her display of artificial fishing flies in the Melbourne International Exhibition of 1880.[B165]

Land in South Kensington was purchased with the profits of the Great Exhibition and organisations like the RSA began to collect objects for future display. The South Kensington Museum was opened in 1857 at what is today the site of the Victoria and Albert Museum, but it focused increasingly on art rather than technology. To help convince government of the need for a home for the mechanical, physical and chemical sciences, a major international exhibition of scientific instruments was organised. Called the Special Loan Collection of Scientific Apparatus, it was assembled in 1876, three years after Ronalds' death, and included 18 of his items.[B94] He was mentioned in at least four of the lectures in the accompanying conference, where his meteorological and magnetic instruments were described as well as his telegraph.[B95] Through the efforts of a few people over many years, much of this Special Loan Collection was able to be retained and it formed the basis of the Science Museum when its doors finally opened in the late 1920s.

With later additions (and a few de-accessions) there appear currently to be 27 items pertaining to Ronalds in the Science Museum. In the 1980s, a significant number of these objects were on display. The showcase housing his electric telegraph apparatus was accompanied by a schematic of the overall set-up. His large barograph and horizontal force magnetograph (Chapter 15) were assembled, complete with sample results. Another glass cabinet commemorated him as a scientist and contained his portrait, drawings of his two kinds of perspective tracing instruments, his Ideas Book, a volume from his library, as well as memorabilia from his Grand Tour including his Constantinople passport, a tailor's receipt from Malta and his sketch of Catania. Since then the museum has changed dramatically with the move to interactive exhibits and a focus on a younger audience, and nearly all his items have been relegated to storage.

With that introduction, we will now visit the early exhibitions to view some of Ronalds' mechanical devices.

12.1 Perspective Tracing Instruments

Ronalds' contribution to the inaugural exhibition of the National Repository in 1828 was his perspective tracing instrument of the first kind, described in

Sec. 10.1. Not unexpectedly, given his diffidence, it was the retailer Holtzapffel rather than he who submitted it.[B32] Luke Hebert, a strong supporter of the exhibition, determined to write up items he saw there that 'may be deserving of particular notice' in his *Register of Arts and Journal of Patent Inventions*. The tracing instrument was one of the first he chose, and he devoted three pages and two diagrams to it.[A31]

The drawing machine was evidently a popular exhibit because it still featured in the third annual National Repository exhibition in 1830. By then it had been joined by several more of Ronalds' devices. One was the second kind of tracing instrument, which was the first invention Hebert described in his magazine that year, along with the tool to draw circles and ellipses in perspective.[A36]

The Polytechnic Institution also exhibited some of Ronalds' tracing instruments in its early years. A substantial drawing (60 cm by 43 cm) of a vase survives, which is annotated: 'by FR, with a large Ins^t, Drawn for the Polytechnic Institution in about 1837'.[J53] His 'inverse' machines for drawing magnified pictures on walls were also displayed there, as mentioned in Sec. 10.8.

12.2 Horse-drawn Carriage Universal Joint

The 1830 National Repository exhibition included an invention with the title 'Improved Four-Wheeled Carriage'.[A37] Hebert's two page article started:

> To construct a carriage that shall be less liable to overturn than the common sort, and that shall uniformly adapt itself to the most rugged and uneven surfaces, every body must admit to be an object of the greatest interest and value. A simple method effecting this object occurred to Mr. Ronalds…The banks, and other ordinary obstacles over which the carriage is represented as running *[in Fig. 12.1]*, are not so great in proportion as those which we caused the model to surmount on the table of the

Fig. 12.1:   Idealisation of carriage with universal joint. Ref. A37.

Repository. The ability of the carriage to run through ditches, and over banks and milestones, cannot be doubted, however unpleasant and unadvisable it may be to take rides of that kind. Yet the reader will allow, that, were he to attempt to run over a *mile*-stone in a common carriage, he would most likely have a *grave*-stone placed of him shortly afterwards.[A37]

It is apparent that Ronalds' working model was a hands-on exhibit and would have been quite entertaining.

In describing how the device worked, Hebert advised that 'an universal joint… allows of the axletrees of both the fore and hind-wheels to assume any angle with the ground, or rather horizon, without materially disturbing the vertical position of the carriage body that might be suspended above.'[A37] At that time, the fore-bed holding the front axle of the carriage was connected to the main body by a central kingpin or perch bolt to allow rotation of the axle in a horizontal plane when turning corners. The kingpin may be surrounded by horizontal circular framing called the fifth wheel to form a bearing surface for the upper body. In Ronalds' invention the kingpin was replaced by a universal joint that enabled the axle also to rotate in a vertical plane independently of the main carriage. The concept appears in the Ideas Book around April 1828 (Fig. 12.2), with the note that 'by these means a tight pleasant motion is obtained and the necessity of a loose rattling motion of two circles or fifth wheels held together by a loose catch pin avoided'.[J14] He later jotted down: 'Model made Dec[r] 1828 & shewn to M[r] Cha[s] Field &c'.[J14] He sought the views of respected family members and friends on his ideas, in part to gain reassurance that they were of value.

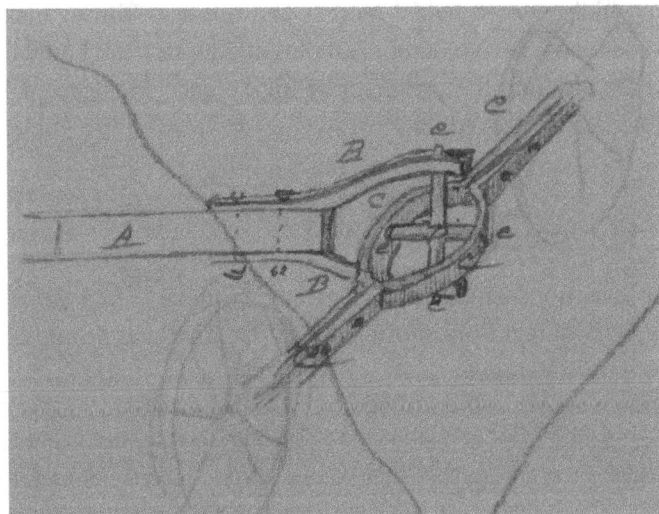

Fig. 12.2:   Original sketch of fore-bed universal joint (1828). IET 1.6.136.

Fig. 12.3: Alfred's drawing of a phaeton at Croydon, made with Ronalds' perspective tracing instrument (c.1829). IET 1.7.41.

A sketch of a phaeton at the family home survives from this time (Fig. 12.3). It was drawn by Alfred and is notable for the detailed delineation of the under-carriage. It may be presumed that it illustrates how the concept was turned into reality — in which case Ronalds' mother had allowed him to make significant alterations to her carriage. No other examples of adoption are known. The original model found its way into the precursor of the Science Museum in 1867 but was 'De-accessioned & Destroyed' in 1921.

12.3 Window-mounted Sundial

Three further inventions appeared in Herbert's write-up of highlights of the fourth exhibition at the National Repository in 1831. The first was a sundial, from which he quotes from the catalogue:

> As a check upon the proceedings of the domestic time-measurers, many of which are too apt to set up a standard of their own, Mr. Ronalds has devised a simple, economi-cal, and elegant little apparatus, which can be set up without calling for the mathe-matics of the schoolmaster, in any dwelling having a window with a southerly

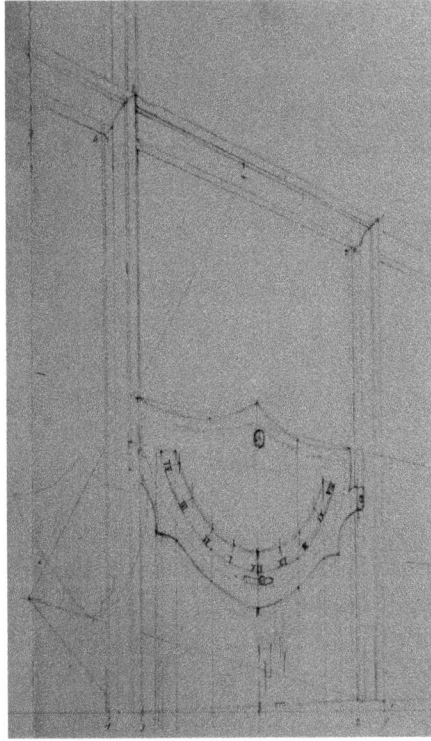

Fig. 12.4:   Window-mounted sundial, drawn when studying perspective. IET 1.6.118.

aspect. A wire fixed angularly into one of the bars of the window frame on the out-side serves as the style or gnomon: the shadow of this is received on a shield-shaped plate of thin copper, or tinned iron, attached on the inner side of the corresponding pane of glass. To allow the shadow to be seen inside the room, an arc of a circle, about an inch broad, is cut out of the lower part of the shield, and the opening is covered with a plate of talc, rendered semi-translucent by grinding one of its sur-faces. The shadow of the style may be traced on the talc plate, with a black-lead pencil, at its hourly stations with the subdivisions at pleasure, in correspondence with any good watch or clock, and the plate being moveable on the centre of the arc by the aid of a small stud, can be adjusted afterwards to a sufficiently near coinci-dence with mean time for all ordinary occasions.[A39]

A drawing survives in the Ronalds Archive (Fig. 12.4). An earlier version of the sundial appears near the front of the Ideas Book when Ronalds was still pursu-ing electrical science. In it, the window pane was replaced with thin ground plate glass, with the note: 'The Ground Surface of the Glass must be placed outwards in order that the shadow may not be refracted'.[J14] The improvements would have been determined after purloining one of his mother's windows.

12.4 Rigging or Tackle Block

Hebert, who could sometimes be very critical, found some fault with Ronalds' rigging block (Fig. 12.5):

> The usual (but not the uniform) method of constructing pulley blocks, is, to let the sheaves turn loose upon their centre pins, the latter being fixed; the common result of which is, that the holes in the sheaves *gulley*, by reason of the superior hardness of the pins; to remedy this defect, Mr. Ronalds very judiciously fixes, or casts, the pin to the sheave, the extremities of which are turned for the axes of rotation, leaving bosses or shoulders next to the sheaves, to prevent the latter rubbing laterally against the confining cheeks. Now, although this plan is a good one, and no doubt original to Mr. Ronalds, it is commonly applied to several purposes by the Birmingham manufacturers…
>
> There is another peculiarity in Mr. Ronalds' plan, which more immediately concerns the ship's-block makers. Instead of cutting the whole framing out of one solid piece, he makes them in separate pieces, which we suppose must be the practice in those of a large size, as it is productive of a great economy in labour and materials[A38]

Nothing more is known of this invention.

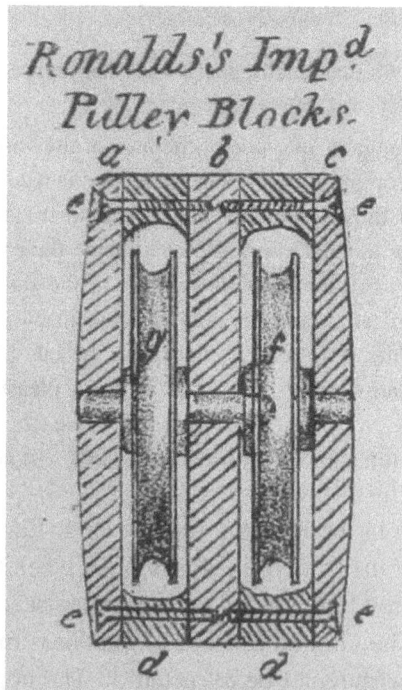

Fig. 12.5:   Rigging or tackle block. Ref. A38.

12.5 Stable Tripod (or Hexapod) Stand

The third description from 1831 is of the tripod (or hexapod) stand discussed in Sec. 10.4:

> By his intimate conversancy with geometrical and mechanical principles, Mr. Ronalds has produced, with almost the least quantity of material, a solid structure. Every thrust, strain, or twist, to which the apparatus in its applications is exposed, is met by the best possible disposition of the parts which compose the framing. It is scarcely possible to conceive how more appropriate strength can be derived from an equal quantity of matter.[A40]

These last three entries in Hebert's journal are noted as being 'By C. RONALDS, Esq. of Croydon'. Charles must have submitted the items to the National Repository as part of his business support to his brother. He may also have helped draft the catalogue descriptions as their tone seems more assured than Ronalds' usual style. No further objects are described: Hebert's magazine ceased publication in 1832 and later catalogues have not been found.

12.6 Teleographic (or Telegraphic) Clock

The teleographic clock mentioned in the autobiographical letter is also listed (as a telegraphic clock) in the Ideas Book and would date from around 1830. Unfortunately, no additional details are given. It may have been a more sophisticated manifestation of the Time Ball, which was first trialled at Portsmouth in 1829. The time ball comprised a large painted sphere that dropped down a post at a predetermined time to enable ships to verify their chronometers. In the 1850s, the Electric Telegraph Company began to operate time balls remotely using a telegraph circuit from an accurate clock at the Greenwich Observatory. The system was a collaboration between George Airy*, Charles Walker* and Latimer Clark*. Ronalds' early version of the apparatus apparently had three large balls that varied in height in a manner that indicated the time continuously, and was possibly also driven by electric telegraphy.

12.7 Turning

Turning was a popular pastime for gentlemen in those times. The most renowned lathe manufacturer, *Holtzapffel & Co*, began displaying a 'gentleman's turning lathe'[B32] in the first exhibition of the National Repository. Later models were displayed at the Polytechnic Institution and at the Great Exhibition of 1851, with the last winning a medal.

Ronalds acquired his first Holtzapffel lathe in September 1813, as soon as the family moved to Hammersmith. It was a five inch centre lathe costing £21 and had the company's number 922 (all Holtzapffel lathes were numbered sequentially).[B167] Warren Ogden's *Notes on the History and Provenance of Holtzapffel Lathes* (1987) advises:

> Owners just died, leaving no instructions or guidelines...In the case of No. 922 for instance, the remains were found under a pile of rubbish which took two days to clear. The date of burial must have been long ago, and the location was a lonely farm in Wales.[B167]

It can be surmised from this that Ronalds gave his lathe to Alfred when the latter moved away from home and it was left in Wales when he emigrated from there to Australia.

Ronalds bought a replacement lathe (number 1500) in July 1830, soon after Alfred had left home. His third lathe (number 1947) was purchased in November 1849 for use at the Kew Observatory. It cost £40 with a 10 per cent discount for cash. It receives mention as belonging to him in an inventory of observatory equipment made in 1871,[S6] but would later have been discarded.

The lathe was an ideal machine for him. To deploy its full range of capabilities required both practical skills and a strong understanding of geometry. He turned wood, various metals and also more distinctive materials such as ivory, tortoise-shell, mother of pearl, alabaster, marble and 'Potter's earths'.[J54] It would have been used for utilitarian purposes — including making his electrical and mechanical apparatus and items around the house — as much as for decorative work. Alfred Frost advises in his biographical memoir of Ronalds that he 'appears to have been an excellent workman, turning being his favourite pursuit. He has left behind several small specimens of his work, which show a high state of proficiency in the art.'[M5] The specimens are now lost. Ronalds and Alfred were not the only members of the family to enjoy the craft: a chess set turned by their brother Edmund was passed down through his descendants.

Ronalds developed various new tools to assist his lathe-work, with one of the first being an addition to the slide rest (Fig. 12.6). The primary purpose of the slide rest is to support the cutting tool while allowing it to be moved along the spinning work piece. His modification automated the simultaneous movement of the tool transverse to the work to form any particular curve. This is done with the aid of a small steel roller that runs along a template of the desired shape attached to the slide rest. A sketch of the device appears in the Ideas Book in April 1829, with the note that 'any number of pieces of work as handles for instance may be turned to precisely the same size & pattern beautifully easily & rapidly.'[J14]

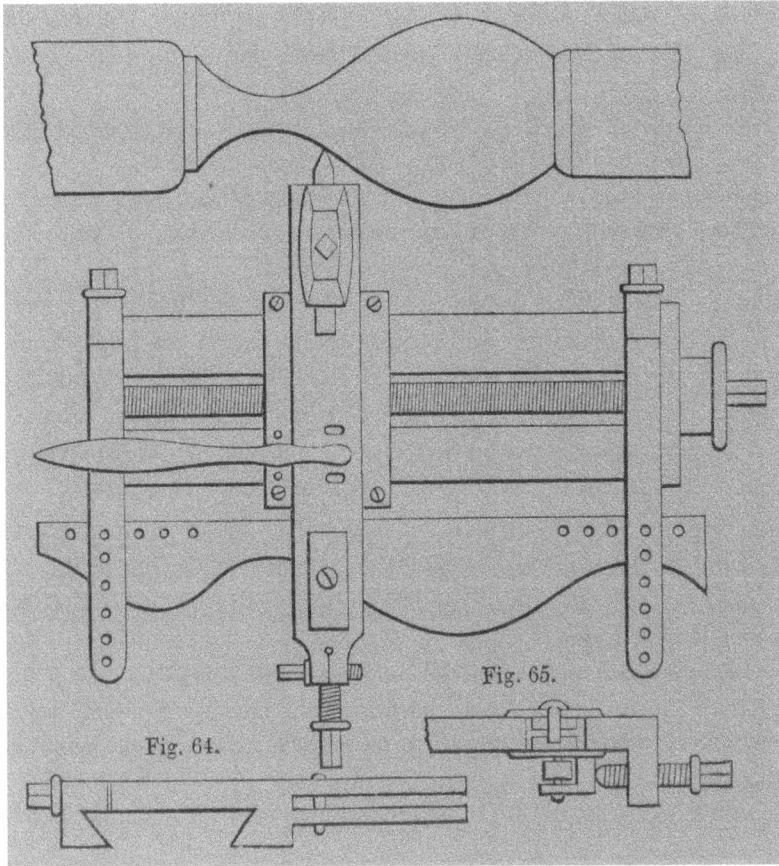

Fig. 12.6:   Curvilinear turning attachment for lathe slide rest. Ref. B50.

Charles Holtzapffel's negotiations in June 1834 concerning the manufacture and sale of the slide rest survive. In the first letter he advised Ronalds:

> we cannot really make any deduction from the charge on the receptacle slide &c[?] 5– as our profit on it was extremely small...on account of the trouble of making so many slides all to agree...but at the same time we are desirous of meeting your wishes as to the amount of reduction...but it is only on the understanding of its being so much allowed in favour of your giving us the Curvilinear Apparatus which we do with much pleasure [C69]

He continued with what he thought was a sweetener:

> we are on the point of printing a new Catalogue and as we are anxious to give every man his just need[?] of praise we shall in noticing the curvilinear apparatus inform the world that was invented by you [C69]

His suggested wording was included for Ronalds' comment.

Ronalds evidently pushed back (perhaps with his brother Charles' influence) because the tone of Holtzapffel's second letter four days later is somewhat different:

> we beg to say that in proposing to advertise the Curved Guide as your invention, we merely viewed it, abstractedly, as an accession to our apparatus...
>
> We can however only conclude from the general tenor of your letter, that you do not desire the introduction of your Instruments in our Catalogue; we shall therefore withdraw them from it.[C70]

The matter must have been smoothed over because the curvilinear apparatus did appear in the 1834 Catalogue.[B39] Ronalds' contribution was still mentioned in early 20th century Catalogues, even though the device had been modernised well before then. The price was then 'from' £10.[B118] The apparatus formed part of Holtzapffel's lathe display at the Polytechnic Institution, the Great Exhibition and perhaps even as early as the National Repository.

Over 50 years after the original invention, the apparatus was described and illustrated in the revered text *Principles & Practice of Ornamental or Complex Turning* (1884), the fifth volume of the Holtzapffel family's book *Turning and Mechanical Manipulation*:

> The *curvilinear apparatus*...contrived by the late Mr. Francis Ronalds about the year 1830, — the figures being reproduced from that gentleman's original sketch, — translates the rectilinear into a curved traverse in a very effective yet simple manner, and the scheme of this apparatus is further worthy of record inasmuch as this application of a guide principle, originally proposed by an amateur for the purposes of ornamental turning, contains the germ of all the turning and carving machinery since so extensively employed for the mercantile production of numerous *fac-similes*[B50]

Mass production of furniture and such items using powered machines had begun around mid-century. Although his contribution is little known, the curvilinear apparatus turns out to have been one of Ronalds' most influential inventions. He probably received little or no financial benefit from the many sales of his device, and certainly none from its later much wider adaptation. No example exactly matching the original design has been found, but slightly later versions exist in various collections including the Science Museum and are sold on occasion with antique lathes. Templates of similar type are used in ornamental turning today.

A second area of focus for him with the lathe was decorative cutting, where he contrived what he called the Turner's Pen. When he had built his geometric pen for drawing on paper, described in Sec. 10.5, he realised that it created curves similar to those produced with a geometric chuck on the lathe. His invention could be adapted to make a revolving cutting frame for surface ornamentation of wood: instead of the cutting tool being stationary and the work revolving with the geometric chuck, it would be the other way around.

John Holt Ibbetson had developed the geometric chuck in about 1815. He published many examples of the fancy epicycloidal and other patterns it produced but remained very secretive about its workings. He advised in his 1833 book: 'I have never made any particular communication of the mechanism of this Instrument…and Messrs. Holtzapffel have only recently become acquainted with its mechanical contrivance.'[B37]

A draft letter survives from 1830 in which Ronalds seems to be responding to a request to design a cutting machine to make curved patterns. Interestingly, this is the very time at which Ibbetson was negotiating with Holtzapffel to manufacture and sell his geometric chuck. The tenor of Ronalds' letter is quite different from his usual deferential tone to the many people he perceived to be his betters: he is providing consulting advice here and regards himself as the senior partner. The concluding sentence illustrates this clearly: 'Now I will expect to hear from you by return of post to Hastings *[where he was holidaying]* to say when you will come to Croydon *[the family home]* and when I know this I will go up to meet you. Perhaps you can get some wheels of 48, 36 & 24 1 inch'.[C63] The recipient would have been Holtzapffel or possibly one of their competitors.

Nonetheless, this business style does not come easily to him and the draft letter is heavily edited. By way of example, one sentence read originally:

Fig[s] 1 & 2 are intended to shew you how I proposed to diminish these apparent evils.

It was later revised in pencil to a much more positive statement with fewer qualifiers:

Fig[s] 1 & 2 will help to explain the manner in which I have succeeded in diminishing at least these evils and in procuring engraved or etched figures exactly resembling epicycloidal turning.[C63]

He was perhaps being coached by family members to avoid the tendency of his modest wording to suggest he had not achieved success.

He went on to describe one manifestation of the device with five wheels. The innermost one *a* is 1 inch (25 mm) diameter with 96 teeth and, moving out radially, the next wheel *b* is ¾ inch (19 mm) diameter with 72 teeth and the following two *c* and *d* are ½ inch (13 mm) and 48 teeth. These are surrounded by the outer driving wheel in the same manner as with the geometric pen. A drawing survives that shows this configuration (Fig. 12.7); it is set up for drawing with pencil on paper however, rather than as a wood engraver. He was already contemplating further adjustments:

A cleaver watch maker here says that the size of the teeth we talked of would do very well but I think they would be better if larger and if 3 or 4 wheels only are used, the same diameter of wheels might be used with half the number of teeth in each thereby doubling the strength.[C63]

Fig. 12.7:   Turner's pen — version 1 (1830). IET 1.6.115.

A third configuration had a solid outer drive wheel, with the small intermediate wheels inserted through holes arranged almost circumferentially rather than radially (Fig. 12.8):

> You will see that this reduces the <u>size</u> enormously without reducing the wheels, the teeth or the number of changes *[possible permutations]*, indeed the latter is increased I believe.[C63]

He summed up:

> I apprehend this or something like it to be the plan of the <u>great</u> geometric chuck but I think that a <u>pen</u> would if made very solid answer as good a purpose nearly and <u>sometimes</u> better.[C63]

A folder has been retained entitled 'Specimens executed by means of The Turner's Pen'[J43] and noting the contents to be epicycloids and conchoids (Fig. 12.9).

Edward Clive Bayley wrote to Uncle Charles Field in 1839 from St Petersburg: 'Pray explain to Mʳ Ronalds, with my best regards, that I am not just now in fettle to attend to his commission for a specimen of work done on my Epicycloidal Chuck,

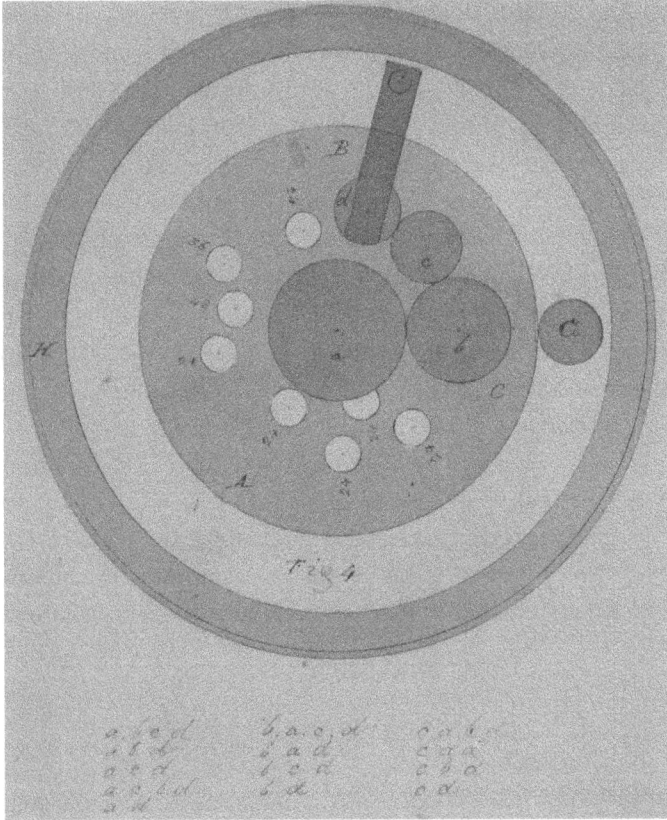

Fig. 12.8:   Turner's pen — version 3 (1830). IET 1.6.103.

Fig. 12.9:   Ink stamp created using the epicycloidal cutting tools. IET 1.5.43.

but I will not lose sight of it hereafter.'[C86] Ronalds had earlier been requested by Bayley to explain the workings of his new device. The letter suggests that the cutting machines Ronalds developed worked sufficiently well for him to not purchase Ibbetson's chuck himself. However, no direct evidence has been found of their sale by Holtzapffel or others. The epicycloidal cutting frame described in Volume Five of Holtzapffel's book was contrived by a Mr W. Pomeroy in about 1870. After that time it became common for intricate patterns to be generated by revolution of the tool rather than the work, as done in both Ronalds' and Pomeroy's apparatus.

Ronalds also developed an Epicycloidal Cutting Apparatus in which both the tool and the work piece revolved simultaneously to enhance the array of patterns that may be produced (Fig. 12.10). This could be achieved by a single drive mechanism — the rotation of one element was transferred to and from an eccentric shaft via two toothed wheels at each end with a chosen gear ratio to drive the second element. An alternative would be to use an overhead shaft to rotate both elements. The first approach was reinvented and patented in 1893 by Alfred Edward Beddow in his 'Epicycloidal, Rose-cutting, Eccentric-cutting, Drilling, Fluting, and Vertical-cutting appliance'[B112] and sold by lathe-makers Britannia Company.

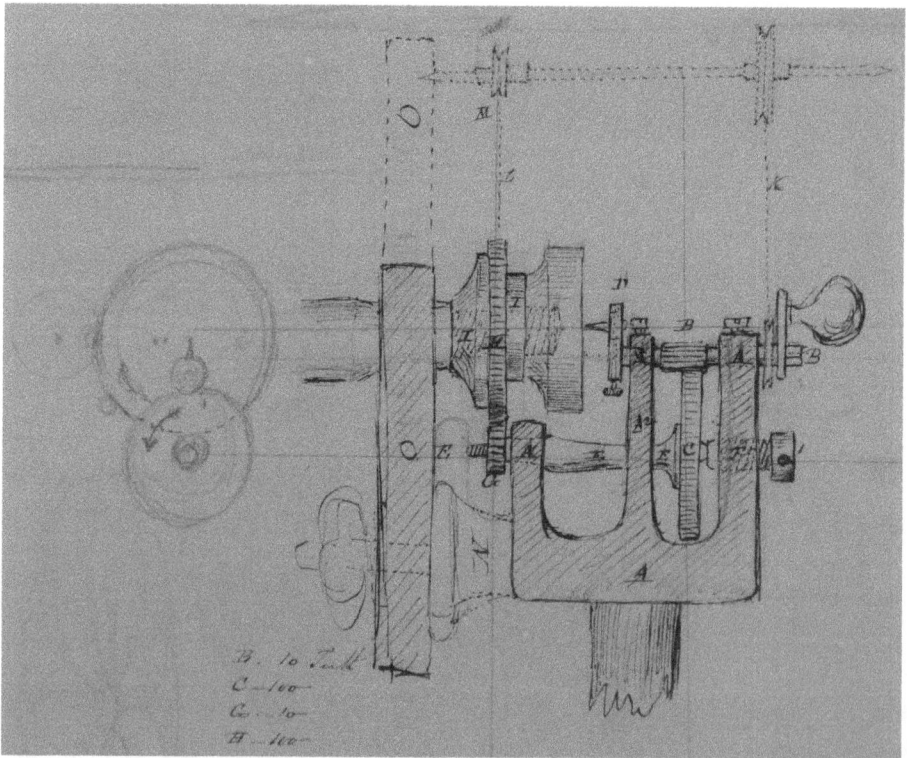

Fig. 12.10:   Device to carve intricate patterns by movement of tool and work piece (*c.*1830). IET 1.2.1.

There are hints of various other rotating cutters but little information is given on them. Ronalds developed devices for 'Rose (or rather Conchoidal) Turning,'[J54] either on one centre or between centres, that were adopted by friends and associates. The Ideas Book mentions a 'Slide rest made to perform rosette work.'[J14] He compared the action of various lathe attachments with one of his devices in a worksheet and concluded:

Advantages viz Combination of the rose engine[1] & elliptic Chuck

Disadvantages viz. Irregularity of the rosette. Now corrected a little [J44]

There is in addition an elegant drawing of a lathe fitted with a division plate and index (Fig. 12.11); this is used with a revolving cutter to produce segmental ornamentation. The index point is inserted in turn into various holes in the disc to hold the work piece at regular circumferential orientations.

The Ideas Book also lists various other concepts relating to slide rests, but with no further detail; these include: 'Vertical position of slide rest'; 'Motion communicated to slide rest from the mandrel of a hand wheel which works the lathe'; 'Circular saw & machinery used with slide rest' and 'Rolling rest & Hollow mandrels for cutting cylindrical sticks'.[J14] Clearly he enjoyed experimenting with new capabilities on his lathe.

Fig. 12.11:   Lathe with division plate and index drawn with perspective tracing instrument (c.1837). IET 1.5.51.

---

[1]A type of lathe that achieves intricate shapes through the action of a rocking headstock, its motion being managed by a rubber moving against a shaped cam mounted on the spindle.

The appeal of turning was such that he decided to write a book on the subject, much as he had done a few years earlier with engraving and printing. His perceived need for a 'Turner's Manual'[J54] is explained in his draft preface written in 1837:

> It has to many noblemen, gentlemen & authorities of great respectability appeared extraordinary that in a country where the beautiful Art of Turning has been carried to a greater degree of perfection than in any other...no <u>English</u> treatise on the subject has yet been devoted principally to the convenience &c of the Amateur...

> Most Amateurs are acquainted with the elaborate and expensive work of M *[Louis-Eloy]* Bergeron founded originally on that of M *[Charles]* Plumier & they will see that we have derived much valuable information from it. The more recent French work of M Dessaignes has also been a source of much assistance...We have also consulted not without profit M *[Menut]* Dessables' Manual...

> We presume to hope that if the mode of classification which has been adapted should be thought somewhat prolix, our indulgent brother Chips, seeing that we had but a choice of great difficulties, will <u>try</u> to admit that we have not stumbled upon the greatest...

> The Subject is divided into 4*[?]* parts...

> neither time nor expense have been spared to render our attempt at least comprehensive. We may also <u>confidently</u> appeal to several gentlemen who have kindly lent their aid to attest that opportunities have not been omitted of gleaning verbal information...Our grateful acknowledgement of this kind of friendly assistance are particularly due to *[missing]*.[J54]

Interestingly, the notes are written in the plural; they are in Ronalds' hand and style but there are several small edits that appear to be in another's handwriting. His co-worker is not identified in the papers.

The detailed Table of Contents shows ornamental techniques to have been a significant focus, with Part Two covering eccentric, oblique, elliptical, epicycloidal and rose turning, as well as the geometric chuck. Part Three describes approaches where the tool revolves and the work is stationary, or both revolve, while Part Four covers portrait turning and the square or line engine. Devices not previously published are included, including Ronalds' curvilinear slide rest, his rose engine and conchoidal turning inventions, and the epicycloidal cutting apparatus.[J54] A section was also promised on:

> Precautions against accidents & injuries
>> To the eye
>> To the lungs
>> To the fingers &c &c (to the temper) [J54]

as well as:

> <u>An appendix</u> containing a <u>vocabulary of terms used by Turners</u> &c in English French & German and <u>Comparative tables of English French & German Measures of Dimensions, Weights, & Quantities</u> [J54]

Finally, Ronalds made mention of a range of 'prime movers of Lathes', including the foot, hand, horse, water, wind, steam and 'electromagnetism'[J54] — envisaging a future age of machinery powered by electricity.

It is likely that one of Ronalds' collaborators was Charles Holtzapffel, who published the first volume of his family's book *Turning and Mechanical Manipulation*[B50] in 1843. Points addressed in Holtzapffel's preface are remarkably similar to those in Ronalds' draft. Each would have valued the contributions the other could make to a joint publication. Being about 30 at the time and 18 years younger than Ronalds, Holtzapffel would have admired his elder's experience and achievements. Holtzapffel in turn had not only learnt much from his father, his firm's founder, but was talented in his own right: Ronalds had acknowledged his advice regarding his tracing instruments in *Mechanical Perspective*.

Work on the book advanced to the stage of typesetting[J54] but it was not published. Perhaps the collaboration waned during one of their business negotiations. If so, Ronalds' work may be regarded as an early forerunner of the Holtzapffel family's seminal turning text. Holtzapffel made little intimation to support this conjecture; he explained regarding his book 'that he had made some beginnings in conjunction with his late much-respected father; but that after the death of the latter in 1835, he recommenced his labours on a new plan'.[A64]

Although Ronalds' text was not completed, it is apparent that he had become a respected expert in turning methods and equipment development. He continued to use his lathes for many years, particularly in making equipment for the Kew Observatory.

12.8 Letter Stamps, Printing and Typewriting

A quite different area of interest for Ronalds was printing. We saw that he developed expertise in printing drawings when he invented his tracing instruments and, as mentioned in Sec. 10.7, he quite likely built his own presses. He also began to dabble in the printing of text. An early concept, entitled 'Letter stamps',[J14] appears in the Ideas Book in 1825. It comprised a small cylindrical vessel with holes in the base corresponding to the characters to be stamped. A fitted piece passing through the top carried numerous slightly conical pins which dropped through and a little beyond the base at the location of holes. When the stamp was pressed onto the paper the projecting pins rose slightly, allowing the ink inside the vessel to flow through on to the paper.

At this time the letter stamps used at post offices were handheld and carved from cork or wood, with more accurate steel stamps first being introduced around 1830. Many incorporated slugs that enabled the date and time to be altered.[B121] Ronalds' concept was perhaps more easily adjustable and avoided frequent re-inking of the stamp, but appeared to have other disadvantages.

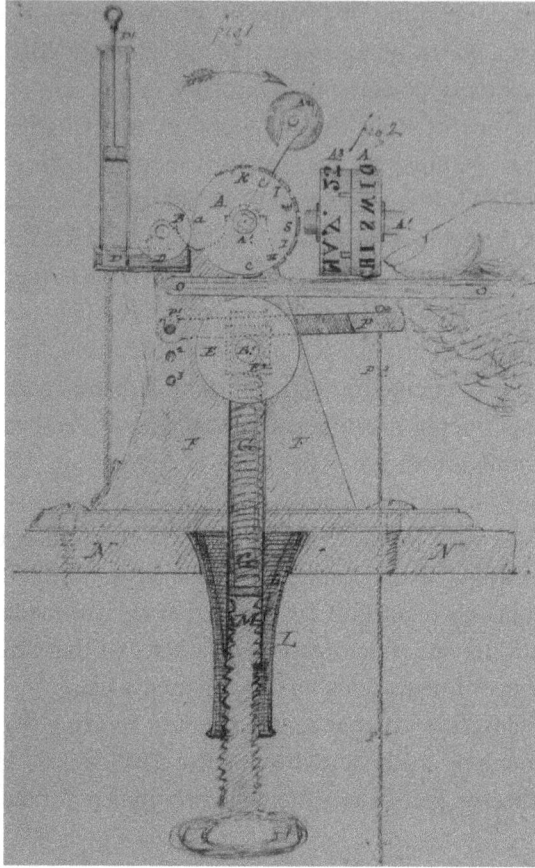

Fig. 12.12:   Letter printer and typewriter (*c*.1835). IET 1.6.119.

A second sketch survives of a more sophisticated lettering device where the characters are placed on a hand-driven roller (Fig. 12.12). The characters shown are 'Chiswick May 23', indicating it was drawn after the family moved homes in 1833. The paper lies on a plate and passes to the left in the diagram when pressed between the rotating roller and another below. A third roller dipped in a well provides ink to the characters. The printing pressure is modulated by means of a foot pedal acting on a spring and, when removed, the plate may be adjusted up and down by hand. The device by this means appears to be able to serve two distinct functions. The first is to simulate a simple printing press by printing any multiline statement numerous times, for example, to postmark many letters conveniently. Alternatively, it can print a sequence of individual characters in the manner of a golfball typewriter by applying the pressure only when the roller has turned to the next desired letter located around the discs.

A modern version of the second application is the 'Dymo' labeller, in which raised characters are arranged around a wheel. The wheel is turned to the desired character and it is then forced against an embossing tape by pressing a trigger. The device also has some resemblance to the 'typographer' patented by William Austin Burt in America in 1829 and regarded as a forerunner of the typewriter. It is additionally similar in concept to a printing telegraph Charles Wheatstone* patented in 1841 where electrical circuits actuated both the rotation of the dial on which the letters were arranged and the printing pressure.

Ronalds later developed plans for a more complex printing device, retained in the Ronalds Archive, but its precise working is unclear.

12.9 Surveying

We know of Ronalds' surveying ability only from his plans of the Carnac megaliths. A survey plan made by Alfred of part of Geelong also survives, and it is easy to imagine Hugh deploying similar techniques in his new community of Albion, Illinois. Ronalds, and presumably his brothers, would have enjoyed the blend of practical measurement techniques and geometry offered by surveying. Its principles and processes are akin to those he honed to achieve precision in his drawings and that all three used when navigating on their travels. It seems that for them simple surveying and map-making was a natural extension of their schoolboy education and hobbies.

It is not altogether unexpected that Ronalds would contemplate some new and improved instruments for surveying. One was a device to measure the angular separation of two objects. Doubly-reflecting instruments had been invented in the 18th century, the best-known being the sextant and octant (or reflecting quadrant). Such instruments comprise a fixed, half-silvered horizon glass and a second index mirror that rotates about its centre and is attached to an index arm. When held so that an object is sighted directly through the clear portion of the horizon glass, it is also seen in the mirrored part (through reflection from the index mirror) when the index arm is aligned with zero. The index arm is then moved until the second object comes into view in the mirrored segment of the horizon glass, which is achieved by the arm and its mirror rotating through half of the angular separation of the objects. The true angle is therefore read from an arc graduated to fit 90° in a 45° arc or 'octant'. Figure 12.13 is Ronalds' illustration of the geometry of this type of instrument.

In 1811, Sir Howard Douglas patented an ingenious but intricate variant to the device. The index mirror now rotated about an offset point and the horizon glass also rotated through a linkage so that it always showed an image of the moving index mirror. By this means the index mirror is able to rotate through the total

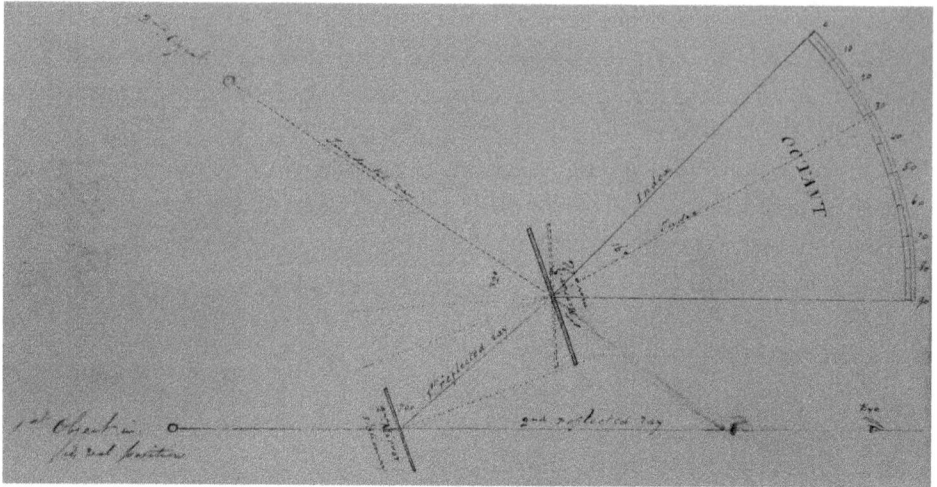

Fig. 12.13:   Optics of the double-reflecting octant (*c*.1829). IET 1.6.107.

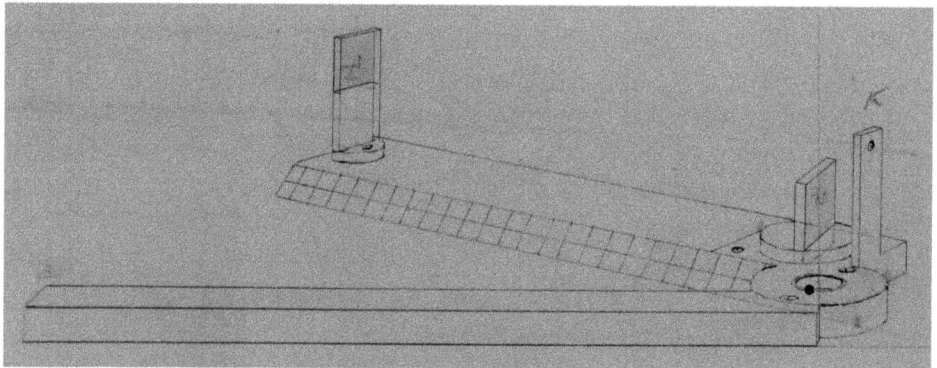

Fig. 12.14:   Doubly-reflecting pocket surveying instrument (*c*.1829). IET 1.6.108.

angular separation of the objects. The advantage of the arrangement is that the index arm can then be used to draw a line at the correct angle directly onto a plan without the intermediate steps of reading off the angle, noting it and then using a protractor. Douglas' invention was communicated in Tilloch's *Philosophical Magazine* by the optical-makers Cary and was manufactured and sold by them.[A10]

Ronalds purchased the instrument and would have enjoyed experimenting with it. It soon occurred to him to take the opposite approach, stripping out the additional elements and achieving the same goal by simply ensuring that the mirror rotated through half the angle of the arm. He adapted the common sector to the purpose by placing the index mirror near the hinge connecting the two rulers (Fig. 12.14). The two revolving elements were linked by 'a piece of strong watch spring'[J39] (Fig. 12.15) and the barrel supporting the mirror was twice the diameter of the

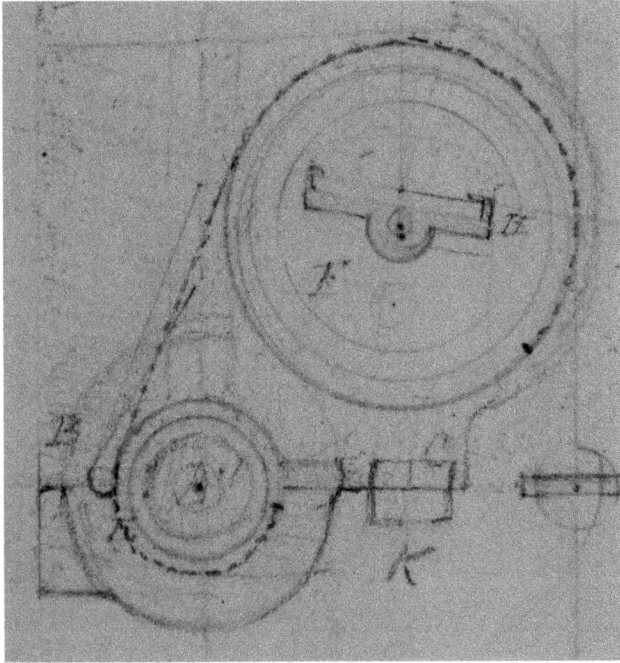

Fig. 12.15:   Doubly-reflecting pocket surveying instrument — sector hinge and rotating mirror. IET 1.6.108

hinge to give the required angular ratio. A rough sketch and description appeared in the Ideas Book in March 1829. The 'Pocket Surveying Instrument'[J39] apparently progressed to manufacture and sale, with Ronalds drafting promotional material:

> The proposed object was to produce a more compact, smaller & cheaper instrument for serving the office of both quadrant & protractor than seems hitherto to have been made known. This is principally usefull for rapid plan sketching on the spot but will also *[be]* found perhaps superior in point of accuracy to a quadrant or sextant of equal size as there exists no necessity to read off by divisions.[J39]

No mention of the instrument has been found in the literature.

Ronalds and Blair noted in their detailed survey of the Kerlescan Enclosure at Carnac in 1834 that 'angles…were measured by Sir Howard Douglas's reflecting semicircle, (a most useful little instrument to antiquarian travellers,)'.[R11] It is not clear whether Ronalds reverted to a larger and more accurate instrument for this work, or whether he simply used familiar terminology to denote the type of instrument employed — and modestly avoided advertising his own device. He also seems to have made use of a plane table in his surveying, which provides another means for bearings to be marked directly on paper in the field. A final point is that the date of his pocket surveying instrument suggests that he and Blair may have been surveying ancient monuments in England as early as the late 1820s.

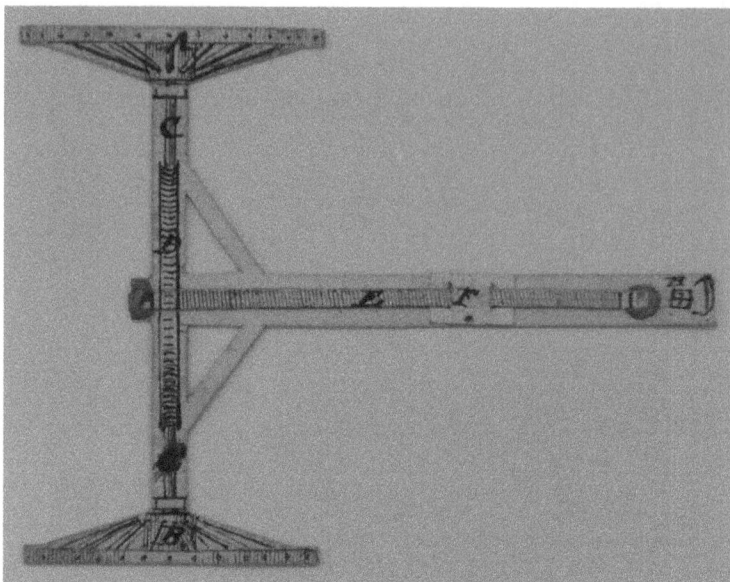

Fig. 12.16:   Cardboard model of Trochiametrograph (1827). IET 1.6.104–105.

Ronalds developed what he called a 'Trochiametrograph'[J14] in early 1827. It was inspired by the Odometer or Trochiameter developed by William Wollaston's* brother Francis, which was an instrument attached to the wheel of a carriage to count the revolutions, from which the distance travelled could be computed. The purpose of Ronalds' device (Fig. 12.16) was to record distances in graphical form in surveying. The apparatus had two large wheels with a worm D on their axle that meshed with a toothed wheel to drive another transverse screw E that carried a slider F. A pencil on the slider recorded the scaled distance travelled along the screw on an attached drawing board (not shown in the cardboard model). Ronalds chose a diameter of 80 cm for the large wheels. With the smaller wheel having 80 teeth, and the screw eight threads per inch, the resulting scale was one inch to one mile. The scale was readily adjusted by inserting a wheel with a different number of teeth. Reorienting the drawing board when starting a new traverse would enable a plan of the surveyed area to be produced automatically.

Ronalds had another idea pertaining to Wollaston's trochiameter, which he called an 'Oil feeder to Machinery'.[J14] His concept was to combine the instrument with an oil can that would automatically put a drop of oil on the bearing of any rotating machinery at given intervals.

One further innovation in the field of surveying is outlined in the Ideas Book in 1827. It was an 'Improved method of adjusting the level & telescope of a theodolite, adopted by Mr Owen of Shrewsbury in his Cambrian level'.[J14] The theodolite was held by pivoting supports, one of which may be raised or lowered by turning a screw to make fine adjustments to the level.

12.10 Large Telescope Supports

The telescope was a favourite instrument of Ronalds' and new uses for them and novel mechanisms for their support appear throughout his papers. In 1833 he configured an equatorial mount for a large astronomical telescope. In doing so, he became involved peripherally in a notorious scientific dispute, replete with characters described politely as 'difficult'[A171] by Dr Michael Hoskin in a detailed article on the saga written in 1989.

An equatorial mount comprises supports and a clock mechanism that enable the telescope to follow the apparent nightly motion of the stars around the earth. The inclined polar axis of the machine spans between bearings near the dome of the observatory and the floor and rotates slowly about its longitudinal axis. The telescope is supported in the polar axis in a way that enables rotation about a perpendicular axis to the point of interest in the skies.

The lead character in this story is Sir James South*, who purchased an object-glass with 30 cm aperture and a focal length of 5.8 m for £1,000, with plans to build the largest refracting telescope in the world. The esteemed instrument-maker Edward Troughton FRS was entrusted with the construction of the equatorial mount. Little is known about Troughton's design beyond stylised sketches later published by Revd Richard Sheepshanks FRS.[B65] These show it to be a simple arrangement with the polar axis comprising two pairs of curved wooden members each linked at their centre by a St Andrew's cross on which the telescope pivoted.

The mount was installed in late 1831 but the telescope was found to vibrate in use. Troughton's attempts to remedy the problem failed and he also refused South's requests to start again with another design. With Troughton then being 79 years of age, his friend Sheepshanks decided to get involved in late 1832. Sheepshanks was an outspoken person and he and South already had an acrimonious relationship from their roles as Secretary and President of the Royal Astronomical Society. Sheepshanks and his close friend George Airy, then Professor of Astronomy at Cambridge, attempted various repairs including adding diagonal bracing to prevent twisting, which were partially effective by June 1833 and improved further over time. By then South was not interested in their solution; he was focused on attributing blame and had started to collect formal opinions from technical experts.

It is at this point that Ronalds enters the story. We know from his diary[J45] that South and his friend Joseph Gwilt, an architect, dined at his home in Chiswick on

13 May 1833. South followed up two days later with an 'invitation'.[C67] Ronalds replied advising that he was unable to attend the event, but was sending his model of an alternative equatorial mount:

> The bearer will I hope bear ~~this~~ your model safely & the note before pudding time.
> If the frame should be honoured by the inspection by <u>other</u> grave authorities will you
> be so kind as to represent it as a mere <u>sketch</u> deficient in proportions & unprovided
> with several essentials [C67]

He finished by noting his 'desire to contribute' to South's 'superb undertaking (but of course wishing to avoid all undue interference with others).'[C67]

In one version of Ronalds' design, each of the two transverse pivot points of the telescope is positioned at the apex of three longitudinal triangular frames in two perpendicular planes (Figs. 12.17 and 12.18). The frames are connected together at the top and bottom bearings as well as midway at the telescope support by gusset plates. Sketches of an alternative triangulated configuration also survive (Fig. 12.19).

The situation came to a head when Troughton sued South for payment. The case went to legal arbitration in 1834, where Ronalds' model would have been studied by witnesses on South's side (including the latter's 'dear friend',[J46] astronomer Thomas Romney Robinson*) as well as Airy and Sheepshanks. The judge ruled four years later that the invoice must be paid. Enraged at the verdict, South smashed up Troughton's polar axis and auctioned off the debris. Posters advertising the sale included prominent mention of the firm and 'Botchings cobbled up by their Assistants, MR. AIRY AND THE REV[D] R. SHEEPSHANKS'.[A171] Sheepshanks published a 92-page letter in 1854 spelling out his side of the story, and showing he was proud of his role in precipitating South's humiliation:

> my conduct on the trial was more provoking than ever...
> I took a very active part in plucking off the daw's stolen feathers [B65]

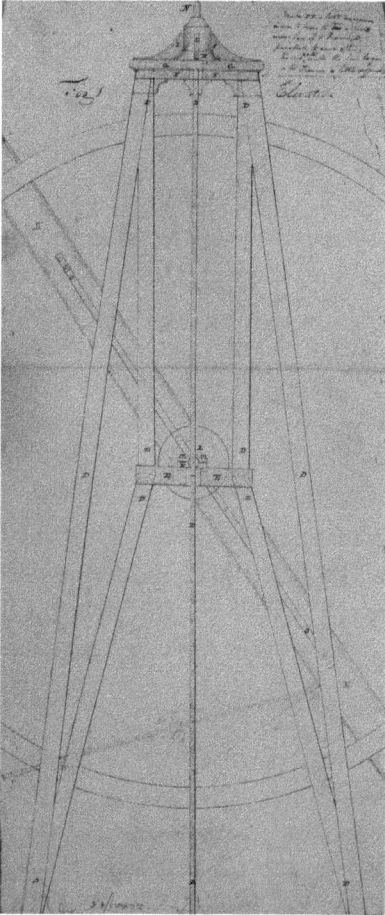

Fig. 12.17:   Equatorial mount for large telescope — model 1 — elevation (1833). IET 1.5.5.

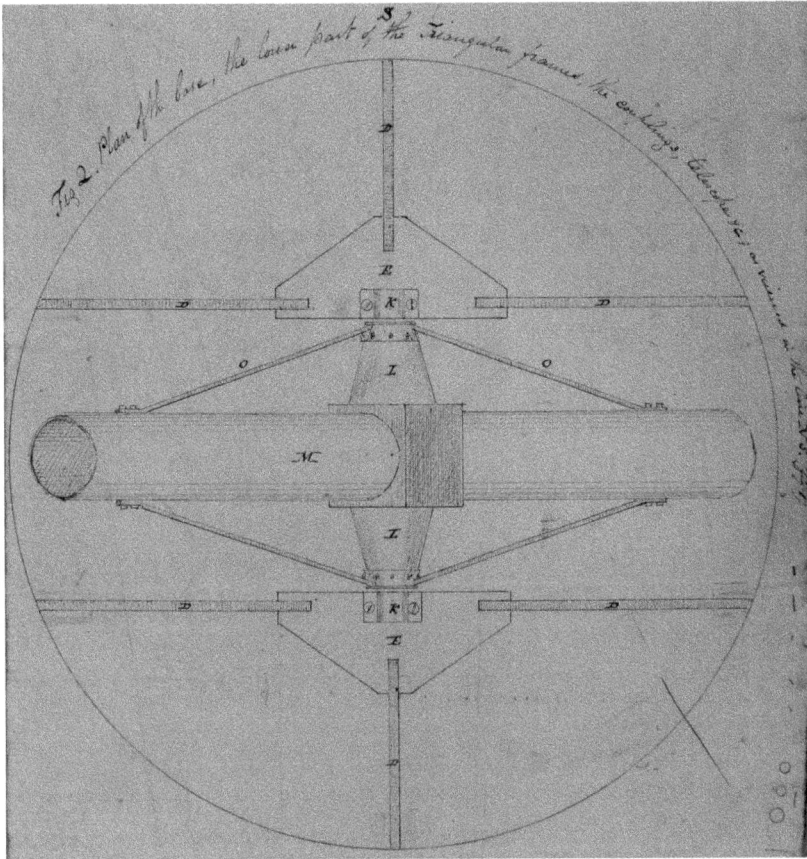

Fig. 12.18:   Equatorial mount for large telescope — model 1 — plan (1833). IET 1.5.5.

Sheepshanks died the next year and South's last published retaliation was a letter of reply to his obituaries. Ronalds' name was never mentioned in the many years of public mud-slinging, although he did follow events; Airy's subordinate James Glaisher* wrote to him in 1846: 'Should you come *[to Greenwich Observatory]* be so good to bring with you Airy–South's friendly correspondence'.[C173] Their acrimony was being played out in a series of letters to the newspapers.

Airy had been presented in 1833 with an object-glass the same size as South's by the Duke of Northumberland. He published a detailed account of his telescope's construction (complete with 67 diagrams) in 1844 where he noted, not surprisingly, that 'The selection of a form of equatoreal *[sic]* mounting was a matter of great anxiety.'[B52] Airy however was an expert in solid mechanics: in 1862, he described what is now known as the Airy stress function for determining the strain and stress field within a beam. He decided to support each of the two telescope

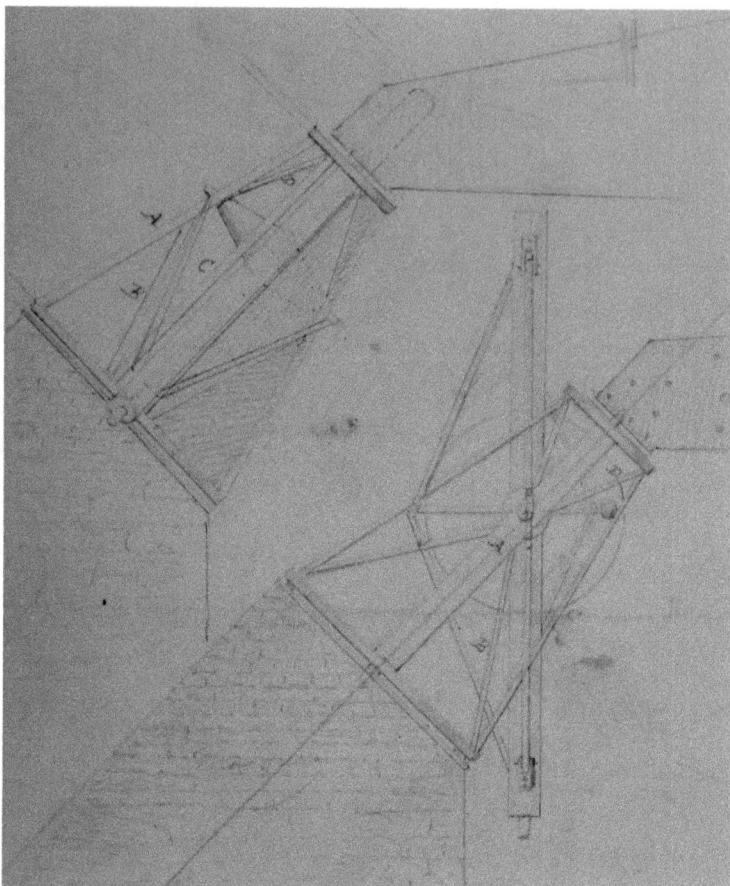

Fig. 12.19:   Equatorial mount for large telescope — model 2 (1833). IET 1.5.2.

pivots in a long, framed pillar (Fig. 12.20). Each pillar comprises three longitudinal members interlinked with cross bracing and lateral ties in the three faces and at the top and bottom bearings.

Ronalds summarised his involvement in the overall episode in his autobiographical letter:

> At the time of Sir Ja[s] South's dispute with old Troughton, about a very large & costly Equatorial, I presented, to the former, a model of an Equatorial formed on the principle of the...hexipod staff *[in Section 10.4]*...This was placed before certain Arbitrators, on the occasion, & afterwards, the Northumberland equatorial, at the Cambridge Observatory, was constructed on a plan very much resembling that of this model: not improved, I think; but I may be (naturally) prejudiced.[C375]

Ample reason for Ronalds' possible bias against Airy will become apparent in Chapters 14–16 concerning the Kew Observatory.

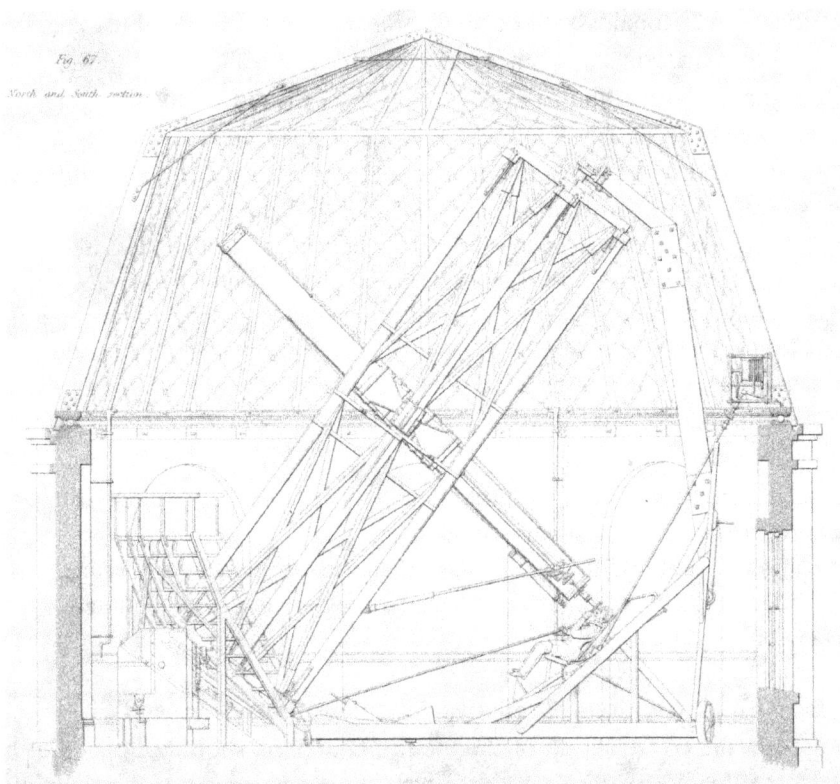

Fig. 12.20:   Equatorial mount for Northumberland telescope. Ref. B52, University of Cambridge, Institute of Astronomy.

There are similarities and differences in Airy's and Ronalds' designs. Both used the stiffening effect of triangulation to support each end of the telescope's transverse axis. Ronalds introduced the triangles at least in part through the longitudinal shape of the polar frame whereas Airy chose parallel legs and positioned and braced them triangularly. Ronalds' configurations are simpler structurally, although have the disadvantage of a larger base size.[2] Airy's design in contrast was actually built and continues to demonstrate its suitability. It is now regarded as a prototype of the 'English form' of equatorial mounting.

Ronalds later designed the mount for another outdoor telescope of significant size. The request came from manufacturer John Cary (who was also involved in South's case), for his client Mr Kilner. The friendly tone of the correspondence between Ronalds and Cary suggests a warmer relationship than Ronalds managed

---

[2]Personal Communication, Mark Hurn, University of Cambridge, Institute of Astronomy, July 2015.

with Holtzapffel. Ronalds' letter also demonstrates his considerable experience in the strength of materials and other aspects of mechanical design.

In Ronalds' words, the brief was: 'you want a solid cheap serviceable stand and you do not care either about appearances or great portability.'[C96] His solution was a tripod with hollow metal legs of triangular cross section. He built a model of his suggested design to accompany his letter to Cary:

> with the hope of assisting you to construct a stand which shall have less tremor than any <u>portable</u> affair that I ever saw or heard of I have made a paper model which clumsy as it is will serve Squire Kilner better than drawings and will save you some trouble in explanations as he has only to use the <u>scale sent</u> with it as he would a ruler to measure every part, the small divisions are inches the large feet.[C96]

He recommended the use of '<u>tin'd plate</u>', noting: 'I borrow the idea from Cap[n] *[Joseph]* Huddart's Equatorial at Sir Ja[s] South's'.[C96] The polar axis of Huddart's instrument comprised two longitudinal members shaped from tinned iron plate.[A26] The rest of Ronalds' design was original:

> The legs...should be filled with water when used for which an aperture should be made in the top of each to let it in and a cock or cork at the bottom to empty them (as you do not want to move it after, you may leave this part of it out of doors).[C96]

The Ideas Book mentions alternative fillings of 'gravel or sand'.[J14] Wood was also used in the construction of the mount. One part 'should be made of beech & at least 2 in *[5 cm]* thick'.[C96] Another 'should be made of 3 pieces of well seasoned deal so that the grain may run from the centre...towards the angles in order to obtain the greatest strength of the wood'.[C96] Yet another 'should be made of 2 or more parts the grain of one...running as shewn in the model and the grain of the other at...right angles to it'.[C96] This appears to be an early example of composite timber construction. He also concerned himself with design details: 'I intend to make the Handle a nut which is used for the horizontal motion stand out in front instead of the side which will be much more convenient to reach.'[C96]

The idea was sufficiently novel that he felt he needed to twist Cary's arm to try it. He was also interested himself to see how well it would work:

> Pray try the tin legs...If you should imagine that I want the tin legs tried for the sake of my own convenience or curiosity you would not be much mistaken but I calculate upon your own convenience also. It cannot fail of being a good solid stand and not a dear one...Now pray send for the tinman, put the tin legs in hand before you put this epistle out of your head[C96]

Cary did not need much convincing. He replied: 'I like the Stand <u>very much</u> and shall put it in hand immediately.'[C97]

12.11 Window of Vision in Fishing

Ronalds applied his knowledge of physics in an entirely different way when he provided assistance to Alfred regarding fly fishing in about 1835. Their nephew John Carter made mention of Ronalds' 'help in the scientific parts of the book'[B42] in his preface to the tenth edition of *The Fly-fisher's Entomology*. One of his contributions concerned the effects of refraction on the fish's window of vision from water to air. Vincent Marinaro, in his classic book *In the Ring of the Rise* (1976), noted that Alfred was the 'first fisherman-writer to call attention to this strange physical phenomenon'[B159] and also reproduced the relevant plate (Fig. 12.21) and accompanying text from the original book.

That refraction alters the apparent position of the fish in the water is well known. One of the less obvious consequences explained in Alfred's book is that a fly fisher should stand in the shallows because the fish cannot then see them clearly. They are in part invisible as light from the portion of their body just above the water will largely reflect on the water surface. The upper part of their body will be indistinct and apparently greatly separated from their feet

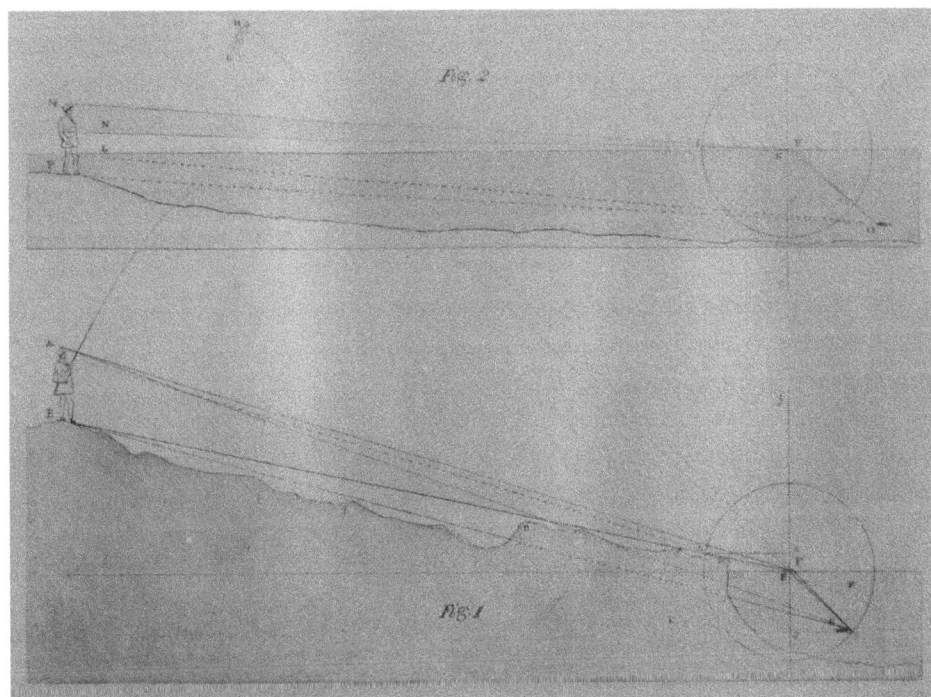

Fig. 12.21:   Window of vision in fishing (*c.*1835). Ref. B42.

due to the large angle through which the light bends on entering the water to the fish's eye.

Alfred would have observed the effects of refraction during his detailed entomological studies, with Ronalds providing its numerical and pictorial quantification. This is just one illustration of Ronalds' understanding of optics. Perhaps the first example was when he correctly deduced the origins of the 'extraordinary phenomenon'[J20] of the mirage he saw in Egypt in 1820 as being the refraction of light due to the warmer air near the ground having a lower density. He deployed his optical knowledge again in the 1840s in developing early cameras to photograph meteorological phenomena (Chapter 16).

12.12 Steam

Ronalds devoted considerable time and money in 1837–38 to the topic of steam propulsion of boat-trains on canals. His target was Birmingham Canal Navigations, the company operating the canals around Birmingham, while the catalyst would have been a conversation with his brother-in-law Samuel Carter, who had intimate knowledge of transport trends in the area through his occupation.

Both the timing and intended application were favourable for such a venture. The world's first trunk railway was opened in 1837 to link Birmingham with Warrington and Liverpool, and a second line connecting Birmingham with London was to open in 1838. Samuel's legal firm *Corrie & Carter* had been appointed Solicitors to the London & Birmingham Railway back in 1831 (he apparently saw no conflict of interest in also assisting Ronalds). Partly because of the looming threat of competition from rail, but also because of the system's prior success, a series of improvements to the local canals was also completed in 1838 following recommendations by Thomas Telford. The main canal was straightened, widened and rerouted to avoid many of the locks as part of this work.

Rail had inherent advantages over canal transport in speed and manpower needs per unit of cargo that more than overcame the greater cost of the rolling stock. In Birmingham these advantages were less acute. The area supported a large network of short canals with the primary cargo being bulky and low cost coal. Ronalds wrote in a draft reply to Samuel: 'I had no idea that so large a quantity as 65 to 75 loads of 24 Tons each or about *[blank]* Tons of coal arrived in Birm^m daily…from so small an average distance as 10 Miles.'[C80] Ronalds' idea was tailored to these attributes.

It is interesting that he chose to offer assistance to the canal company rather than the railway. This was the exact time at which Cooke and Wheatstone conducted the first trial of their electric telegraph for the London & Birmingham Railway, which was quickly deemed by the client to be too complicated. If

Ronalds had dusted off his telegraph work and approached the rail company, history might have been very different.

The first commercially successful steamboat built by Robert Fulton was in operation by 1807 on the Hudson River in New York State. It was propelled by two paddlewheels. Since then, steamboats had begun plying the Thames estuary and the English Channel. It was quickly realised that steam travel would be challenging on narrow canals because the wash from the paddlewheels would damage the banks. Ronalds' preferred solution to avoid churning the water was to propel the boat using a rack and pinion arrangement (Fig. 12.22). Two large, horizontal steam-driven pinions on the deck of the boat engaged with a cogged timber rail, with trundles on the far side of the rail maintaining contact at a height and orientation that adjusted to accommodate geometry variations. The rail was pinned in place with piles but could move up and down with the water level. He also developed concepts using a serrated wheel and a plain rail that ran along the brick wall forming the side of the canal. The efficiency of the overall operation would be improved by towing a series of boats. He mentioned another advantage of the scheme as obviating 'extreme torture'[C82] of the horse.

He developed considerable enthusiasm for his concept, taking out a Caveat on 11 October 1837. Samuel explained how it would work:

Dear Frank,

It seems that whenever any one applies for an Improvement in Steam Engines you will have a notice to oppose…& then the Attorney General will…privately question you & the other man — & decide whether your affairs clash [C76]

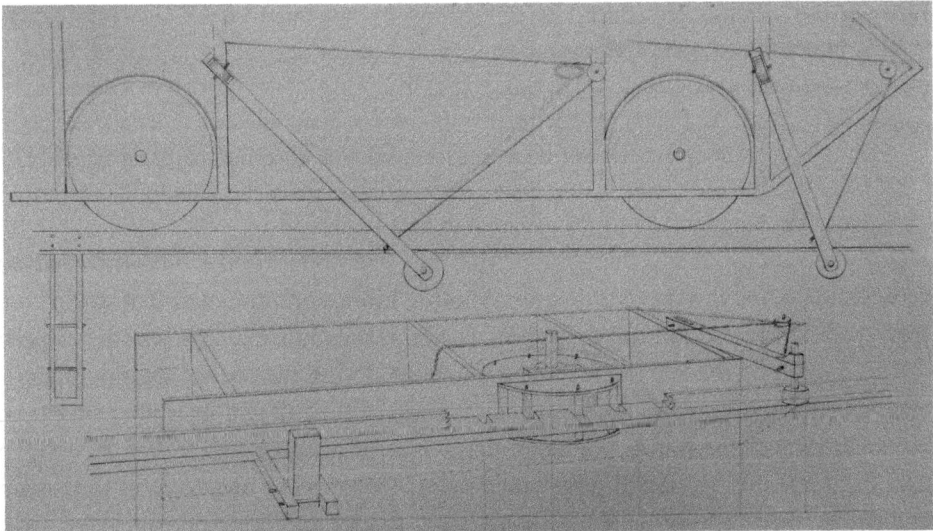

Fig. 12.22:   Canal tugboat with rack and pinion propulsion (1837). IET 1.4.51.

Ronalds would be given priority because of his earlier lodging date if the applications were found to cover the same innovation. Patents in the broad field of 'Steam Engines' came thick and fast at that time and he raised an opposition to at least one. His lawyers advised when various other notifications came through: 'if we are to oppose all these patents you will spend no end of money.'[C79]

Ronalds worked simultaneously on his commercialisation plan:

1st   To make a working model at my own expense (moved by clock work)

2nd   To exhibit it to the Company of the Birmingham Canal only

3d   To superintend the construction of an experimental Aquatic Rail & the necessary machinery & boat to be tried on any fitting situation that the Company may approve[J55]

The bottom of the page has been torn off so any subsequent steps are missing. The model canal was probably built in his mother's back garden at Chiswick. He updated Samuel in October 1837:

The boat carriage now at work on my Canal is supported almost wholly by the water, the pressure of the wheels on the rails being only sufficient to cause adhesion enough to tow the coal boats which follow it & that pressure can be regulated at pleasure. In this part of the affair I succeed far better than I expected.[C77]

He also mentioned his ideas for a large-scale trial:

I should want only the hire of a lot of timber (it would not be spoiled) of a few old monkey boats & of a one horse smoaker [C77]

The last point might have panicked Samuel, because Ronalds' next letter notes:

Your advice about making exp^ts & enquiries before incurring further expense is exactly the needful & shall be followed implicitly.[C80]

Using the test results, Ronalds performed calculations on the financial feasibility of the scheme. He started by describing his scenario:

It is inferred from the experiments made on a scale of one inch to a foot, & from other data, that a steam engine, working up to five-horse-power, placed in a tug boat, and acting upon the floating cogged rail, will tow Ten boats attached to it and to each other…at the rate of 3 miles per hour. Each towed boat carrying the usual quantity of Coal, viz 25 Tons, and the tug boat carrying the quantity required by the Steam Engine. The length of the trip is assumed to be 17 miles.[J55]

He was assuming in his calculations that his steamboat train would travel at no more than double the speed of horses. Another key assumption was that seven boat trains would operate 312 days per year. The largest capital expense was the cogged rail for which he made a detailed estimate considering the cost of the raw materials, their transport, and fabrication and erection of the rail and its timber supports. His operating cost estimate included the wages of an engineer, stoker, steersman and trimmer plus labour to assist in passing locks; wear and tear on the steam engine, toothed rail, tug and coal boats; coal usage; and other consumables such as oil. The resulting haulage cost per ton per mile was just a fraction of prevailing rates. A key question was whether an investor would accept the payback period of several years required to recover the significant upfront costs.

Noting 'we fear to talk too much before taking a patent,'[C80] he shared his idea with several experts, and their opinions were mixed. Railway engineer George Rennie, brother of renowned civil engineer Sir John*, saw technical and financial difficulties with the scheme.[C81] Ronalds had earlier contacted John Benjamin Macneill FRS, who had worked with Telford, had road and railway experience and was an inventor;[A53] he 'thought it worth a patent'.[C80] Another opinion came via Ronalds' patent lawyer:

> My friend Lee considers there is no doubt in the world that your innovation would cause a saving in working Canal boats…& then he s^d there were all Kinds of improvem^ts that Canal Comp^s might adopt but they were the most backward people going & they never wo^d adopt any thing new [C79]

Nothing more about the story is known except that he did not take out a patent and his scheme was not adopted.

So what actually happened in Birmingham? The London & North Western Railway Company (LNWR) took over Birmingham Canal Navigations in 1846. LNWR's motivation was that two of the railway lines it owned could then be joined up using Telford's main canal route, avoiding lengthy and expensive negotiations with landowners. The company came to realise that the railway and canals could coexist beneficially, with the subsidiary canals providing goods to the railway spine. As a result, much of the Birmingham canal network has survived to the present day.

Ideas resembling aspects of Ronalds' scheme saw later proposal, trial and use in other places. The committee of the longer Birmingham and Liverpool Junction Canal received a report in 1841 recommending the use of a train of six boats. It would be pulled by three horses and also require fewer operators, thereby reducing costs significantly. The report's author acknowledged however:

> I am well aware that the plans by which I am about to propose to meet the railway competition are not such as the canal carrier would wish to see adopted — nay, that they may even provoke his active opposition — his view being to carry on the struggle with his present establishment and on his present system [B46]

LNWR took over the canal in 1847.

Steam was deployed on canals in America, Europe and the UK in various ways later in the century. In one approach, a series of boats was towed by a steam locomotive running along a rail on the towpath. Macneill experimented with this scheme in 1839. Subsequent evolution to electric and then diesel engines and perhaps a road roller enabled the method to be used until modern times.

Ronalds also dabbled in other aspects of steamboat propulsion. In May 1828 he built a model of a paddlewheel to which was attached a spiral spring housed in a box.[J14] Current acting on the wheel when the boat is moored wound up the spring, while in motion the spring would partially drive the wheel. He contemplated a similar idea for storing excess energy in windmills, perhaps using a weight rather than a spring. In this case the purpose would be to smooth the supply of energy rather than to alleviate fuel usage. His concepts are very familiar today. Hybrid vehicles combine two forms of energy to optimise fuel consumption, while energy storage mechanisms are used to help match demand and supply of intermittent sources such as wind.

A rather different concept in the Ideas Book is the 'Steam engine jet'.[J14] Ronalds' brother Hugh took out a US patent for a water jet propeller for steamboats in 1840 (Fig. 12.23).[P4] Propulsion was achieved by forcing water out of

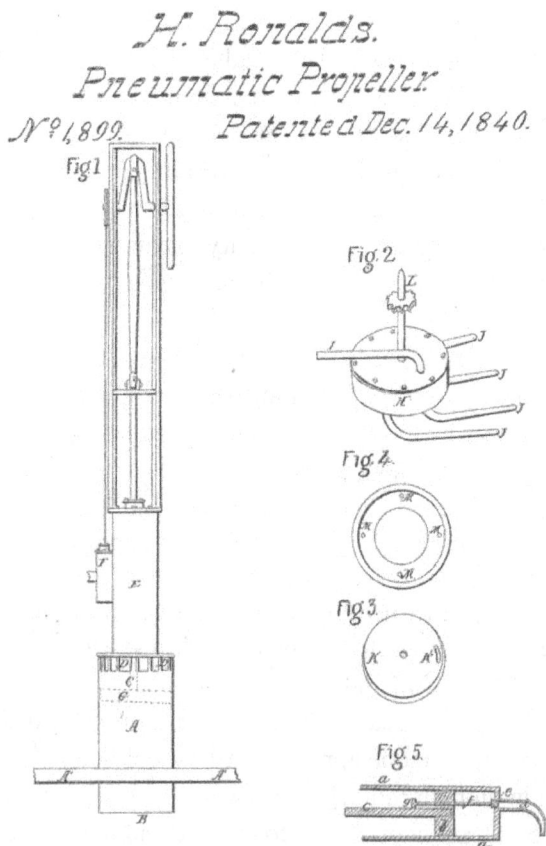

Fig. 12.23:   Hugh Ronalds' patented water jet propeller (1840). Ref. P4.

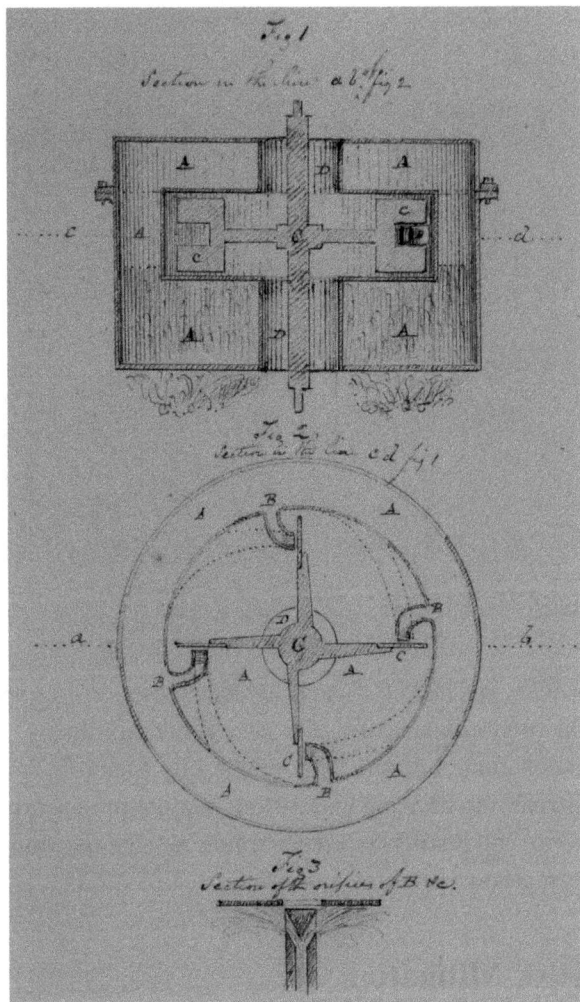

Fig. 12.24:   Steam-driven rotary motion (*c*.1835). IET 1.6.119.

open-ended steam-driven piston cylinders aligned near the stern. The steam was then released and the partial vacuum, together with the head of water, generated water inflow and the piston's return cycle. The system was particularly suited to the seasonally shallow water of the rivers near where Hugh lived.

Ronalds additionally explored steam-driven rotary motion. Figure 12.24 shows a vessel partially filled with water and bolted airtight. On heating, steam flows through nozzles into an inner drum where it strikes blades and revolves the shaft to which they are attached. The steam returns to the outer chamber through another pathway. Although reciprocating steam engines had been developed in the 18th century, invention of the practical steam turbine is generally credited to Sir

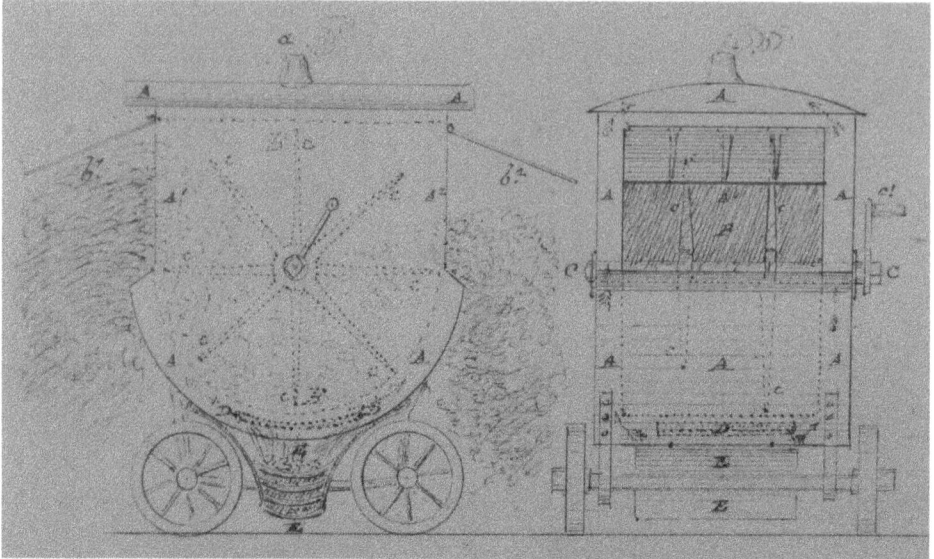

Fig. 12.25:   Sketch of portable apparatus, possibly to make hay using steam. IET 1.6.120.

Charles Parsons FRS in 1884. He used the rotary motion to generate electricity and it remains the predominant form of power generation today.

Finally, Ronalds contemplated 'Hay-making by Steam or hot air'[J14] in 1829, mentioning the possibility of using under-floor heating in the barn for the purpose. He also sketched an interesting-looking portable device that might have had the same goal, or been a powered threshing machine (Fig. 12.25).

12.13 Fire Risk Mitigation

In early 1840, Ronalds turned his attention to the serious challenge of fire. Although it was an ever-present threat, there had been renewed interest in the topic since the conflagrations of the Palace of Westminster (the home of Parliament) in 1834 and the Royal Exchange in 1838. One of his motivations was statistics given in *The Times* in the late 1830s that on average two fires broke out each day in London; they consumed £1 million worth of property per year of which only half was insured. Perhaps family or friends had also suffered a recent fire. The family cheesemonger business being destroyed 20 years previously (Chapter 2) would additionally have been etched in his conscience. The account of that fire in the *Gentleman's Magazine* had noted that 'owing to a great scarcity of water, in consequence of the frost and the water being turned off, the flames extended to several other houses.'[A21]

Ronalds mentioned in his draft book on electricity (Sec. 6.6) that he had written 'a dissertation on Fire'[J16] back in 1817. Returning to the subject, he scoped a pamphlet on 'Preservation of life & property from Fire'.[J56] It set the scene by giving information on recent fires, their common causes of starting and spreading and typical losses, and went on to address the issue under four headings: Fire Prevention, Early Warning, Damage Minimisation, and Rescue. Each category was packed with useful suggestions.

Under Fire Prevention, he listed precautions against incipient fires in different types of buildings, with particular note of spontaneous combustion in flues. He also offered suggestions used elsewhere such as 'no smoking allowed in Vienna after a certain hour' and 'curfew' and, interestingly, 'Precautions…as to the character of Insurers'.[J56]

Means of Early Warning included natural signs like 'Smell, Indications by Animals, Smoke', schemes adopted in different cities such as 'Watchmen' and 'Watch tower for fires at Vienna', as well as 'warning machines' and, pointedly, 'Electric telegraphs (hint)'.[J56]

He considered damage minimisation factors to be pertinent before, during and after a blaze. He began with 'the use of incombustible materials in buildings &c as stone stair cases, tiled floors, iron roofs &c', '(So called) Fire proof cements' and iron safes set in the wall. He also noted the effect of opening doors and windows in fuelling the fire. Concerning fire extinguishing, he advised on options readily to hand and the possibility of 'a private engine kept allways ready charged with water & with condensed air', before addressing public firefighting systems and their required infrastructure including 'Parish engines', 'Fire Plugs', suitable dress for firemen and 'their character for bravery steadiness & experience'. He finally touched on the possibility of 'Wanton mischief, obstinacy of mob, Robbery', the roles of neighbours during and after a fire and, fascinatingly, 'Beer'.[J56]

The topic of Rescue covered 'Modes of personal Escape & of withdrawing property'. It included advice such as remaining calm, 'crawling to avoid suffocation' from the rising smoke, keeping a means of escape at hand for an emergency, such as a ladder or 'rope slung between 2 neighbouring houses', and possible public infrastructure to aid rescue.[J56]

In total the document is a remarkably holistic overview of the gamut of likelihood and consequence factors, at a time when a system-wide appreciation of fire risk management was virtually absent. It is also illuminating concerning the technology and culture of firefighting at the time. Fireman James Heathman covered many identical points in his 1882 book (with the same title as Ronalds' pamphlet) because, sadly, little had improved.[B104]

It may be useful to provide some context on the situation prevailing in fire-fighting, personal safety and notification to help understand Ronalds' remarks. The fire hydrant or 'plug' — a key element of firefighting infrastructure — had been invented early in the century. The London Fire Engine Establishment (LFEE) was formed in 1833, but it was a private enterprise, funded by insurance companies to reduce their outlays. One of their jobs was to remove valuable objects from the burning building (if they were insured). Outside the City of London there was still competition between parish fire brigades. Both systems retained risks of less than ideal behaviour that had been seen in the past. A large fire could burn for a day or more, and beer was often made available to refresh those fireman and spectators who manned the hand pumps of the fire carts.

The priority of the LFEE was the insured building and its contents, not its occupants. In 1836, the Society for the Protection of Life from Fire was established from voluntary subscriptions and with the Lord Mayor as president. In the next few years it developed an escape ladder and they and their operators were dotted around the streets of inner London at night. Amazingly, it was not until 2005 that all buildings were required to have safe means of escape from fire, although new and modified buildings have been so regulated in the UK since the 1960s.

Cities in Ronalds' time might have had roving watchmen who rang a bell if they saw signs of a fire at night. The first telegraph-based alarm systems were installed in the early 1850s. The Boston scheme, developed by Dr William Channing, comprised alarm boxes in the streets which when triggered notified the station of a fire and its approximate location; alarm bells were then sounded. Channing's description of the scheme is in the Ronalds Library. Siemens introduced a comparable set-up in Berlin, while London introduced the telegraph and street alarm posts in 1879. These systems were manually operated: automatic fire detection systems were first incorporated late in the century and worked through exposure to heat. The convenient battery-powered smoke detector and alarm for the home did not appear until the 1970s.

Returning to 1840, Ronalds visited the Town-Clerk's office to obtain details of the types of escape ladder being investigated as part of the Lord Mayor's new Society for the Protection of Life from Fire. He received little assistance there: 'there seemed to exist a decided wish to get rid of me as speedily as possible'.[C99] He changed tack and wrote directly to the Lord Mayor about an idea he had: 'Your lordship is probably aware that the curious old work of...'[C99] Whether Thomas Johnson was aware of Roberto Valturio's book *De Re Militari* which had been published in 1472 is unknown, but Ronalds' point concerned the 'Ancient machine for scaling the Walls of a fortified town'[J56] depicted in it. He included his own sketch of the device (Fig. 12.26), which is what we would today call a scissor-lift and had the advantage of folding down into a compact wheeled base for transport.

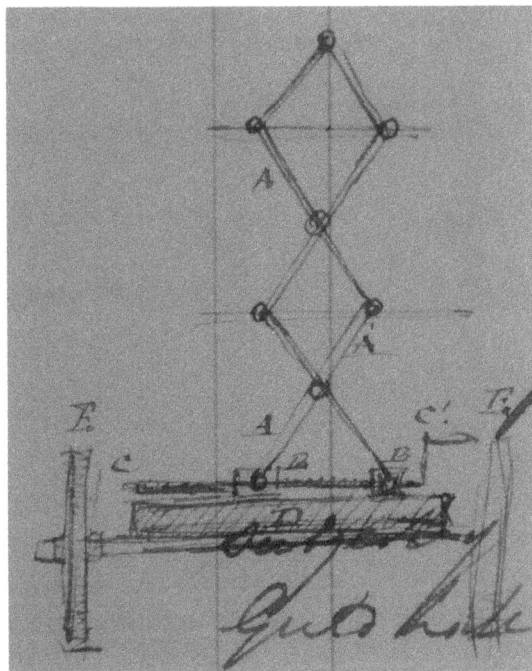

Fig. 12.26:   Fire escape ladder (1840). IET 1.5.14.

He noted in the letter that he was 'resolving…to have one constructed'[C99] if better ideas were not yet available. No reply survives and the Society went on to adopt a long ladder on wheels.

He also pursued several new technologies in his category of early warning devices (these did not include the telegraph). The first was a self-actuated fire 'alarum' (as alarm was spelt in those days) 'to procure timely notice of incipient fire in dwelling houses &c'.[C94] It comprised a U-shaped enclosed glass tube that rested on pivots close to its centre of gravity (Fig. 12.27). The tube contained mercury, with a vacuum at the top of the thin right leg. When the air in the spherical bulb on the left leg expanded under rising temperature, the mercury moved to the right, altering the centre of gravity until the device toppled from its support. The action pulled a cord and triggered a remote alarm bell. The device, like his other fire work and indeed most of his inventions, was not intended to be a money-making activity: he noted in a draft letter to *The Times* that 'It may be made by any mechanic indifferently versed in such like matters.'[C94]

He sent the fire alarm to the Polytechnic Institution in May 1840. There is just a small hint in the catalogues of the complaint he makes in his autobiographical letter about the labels being switched between two different alarms. The 1845 catalogue lists Mr Ramsey's alarm in Case O and 'A New Fire Alarum. Deposited

Fig. 12.27:   Self-actuating fire alarm (1840). IET 1.5.18.

by Mr RONNALDS'[B54] in Case Q. The previous year, Case Q's alarm (but not that Case O) had been unidentified. Other fire alarms would have been installed at home in Chiswick and with family and friends who wanted them. As noted in the autobiographical letter, he set up one at the Kew Observatory.

Ronalds also expended significant effort in devising a scheme to pinpoint accurately and quickly the location of a fire, while it was still small, to assist in directing the fire crew. He called it a 'Fire Finder'.[J57] Its basis was a watchtower fitted with a theodolite, which was used initially to measure and record bearings and vertical angles from the horizon to numerous places around the tower. When a fire was observed by the watchman, the theodolite would again be used to confirm its angular position, from which the street and local landmarks could be identified even in the dark. The alarm would then be raised.

He described in detail three different ways in which the measured location data could be documented to enable rapid retrieval at the time of the fire. The first was a multi-paged table in an indexed parchment ledger, with columns labelled

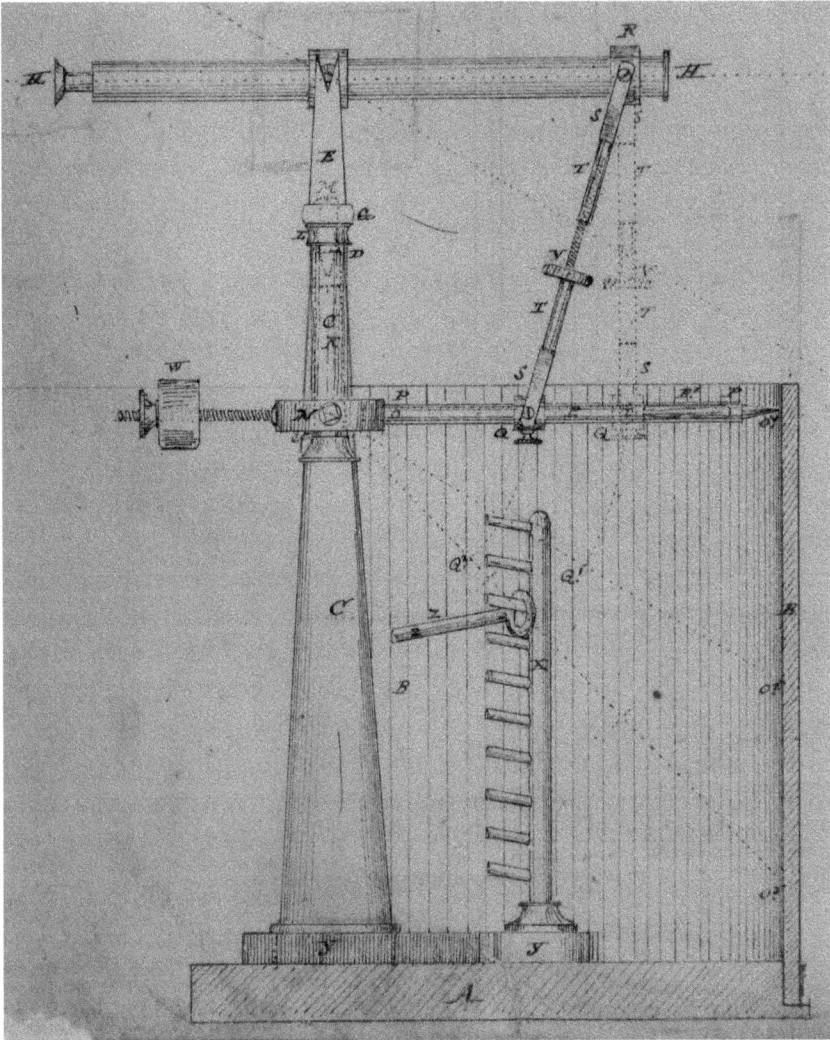

Fig. 12.28:   Fire-locating telescope with panoramic map (1840). IET 1.5.9.

1° to 360° horizontally and rows 1° to (say) 30° vertically. The name of each place was recorded in the box defined by its bearing and angular distance from the tower.[157] He had used the same concept for the message grid for his electric telegraph described in Sec. 7.1.

The second method was based on the cylindrical projection device he developed at this time to draw panoramas; it is outlined in Sec. 10.8. The positions of places were marked on the cylinder by a long pencil connected below the observing telescope (Fig.12.28). It differed from the earlier drawing apparatus in allowing the inter-linkage to be set non-perpendicular to the telescope and pencil when these were horizontal. In this way the pencil rotated through a greater angle than the telescope and stretched the marked locations down the cylinder to give greater clarity. He noted that 'The picture will (of cource) be no longer a <u>correct</u> panorama but this matters not for the <u>present</u> application.'[J57] He also added a pillar with pegs on which a ring-ended stick may be hung. It was designed as a maulstick on which the hand could rest and be steadied when using the pencil to either mark the cylindrical drawing or identify a fire's location.

His third attempt incorporated a circular table at the base of the instrument and a mechanical arrangement by which places observed through the theodolite could be marked and then labelled on it as a type of plan.[J57]

Watchtowers were first introduced in London by the LFEE's successor, the Metropolitan Fire Brigade, in the 1870s.[3] A simpler version of Ronalds' theodolite scheme was invented in 1911 by William Osborne, who worked for the US Forest Service. Remarkably, he gave it the same name as Ronalds did. The Osborne 'Firefinder'[A186] uses a scope or other sighting device (depending on the model) to determine the horizontal and vertical angles to the fire. The device swivels on a horizontal plate of up to 75 cm diameter on which a map centred on the lookout tower is placed. Some models also enable the topography followed through the eye piece to be traced by a pencil to give a panoramic drawing around the rim of the plate. Thousands of the instruments have been sold and they are still used on lookout towers in various countries in combatting forest fires.[A186] It seems that with fire management, as in several other areas, Ronalds was well ahead of the prevailing views of his time.

12.14 Tracing Glass

The Ideas Book is packed with simple devices that Ronalds would have made and found useful. In January 1827 he described a tracing glass. Reflectors were placed at an optimum angle with respect to each other and the illumination from a window or lamp to reflect 'a strong concentrated light'[J14] through a horizontal pane of glass supported above them. He often traced his drawings before sending off the originals — Jean-André de Luc* had commented on the 'ingenuity'[C11] of his approach as early as 1815.

12.15 Tensioned Bow to Straighten Rods

A 'New method of straightening long rods or bars without turning or planeing'[J14] appears in the Ideas Book in 1827 (Fig. 12.29). The rod is heated while being tensioned in a strong bow that is positioned to the side of the fire. The concept is a much larger-scale version of his approach to making straight wire indicators for

---

[3]Personal Communication, Elena Payami, London Fire Brigade Museum, November 2013.

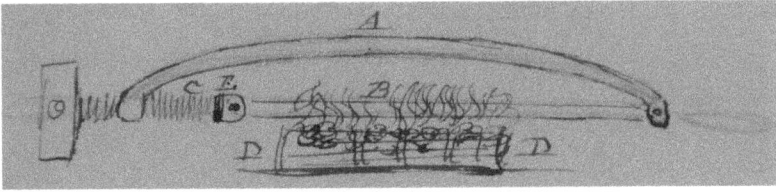

Fig. 12.29:   Tensioned bow to straighten rods (1827). IET 1.6.136.

electrometers, which formed part of his publication of an improved apparatus for observing atmospheric electricity in 1817 (Fig. 6.6 and Sec. 14.2).

12.16 Watch Alarm

He also described a watch alarm in 1827.[J14] The desired time for the alarm was set by inserting a pin into the corresponding hole on the perimeter of a numbered wheel fitted on the watch winding square. When the pin reached a lever (E in Figure 12.30), it pressed down on one end and the other end of the lever swung up and disengaged from the peg G on the spring box F. The box therefore unwound

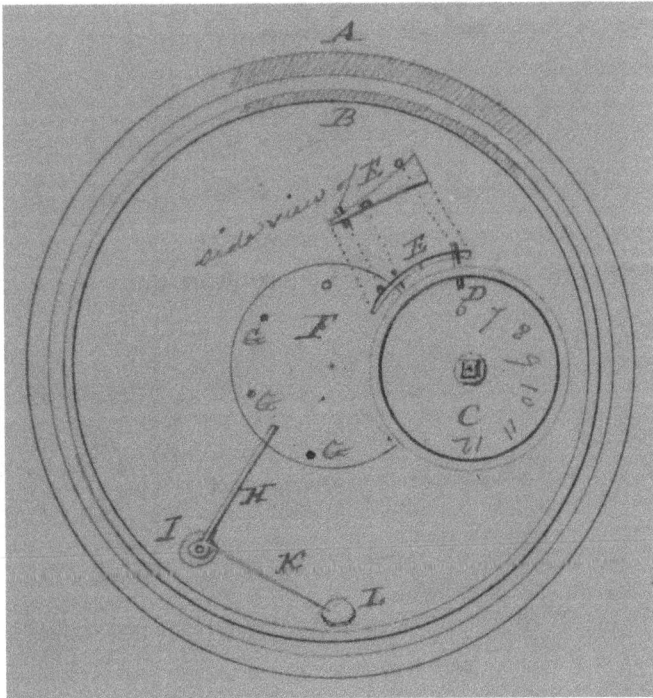

Fig. 12.30:   Watch alarm (1827). IET 1.6.136.

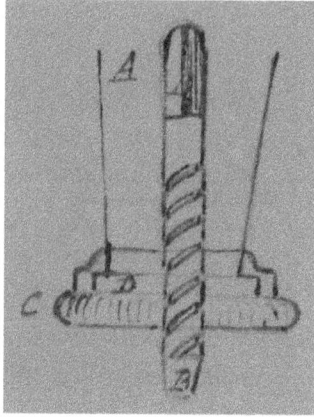

Fig. 12.31:   Table leg level adjustment (1824). IET 1.6.136.

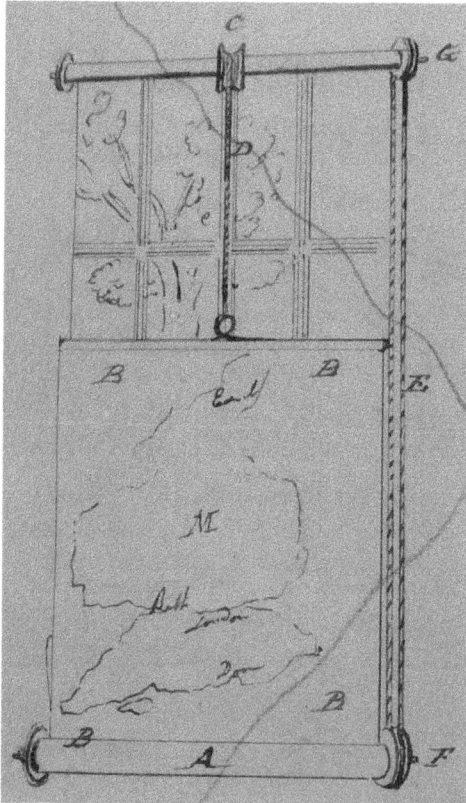

Fig. 12.32:   Upside-down window blind (1826). IET 1.6.136.

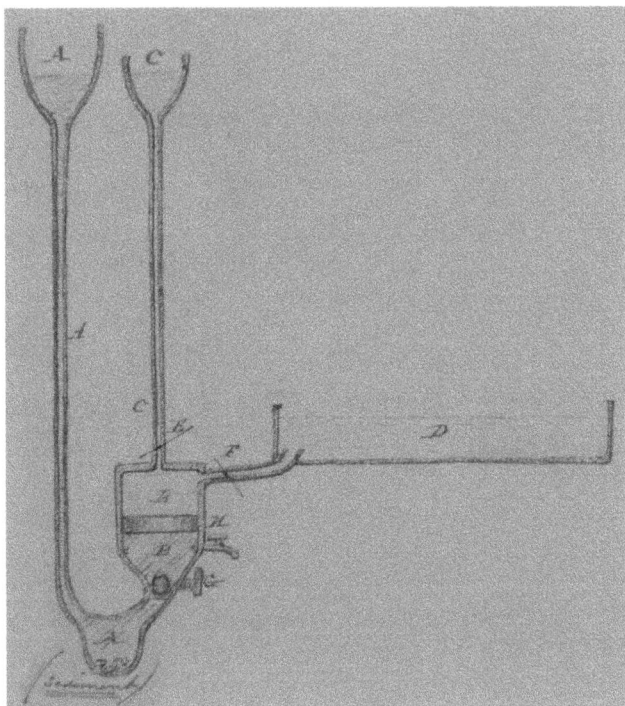

Fig. 12.33:   Plumbing for multi-storey house (1829). IET 1.6.136.

quickly and other pegs *G* caused the clapper *L* to repeatedly hit the outer bell. The device would have helped remind him of upcoming commitments when he was engrossed in his work.

12.17 Items for Family and the Home

The Ideas Book additionally contains numerous simple concepts to improve life for the family. One that is still used today is an adjustable levelling screw fitted to the base of a table leg (Fig. 12.31). As an alternative, he suggests that 'The milled headed screw may be attatched to the middle of the leg instead of the bottom for greater convenience & the spindle prolonged.'[J14]

Another idea was a pulley arrangement that allowed a blind to be drawn up rather than down (Fig. 12.32). For deep windows this would limit visibility into the room while still letting light inside and allowing occupants to see out. The sketch appears in the Ideas Book in 1826, not long after the family had moved to Heath Lodge in Croydon with its floor-to-ceiling windows on the ground floor.

Fig. 12.34:   Temperature-controlled ventilator (1825). IET 1.6.127.

He also developed a plumbing scheme for Croydon in the summer of 1829. In those days water was carried upstairs for ablutions. To avoid this, he envisaged two vessels on the upper floor, with pipes down to the ground floor where they were connected together via a vertical cylinder with a piston (Fig. 12.33). Clean water is taken from one vessel and, when used, deposited in the dirty water vessel. The water pressure differential during this cycle lifts the piston and the clean water rises again to its former level. When the cylinder becomes full of dirty water, the system is recharged by emptying the dirty water through a tap while clean water from the reservoir on the ground floor simultaneously refills the cylinder.

A drawing survives of a 'thermometric ventilator'[J34] that opened automatically at higher temperatures (Fig. 12.34). The concept worked on a principle similar to one of the methods for regulating his battery-operated clock against temperature effects (Sec. 6.3), and included a weight wound around a wheel on the ventilator and resting on a column of mercury. Around the drawing he listed the questions he needed to address in designing the device:

1   What size is the aperture required to be when quite open
2nd What weight would be sufficient to overcome the friction of the wheel and open the ventilator
3d   What weight would be required to overballance *[sic]* the same and shut it…
4th What area of Mercury would be capable of sustaining the said Wt
5th What capacity must the Thermometer have [J34]

He also addressed the stability and operation of candles through an 'Improved Candle-stick'.[J14] It had a 'spring nozzle'[J14] to hold the candle firmly in place even if it was not quite round and the required size for the holder, and a screw to lower the candle when snuffing it out or to raise it in use.

The Ideas Book mentions two toys that he presumably built for his nieces and nephews. One was a mechanism he called 'The Artificial Grinner'[J14] for making amusing faces. A thin piece of india rubber is marked with a face using stretch paint (made with a solution of rubber) and then positioned over the top of a snuff box. The box contains various levers, screws and knobs 'to pull or push the face… into all sorts of funny grimaces',[J14] and is set in motion by winding a clockwork mechanism. The idea is dated 20 August 1829; the next day his brother Charles' wife Catherine died, leaving two toddlers.

The second concept dates from the period of his work on dry piles (Chapter 6). The terminals of a small box of piles alternately attract each end of a painted see-saw complete with little figures. The piles are 'screened from view by painted representation of earth covered with verdure Bushes &c'[J14] and the whole apparatus could be housed in a deep picture frame for safety. He called it 'An Electrical Toy',[J14] perhaps channelling Singer's* remarks about his electric clock.

12.18 Concluding Remark

In total, Ronalds devoted considerably more time to 'mechanical' activities than to his earlier electrical or later meteorological and photographic work, or his art. In doing so he developed a greater number of more diverse devices in that broad category, some of which were highly innovative — yet his attainments as a mechanical designer and engineer are unrecognised. It could be argued that this is due to the limited take-up of his mechanical inventions, but this is an incomplete view given the popularity of his perspective tracing instruments, the enduring success of his tripod stand, the importance of his modified slide rest to both amateur turning and the machine age, the use of other devices by friends and associates, and the later adoption of inventions like the fire finder. A key factor is that he did not seek and even actively avoided public attention through this period whereas his earlier and later work was published. It suited him to be just one of the growing band of mechanical engineers from various walks of life who were together enabling the 'modern' industrial world.

We can sense however that he eventually tired of applying mechanical design almost randomly for himself, his family or others. After 25 years, he found the courage and a fitting opportunity to return to his first love — the combination of electrical and meteorological science — by re-establishing the Kew Observatory. We travel to Kew in the next chapter.

Chapter 13

KEW OBSERVATORY 1842–55
AND BEYOND

Development of Kew's central role in international meteorology and geomagnetism, highlighting Ronalds' contributions as inaugural Director and inventor and the extended struggle for survival.

This chapter, and the next two, concern Ronalds' activities at the Kew Observatory. Robert Scott FRS, Director of the UK Meteorological Office, wrote in his 1885 history of the observatory:

> much of Mr. Ronalds' attention was directed to the subject of atmospheric electricity...
> At the same time the Hon. Superintendent was far from neglectful of other branches of physical research, and he devoted much of his rare mechanical energy to the invention and perfecting of the photographic process for the registration of meteorology and terrestrial magnetism, with which the name of the Kew Observatory has been permanently associated.[K9]

In effect, Ronalds created three substantial roles for himself. The first was being the inaugural Director — developing the observatory's activities, credibility and sustainability, and managing its affairs — and is a focus of the present chapter. His second goal, described in the next chapter, was to re-establish the science of atmospheric electricity; he invented a suite of new equipment that was used to generate extended datasets, first at Kew and then internationally. In his third area of attention (Chapter 15), he helped shape the new invention of photography in creating a series of self-recording instruments. He had considerable success in all three roles, although juggling them stretched him severely. Kew grew to become arguably the preeminent meteorological and magnetic observatory worldwide. Atmospheric electricity observations continued up to the observatory's closure in

1980, while his photographic machines were deployed around the world as one of the most important innovations in meteorology of the era.

A key organisation in the early history of the Kew Observatory was the British Association for the Advancement of Science (BAAS) — now called the British Science Association. The Association had been founded in 1831 by a group including Sir David Brewster*, Professor John Phillips*, Revd James Yates and geologist Roderick Murchison FRS to coordinate and fund prioritised scientific endeavours. Quickly becoming an influential forum, it attracted around 1,000 attendees to its annual conference and associated events including formal dinners, evening talks, promenades, excursions, private parties and perhaps a ball. Harriet Martineau, who attended the 1838 conference, described another side not unusual for such events: 'the occasion is really so sadly spoiled...by the obtrusions of coxcombs, the conceit of third-rate men with their specialities, the tiresome talk of one-idead men...and the disagreeable footing of the ladies'.[B97]

From the beginning, meteorology was a priority for the BAAS. Weather affected everything from agriculture to maritime transport to human health and yet scientific knowledge of climatic processes remained scant. Sir John Herschel*, whose view was always sought and heeded, described meteorology as 'one of the most complicated and difficult, but, at the same time, interesting subjects of physical research'.[A50] An associated interest was the continual perturbations in the earth's magnetic field. Making progress in understanding these global scale phenomena required a global effort. The first step was consistent observation in geographically dispersed locations and over extended time periods. The hoped-for later steps were to discern patterns in the data and then to develop mathematical and physical laws explaining their basis. It would be the 20th century before such theoretical models became credible. A significant advance was made in the late 1830s: Britain set up observing stations at key points across her Empire and joined observatories in Europe and Russia in the effort — it was the first truly international science project. The scale and reach of the BAAS made it an ideal organisation to coordinate the overall response. The Committee overseeing 'the system of Simultaneous Magnetical and Meteorological Observations'[S1] was chaired by Herschel while the Association's General Secretary, Colonel Edward Sabine*, was at the centre of the 'Magnetic Crusade'[A166] for many years.

Another pivotal character in the story is George Airy*, whom we met in Sec.12.10 as Professor of Astronomy at Cambridge. Appointed Astronomer Royal in late 1835, he now led the famous Royal Observatory at Greenwich, to the east of London, which had been founded in 1676 and had a rich history in astronomy. His first priority in the job was to get Greenwich on the magnetic and meteorological bandwagon. A new building for the purpose was soon in place funded by the Admiralty.

The Royal Society had also long studied the weather. In June 1840 the Committee of Physics and Meteorology persuaded the Society's Council to establish their own Magnetic and Meteorological Observatory near London as part of the international effort. Airy was outraged, but quickly turned the decision to his advantage. As he put it in his autobiography:

I wrote to the Council on July 9th, pointing out what the Admiralty had done at Greenwich, and offering to cooperate…Wheatstone looked at our buildings and was satisfied. My estimate was sent to the Admiralty, viz. £150 outfit, £520 annual expense…Most of the contemplated observations were begun before the end of 1840[B115]

There was another Royal Observatory near London too. This observatory had been built in 1768 by King George III, who had a keen interest in astronomy. The new Queen did not seem to share her grandfather's interests and by 1841 it was disused and empty apart from the caretaker Mr Cripps and his family. Just 12 months after their initial proposal, the Committee of Physics and Meteorology recommended that the Royal Society 'secure' this alternative Royal Observatory 'for various scientific purposes'.[K9] This time the Society's Council wisely asked 'what specific scientific purposes' and 'what would be the probable annual expense'.[K9] The Subcommittee comprising Herschel, Sabine and Charles Wheatstone* reported back in February 1842 that it had changed its mind.

By then these gentlemen had an alternative plan. They and others mentioned in these pages, including John Daniell*, John Gassiot*, George and Sir John Rennie*, Thomas Graham*, William Buckland, John Taylor, Henry Fox Talbot and Murchison, contributed to a Subscription for 'a Physical Observatory'.[K9] With this seed funding of around £200, the BAAS agreed to take on the building. The President announced the decision in his commencement address at the annual conference in June 1842, describing the observatory succinctly as providing for the Association: 'a fixed position, a place for deposit, regulation, and comparison of instruments'[S1] and then, less convincingly: 'I am informed that the purposes to which this building is readily and immediately applicable, are of an importance which none but men advanced in science can appreciate.'[S1]

The lack of planning behind the BAAS's decision is remarkable when seen through today's eyes. Issues that were to prove so troublesome for Ronalds had not been thought through: the role of the establishment (and especially its differentiation from Greenwich); its governance; stakeholder buy-in; and the ongoing provision of adequate funding. The BAAS had not even made an estimate of running costs. These issues and the progress made on them during Ronalds' tenure are central themes of this chapter.

Much of the story concerns the struggle for supremacy between the two Royal Observatories. Airy's first foe was an unsuspecting Ronalds but, as Sabine became more prominent at Kew and Ronalds departed, the combat shifted to two opponents much more evenly matched in their trickery — Sabine and Airy. Skirmishes continued for over 30 years.

Airy expressed some understanding of the proposed distinctive roles of the two observatories as early as September 1843 in a letter to Faraday*. The BAAS's observatory would develop instruments, which Greenwich would use for the regular recording of phenomena:

> Their object is, to indicate time of observation and to devise apparatus adapted to it. Mine is, to work myself and to set my people to work like horses in a mill, when the proper observations are indicated.[C125]

A decade later, Kew's focus was outlined in an article in *Annales de Chimie et de Physique*:

> The British Association for the Advancement of Science has founded at Kew…a meteorological observatory led by Mr. Ronalds, which is occupied less with regular observations than in improving observation procedures and conducting meteorological experiments, that is, research which because of its difficulty cannot be trialled in most observatories.[A85]

These ideas continued to change over time to suit individuals' purposes — powerful people like the flexibility of ill-defined rules. In an article Airy and his First Assistant Robert Main FRS wrote in 1851 entitled 'The Observatories of London and its Vicinity' they slipped in that Greenwich was England's 'head quarters of science'.[A84]

If all this rivalry sounds incongruous with the person we have known up till now — it was. Ronalds was not interested in turf wars and strove to enhance the science conducted at both institutions. Sabine warmly acknowledged his achievements at the time and even years after he left, while Airy was the opposite as we will see in these three chapters. Wheatstone proved to be another minor irritant to Ronalds through the period. Ronalds also had many influential supporters in the BAAS ranks. The early history of the Kew Observatory is an important illustration of science leadership in the mid-19th century and has many more characters than previous chapters of this book.

13.1 The Observatory Building

An 1866 article on the BAAS's observatory started, with unusual levity for an engineering journal: 'KEW OBSERVATORY is so called because it is much nearer Richmond than it is to Kew, on the same principle that Primrose-hill is so named because no primroses grow within two miles of it.'[A108] Although the official title had been the King's Observatory or Royal Observatory, it was known as the Kew Observatory throughout the years it was associated with meteorology and magnetism. The misnaming caused significant inconvenience for Ronalds — it delayed mail and baffled visitors.

The observatory's location is 17 km west from London on a bend of the River Thames; it was at that time surrounded by a farm. An isolated situation had the scientific benefit of being remote from human activity that might influence measurements. The disadvantage was difficulty of access at a time when transport was slow and uncomfortable. For Ronalds, the location was quite convenient, being a pleasant 7 km walk from his mother's home at Chiswick (Fig. 3.1). Very soon after starting there he wrote to family friend William Aiton Jnr who ran the Kew Gardens to the north:

> I (naturally) wish to diminish the tedium a little by curtailing the walk from Kew Bridge, which could be done by a distance of about three quarters of a mile (I think) if I could be allowed an entrance at the Botanic Garden Gate, and an exit by a Gate which I think exists nearly in a line with it and the Observatory.[C114]

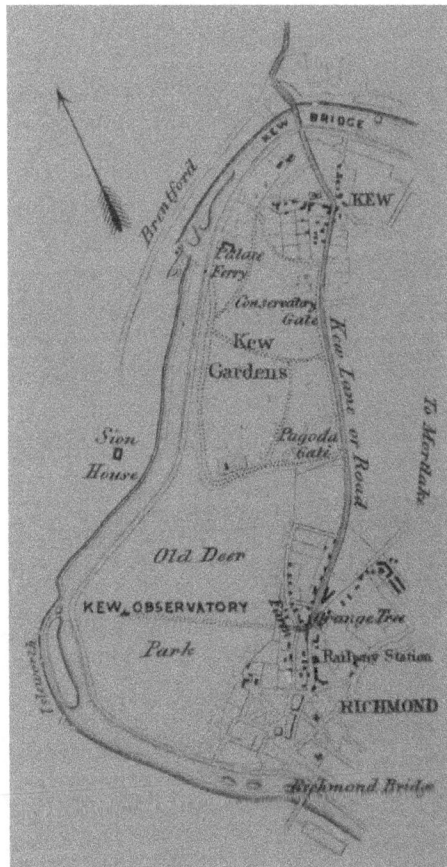

Fig. 13.1:   Ronalds' map of Kew Observatory environs (*c*.1850). IET 1.9.1.

Aiton obliged by giving him a key, and shortly afterwards visited him at the observatory.

Ronalds made a small map to help people find the building (Fig. 13.1), but it was not always successful. When Sabine's assistant Charles Younghusband first visited, he arrived at Kew Bridge to find: 'every one most obstinately confounding the conservatory with the Observatory…then I found that every one omitted the first syllable in pronouncing either word'.[C224] The spectacular glass and iron Palm House had just been opened at the Kew Gardens. Another visitor who had got lost wrote in 1848: 'Your excellent observatory seems to be more known anywhere else than its very neighbourhood.'[C222] When friend Samuel Sharpe* visited, Ronalds provided detailed instructions on how to reach the observatory from the City, giving him two options: 'a Richmond Omnibus'[C206] or a steamboat up the river. Regarding the route home:

> I will now tell you how you must get away from Kew.
>
> You must get into an Omnibus as before, or walk through the gardens with me and get to Chiswick, where you must get a piece of cold Beef & a glass of Wine [C206]

Sharpe replied: '*[I] will meet you at the observatory by 4, unless I lose my way in the Park.*'[C216] By the end of the decade, transport was much simplified: visitors could arrive by train from Waterloo and then walk down a long entranceway to the observatory from just near the Richmond station.

The building itself is three storeys high, topped with a roof terrace and a central dome (Fig. 13.2). It is surrounded by a mound supported by subterranean

Fig. 13.2:   Kew Observatory in 2011.

vaulting to reduce the effects of flooding. The middle rooms are octagonal in plan with walls lined with glass cases and, before long, these were used to display instruments from the Royal Society as well as the BAAS. In 2011 one of these rooms still held a large stone tablet commemorating the various observatory superintendents.

The Commissioners of Works had apparently advised the Queen in 1840 that the building was in a dilapidated state, and her initial decision was that it be pulled down.[J105] The Subcommittee in contrast had informed the Royal Society Council that it was 'in excellent repair'.[K9] Ronalds found that there was significant dry rot, including in the display cases, the kitchen, water closets and floor joists, and also increasing levels of water ingress that was damaging the ceilings. Petitioning the Commissioners of Works to make essential repairs to the building fabric was an ongoing role for him during his tenure. It is now Grade I listed.

13.2 The First Years

There is no record of Ronalds participating in the machinations that led to the BAAS taking on the observatory. He did not contribute to the private subscription. It seems that he had some input to the accompanying Prospectus, however, drafted by Wheatstone, because his ideas are apparent in that short document. One specific item was:

> A complete series of apparatus for experiments on atmospheric electricity. For such investigation the locality is peculiarly adapted. Nothing of the kind at present exists in England, and yet there is no subject in meteorological science for which so much remains to be done.[K9]

This proposal must have been well received because it was where the collected monies were applied. Also mentioned in the Prospectus were objectives akin to his interests in improved instrumentation and precision measurement:

> A repository and station for trial of new instruments…
> a station to which persons…may bring their instruments for the purpose of comparison with the standard instruments there deposited…
> any person desirous of doing so may understand their construction and acquire their use…
> As complete a collection as can be gathered together of measuring instruments…for the purpose of obtaining accurate quantitative results[K9]

Another item had Wheatstone's name on it:

> a universal meteorograph, which will accurately record half-hourly indications of various meteorological instruments, dispensing entirely with the attendance of an observer[K9]

Ronalds was able to achieve all of those broad goals within a decade. In doing so he utilised and built on equipment and experience he had gained 30 years earlier, particularly in the novel areas of atmospheric electricity and self-recording (Sec. 6.4).

The first records of his involvement at Kew are dated November 1842. He and Wheatstone had already inspected the observatory together by this time and Wheatstone had viewed his meteorological instruments at Chiswick. Wheatstone wrote: 'Do not forget to send me, at your earliest convenience, your plan for the atmospheric electric apparatus…Let me also know what part of the apparatus you intend to provide.'[C107] Ronalds seized the initiative and sent a project plan running to 12 pages plus drawings. He had designed the necessary alterations to the extent of describing to Wheatstone the form of reinforcing ring to be placed around the hole cut in the roof of the dome for the conducting rod. His plan also encompassed the broader needs of the facility, including general repairs and personnel. He suggested the value of having two employees — an observer and a part-time 'jobber in fiddle faddles' to avoid having 'to send an instrument to London…for a screw.'[C108] Ronalds would focus on instrument development and training — he later called himself the 'experimentalist'.[J63]

At the June 1842 BAAS conference, £200 had been 'placed at the disposal of the Council for upholding the establishment in the Kew Observatory'[S1] for the first year. Part of this was to employ Sergeant John Galloway, who came from the Royal Artillery, as observer. Another £50 was granted to Wheatstone for his meteorological recorder. These were significant outlays for the Association, although only a fraction of the amount Airy received from the Admiralty for meteorological observations at Greenwich. It was late 1849 before a second assistant was able to be hired.

The initial governance arrangements emerged over the ensuing months. The size of their commitment meant that oversight stayed with the BAAS Council — they wanted to keep a tight rein on expenses. Sabine became 'the organ of Council, in the matter of the Observatory'[C141] although he could not make financial decisions. Next in the pecking order was Wheatstone, who was also on the Council and had his own project relating to the observatory. He had advised Sabine that he was 'willing to undertake the responsibility' for 'specific grants for important objects to be there carried on'.[C104] Of the three prime movers he lived and worked the shortest distance from Kew, in London. Sabine was based at the Royal Artillery Barracks in Woolwich — well to the east of London, past Greenwich (Fig. 3.1) — while Herschel lived in Kent. It was for these reasons that Wheatstone provided the initial instructions to Galloway and also assumed management control for Ronalds.

Ronalds had made it clear in early correspondence that he did not intend to have a leadership role in the facility. His plan submitted to Wheatstone included the phrase: 'knowing nothing about the views of the Society *[BAAS]* as to expenditure, & having no interest to consult beyond that of the science itself…',[C108] while his letter to Aiton noted:

> My whole & sole motive is the advancement of a branch of science which is (all over Europe) daily increasing in Interest viz. atmospheric Electricity. Having volunteered to direct the erection of an apparatus for observation on a principle which I invented many years since (& which your old friend Mr De Luc*[*]* of Windsor highly approved)…[C114]

He certainly did not apply to be Observatory Director, although at least one person did, as becomes apparent later. For Ronalds it was natural to offer his services in an honorary capacity and provide much of the equipment himself in an informal arrangement.

Nonetheless the reporting arrangement must have felt awkward to him. Wheatstone, who was a youth when they first met, was now being hailed around the world as the inventor of the electric telegraph and had positioned himself at the centre of scientific circles. Ronalds meanwhile had all but disappeared from the scene. He seemed to feel their 14-year age difference keenly — his plan submitted to Wheatstone is littered with self-deprecating remarks, particularly concerning the currency of his electrical knowledge:

> you may skip some diffuse remarks without offence; for I well know them to be founded in old fashioned ideas & proceedings…
>
> Many new facts &c have grown up…of which I am totally ignorant no doubt.[C108]

For his part, Wheatstone did not treat his older colleague with particular respect or kindness — his letters are quite brusque. He seemed to like the kudos of being seen as the leader. He was at this time demonstrating the same trait regarding Cooke (with the electric telegraph) and Bain (with the electric clock and the printing telegraph), as discussed in Chapter 7. The ill-fitting connection between Wheatstone and Ronalds was reasonably short-lived but perhaps contributed to the occasional friction between them over the ensuing years.

Not unexpectedly, there were very few visitors in those first months — Ronalds and Galloway were largely left alone to set up the observatory. Ronalds stipulated and managed the works necessary to make the building serviceable. Galloway and his family resided in the basement along with Cripps and a bedroom was fitted out on second floor for when Ronalds stayed overnight. Accounts in Ronalds' hand show that he was soon monitoring all expenditures at the observatory since start-up in June 1842 when Galloway arrived. He was already effectively running the observatory.

He kept Wheatstone fully informed of progress and costs in writing and by visiting him in the City. All expenses he wished to incur needed to be first agreed by Wheatstone, who later reviewed the invoices and receipts. Wheatstone in turn passed the documentation up through Sabine to the Council for approval and payment. The process was longwinded and both tradespeople and Ronalds became frustrated, Ronalds writing:

> I have lately made <u>four</u> impotent journeys to see you Three for the purpose of obtaining your authority for incurring some additional expenses...which I <u>believe</u> you approved...and one for the purpose of getting you to sign the inclosed bill of Messrs Powell in order that they may receive the amount [C128]
>
> I am sure that you so well knowing the excessive vexation of great delay in the progress of observation will know how to excuse my impatience.[C129]

He began to expedite the work by instigating and funding purchases himself. The understanding was that if Wheatstone did not subsequently approve the expense, the property would notionally belong to him rather than the BAAS. Either way, he wrote, 'I shall be content'.[C117]

Even more annoying for him was that his ideas were perceived to belong to another. Wheatstone was eager to show at the 1843 BAAS annual conference that progress had been made in his new area of responsibility by tabling what he referred to as 'my report'.[C116] Ronalds duly prepared for him a written description and drawings of the atmospheric electricity apparatus together with a journal of the first six weeks of observations of the electrometers and meteorological equipment.

The first section of Wheatstone's report, which was a detailed explanation of his own meteorological recorder, ended: 'A description will also be presented to the Association by Professor Wheatstone of an Electrical Apparatus'.[S1] The second section noted that while 'many of the contemplated objects' of the observatory had not yet been carried into effect 'a very perfect and efficient apparatus for making observations on the electricity of the atmosphere has been established.'[S1] Ronalds was mentioned only in relation to two technical details. The impression that the atmospheric electricity equipment was at least partly Wheatstone's innovation lingered for years.

Ronalds realised that he had a choice. He could remain fully focused on his beloved science and stay in the background, or he could step up into a formal leadership role. He chose the second option and attended the 1844 BAAS conference at York himself — although he was very late getting there. Wheatstone wrote to him from the event:

> I have been anxiously expecting you here every day, but as the meeting is now more than half over I have given over all hopes of seeing you. The Report of the... Observatory has been deferred until Tuesday morning the last day of the sectional meetings, and I hope I shall receive it from you against that time...Your nephew has called several times upon me to enquire whether you had arrived *[this would also have been Dr Edmund Ronalds' first meeting]*. Do not fail to send me the Report immediately.[C137]

Ronalds was clearly in a panic pulling together a description of the observatory, the first full year of atmospheric electricity and meteorology results and his early photographic registration curves. It was a very long time since he had submitted a paper to scientific peers and it would have been his first public address of any size. He perhaps also had in mind the potential benefits of allowing little time for Wheatstone's scrutiny of his work — his report started by noting that the electrical laboratory was: 'fitted up and supplied with instruments under my direction, and principally in accordance with a plan which I had in November 1842 stated to Professor Wheatstone'.[R12]

He arrived in time. He shared the rostrum that morning with such luminaries as Herschel, Brewster, Phillips, Romney Robinson*, and chemist and photographer Robert Hunt. It was not an instant success. The *Literary Gazette* in covering the session noted that 'Mr Ronalds' report formed a volume of most clear and admirable registries. It was too long, however, to read; and Mr Ronalds was too unaccustomed to address a numerous assembly to give a verbal account.'[1] He was ill-prepared for the conference format and had got bogged down in the detail. Fortunately others there could appreciate the overall value of his work. The *Athenæum*[2] advised that Romney Robinson went on to underline the novelty of several aspects of Ronalds' apparatus; knowing him through Sir James South*, Romney Robinson had visited Kew a few months before the conference.

By the end of the day, Ronalds had been awarded his own grant of £50 to conduct electrical experiments and £57 for completion of his electro-meteorology observatory. In addition, £150 was allocated for Observatory operating expenses and £30 for the purchase of a 'barometrograph' (Table 13.1). As Observatory Superintendent, he would now be interacting directly with Sabine rather than through Wheatstone. A further resolution was:

> That the Meteorological Observations made at the request of the Association be discontinued, and the instruments transmitted to the Kew Physical Observatory, except...where the observations can be continued gratuitously.[S1]

The last decision especially highlights the esteem in which Ronalds' endeavours were held, but was unwise for stakeholder relations — numerous meteorological scientists had been supported by the BAAS in past years. To top it off, all of this was achieved while seemingly enhancing his relationship with Wheatstone — the latter's next letter was decidedly more polite:

> It is a long time since I have paid a visit to the Observatory. I intend to do so on Wednesday next with one or two friends from Ireland. I shall have great pleasure to meet you there if convenient, but do not put yourself at all out of the way.[C145]

---

[1] *Literary Gazette* 26/10/1844 689.
[2] *Athenæum* 19/10/1844 954.

Table 13.1:   Funds Provided to Kew Observatory.

Year	Operations (from BAAS) (£)	Specific projects			
		Amount (£)	Grantee	Purpose	Donor
1842–43	200	200	Ronalds	Atmospheric electricity apparatus	Private subscription
		50	Wheatstone	Self-recording meteorological apparatus	BAAS
1843–44	200	—	—	—	—
1844–45	150	57	Ronalds	Electrical apparatus	BAAS
		50	Ronalds	Electrical experiments	BAAS
		30	Kew	Purchase of Kreil's barometrograph	BAAS
1845–46	150	50	Ronalds	Self-registering equipment	Royal Society
1846–47	150	10	Kew	Purchase of Romney Robinson's anemometer	BAAS
1847–48	150	—	—	—	—
1848–49	100	50	Birt	Reduction of atmospheric electricity observations	BAAS
1849–50	250	—	—	—	—
1850–51	300	100	Ronalds	Self-registering and other meteorological equipment	Royal Society
		100	Ronalds	Experimental trial of new equipment	Royal Society
1851–52	300	175	Stokes	Friction index of gases	Royal Society
		150	Kew	Development of standard meteorological equipment	Royal Society
1852–53	200	261	Kew	Balloon ascents	Royal Society
1853–54	200	—	—	—	—
1854–55	500	140	Kew	L'Exposition Universelle Paris	Royal Society
1855–56	500	180	Kew	Photoheliograph	Royal Society
		250	Kew	Gas reticulation	Royal Society

After the meeting, Ronalds liaised with the Assistant General Secretary, Phillips, about what was required for the printed proceedings. Phillips replied that his report should be 'not curtailed.'[C140] He added: 'I wish Wheatstone had finished his Instruments.'[C141]

13.3 Honorary Superintendent of the Observatory

Ronalds presented a comprehensive Annual Report to the BAAS conference each year after that. In it he outlined the state of the building and its contents, summarised all observations made, and described in detail each of the new instruments he had developed during the year as well as investigations conducted towards improving operations and equipment. He ended with proposals for new work to attract further funding. As he noted in his sixth report in 1849: 'the number of such-like propositions has accumulated so much faster than the means and time required for their execution, that the catalogue arrives at an almost despairing magnitude.'[R19] His presentations were usually chosen for write-up in popular magazines like the *Literary Gazette* and the *Athenæum*.

He had entered a different world — he was in the public eye and he was working as part of a large team. He needed to lead subordinates, influence scientific peers and take direction from those holding the purse strings. Up until now, his activities had been private. The new world had both advantages and disadvantages. On the plus side, there were now many people who knew about and appreciated his science and aided him in achieving his goals. He could work at a greater scale and quickly deliver benefit (as he always wished to) because development funding was available, as were ready markets for his products. The negatives included loss of the freedom he loved, and being increasingly caught up in management and administrative issues as well as scientific jealousy at individual and institutional levels. He had little experience or aptitude in many of these interpersonal matters.

Ronalds was now interacting regularly with the most powerful scientists in the country — people like Sabine and Airy. This was never his style but fortunately Sabine proved to be a great mentor and friend. Sabine was highly strategic — always thinking ahead to future goals and how to attain them — and could provide sage advice when Ronalds was caught up in the pressures of day-to-day operation. Recognising Ronalds' honorary status and scientific nature, his style was to suggest or ask rather than instruct, despite his military background. He had a warm and welcoming manner — '[I] am always glad to see you'[C246] — and also took a strong personal interest in Ronalds' well-being. Above all, he recognised that Ronalds' genius was a critical ingredient for Kew's success. With their highly complementary skills, their relationship over the next eight years was much more productive and enjoyable than that with Wheatstone. One of the few downsides was the long trip to Woolwich that Ronalds now needed to make frequently. Years later, Sabine acknowledged 'the great inconvenience, (of which you are so kind as never to complain)'[C309] of going all that way to see him.

One of their early interactions showed how Sabine had the confidence to streamline business decisions. Ronalds had requested an abnormal outlay for a better heater: 'I get dreadfully wet &c in walking through the grass & snow.'[C146] Sabine replied immediately:

> as you say the matter is pressing, and as the expense (£7) will not break down our finances for the year, I will take upon myself to request you to be so kind as to get the new kind of grate which you propose at once, and without waiting for the meeting of the Council.[C147]

This would have been a refreshing change for Ronalds. He needed to wait several more years before an efficient governance mechanism was agreed whereby he could submit his annual budget for approval, upon which funds were released to him.

Ronalds' relationship with Airy was also very friendly at this time. He had been visiting Greenwich regularly to assist in the development of atmospheric electricity apparatus there. Airy estimated that the trip between Kew and Greenwich took two and a half hours each way, so it was another considerable commitment for him. Airy had written at Christmas 1843: 'I am very much obliged by…the interest which you take in our Electric matters…would you stop an hour later and dine with my family at ½ past 3?'[C130] He also invited Ronalds to Greenwich's Annual Visitation dinner in June 1844.[C134] The personal interaction indicates the importance of Ronalds' work in Airy's eyes: he did not introduce James Glaisher*, the Superintendent of his 'Magnetical and Meteorological Department', into the relationship for over two years. When he did so, he told Ronalds: 'I have really heartily heard Mr Glaisher's report of his visit to Kew (which I think has made an impression on him that will last to the end of his life)'.[C156]

Ronalds and Glaisher became good friends — they used to dine at each other's homes as part of any observatory visit. The closing of a typical letter from Glaisher to Ronalds reads: 'Mrs G. unites with me in Kind respects to yourself, to Mrs Ronalds and to Miss Field [Jane's sister]'.[C172] Another Greenwich colleague appeared in February 1846 when Galloway advised Ronalds of the visit of 'a Gentleman by the name of Mr Brooke'.[C163] Dr Charles Brooke* was to become Ronalds' rival in the development of photographic self-registration, as discussed in detail in Chapter 15.

Visitors had started coming to Kew as soon as the atmospheric electricity apparatus was erected and now increased steadily. Sir William Snow Harris was one of the first; Harris is known in particular for his lightning conductor design for ships. Early German callers included Professor Justus von Liebig from the University of Giessen (mentioned in Chapter 9), and Berlin physics professors Gustav Magnus and Adolf Erman (son of Paul Erman of Sec. 6.4). Ronalds dined with Magnus, although his guest 'was obliged to retire during dinner with a pain in his guts'.[C371] Another renowned visitor was Hans Oersted, who had discovered electromagnetism (Sec. 7.2): 'D^r Œrsted is very desirous of having the pleasure of making your acquaintance, & of seeing your apparatus'.[C175]

After Ronalds' early photo-recording instruments were built in 1845, the observatory's reputation was made and requests for visits were never-ending — eminent scientists, representatives from most of the observatories and scientific academies in the UK, Europe and the Colonies, and several from the USA. Many expressed interest in having similar equipment made for them and being trained in its use. Ronalds always tried hard to assist them, despite the other calls on his time, but the high instrument costs meant that only a small portion of requests came to fruition. Dr Carlo Matteucci, esteemed Professor of Physics at the University of Pisa, was an early international example in 1844. He was 'very desirous of seeing the Electrical Apparatus at Kew'[C139] and Ronalds went on to provide him atmospheric electricity equipment and instruction.[C213] He seems to have visited again in 1846.

These were the beginnings of Kew's remarkable track record in supplying quality instrumentation around the world. It was late 1850 when Sabine confided to Ronalds: 'My head is full of apparatus for making standard Meteorological Instruments of different kinds — we ourselves at Kew to be the makers; & the instruments to bear our stamp'.[C312] Instruments verified at Kew started to be so marked in early 1852.[S1] The 'Kew-pattern' designation was in use by 1867 and the famous 'KO' hallmark of quality appeared in the late 1870s.[K18]

Other scientists also used the facilities on occasion and provided equipment — particularly, it seems, instruments beginning with 'a'. At the request of the BAAS, Robert Hunt placed his new actinograph (an instrument for measuring the light available to expose a photograph) at the observatory. Dr Samuel Horatio Banks was another early guest who brought his experimental registering anemometer. Follett Osler later offered Ronalds his self-registering anemometer. Over time, sufficient instrumentation was in place to enable comparison of new apparatus with existing 'standards', and this became another of Kew's critical roles. Samuel Burt Howlett, Chief Surveyor at the Board of Ordnance, for example, erected his experimental anemometer at the observatory in 1850 for comparison with Ronalds' instrument. Ronalds prepared a statement of its performance for the head of the Royal Engineers, Field Marshal Sir John Fox Burgoyne.[J84]

Family and friends also visited — he would have been very proud to show off his observatory. Peter Martineau and Dr Edmund were there on 1 August 1844 and

Hammersmith friend James Atkinson Peacock and his sisters visited later that month. Some years later, first cousin George Field Jnr wrote:

My dear Frank

I — a neighbour of mine a M^r *[Robert]* Hudson, a great Botanist, & *[Edward]* Cook*[e]*, the Marine Painter, are going to meet Sir W: Hooker *[who took over from Aiton Jnr]* on Tuesday next to look over the Gardens at Kew. I should like to shew them your wonders, if it would suit you to be at your post on that day. We return to my House to dinner & if you will return with them in my Carriage & dine & sleep at our House we should be much delighted; & I will send you then to Kew the next morning.

I have got some works of Art, which I flatter myself would please you to look at, & I will ask Reid*[3]* to meet you. So pray say you will come [C248]

Ronalds must have been too busy to accept the dinner invitation, because George Jnr later advised:

I cannot express to you how much pleased Reid was with your exhibition yesterday…

My neighbour Gasiot *[sic]* dined with us, & he was talking to Reid much about your works [C249]

These would have been the happiest times of Ronalds' tenure at the observatory. He was very busy but doing what he loved: designing and building new instruments, collecting detailed meteorological data, conducting novel experiments, and assisting others in the scientific community (Table 13.2). In addition to the creative work he was also performing the other roles expected of the Director of such an establishment — setting science direction, attracting funding, ensuring project delivery, facility upkeep, financial management and stakeholder engagement. Galloway focused on the atmospheric electricity observations and some clerical activities, including making copies of Ronalds' important letters. Ronalds admitted to Glaisher: 'Since the commencement of my work at Kew Observatory at the beginning of 1843, I have taken scarcely any other holiday than a partial one of few weeks lately spent at Brighton with my family and almost by compulsion.'[C164] He was sometimes not getting home from work until 10 p.m. Sabine agreed that Galloway's wife Mary should be paid an allowance for providing his meals 'As the observatory has the advantage of such frequent attendance from you'.[C176] Ronalds was also continuing to work in an honorary capacity. The comparison with Greenwich is stark. Airy's annual salary at this time was £800, with an additional £300 per annum from the Civil List and his home provided at the observatory. He always left for the day by 2 p.m. to take a brisk walk before an early dinner. Glaisher in the magnetic and meteorological department was aided by three assistants and

---

[3]General Sir William Reid FRS (1791–1858) was an officer in the Royal Engineers and senior Colonial administrator, with scientific interests in meteorology. He was related through marriage to George Field Snr.

Table 13.2:   Chronology of Key Activities at Kew illustrating Ronalds' Simultaneous Areas of Focus.

Year	1843	1844	1845	1846	1847	1848	1849	1850	1851	1852	1853
BAAS conference	Cork	York	Cambridge	Southampton	Oxford	Swansea	Birmingham	Edinburgh	Ipswich	Belfast	Hull
Personal / Family					Alfred's upset	Prize	Alfred's book	Brentford problems		Jane's estate	Edmund's mill loss
Observatory	Establishment	Early visitors	Closure threat				Decision to close			Kew Committee	
Assistant			Galloway				None	Birt		Welsh	
Atmospheric Electricity:											
Kew Instruments	Set-up	Portable · Night electrs	Coulomb								
Observations						Telegraph	Analysis	Round-house			
Dissemination		Greenwich	Pisa		Bombay		Toronto		Arctic · Trivandrum	Madrid	
Meteorology	set-up	Storm clock			Kite			Instrument screen	Hygrometer	Std thermometers	Balloon
Self-registration:											
Kew Instruments		Resin plate	First generation	Declination Mag	Compensated barograph			HF · VF	Upgrades		
Observations							Toronto		24/7 Trial		
Dissemination							Toronto			Madrid	Oxford
Photography			Talbot paper process		Paper improvements		Daguerreotype	Gelatine prints			
Illumination			Daylight or Argand lamp with olive oil				Daylight or Rumford polyflame lamp				Coal gaslight

several 'computers',[A175] while Galloway was making observations from sunrise until late evening with no backup apart from Ronalds.

This busy but happy period did not last long. At the very next BAAS conference, held in Cambridge in June 1845, one of the agreed recommendations was: 'The Council to take into consideration…the expediency of discontinuing the Kew Observatory.'[S1] Kew was not only the major budget item but also the Association's first recurrent expense. It is not surprising that this could draw disaffection from two quarters — the head of the other observatory near London (Airy) and those who had previously attracted the funding for their own research in such topics (like Wheatstone perhaps?).

The timing is noteworthy. It was also resolved at the conference 'to encourage by specific pecuniary reward the improvement of Self-recording Magnetical and Meteorological Apparatus'.[S1] A prize of £500 was offered by Airy's masters at the Admiralty. Just after the conference Airy and Brooke began to develop self-registration equipment at Greenwich — Airy had forgotten the agreement that they would focus on observation rather than instrument development. Ronalds had already been working on the problem for some time and his first instrument was complete. He had applied for funding at the conference to extend his apparatus to other meteorological and magnetic parameters. Not only did the threat of closure hang over his head all the next year, but he was also starved of funds to continue his instrument development. He made mention of the 'continual source of anxiety'[J63] in his 1845–46 Kew Observatory Journal.

He struggled on, borrowing equipment and probably purchasing items himself. By the end of March 1846, Sabine felt confident that they could attract funding from another source: 'I am delighted to hear of your success in the self-registering apparatus…You shall not be long without…money — but I have to consult my colleagues.'[C166] Wheatstone provided a little more detail the next day:

> Dear Ronalds
>
> Col[l] Sabine and myself propose to apply to the Council of the Royal Society for a grant from the Wollaston[*] fund to assist you…Can you contrive to send me… some specimens of your photographic registering [C167]

Wheatstone and Sabine were both members of the Council, which was very useful in gaining research funding. The application for £50 was successful and Ronalds expressed his gratitude to both for 'procuring for me that "sinew of Science" Cash.'[C168]

Straight after that win, a Committee was formed for the review of the observatory. It included the BAAS's President that year (Herschel), as well as Airy, Sabine, Wheatstone and Graham. They visited Kew on 7 May 1846. The good guys won and their recommendation that the establishment be maintained at its present level was announced at the annual conference in September. Among the generalities, including that it 'has already become a point of interest to scientific foreigners, several of whom have visited it',[S1] the tangible reasons given concerned the quality and success of Ronalds' work in his two priority areas. The Committee highlighted his 'actual progress'[S1] in self-registering apparatus. Concerning atmospheric electricity, they noted:

> that a systematic inquiry into the intricate subject of atmospheric electricity has been carried out by Mr. Ronalds, which has been productive of very material improvements in that subject, and…[S1]

as Murchison, the incoming President, continued in his opening address:

> …to which no higher praise can be given, than that it has, in fact, furnished the model of the processes conducted at the Royal Observatory of Greenwich…I earnestly hope it may be sustained as heretofore by annual grants from our funds, particularly as it is accomplishing considerable results at very small cost.[S1]

Murchison had visited Kew not long before the conference.

There was considerable harbour-side frivolity and the Southampton conference should have been a happy one for Ronalds, but it was not. He had a disagreement with Wheatstone during the event and with Airy afterwards. His Kew colleague Banks wrote to him soon after it had wrapped up:

> you were not at the "<u>blow up</u>" on board the "<u>Excellent</u>" *[sailing gunnery ship]* or at the "<u>blow out</u>" at the admiral's!…
>
> I hope you are more at ease now.[C183]

The first problem was the Presidential speech. As published in the newspapers, Murchison had said that Ronalds was working 'under the suggestions of Prof. Wheatstone'.[4] This credit was removed with Sabine's assistance[J68] for the formal record of the address in the BAAS proceedings. Ronalds was always very sensitive about the originality of his ideas, but his emotions would have been heightened by other events at this very moment — Wheatstone and Cooke were achieving phenomenal financial reward and public recognition with the sale of their telegraph patents to the Electric Telegraph Company (Sec. 7.2). Ronalds' depth of feeling is made clear in a letter to Airy: 'There was not <u>One</u> exp^t or instrument made under <u>his</u> suggestion. Sir Rod^k has kindly consented to correct the error in future prints… <u>I</u> will not be deemed a Charlatan, small as the Inventions themselves may be.'[C185]

Ronalds also appeared to have another altercation with Wheatstone. Glaisher's letter, two days after Banks', started:

> I do feel much grieved at the desponding tone in which have written to me…That you have been subjected to much annoyance and not quite fair play is evident…
>
> It was also annoying that Prof. Wheatstone should have been president of the

---

[4] *Athenæum* 12/9/1846 933–8.

section, as he appears to have acted from a very mean feeling, but such a one he has been accused of more than once.[C184]

Herschel was President of the Mathematical and Physical Science Section but was perhaps too busy to fulfil both that and his overarching role that year. The senior Vice-President Brewster was prevented from attending by an accident. This left Wheatstone, as another Vice-President, to take the Chair. The Section meeting was where Ronalds and Brooke both presented their reports on photographic self-registration, and George Dollond introduced his registering apparatus using pencil and paper. Wheatstone certainly had scope there to influence the proceedings.

The pressing need for self-recording instruments ensured that the Section meeting was well covered in the literature. Details were provided in the *Literary Gazette* and the *Athenæum*, and these were reprinted across various other publications. As the first-named magazine noted in its write-up, the 'progress' of these instruments was 'a point essential to the advancement of meteorological science, and occupying the thoughts of numerous observers'.[5] Brooke's paper was the first read and early in its description the *Literary Gazette* advised that 'Some photographs were exhibited, and were much approved of by the president…and other leading members of the Committee who were present'.[6] Perhaps things went downhill from there. Ronalds would certainly have lost his confidence. He might have been hoping for additional funding to continue his work, which Wheatstone as Section President declined to support. Alternatively, it might have been Wheatstone who recommended against printing Ronalds' full report, where his precedence in photo-recording was made clear. Wheatstone's behaviour is interesting as other evidence in Chapter 15 suggests that Ronalds' photos were superior to Brooke's. Ronalds' speech was the next day and the same article noted of his success to date that 'This must be gratifying to Mr Ronalds, who has from the first so ably devised and conducted the experiments and observations at Kew.'[6] The *Athenæum* made no mention of Brooke's work, although the other two papers were summarised.

Things did not get better after the conference. Ronalds had quoted some flattering remarks made by Glaisher on the quality of his photo-registration and these were reprinted in the magazines. That did not suit Airy at all and Ronalds received a formal knuckle-rapping. He then suffered a bout of illness and was away from the observatory for several weeks. Pressure was starting to build, with his long hours, sustainability challenges and the race he found himself in with Brooke. Up to now he had been assisting Greenwich rather than competing with them and he would have hated the idea of a contest.

---

[5] *Literary Gazette* 26/9/1846 835.
[6] *Literary Gazette* 10/10/1846 874–5.

It did not take long for the observatory to be called into question once more. The same review committee was reappointed in May 1848, again chaired by Herschel. Ronalds was asked to address a range of points in his submission to the committee, including:

- The condition of the observatory,
- What had been achieved since the previous committee's report,
- The process and probable cost of analysing the observations already made,
- If he thought the continuance of the Kew Observatory desirable, a statement of respects in which the observations would differ from those now made at Greenwich,
- Supposing the observations at Kew were not to differ, the reasons inducing him to recommend their continuation at the two stations,
- A fair estimate for the future annual cost of maintaining the observatory, should the Council take the same view as he of the expediency of continuing it.[C213]

He also added 'a complete catalogue of all the property of the Association on the premises';[S1] this would have been a time-consuming task as books, instruments and meteorological reports from other observatories had been arriving steadily for some years. He distinguished whether each item had been funded, was a gift, or on loan.

It was much more difficult to justify an important role for Kew this time around. Greenwich was not only strongly funded by government, but now had its own self-registering equipment as well as a version of Ronalds' atmospheric electricity apparatus. Any other new instruments were less pressing. One of the few courses open to him was to argue the superior quality and novelty of the Kew instruments, experiments and results, which was an unenviable position with Airy on the Committee.

He managed to inject some passion into his nine page submission:

Would it not be of real public advantage to test at Kew, the instruments used by Gentlemen who furnish reports for the Registrar General…Have none of the gentlemen private observers who have visited Kew, derived any benefit thereby? Have two meteorol[l] Instruments coming out of different Shops ever agreed? Does any Shop in London (or Paris) possess an electrometer convenient for meteorol[l] purposes? Does the distinguished and justly celebrated honorary secretary of the BA really grudge £150 p ann[m] for even the prospect of establishing an efficient meteorol[l] proving house, at Kew, whilst he is doing so much good work in his excellent magnetical proving house at Woolwich? Blame me as much as you please, for not having done more than I have, but do not let the establishment fall entirely for want of proper support & assistance.[C213]

Sabine appeared to be at best neutral in the deliberations. He wrote to Herschel:

> I cannot bring myself to think with M[r] Ronalds, that Kew could be maintained & do all that it now does, & all in addition that M[r] Ronalds proposes should be done... upon the present funds, i.e. £150 a year [C214]

He and Wheatstone had come to realise that effective running of the observatory required an annual outlay of about £450,[C212] which would cover operating expenses and employment of two senior assistants — much the same team that Ronalds had suggested to Wheatstone in 1842. Herschel was still hopeful however that a way could be found to continue the work:

> If the funds of the assoc[n] were in a flourishing state, not only would it be desirable to maintain but to extend the course of obs[ns] carrying on...I am of opinion that the association ought not, except on very urgent grounds, to throw up the observatory [C210]

Personally, Ronalds came through the review well, but inevitably it was bad news overall. The committee's report to the BAAS conference quoted extensively from his submission before summarising very positively on Ronalds' technical achievements in his two principal arenas. It continued:

> The question as to the expediency of continuing the present expenditure of the establishment has occupied the anxious attention of the Committee...they have taken into consideration the state of the funds of the Association, and also the circumstances of the establishment itself, which they are of opinion cannot for the future, or even for a single additional year, be carried on in a manner satisfactory to the Association *on so low a scale* of expenditure as that which, by a fortunate conjunction of personal circumstances eminently favourable, has hitherto been found practicable...they see no course open but to recommend its discontinuance from the earliest period at which it shall be found practicable...resigning it into the hands of Government [S1]

Funding at a reduced level was allotted for the interim period until alternative arrangements could be made, together with a grant of £50 'to be employed in the reduction and discussion of the important and unique series of Electrical Observations which have been made at Kew under the superintendence of Mr Ronalds'.[S1] It was agreed that astronomer William Radcliffe Birt would be a good person to perform the analysis; he had performed similar computational work for Herschel and had become very interested in the observatory.

The *Literary Gazette* reported the decision:

> The grant of 100*l.* for the Kew Observatory was read, together with the draft of a resolution by Col. Sabine, expressing the concurrence of the General Committee to the opinion of the Council as to the expediency of discontinuing Kew Observatory... Throughout the discussion that ensued, a general feeling of regret prevailed, and also an unwillingness to risk the disturbance of the continuity of the observations.

It seemed however to be generally understood that no positive steps would be taken before the meeting at Birmingham; so the resolution passed.[7]

It was a difficult situation but the BAAS had not handled it well. The funding quantum to Kew constituted 70 per cent of the overall grants made that year. Nonetheless, it was smaller than those of previous years, which the review committee had regarded as much too low, and the General Committee endorsed to be 'confessedly inadequate'.[S1] The idea of seeking government funding had been dismissed by Herschel even before the conference:

> I cannot give my support to an application to Gov[t] to take on itself the support of the Kew Observatory because I am not sufficiently impressed with the scientific necessity of such an establishment unconnected with the peculiar objects which made it desirable for the British Association as their private Observatory.[C217]

It was not at all clear what other options might be available. Meanwhile, Ronalds was working day and night for no financial reward and even providing some of the needed equipment and materials himself to keep the observatory afloat. Rather than providing relief to him, the BAAS's decision increased both his workload and his uncertainty. Straight after the meeting, Sabine gave Galloway his three months' notice and the atmospheric electricity observations stopped.[C223] He wrote to Ronalds: 'I agree with you that we might make M[rs] Galloway a present.'[C271]

In the lead-up to the conference Ronalds had printed a paper summarising his scientific work at Kew, entitled: *Epitome of the Electro-Meteorological and Magnetic Observations and Experiments, &c. made at the Kew Observatory*. It was based on his submission to the review committee and included a detailed breakdown of costs by year and category. A total of £804 of BAAS funding had been expended over the five years. Galloway's wages was the biggest cost item, followed by 'Instruments' and then 'Coals' for heating.[R18] The summary formed the basis of Ronalds' talk at the conference and he handed out copies to attendees. With this form of dissemination, few copies remain today, although it can be found at the IET, the Met Office Archive and the Wellcome Trust. The *Literary Gazette* wrote of it:

> A retrospective of five years' proceedings at the Kew observatory was not at all required to prove a constant desire on the part of Mr. Ronalds to promote the views of the British Association. The printed epitome, however..., widely circulated, will only increase the regret generally felt at the contemplated abandonment of the "Home of the British Association."[8]

A surviving comment from one of Ronalds' colleagues, Dr John Lee* reads: 'The Epitome is a valuable communication and highly creditable as a work of patience

---

[7] *Literary Gazette* 19/8/1848 555.
[8] *Literary Gazette* 9/9/1848 600.

steadiness and ability, and you should be allowed to enjoy the merit of the discoveries which you have made.'[C231]

On top of all this, Ronalds had family worries through this period — in particular, Alfred's wife Margaret had died. As Betsey's sister records: 'My Cousin Alfred is in deep affliction for the loss of his Wife who died about a month ago after the confinement of her 8th child. She is an irreparable loss to him.'[C198] Alfred decided to seek a new life in Australia but leave baby Hugh with family in England. Ronalds, as the senior male in the family and probably Alfred's closest sibling, would have had considerable involvement in the episode. Edmund had also incurred large financial losses as part of the general downturn in the economy. Ronalds had apologised for the quality of his submission to the review committee, 'having written it under pressure of other engrossing & disagreeable affairs'.[C213]

Things quickly went from bad to worse. Straight after the BAAS meeting, Airy arranged for Brooke to be awarded the Admiralty's prize for self-registering apparatus. His timing was again impeccable. Grisly details of the saga are played out in Sec. 15.10, which show that his actions were planned carefully over several years. His goal was to demonstrate through the award that Greenwich was superior to Kew and Ronalds was the collateral damage. Ronalds' achievements were also formally recognised after eight months of torment, at which time Sabine wrote to him simply: 'It is not so bad a country that we live in after all'.[C252]

Another minor annoyance at this time concerned Ronalds' walk to work. The new Director of the Kew Gardens, Sir William Hooker FRS, changed the locks on the gates to avoid trespass. He had heard that the BAAS had 'given up'[C250] Kew, so did not give Ronalds a new key. It took some weeks of warm correspondence before his right-of-way was reinstated: 'You must put up with a little inconvenience…till I hear from the Board.'[C254]

On the positive side, Sabine swung into action to support him after the conference. He had asked Ronalds to create specialised magnetographs for the Toronto Observatory, which had the knock-on effect of providing additional equipment and funds for Kew from Woolwich. He also remained optimistic, at least to Ronalds, writing for example:

> I hope yet to see you engaged in many jobs at Kew besides mine, some of your own for example [C223]

The BAAS made an approach to government to take over the observatory, with Sabine updating him:

> if it is in the affirmative, the purse strings may be at once loosened.[C223]
>
> I am quite charmed to find you so much on the alert [C223]

The application was not successful. We will see that funding uncertainty continued for several more years.

Sabine gave Ronalds another glimmer of hope in June 1849: 'The council of the Assoc. have agreed to keep on the Kew Ob^y till the Birmingham Meeting'.[C257] The Birmingham conference was three months away and, incidentally, was where Blair was to present their observations on the Carnac megaliths (Sec. 11.3). There the Council announced with 'great pleasure'[S1] the government's award of pecuniary prizes to Ronalds and Brooke. It also advised: 'Mr. Ronalds has drawn up a Report describing the modifications and improvements which he has introduced in the self-registering apparatus during the last year'.[S1] Birt also had his own paper discussing the atmospheric electricity observations. 'Both these Reports will be read to Section A. preparatory to a consideration of any further recommendation which it may appear desirable to make for the continued maintenance of the Observatory.'[S1] These were the first and second papers read in the Section meeting.

The session, including detailed summaries of both Ronalds' and Birt's papers, was written up in the *Literary Gazette* and the *Athenæum*. It is clear that Ronalds was not alone in arguing for the observatory's continued existence. He also did not miss the opportunity to make some pointed remarks:

Mr. RONALDS handed in his annual report 'On the Kew Observatory,' from which it appeared that the observations have, from the deficiency of applicable funds, been discontinued since the meeting in 1848[9]

Following his presentation:

Colonel Sabine, Mr. Gassiot, and others, bore testimony to the great value of Mr. Ronalds' labours at Kew, and to the readiness with which he furthered the views of all who visited the Kew Observatory; and Colonel Sabine especially referred to the instruments constructed there under Mr. Ronalds' directions, and the instructions drawn up by him and sent therewith to private observatories. Great advantage had thence arisen, and on this point alone much inconvenience would result if the Kew Observatory were given up.[10]

The incoming President also provided input:

Dr. *[Romney]* Robinson could not let the subject pass without observing how evident it was that we were now on the point of at least obtaining the laws of atmospheric electricity from systematic observations, and how unfortunate that at such a point, for want of funds, the observations at Kew were obliged to be discontinued.[11]

In the end it was Herschel's remarks that were quoted by the General Committee in justifying a further year of funding:

Sir John F.W. Herschel having reported that the Meteorological Observations made at Kew are peculiarly valuable, and likely to produce the most important results, the

---

[9]*Athenæum* 15/9/1849 934–5.
[10]*Literary Gazette* 22/9/1849 685–6.
[11] *Literary Gazette* 15/9/1849 672–4.

Committee resolved that the sum of £250 be voted for the continuance of that establishment for the ensuing year [S1]

This was the largest quantum allocated to date and again constituted two-thirds of the total financial outlay. The Editor of the *Literary Gazette* cheered: 'We rejoice that Kew has not been abandoned',[10] while Sabine's assistant Younghusband told Ronalds: 'I was very sorry I could not be at Birmingham to hear your paper read, but hope soon to read it without abridgment myself. It excited universally, I am told, the greatest attention.'[C268]

Ronalds was also nominated and elected to the BAAS Council — it was the first time he had taken on such a role. It confirms not only that he was well known and respected by his peers but also that he now appreciated the value of being more influential in the circles of power. He was re-elected each of the next four years until he moved to the Continent. He also served on the Committee of 'Section A — Mathematical and Physical Science'.[S1]

With his friends' help, he had overcome indecision and lack of funds and Kew had risen from the ashes; it had however taken a toll on his health. He was invited to stay with John Lee, but had to leave the conference early: 'you returned home unwell or out of order; and Mrs Lee and I shall be glad to have a short note from you to inform us of your health.'[C267] Sabine also started to worry about him: 'I hope you returned home <u>recovered</u>: I was sorry to see that you looked not cheery[?] so on the last day or two.'[C264] The ongoing struggles at the observatory had started to affect his well-being. In addition to lack of funding and continuity, and Airy's actions, he was feeling pressures from various other directions, which are explained in more detail elsewhere. Sabine's relentless drive had been revealed in urging his timely completion of the magnetographs, and they had proved very challenging technically. After all the effort, the first one was then apparently lost at sea *en route* to Toronto (Sec. 15.9). Ronalds also greatly wished to restart the atmospheric electricity measurements and compare them with geomagnetic and other data, there having been some unusual observations in recent times (Sec. 14.4). He had other things outside the observatory on his mind as well — Alfred had asked him to produce a new edition of his book *The Fly-fisher's Entomology* on migrating to Australia (Sec. 10.7), and he provided assistance to Elise Otté in her translation from German of Alexander von Humboldt's seminal work *Cosmos: A Sketch of a Physical Description of the Universe.*

There was a further resolution at his first Council meeting that 'Mr Ronalds be informed that this Council do not in any way consider themselves pledged to the continuance of this expenditure beyond the meeting at Edinburgh'.[J76] Edinburgh was just 10 months away, so he started to spend. Cripps had died recently so he took on Sarah Perrett as housekeeper. Birt was appointed to a position with residence at the observatory and Richard Nicklin* for short-term photographic and mechanical assistance. At last having a 'jobber in fiddle faddles' in Nicklin, he also purchased a Holtzapffel lathe and brought his portable forge from home as part of building up the laboratory and mechanical workshop he had always wanted. He saw Birt's primary focus as being the reinstatement of the atmospheric electricity and associated meteorological observations, as highlighted by Hershel to justify the increased funding.

Birt's closer involvement turned out to cause further grief rather than the relief he needed. The week after his arrival he started to write long letters of complaint about Ronalds to the BAAS hierarchy. The nub of the issue was that Birt wished to pursue his own interests and to report directly to the Council — he wanted the observatory to be his own. He had been seeking employment as an Observatory Superintendent for some years without success and had in fact applied for that position at Kew back in 1842.[C209] He was also unwell. Others were not blameless. With Ronalds' various pressures he would not have managed the emerging conflict well, while Sabine was becoming so involved in Kew that reporting lines were ambiguous — it is easy to imagine him encouraging Birt to extend his activities to magnetic problems, which were Sabine's first love. Sabine was also seriously ill through much of the period which delayed decisive action.

Birt had offered his services to Kew for a second time in May 1848 when the observatory's future was under review and the importance of analysing the atmospheric electricity results was being discussed. He wrote to Herschel:

> M[r]. R has also informed me that he finds a difficulty in carrying on the necessary observations and experiments under existing circumstances…if my services would be at all available at Kew in continuing (under the superintendence of M[r] Ronald who may be said to be the honorary Director) the Electrical Observations also the photographic registration and in devoting particular attention to the discussion of the observations which M[r] Ronald especially desires I should be most happy to render them upon such terms as should not augment the annual grant from the funds of the Association.[C207]

As soon as he arrived in November 1849, he went over Ronalds' head to advise BAAS Assistant Secretary Phillips (Sabine being indisposed):

> I should particularly wish to receive my instructions…from the Council…The only daily observations that I contemplate making are meteorological at 9 am, 3 and 9 pm for the purpose of transmitting to the Registrar General.[C274]

He also described a 'warm'[C274] dispute he had just had with Ronalds. The next day he told tales about Ronalds to Hunt, before adding:

> Although I have felt myself compelled to speak as I have I do not by any means wish to be unfriendly with M[r] Ronalds but on the contrary I shall be most happy to assist him with his experiments…as time and circumstances in connection with other objects may permit.[C275]

Wheatstone then appeared to muddy the waters: 'I have had an interview with him *[Wheatstone]* and I am glad to find that I have not been mistaken in my apprehension of the position it is intended I should hold here'.[C276] Wheatstone offered to apply to the Royal Society for a grant for Birt to continue his own research, although it was not successful. In the meantime Birt mapped out a plan of activities for himself.

The next month he offered his detailed views to Herschel on the value of the observatory:

> There are also some "internal" circumstances which render its continuance very doubtful to my mind...unless something much more efficient is accomplished in the remaining six months than has been effected in the first three, Gentlemen who enquire "what have you to shew for the money granted?" will certainly have good cause to withhold their support. I ought not I think in such a communication as this to conceal two or three points that bear particularly on my own position here...[C277]

By sheer luck, Sabine wrote from his sickbed to Herschel very shortly afterwards a glowing account of Ronalds' latest achievements in magnetic registration.[C278]

Birt then contacted Sabine himself three times in a fortnight:

> In order to recover from the state of extreme nervous debility under which I laboured I found it necessary to remain for some time in my own apartments which I did from Dec 13 1849 to Feb 25 1850.[C279]

> With regard to my non-resumption of the Electrical Observations I beg to remark that I never considered these or any other observations I may have made here as portions of my official duties...I further beg to remark that Experience has convinced me that it is utterly impossible to carry out any thing efficiently in Electricity and Meteorology under the direction of Mr Ronalds.[C285]

Ronalds had first advised Sabine of the situation in January and, now that the latter had returned to his post, submitted a 14 page report with his view of what had transpired.[J77] Both sides of the story were heard by a committee. The committee evidently sided with Ronalds, because Sabine wrote to him after the meeting: 'Will you have the goodness to see whether the first Minute is worded to your satisfaction'.[C288] He advised Ronalds a month later:

> I wrote to Mr Birt, as you wished, informing him that the Com^{ee} w^d be at Kew on Monday, & that I thought they w^d expect to find the *[electrical]* Frequency observations in progress, & merely remarking incidentally that by the Resolutions of the Committee he must have seen that it was from you & not from me that he was to receive directions for those observations.[C289]

Birt meanwhile was updating Herschel on private observations he had been making over the last four months on the Zodiacal light.

Three weeks later, Birt sent a doctor's certificate dated 3 June 1850 to Sabine:

> 'I certify that Mr William Birt is in a weak & nervous state, and unfit for his present early duties.'[J78]

The accompanying letter advised:

> I am almost too ill to write...He *[the doctor]* tells me to-day that it is indispensable
> for my recovery that I must leave the house immediately for the sake of change of
> air and scene and that I must not attend to any instruments for some time to come...
> I should be unwilling to undertake the duties of the Observatory another twelve
> months <u>under present arrangements</u>, these arrangements are not satisfactory to me
> nor are they such as I expected would have been made.[C291]

Another month later, Birt contacted surgeon and BAAS Councillor Professor
Forbes Royle:

> I don't know that on any previous occasion I was more astounded than when I found
> that a note written under the influence of extreme weakness and debility...should
> without any knowledge...on my part have been...viewed as a <u>resignation</u>...I trust it
> will be readily acceded that I have not resigned the appointment[C294]

When the resignation was upheld, Birt determined that he would remain in residence at the observatory even though he was not working:

> I beg to say that I expect my salary to be paid...I entered the Observatory on Nov 2
> 1849 so that the Twelve months...does not expire until Nov 2nd 1850. At the same
> time I do not wish to occasion the Committee any inconvenience by remaining in the
> observatory, and shall be happy to vacate the rooms upon a settlement with me[C301]

He also sought Airy's assistance in presenting his views on the state of Kew
to the upcoming BAAS meeting:

> I am extremely desirous that a thorough and <u>impartial</u> investigation should be made
> into the management of this observatory...I have laid the entire correspondence
> between the Visiting Committee and myself before the President Sir David Brewster
> and sincerely hope some steps may be taken to set the matter of the observatory in
> a proper light before the members.[C296]

Airy passed the letter on to Professor James Forbes, now the Chairman of the
Section, suggesting 'it may be worth keeping in mind if you visit Kew.'[C306]

Ronalds was hoping that Birt would have left when he returned from the
Edinburgh BAAS conference in August. That was not to be. Royle then coached
him:

> we are likely to have some trouble with Mr Birt, who seems as unwise in all his
> proceedings as in his resignation...I think it very desirable that you should screw up
> the doors of all such rooms as are not in use & get new locks for such as are in use...
> I think with Mr Gassiot that people coming to the Observatory should by some
> means be informed that Mr Birt though residing in the Observatory is not in the
> services of the Association as his resignation may be seen noticed in the report of the
> Committee read at Edinburgh & published in all the Newspapers. How this is best to
> be done I trust however to your discretion[C303]

Gassiot achieved a breakthrough in late September:

> I have arranged with Birt to give him (as on my own responsibility) £25 on his giving
> possession which he promises to do on Wednesday...knowing the character of my

man I have taken the precaution of drawing up a minute of our conversation which was read to him in presence of Nicklin and acknowledged to be correct.[C308]

Birt requested £50 from the BAAS, which was declined, the minutes noting: 'Mr Birt ceased to perform any duty whatsoever on the 6th of June.'[J83]

The last surviving letter on the saga was a rather confused one from Birt to Sabine nearly a year later:

> I ask what would Mr Ronalds apparatus have been without my discussion? Surely it might have been enough to have left Mr Ronalds and myself to have fought our own battles…I am willing at any time…before an unprejudiced Committee to have the whole matter fairly and fully investigated. Is he as willing?…I have communicated my views on these subjects to the Astronomer Royal and hope the investigation I sought eleven months since will be no longer delayed.[C325]

For many, the ongoing threat of closure over most of the last five years, let alone these other factors, would have been debilitating. Ronalds did admit to Sabine that he felt 'cruelly hampered'[C302] by Birt's behaviour. He found the mental strength however to remain focused and creative. In particular, he made significant innovations in photography and instrumentation at this time that brought self-recording apparatus to a new level of precision.

It was clear however that something needed to be done to move beyond the drip-feeding of funding. The first step was an improved governance structure, with a committee of Council being appointed for the first time to oversee Kew. Benefits included a formalising and widening of the support base for the observatory, which had proved so useful at the Birmingham conference and in dealing with Birt. It would also facilitate the attraction of additional funding from other sources and provide backup in case Sabine was again laid up. It was timely too to revisit the issue of differentiation of Kew and Greenwich activities.

In its early days, the 'Kew Committee of the British Association'[S1] comprised Wheatstone, Royle, Reid, Gassiot and Colonel William Sykes FRS, who was later Chairman of the East India Company. Ronalds attended in his role as Director and BAAS Secretaries Sabine and Phillips also followed proceedings closely. Herschel was another member although he stressed he could not be part of a 'real working Committee'.[C332] Sykes and Gassiot were therefore the early Chairmen. Committee minutes show the group to have been both formal in its business and most enthusiastic about the observatory and its potential.

By the time of the Edinburgh conference (Fig. 13.3), a team of people were working together in a coordinated way to further the Kew Observatory cause. The public relations offensive began with an update through the Council:

> Kew Observatory…has given to science self-recording instruments for electrical, magnetical, and meteorological phænomena, already of great value[S1]

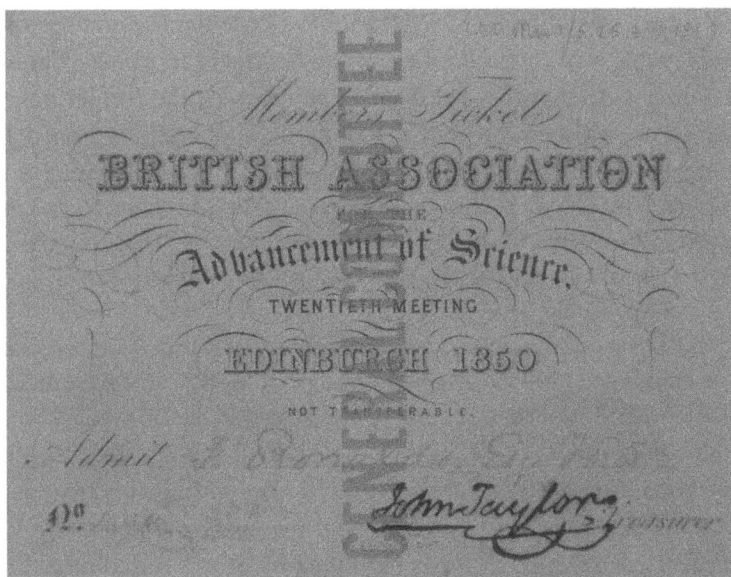

Fig. 13.3:   Ronalds' ticket for 1850 BAAS annual conference in Edinburgh. IET 1.5.25.

The update then addressed continuity and the observatory's purpose, noting that a full suite of observations using Ronalds' instruments:

> would be quite incompatible with the limited funds at the disposal of the Association, and inconsistent with the general intention of the establishment — which is an *Experimental Observatory*, devoted to open out new physical inquiries, and to make trial of new modes of inquiry, but only in a few selected cases to preserve continuous records of passing phænomena.
>
> It is on this view of the character of the Observatory that the Committee found their opinion, that it may be maintained in a state of efficiency, and kept always ready to take its proper share of the Advancement of Science, by means of a moderate annual grant from the Association.[S1]

These sentiments coincided with Ronalds' arguments in his review submission two years earlier: 'I have always felt, & said, that we should rather make Registrars, animate & inanimate than be Registrars ourselves.'[C213]

Brewster, the incoming President, continued the theme in a very promising way:

> I regret to say, that in consequence of our diminished resources, the Association, at its meeting in 1848, came to the resolution of discontinuing the observations at Kew…I trust, however, that means will yet be found to maintain the Observatory in full activity…Having had an opportunity of visiting this establishment a few weeks ago, after having inspected two of the best conducted observatories on the Continent…I have no hesitation in speaking in the highest terms of the value of Mr. Ronalds' labours, and in recommending the institution which he so liberally superintends to the continued protection of the Association[S1]

Brewster's timely invitation to Kew would have been the Committee's idea.

Ronalds' paper was again the first in the Sectional meetings. The *Athenæum* and the *Literary Gazette* provided detailed summaries, the former describing his written report as 'voluminous and elaborate'.[12] Supporters weighed in to help underline the key points, starting with the Assistant General Secretary:

Prof. Phillips bore testimony to the ingenuity and zeal of Mr. Ronalds…He invited… all members to go down to the observatory to witness the beautiful and accurate processes in operation, of which, without seeing, no adequate idea of the kind of work done at Kew could be formed.[13]

Prof. PHILLIPS then gave a sketch of the observations which had been established under the unpaid and invaluable superintendence of Mr. Ronalds since 1842–3…

Sir D. BREWSTER wished to suggest to Mr. Ronalds that…[12]

Ronalds' diary of the event survives and provides interesting insight into the conduct of a major conference at this time. It also outlines his own activities — with whom he conversed and how he spent his spare moments. He was now a more confident contributor, and was prompted by others for comment as a meteorological expert. Throughout the conference however he worried about the future of Kew.

The *Athenæum* records in Section A:

Dr. *[Charles]* MARTINS addressed the meeting in French 'On the Six Climates of France.'…Mr. RONALDS inquired whether the state of the dew point had been attended to in this classification, as in his opinion that was one very important element.[14]

From Ronalds' diary, at the same session:

Mr Phillips read a paper…of a committee which had visited a tree about 4 miles from Edinb^g which had been struck by lightning. Much discussion ensued & I was asked by the President Professor Forbes to speak on the subject.[180]

Ronalds even attempted some horse-trading with Airy, who was clearly still his primary adversary and was in a powerful position as President-Elect:

I expressed my willingness to assist him if he would erect an electrical apparatus… with an understanding that he would countenance or support the Kew establishment as an experimental observatory (not as one for regular continuous observations). He said that if a system of registration for electricity for 24 hours without attendance could be adopted he would…support Kew under a smaller expenditure than the present [180]

Airy did not get his way. The decision was to increase the funding for Kew to a new high. Ronalds indicates in his diary how the process played out through the committee structure. On Thursday 1 August:

---

[12] *Athenæum* 10/8/1850 839.
[13] *Literary Gazette* 10/8/1850 568.
[14] *Athenæum* 17/8/1850 873.

Attended Council at 2. It was resolved to recommend £250 grant for Kew[J80]

Monday 5 August — after his presentation:

Mr Gassiot came to me and said that a unanimous resolution had been passed in the Committee of the Section A that morning for recommending that £300 should be granted for the Kew Observatory this year.[J80]

The next day:

Went to the Royal Institution, at about 3, and waited untill the Committee of Recommendations had broken up, when Sir D Brewster told me that I had got the £300 & Col Sykes also.[J80]

And the next:

Attended General Committee where it was resolved amongst the other resolutions concerning Kew that a committee should be appointed to endeavour by application to the Royal Society & if necessary to the Government to relieve the Association of the expense of maintaining the Kew Observatory.[J80]

Concerning the last:

Never mind <u>that</u> said Mr Gassiot to me if you can get some good work done.[C302]

Ronalds was understandably shaken by his experience with Birt and took great care in ensuring his next assistant John Welsh would be suited to the job. Welsh resided in Edinburgh and they spent considerable time together during the conference. Ronalds put together a duty statement for which he obtained the approval of Committee members and Welsh's agreement. He also visited his mother: 'Called at Mrs Welsh's and agreed with her that I would on my own responsibility order an iron bedstead & mattress and a little other furniture for her son at Kew'.[J80] Committee members had also learnt their lesson and continued to reinforce to Welsh and others that he reported to Ronalds, notwithstanding their increasingly hands-on attitude to the observatory. In the absence of funding continuity Welsh, like Birt, was offered a 12-month appointment with a salary of £100.

Ronalds participated in few of the conference's organised social events, preferring instead to work, dine with a friend or pursue other scientific interests. On the first Saturday he 'Worked nearly all day on Account Book.'[J80] He spent much of the intermediate weekend at nearby Craigie Hall, probably with his uncle Charles Field and cousin Thomas Field Gibson who also attended the conference. He also bought gold pens as a gift for his brother Charles, sought books for his electrical library and took considerable interest in a new form of mounting for a reflecting telescope.

After the conference, he took the luxury of a short break with Uncle Charles on his way home. He confided to Sabine that he was in need of it 'allmost as much as last year'[C302] when he was ill. Royle wrote incredulously to Gassiot: 'Mr Ronalds is taking a holyday of a week!! which he begged I would consent to!'[C297] Sabine also passed comment: 'My two ladies *[his wife and mother-in-law]* regret that you only allowed yourself a ten-days holiday'.[C307]

Table 13.3:   Ronalds' Vacation in Scotland and Lake District (August 1850).

Day	Destination	Travel/Activity
12th	Edinburgh	Booksellers & Library, time with Welsh
13th	Callander	By train via Stirling
14th	Glasgow	By public coach along Loch Katrine, steamboat & public carriage down Loch Lomond to Dumbarton, then up Clyde
15th	Glasgow	Booksellers, Theatre
16th	Carlisle	Visited Engineer to Clyde Trust and viewed tide registering instrument, then to Carlisle
17th	Birmingham	Via Kendal to Windermere, steamboat down lake & back, to Oxenholme, then by train via Preston
18th	Birmingham	Visited Osler to examine anemometer, evening with Carter family [Ronalds' sister]
19th	Chiswick	Travelled home

It was certainly a whirlwind trip. He spent just one day at the Scottish lochs and half a day at Lake Windermere in the Lake District (Table 13.3). It was very different from the leisurely pace of travel he had enjoyed on his Grand Tour, which was aided now of course by the advent of the railway. Another notable aspect is that he mixed business with pleasure even on this very short break, viewing self-registering instruments for tide and wind along the way. His diary gives just a hint of the old Ronalds with his artistic contemplation on Lake Windermere: 'Weather very fine. Breeze upon the Lake refreshing. Views much improved & diversified by the shadows of heavy clouds on the mountains and incidental light &c. Beautiful sailing Boats.'[J80]

Airy, on his post-conference vacation in the Scottish Highlands, continued to mull on the issues and soon tried another angle with Kew Committee Chairman Sykes:

> it appears to me that the continuous observations made there *[Kew]* may be expunged without sensible loss to science <u>except the electrometer observations</u>…These… might, I think, be transferred to Greenwich (the very instruments being shifted if necessary) provided the indications can be registered photographically. This last condition…is indispensable to us: but I believe that Mr Ronalds has so far managed it that we might reckon upon it.[C300]

When Sabine heard of Airy's proposal, he reassured Ronalds from Amiens, France:

> dont let it trouble you…we have no power whatsoever to <u>transfer</u> to any other establishment an apparatus of which the cost was defrayed by private subscription expressly of <u>members of the B Assoc. for its establishment at Kew</u>.[C309]

Sykes replied formally a few months later advising why transfer of equipment to Greenwich would not be possible. Airy followed up with Ronalds:

> I have just heard fully from Col. Sykes in regard to the Kew Observatory. His letter affords a most satisfactory answer to the question which I put to the Committee at Edinburgh "What was now done at the Kew Observatory" — a question which it was necessary for me to put when the suggestion of suppressing or modifying the Observatory was in consideration, and which some injudicious person reported to you as if it had been a reflection upon the Observatory. This I explained to you verbally [C319]

He went on to request Ronalds to assist in the development of improved atmospheric electricity apparatus at Greenwich as discussed in Sec. 14.5.

During the year, Kew received an additional £300 in three grants from the Royal Society, which had just received a new source of research funding from government. Survival was now looking more likely. The first enabled Ronalds to develop new magnetographs. Sabine updated him on its award:

> We had a narrow squeak yesterday for our £100...Mr Wheatstone did not attend the Committee & it was stated by some gentleman from him that for some reason, not stated, he did not wish his name to appear in it, the meaning of w^h I do not know... My name was substituted for Mr Wheatstone's for the reason above named & Colonel Sykes, Dr Royle and Mr Wheatstone are a Committee to see that the funds are appropriated according to the grant, (which gives them no control in other respects). [C290]

Ronalds' second grant built on the recent reaffirmation of one of the originally stated purposes of Kew in trialling new instruments. Sabine had written to him:

> I think the proposition for an <u>experimental trial</u> during six months of your Magnetographs likely to be of great service to Kew, to Magnetical Science, to the instruments themselves, & to the Inventor of them...Had you not better call it an <u>experimental trial during six months</u>, than <u>observations for six months</u>. [C315]

The third award was to George Stokes, who wished to conduct experiments at Kew. Stokes was later President of the Royal Society and remains famous to physicists and engineers today for the Navier-Stokes equations in fluid dynamics.

The Kew Committee also started to invite strategically important visitors to the observatory. Sabine advised: 'Capt. Smyth & Mr Galloway have arranged to go down with me to the Ob^y on <u>Monday</u> next at the usual hour *[there was now rail transport to Richmond]*. One after another we shall convert all the influential people into warm supporters.' [C318] These two guests were Captain (later Admiral) William Henry Smyth FRS and Thomas Galloway FRS who were elected officers of the Royal Society and various other learned bodies. Smyth's son Charles Piazzi Smyth was introduced in Sec. 10.6. After Airy's inspection, Gassiot mentioned to Ronalds: 'I was glad to find that Mr Airy was so satisfied with the <u>work</u> at Kew.' [C324] It was then less than a fortnight to the 1851 BAAS conference and Airy's views as incoming President were critical. Hot on his heels, Sabine was the new President-Elect!

The BAAS Council advised the Ipswich meeting of Kew's funding successes from the Royal Society and made a clear recommendation:

> taking into account that the institution works well under its present arrangements and on its present footing, and believing that its continuance will be conducible alike to the advancement of science and to the credit of the British Association, they recommend that the grant of £300 to the Kew Observatory should be continued for the next year.[S1]

Airy was less effusive in his Presidential Address than previous incumbents, saying simply: 'it is hoped by the officers of the Association that the Kew Observatory will be made really efficient for the testing of new instruments.'[S1] He also managed to slip in a technical criticism of Ronalds' photo-recording machines. However, he was powerless to change the positive sentiment of the meeting.

Ronalds delivered a bumper report that year, with the printed version running to 36 pages plus six plates of diagrams. It gave detailed updates on his various photo-recording, meteorological and atmospheric electricity instruments and their results and also covered Welsh's activities. Sabine and Welsh both gave presentations on the trial of Ronalds' magnetographs and Welsh also outlined a new sliding-rule he had developed to facilitate the reduction of hygrometer observations. The team was working well.

There was another big event in 1851 — the Great Exhibition at Crystal Palace (see previous chapter). It would not have occurred to Ronalds to submit his instruments to the exhibition and he had no time for it anyway. It is a shame because, if he had, he would almost certainly have won one or more medals.

Concerning meteorological instruments, the *Reports by the Juries* lamented:

> To judge from those exhibited, it would appear that as little attention is paid now to the construction of meteorological instruments in London…as a few years ago, before the commencement of the systematic researches in meteorology at present being carried on[B61]

This could be read as a hint that the Kew Observatory should have exhibited its new devices. The Chairman of the Jurors in this Section was Brewster, another Juror was Herschel and Glaisher was the Reporter — all of whom knew and admired Ronalds' meteorological inventions. Indeed Brewster had made those flattering remarks the preceding year in his Presidential Address.

Atmospheric electricity instruments came in for special criticism:

> The electric instruments are few in number, and there is not one adapted for the purpose of determining the quantity of atmospheric electricity for meteorological purposes. This is a matter to be regretted[B61]

Certainly this one was a direct dig at Ronalds. Perhaps the only mention of his electricity equipment at the exhibition was in instrument-maker John Newman's

Catalogue in the Exhibition Prospectus — he was now making the instruments for sale.

The situation was even worse for photo-recording apparatus — medals were awarded for both Brooke's and Dollond's 'beautiful'[B61] self-registering instruments. Robert Hunt kindly made mention of Ronalds' contributions in his *Hand-book to the Official Catalogues*,[B60] although Glaisher did not in his lectures concerning the exhibition.[A89]

Buoyed by their recent funding successes, and perhaps goaded by the Juries Report, the Committee started this year to become more involved in shaping business at Kew. As Scott notes in his history of the observatory: 'In 1852…the Kew Committee first appeared as actually taking control of the establishment'.[K9] Its first focus was the design, manufacture and testing of thermometers and barometers, which would become another area of renown for Kew. On request, instruments were created for particular applications or others' devices were verified against accurate 'standard' instruments established at Kew. It was Welsh who had special expertise in this domain but Ronalds designed and built some of the equipment for making and calibrating the instruments.[J84] Related activities had commenced several years earlier; the Paris standard thermometer was brought to Kew in 1849 for comparison with the observatory's instrument, and Sabine informed Ronalds the next year:

> I have promised L^t *[Joseph]* Dayman that as soon as you have a standard Barom^r ready, you will compare for him his two Barometers which are just returned from a voyage of circumnavigation.[C317]

This was the historic voyage of discovery to Australia and New Guinea undertaken in 1846–50 by HMS *Rattlesnake*. One of the early specialist thermometers made at Kew was for Ronalds' nephew Dr Edmund in 1852 — it extended to 220°C.

The BAAS conference was held in Belfast, where both Ronalds and Dr Edmund presented papers. The best surviving summary of Ronalds and Welsh's activities at that time is the review of this presentation in the *Literary Gazette*.[15] The primary emphasis was the provision of equipment and advice to other observatories and the Council again wrote very positively on these achievements.

13.4 The Decision to Leave Kew

It was at this time that Ronalds started to reduce his more than fulltime commitment to Kew. As is often the case, multiple reasons contributed to his action but the trigger was the death of his elderly mother in March 1852.

---

[15] *Literary Gazette* 11/9/1852 698–9.

Jane's passing created a large workload for him. She had chosen three executors for her complex estate: Ronalds, Uncle Charles Field and family friend Thomas Berry Rowe who ran the Thames Soap Works in Brentford. Rowe formally 'renounced'[J12] his role immediately and died himself within a few weeks. Uncle Charles had already written his own will (a sign in those times of ill-health) and passed away in early 1854 — he was probably of limited assistance. Much of the burden fell on Ronalds. His first task was to prepare for sale Jane's three properties in Chiswick, Upper Thames Street and Highbury. His sisters would have helped clear out the family's possessions, but he also had to pack up his own life — and sorting through a lifetime of work and memories is always an emotional task. By late in the year, the properties and their contents were sold, Ronalds' most valued possessions were in storage at the Pantechnicon and he was staying at the Golden Cross Hotel at Trafalgar Square. We have his care at this time to thank for the survival of his key papers in the Ronalds Archive.

He then needed to distribute Jane's assets across the family according to her will. For his overseas brothers Hugh and Alfred, unmarried sister Emily, and Charles' daughter Julia, this necessitated him investing in stocks from which they would receive regular dividends. There would have been a desire for urgency in this as his brothers were in financial difficulty. Edmund had just lost considerable money in the silk mill in Derby but wanted to help establish his sons in New Zealand, while Charles was incapacitated in some way. Alfred immediately used his inheritance to establish a large nursery in Ballarat and Hugh similarly purchased farmland near Albion.

There was other sadness in the family too. Hugh's wife Kate passed away the same month as Jane, while Edmund's wife Eliza died the next year. Two of the Brentford cousins had died in recent times and Betsey was very ill — the nursery was in rapid decline[A197] and Dr Henry's widow had just been hospitalised near Brentford as a 'lunatic'. Whether there were still more issues is not known but Ronalds hints in several letters that it was a very difficult time for him: 'I will not trouble you with a recital of the many distressing circumstances and imperative private duties &c, which have of late absorbed the whole of my time and attention.'[C340]

The positive side of the new situation was that he was no longer tied to Chiswick (and Kew). His first use of that flexibility was in enhancing the assistance he provided to other observatories and, in particular, supervising the installation of his photo-recording apparatus at Oxford University.

He gave several other reasons for his gradually decreasing commitment to Kew in letters written over the following years. One concern was his reduced scientific autonomy now that the Kew Committee had become so active. He admitted to Airy:

I have been painfully forced to believe that the conditions which I have deemed essential to the favourable prosecution of my humble labours at Kew are somewhat modified [C338]

That was written in early 1853. Two years later, he told Charles Piazzi Smyth:

The Kew Committee having completely superseded the kind of duties which I had performed gratuitously for 9 years I do not feel the same interest now as formerly in Kew proceedings & in fact I do not desire to share in the responsibility of all of them. [C364]

He was even more explicit to Romney Robinson in explaining that 'Thermometer-making and such kind of operations' are not 'consonant with my tastes and inclinations'. [C365]

Funding difficulties were another factor for him in his withdrawal — be it the lack of steady support, his need to contribute personally, the strings attached to external funding when it did come, or the selfish power plays it could induce. He wrote to an associate in Turin in 1853:

Our Kew Observatory was first appropriated meteorologically by private means and I hope that…it will receive support from our government in larger measure than hitherto. But there is always danger that government patronage may engender considerations of private advantage which have very often obstructed rather than stimulated progress [C340]

He was also aware that he himself had limited funds left for meteorological science, having told the Kew Committee several years earlier:

I have now for at least seven years laboured gratuitously & diligently in the service of Kew and have actually spent a large sum in personal & other costs attendant on that service…I am neither rich nor young nor likely to become rich by magnetic pursuits, or any other. [J79]

The funding situation had also changed the relationship between Ronalds and Sabine. In earlier years, Sabine simply provided oversight and support to Ronalds, but that began to change in 1848 when he requested magnetographs for Toronto. Now Sabine was also Ronalds' client upon whom he was dependent for funding through the period when Kew was threatened with closure. Over time, Sabine developed a sense of ownership of the observatory and its direction and his tone altered from suggesting, to requesting, and even pestering. In late 1850, for example, concerning the vertical force (VF) magnetograph: 'the first thing should be to try your V.F.' [C317] The next week: 'Will you soon have the V.F. mounted?' [C318] And a fortnight later: 'I long to see the V.F. up and working — After that you will remodel the Declinometer *[something Ronalds was keen to do with experience gained from his recent machines]*.' [C320] Sabine was impatient to execute his plans while Ronalds was focused on quality — and fighting influenza. He had lost his freedom. Years later he described himself to Sabine as 'your old Kew magneticmechanic'.[C379] It was written with affection but illustrated what the relationship had evolved to in his mind.

An enthusiastic governing committee with lots of ideas needs a salaried superintendent to carry out its wishes. Ronalds was essentially a 'gentleman scientist', used to pursuing his own interests, and the arrangement with Kew had been successful up till now only because these interests were coincident with the BAAS Council's loosely defined goals for Kew.

The final straw for Ronalds concerned his favourite atmospheric electricity. It had become clear to him that even with Welsh on board the Kew Committee would not agree to the resources required for daily observations to be reinstated. The only way he could achieve his electrical goals was through other observatories. Captain John Lefroy* had developed a strong interest in atmospheric electricity and sought Ronalds' assistance with equipment for the Toronto Observatory, but he was on a tight budget. Ronalds' idea was to provide him with some of the disused apparatus at Kew. Ronalds carefully preserved an excerpt from his diary describing what transpired at the Kew Committee meeting in July 1853:

> The proposal was met by objections by Col[l] Sabine and M[r] Gassiot…
>
> The affair was finally refered *[sic]* to the (self elected) Kew Committee viz Col Sabine, M[r] Gassiot & Col[l] Sykes. I am therefore deprived of…an opportunity of doing a disinterested service to Meteorological Science. I even offered to pay Ten Pounds toward the completion of the apparatus and to take my chance of being repaid by the American Colonial Government. This is disgusting.[J85]

All his frustrations had built up and he needed to leave. The depth of his feeling will be better understood in reference to the next chapter.

Gassiot, as Chair of the Committee, followed up in a letter in early August 1853:

> There appears to be a very general feeling that no portion of y[r] apparatus ought to be removed from the Observatory — it is all so peculiarly & so intimately connected with your original researches at Kew that we should all regret to see it removed [C345]

He continued:

> As the time of the Meeting of the British Association is fast approaching will you have the kindness to send me the Dft of any report you may have to make — I have requested M[r] Welsh to prepare any items he may wish to insert.[C345]

This was the first time that the Committee rather than Ronalds organised and signed off on the annual report — a process that would continue in subsequent years. Indeed Kew Committee Reports were issued to the turn of the century.[S6] Although he was still notionally the Director (Welsh continued to seek direction from him and Ronalds noted him to be 'the Assistant'[C346]) Ronalds was spending very little time at Kew. He took a break from his activities at Oxford to attend the BAAS conference in Hull in September. He was re-elected to the Council but it was probably his last face-to-face interaction with his colleagues on Kew matters.

He was now 65 years of age and the family events of the previous year had reminded him of his mortality. There was still much he wished to achieve. It was also the perfect time to leave Kew. The observatory now had a strong reputation, with many supporters in Britain and international reach. A committed and energetic oversight Committee was in place. The scale and consistency of funding was sufficient to support a salaried superintendent and there was a very capable successor in the wings in Welsh. Ronalds could be proud of what had been achieved.

There was no precise date of departure, just as there was no specific arrival date at Kew. Years later he wrote of 'the Kew Observatory, of which I relinquished the hon[y] direction in 1853.'[C376] His 1860 autobiographical letter outlined his arrival and departure as follows:

> In 1842 I determined to renew electrical & meteorological pursuits, & accepted the honorary direction of the hardly more than projected Meteorological Kew Observatory under the auspices of the British Association.[C375]

> In 1852 I was annoyed & oppressed at Kew, left Chiswick, on the death of my Mother...and quitted a flourishing establishment of which I have been called the founder. This is not strictly true. In 1842 it was in <u>dissolving</u> embryo: but it <u>is</u> true, as General Sabine has said, that it would not have flourished but for my exertions. I maintained it against the strong opposition of many influential members of the British Association (& particularly of one who became & is now one of its most strenuous supporters *[Airy, who in later years supported the concept of Kew as an experimental facility]*). The system now pursued at Kew is precisely that which I long since in a document, received by a committee, strenuously advocated. For the constantly kind, yet perfectly disinterested, & powerfull *[sic]* support of Sir John Herschell Bar[t] and several other gentlemen I shall ever feel grateful. The Royal Society has been very kind & liberal to me at Kew & its late President Lord Ross *[sic]* particularly so.[C375]

13.5 Kew's Later Years

Ronalds' next surviving letter to Gassiot, in December 1853, is written from Paris. He is still smarting:

> I am very sorry that the Kew Committee has been obliged to use M[r] Welsh's less convenient room instead of mine, but I hope to be able to attend its next meeting, and to determine then upon arrangements suitable to future prospects &c.[C351]

He might have done a good job of packing up Chiswick, but he had not even started at Kew. Quickly recovering his cheeky sense of humour, he continued:

> These affairs *[Ronalds' personal scientific equipment]* are valued as being, for the most part, very old friends, particularly some De Luc's Electric Columns, made by Singer*[*]* nearly <u>Forty</u> years since; which...are still active. If you, a brother electrician, and member of the superintending Committee, will refrain from scolding me

over much when we meet, and if you think them interesting…I shall beg your acceptance of one of them (mounted by its electrodes) upon a pair of electrometers, in order that you may, at any time, compare (approximately) the progress of its decay, with your own.[C351]

Gassiot had made important inroads into the long-standing question of whether battery operation was driven by contact between the metals or decomposition (see Sec. 6.2).

Gassiot replied immediately:

I assure you it gave me great pleasure to see y[r] writing again, I had almost come to the conclusion that you had <u>cut</u> us for ever — I certainly had not the slightest Idea of y[r] being in the gay City of Paris…

I also hope when you return you will attend our Committee and if you cannot spare so much time as formerly we shall at least have the benefit of y[r] advice…

if on your return there are any Experiments you would like to undertake, I am sure there would be no difficult [sic] in obtaining any sum you might require[C352]

This did not occur — despite his apparent intentions Ronalds did not return to England for nearly a decade. He liaised from Paris with Welsh who kindly sorted through his many belongings. Ronalds' equipment and papers remained at the observatory for many years. An inventory in the Kew Committee report of 1871, for example, lists his lathe, perspective tracing instrument and 'Various pieces of Electrical Apparatus'.[S6] The 1875 Kew Committee report notes:

A collection of apparatus, principally electrical, which had belonged to the late Sir Francis Ronalds, and had remained in store at Kew from the time of his resigning the superintendence of the Observatory, was most kindly presented to the Committee by his executors[S6]

The next year, the Committee donated his electric generator and some of his personal meteorological instruments to the Science Museum. They kept others for display.[A139] His annual Journals containing all observations as well as his full reports were also carefully preserved and are now in the Meteorological Office Archive in Exeter.

Ronalds retained his strong interest in meteorological and magnetic matters and remained an annual subscriber of the BAAS for some years. His next major interaction with Kew was probably that associated with *L'Exposition Universelle* in Paris in 1855, 'It having been represented to the *[Kew]* Committee that Her Majesty's Government were anxious that magnetical and meteorological instruments, showing the state to which they had advanced in this country, should be exhibited' there.[S6] Gassiot advised him of the desire to include a number of his inventions in the display. He replied from Paris:

I must therefore rely upon your kindness in taking proper steps for rendering their appearance amongst their many highly polished competitors for admiration tolerably respectable. They must, at present, cut rather forlorn figures I think.[C358]

The display included a total of 16 of Ronalds' inventions in atmospheric electricity, meteorology and photo-recording. He noted in his diary in July 1855: 'I learnt at the Palais de l'Industrie that the Jury Sir David Brewster &c, had examined my Instruments, and had spoken highly of them.'[J85] It probably compensated in some way for the debacle of the Great Exhibition. He also would have enjoyed catching up with family, old friends and colleagues who were jurors, exhibitors or spectators, including Welsh.

There were several write-ups of the display. The guidebook *Visite à L'Exposition Universelle de Paris* was edited by Henri Tresca (who remains famous to materials engineers for his plasticity theory):

> France has relatively few meteorological instruments in the Exposition. It is from England that we see the most interesting devices.
>
> The British Association for the Advancement of Science has established a meteorological observatory at Kew, near London, for the verification of precision magnetic and meteorological instruments; various of these devices are exhibited, most remarkable both in their detail and as a whole. It is very useful, especially for the meteorologist, to visit this beautiful collection of instruments[B70]

The French science magazine *Cosmos*, edited by Abbé Moigno*, wrote of the effort:

> Instruments from the Kew Observatory were exhibited…the expense of the exercise exceeded 5,000 fr.; and, to make matters worse, they were awkwardly placed remote from the competition, so that Mr. Welsh, its director, and Mr. Ronalds, the inventor of the beautiful self-registering machines, received no medals or rewards.[A97]

The instruments were positioned on the south east side of the upper gallery.

Strong interest in Ronalds' machines was shown by numerous European observers. The wide recognition was timely for him personally as it coincided with Wheatstone and Cooke's very public battle over who was the inventor of the telegraph. To help satisfy demand, he arranged for a collection of his BAAS reports to be printed as a volume to give to enquirers. He told Romney Robinson: 'finding that some of my Instruments (particularly the Photobarograph, and Thermograph) are beginning to be in request on the Continent, I have spent much time in concocting a french account of them'.[C365] The booklet he prepared was entitled *Descriptions de quelques Instruments Météorologiques et Magnétiques*[R24] and ran to 68 pages of text with 14 plates of illustrations. It gave detailed descriptions of 18 of his most important meteorological instruments, including those at the exhibition, 'for the use, principally, of other (continental) observatories & private observers'.[C367] It took considerable preparation time, going through at least seven proofs, and was not complete until 1856. Although he had communicated in French for many years, this was his first attempt at a formal publication in another language. He acknowledged the assistance of Moigno and his nephew — Dr Edmund had studied in France towards his PhD. Dr Edmund later admitted to him: 'I have sometimes thought you had not been pleased with the criticisms upon the French composition which…I perpetrated upon it'.[C370]

Moigno was delighted to be able to at last advise of the publication:

Mr. Francis Ronalds…has just published the first part of a French description of the meteorological instruments invented or perfected by him…We will say more soon about these wonderful instruments, which are not in planning, but…are good and beautiful realities. Their price makes them accessible to all observatories, they are easy to handle, and their biggest test, the full endorsement of the Royal Society of London and the British Association for the Advancement of Science, prove that they are the perfect substitute for the eye in most applications, they represent the phenomena with the full accuracy that is achievable, and operate continuously, which…is an essential condition for any proper observation of nature and the needs of pure and applied science. We learnt from our noble friend that these two pamphlets, printed… without commercial intentions, will be deposited and sold in the *Cosmos* offices.[A93]

Ronalds immediately sent copies to UK colleagues including Sabine, Gassiot and Romney Robinson. The combined book and/or its two parts are retained in several European institutions including the Universities of Aix-Marseille and Strasbourg (both of which have observatories), the *Conservatoire National des Arts et Métiers*, the *Bibliothèque nationale de France* and the *Accademia di Scienze, Lettere ed Arti* of Padua. The Ronalds Library and the Met Office Archive also have copies.

He went on to advise various observers on their equipment requirements, assist Continental instrument-makers with their manufacture and superintend their installation, as discussed in the following chapters. In terms of delivering broad benefit in his areas of interest, he had made the right decision to leave Kew.

Meanwhile, back in the UK, changes were afoot, which he was also watching with interest. He had mentioned to his colleague Manuel Johnson* in 1853: 'You have probably seen Lord Wrottesly's capital speech about marine meteorological Observations &c made yesterday in the Lords'.[C342] John Wrottesley FRS was advocating participation in a new collaborative scheme to improve the consistency and coordination of meteorological measurement around the world. It was agreed to start off by using ships at sea: this was a relatively fresh arena compared with land-based meteorology, where numerous cherished observational approaches had been developed over the centuries. It also promised in time to lead to improved maritime safety and efficiency.

Ronalds immediately conceived a new instrument that could be used on board ship, continuing to Johnson:

It seems to me that the Aneroyd *[sic]* Barometer might be so modified as to be rendered applicable to self-registration — I am now going to M[r] *[Frederick]* Rippon *[instrument-maker]* to broach the matter to him. If requisite, the motion might be rendered rectilineal, and the whole affair be but very little enlarged.[C342]

The first practical aneroid barometer had recently been developed and the concept was potentially less sensitive to ship motions than traditional varieties. In fact a modification of the usual barometer was chosen, designed by Welsh, and the instruments were tested at Kew before and after each journey.

The British marine meteorology effort was managed by establishing a Meteorological Department under the direction of Robert FitzRoy FRS. He had used the barometer skilfully in the navy and in 1858 published the booklet *Barometer and Weather; How to Foretell Weather*.[B74] He began to use data telegraphed from various coastal stations to issue warnings of impending storms and, from August 1861, published his 'weather forecasts' in the press. The story is taken up in the 1862 report of the Kew Committee: 'Rear-Admiral FitzRoy having been informed of the existence at the Observatory of a Barograph invented and used by Mr. Ronalds' wrote to the Committee requesting to be furnished with 'at least one complete year of continuous record'.[S6] The trial was a success and Ronalds' instrument continued to be used in developing the earliest official weather predictions in the UK.

In the same period Sabine, who could always see the long game, requested the Royal Society to fund a second formal trial of Ronalds' self-registering magnetographs. The results were used to justify, build — and put into operation — an improved suite of instruments.

Kew had evolved into a recorder was well as a developer of instruments. Continuous photo-registration of magnetic and meteorological data was flying in the face of the agreement between Kew and Greenwich, and Airy was going to be upset. Sabine was ready for it. Although now in his seventies, he had recently commenced his term as President of the Royal Society and was more powerful than ever.

Airy proposed at the 1862 BAAS conference that the magnetic observations at Kew be discontinued. What happened next is recorded in his autobiography in the third person:

> a difference arose between Airy and Major-General Sabine, in consequence of remarks made by the latter at…the British Association. These remarks were to the effect "That it is necessary to maintain the complete system of self-registration of magnetic phenomena at the Kew Observatory, because no sufficient system of magnetic record is maintained elsewhere in England"; implying pointedly that the system at the Royal Observatory of Greenwich was insufficient. This matter was taken up very warmly by Airy [B115]

Airy arranged for Sabine to be removed from his Board of Visitors, which the latter then chaired. A period of detailed comparison of the observations at the two laboratories followed, after which major changes were made to the Greenwich magnetic instruments in 1864.[S4] The meteorological self-recording instruments and atmospheric electricity apparatus at Greenwich were also inferior to those developed at Kew, as becomes clear in the next chapters; not being Sabine's focus at the time, they retained their existing form. It seems that competing with Kew was more important for Airy than quality meteorological and geomagnetic measurement.

FitzRoy's weather forecasts were also attracting criticism. Attempting 'to prophesy the weather'[16] was denounced in Parliament as absurd, and many in the scientific community believed the approach could not be theoretically sound. Pressure mounted and FitzRoy committed suicide in 1865. Following his death, the Royal Society was consulted as to the future form of the Meteorological Department. The time was now right for Sabine to recommend extending the Department's remit to land as well as sea. Remarkably, given that 14 years had passed since Ronalds was at Kew, the reports Sabine read to the Society in 1866 made strong reference to his work in justifying the desired approach.

The basis of Sabine's logic was that knowledge of the spatial and temporal variation of meteorological parameters was essential to understanding the weather. The benefits of a distributed network of observations could only be realised however with uniformity in measurement and this required central coordination. He continued:

> The most hopeful way of removing the difficulties which impeded the adoption of a system in which all might willingly unite was obviously the introduction of instruments which should be continuously self-recording...The importance of such a substitution had been recognised at the Observatory of the British Association at Kew at a very early date, even before the Cambridge Meeting of the British Association in 1845; and at that meeting the satisfactory performance was announced of a photometrically self-recording barometer, self-compensated also for temperature *[this is incorrect in detail]*, devised by Mr. Francis Ronalds, the Honorary Director of the Kew Observatory, — the same to whose merits in regard to the instantaneous transmission of messages by means of electricity, at the very early date of 1823, attention has recently called in the public journals *[the call for his knighthood had just begun]*[A110]

> Mr Ronalds added *[at the BAAS conference]* that it was his "intention to provide during the ensuing year complete apparatus on like principles for registering as many of the other meteorological and magnetical instruments as funds will permit."[A109]

> self-recording instruments are of eminent local and international utility. The establishment of a series of them in England would confer a wide benefit. They would give precision and fulness to the charts of our own weather; they would set an example that foreign governments would soon follow[A109]

---

[16] *Hansard*, Mr Augustus Smith 12/5/1864.

The President and Council suggest that the observatory of the British Association at Kew might with much propriety and public advantage be adopted as the central meteorological station…the Kew Observatory possesses all the instruments required in a complete system of continuous self-recording meteorological observation.[A109]

It was not the first time that a scheme of this type had been suggested, but Sabine had the means to bring it to fruition in a way that would ensure its longevity. The new arrangement was overseen by a Meteorological Committee of the Royal Society whose members included Gassiot and Wheatstone as well as himself. FitzRoy's replacement at what was now called the Meteorological Office was Scott, whom we have already quoted.

Airy had not been involved in the deliberations and there was no mention of Greenwich. Predictably, he was not happy. He was advised formally, and not quite truthfully:

The President and Council desire to assure M[r] Airy that nothing in their communication to Her Majesty's Government was intended to imply any disparagement of the Meteorological Department of the Royal Observatory; and as little did it enter into their mind to exalt the Observatory at Kew to the disadvantage of Greenwich.[C385]

Behind these words was a rich correspondence from Airy to the Society's Secretary George Stokes:

The dignity of the most venerable and most respectable Society in the world has been prostituted…in the desire to throw a false glory round an institution of which General Sabine is the principal author.[C384]

Airy also queried the value of Sabine's plan in his next annual report to the Greenwich Board of Visitors. After mentioning that 'so many meteorological observatories have suddenly sprung up',[S3] he mused: 'Whether the effect of this movement will be that millions of useless observations will be added to the millions that already exist, or whether something may be expected to result which will lead to a meteorological theory, I cannot hazard a conjecture.'[S3]

Richard Inwards, in his 1896 Presidential Address to the Royal Meteorological Society, advises:

At the date of the reorganisation of the Meteorological Department of the Board of Trade, under a committee of the Royal Society, the Ronalds system of photographic barographs and thermographs was adopted for all their observatories [K12]

Kew's overseeing role for the observatories included the manufacture and erection of the instruments, personnel training, quality assurance of observations, and collation and publication of the results. For many years, 'London' weather was that observed at Kew.

Sabine and the Meteorological Committee did not wish to reinstate weather forecasts. Sykes, who was now an MP, argued in Parliament and at the 1867 BAAS conference that stopping the valuable storm warnings was 'a pedantic affection of science'.[17] Others agreed and under pressure a 'system of telegraphy of *facts* as contrasted with *prophecies*'[S7] was instituted the next year using measured data. *The Times* started in the mid-1870s to publish weekly graphs produced from the barograph, thermograph and other meteorological instruments as part of its weather reports. Forecasts themselves were later reinstituted after a 13-year break and the Met Office has continued to issue them ever since.[B195]

Long-term financial security for Kew was finally within reach in 1871 when Gassiot gifted £10,000 to the Royal Society 'for the purpose of…continuing magnetical and meteorological observations with self-recording instruments…in the Kew Observatory'.[K9] At this time responsibility for Kew transferred to the Society. Airy took over the Presidency of the Royal Society from Sabine that same year. Before long he moved from the Chair three resolutions to the Society's Council:

> That it rests within the competency of the President and Council at the present time to make such inquiry regarding the necessity of retaining the Kew Observatory as the central meteorological station[B100]
>
> That, in the opinion of the President and Council, it is desirable that the charge of the daily meteorological observations now made at Kew be transferred from the Observatory of Kew to the Royal Observatory of Greenwich.[B100]
>
> That a copy of the last Resolution be transmitted to the Board of Trade.[B100]

No one was willing to second even the first resolution, so 'the President did not deem it expedient to move the Resolutions 2 and 3'.[B100]

There the battle of the London observatories rested. Ill-feeling continued for some years.[B189] At the conferences held in conjunction with the Special Loan Collection in 1876 in the South Kensington Museum (Chapter 10), Brooke took the opportunity to say that Ronalds' magnetograph was 'impractical'[B95] while Scott mentioned why Ronalds' barograph was superior to Brooke's.

The increasing sophistication and interconnectedness of meteorology spurred global demand for Kew's photo-recording instruments over the rest of the century. This was aided when the precursor of the International Meteorological Organisation resolved in 1874 that an observatory was a 'first-order' station when it produced continuous automatic records.

Kew had reached the pinnacle of international observational science. Lord Kelvin wrote in 1871:

> the Kew Observatory…has been of immense service already, and has given England a name and…facilities for accurate observations in all branches of natural science which we never had before. It would be a national calamity if the Kew Observatory were to be abolished[B100]

[17] *The Times* 11/9/1867 10c–f.

By 1893, Dr Frank Waldo was able to write in *Modern Meteorology* that Kew 'is known all over the world by the certificates it has issued with instruments for use in nearly every land'.[B110] In a 1922 history of the BAAS, the Association Secretary described 'the famous little building'[B136] as 'one of the most important scientific foundations in the world'.[B136] More specifically, Sir Francis Galton FRS, in *Memories of My Life* (1909), denoted Kew as 'the Central Magnetic Observatory of the world. It held an almost equally strong position in respect to…meteorological instruments generally.'[B129] Such prominent positioning was seconded as late as 1961 by Dr Robert Multhauf of the Smithsonian Institution, in his review of meteorological instruments: 'In 1842 it *[the BAAS]* initiated observations at the Kew Observatory, which has continued until today to be the premier meteorological observatory in the British Empire.'[A154] Various articles written in different eras on aspects of the history of Kew and its achievements are listed in the Bibliography.

The Met Office closed its operations at Kew in 1980 — now surrounded by suburbia it was far from ideal for observations. The building itself has undergone recent repairs for use as a private residence.

13.6 Concluding Remarks

It is almost certain that, without Ronalds' efforts, Kew would not have survived its early years under the BAAS umbrella and gone on to achieve international renown. As the inaugural Director, he set up the observatory and moulded its successful mission, which was based on his own lifelong interests in instrument development, precision measurement, and wide dissemination and assistance, and focused particularly on atmospheric electricity and self-recording of natural phenomena. The excellence of his instruments brought early prestige and many supporters to the institution. Other contributions in forgone salary and his own equipment were critical in building up the facility and maintaining it when funding was severely limited. Just as important was his steady belief in Kew when others were wavering or, worse, undermining his efforts. Over time his strongest supporters saw the true value of the endeavour and used their influence to promote it. The revered Herschel backed the observatory and Ronalds himself at key junctures; Sabine sourced funds, played the necessary political games over 40 years and also gave vital personal mentorship; and Gassiot provided the cash injection that assured its sustainability.

Ronalds' years at Kew were thus a fitting climax to his scientific career. For the first time, he was at the centre of the British science world, well known and esteemed by his peers and shaping science at scale. While there, he invented 60 new meteorological, electrical and magnetic instruments; examples of 16 of these survive in the Science Museums in London and Oxford. Some were used to create unique datasets at Kew, while he provided others to various UK and international observatories. Later superintendents continued to supply updated versions of his original equipment around the world for another 40 years. It is little known that Ronalds contributed actively to this dissemination after he left Kew — he had found that he could achieve more towards the observatory's mission outside than inside the organisation due to the priorities of the new Kew Committee at that time. One of the greatest impacts of his life's work also emerged in this period, when his two most important inventions — the barograph and the telegraph — came together in helping to generate the earliest official British weather forecasts.

The next chapters explore Ronalds' technical achievements at Kew in more detail.

Chapter 14

ATMOSPHERIC ELECTRICITY AND METEOROLOGY: INSTRUMENTS AND OBSERVATIONS

Ronalds' numerous inventions to measure meteorological parameters and especially atmospheric electricity, and how he and others applied them

While the previous chapter outlined the overall evolution of the Kew Observatory, these next two chapters focus on Ronalds' scientific work in two particular areas — meteorology (and particularly atmospheric electricity) and photographic registration. We start with details of the Electro-Meteorological Observatory he established. The long Secs. 14.2 and 14.3 on Electrometers and Meteorological Equipment, which cover the many instruments he invented for the observatory, may be passed over if desired. Next follows descriptions of the observations taken and the results and ideas deduced from them. He also put much effort into encouraging and assisting others to conduct atmospheric electricity studies, both in the UK and abroad, to help delineate the global features of the phenomenon and to explore its relationship with terrestrial magnetism; these endeavours are outlined in the penultimate section, prior to some concluding remarks.

That electrical instruments responded to changes in the atmosphere and indicated positive charge in fine weather was discovered in the 18th century and denoted by the term 'atmospheric electricity'. The nature of atmospheric electricity was always a favourite topic of experimental study for Ronalds — we saw in Sec. 6.4 that he measured it at Highbury from around 1810, at Hammersmith from 1813 and in Italy in 1819. In his more than 20-year break from the subject little progress had been made. James Forbes FRS, Professor of Natural Philosophy at Edinburgh, said of atmospheric electricity in his major review of meteorology at the 1840

British Association for the Advancement of Science (BAAS) conference: 'almost everything remains to be done on that subject; he who proposes to enter on the field must be prepared to cope with the difficulties of *original* investigation.'[S1] Shortly afterwards, when Sir John Herschel*, Edward Sabine* and Charles Wheatstone* were plotting how the Kew Observatory might be acquired, Sabine admitted an important gap in his Magnetic Crusade: 'How much is it to be regretted that we have not simultaneous electrical observations at these times of great magnetical disturbance.'[C102] Demand had caught up with Ronalds' longstanding interest and he was lured back into the field.

14.1 Atmospheric Electricity Observatory and Apparatus

Ronalds' original project plan for his electro-meteorological observatory at Kew survives in a long letter to Wheatstone written in November 1842. He framed the subject under ten headings: 'The Observatory'; 'Collecting & Insulating Apparatus'; 'Electrometers &c'; 'Additional Electrical Apparatus'; 'Various Meteorological Apparatus'; 'Laboratory'; 'Pagoda & Exploring Wires'; 'Preliminary Operations'; 'Experiments'; and 'Objects of Research'.[C108] It is similar in concept and detail to the first annual report he presented to the 1844 BAAS conference.

He gave little information on his later headings in this first document, noting that 'these (the last particularly) seem, on reflection, rather too important items for an Electrician of bye-gone-times to discuss in this kind of manner.'[C108] He still had very low self-esteem at this time. By a 'Laboratory', he was referring to a clean workshop to make and repair equipment and conduct experiments. It was to include a lathe and various other tools, similar to his laboratory at home. The importance of such a facility for him is illustrated in a comment he made years later regarding the science faculty at Marseilles University: 'I was very glad to see that an extremely well organized workroom (or physical laboratory) forms a part of his arrangement.'[J85]

The Exploring Wires concept entailed a large grid of wires on an adjacent field, connected to electrometers. He raised the idea several times in his early years at Kew but was not able to access the land. He later realised he could utilise an arrangement somewhat akin to it in the emerging network of telegraph lines. Recording of atmospheric electricity over an isolated flat surface rather than at a point near a building, thereby reducing distortion of the electric field, was instituted at Kew at the end of the century.[K13]

Ronalds' project plan started by noting the suitability of the Kew facility for atmospheric electricity observations: 'The neighbourhood of the river and the rather marshy state of the land near the building cause sometimes very dense and interesting fogs...fogs present remarkable electric phænomena.'[R12] The building itself was also well configured — the electrical observatory would be set up in the

Fig. 14.1:   Kew Observatory; Ronalds' original drawing using perspective tracing instrument. Ref. J61.

A Principal conductor with collecting lantern, *B* Wind vane, *C* Balance anemometer, *D* Pluvio-electrometer, *E* Electrical frequency apparatus, *F* Lightning conductor strap.

domed room at the top of building while weather observations could be made from the roof terrace surrounding it (Fig. 14.1). He mentioned the advantages of the sheet metal roof in alleviating charge build-up in the confined space, which would affect the electrometer readings. He also wished to apply a tinfoil lining but this was not carried out.[C108,C126] (The benefits of a so-called Faraday* Cage[1] had been demonstrated by its namesake in 1836 but had also been described by Franklin in the mid-18th century — Ronalds referred to 'Franklin's Can'[J85] in his writings.) He soon noticed that 'if 2 or 3 persons are in the dome at Kew, the amount of Electricity is immediately very much less'.[C157]

Ronalds suggested to Wheatstone: 'An electro magnetic Telegraph might be so constructed by you as to give notice to the Sergeant *[observer John Galloway]* below (& in Bed perhaps) of charges above those usually occurring in serene weather'.[C108] The telegraph did not appear and Ronalds developed an updated version of the bell arrangement he had used in Highbury to signal unusual events.

To Wheatstone's early question as to what apparatus Ronalds intended to provide himself, Ronalds had answered 'all that I can'.[C108] In addition to electrometers

---

[1] Metal shield around equipment to neutralise electric fields.

and meteorological instruments, he brought miscellaneous electrical equipment such as his electrostatic generator, dry piles and Leyden jars; tools for the workshop; useful supplies of brass, glass, etc. for making new devices; his sturdy tripods; furniture; a heating stove; and his fire alarm.

Principal conductor

Ronalds did not bring the insulated conductor he had invented and built around 1815 and documented in 1817 (Fig. 6.6): 'because I think that the first strong wind from the south would transport the fishing rod to the neighbouring abode of the fishes'.[C108] The Kew set-up was on the same plan but considerably more substantial (Fig. 14.2). The principal conductor was a long tapered copper tube that extended through a capped hole to a height of nearly 5 m above the dome — it was then about 24 m from the ground and above the tallest trees in the vicinity. The conductor was supported on a strong trumpet-shaped glass pillar, which was heated constantly inside via a chimney from a small lamp below to ensure it was a perfect insulator. The insulator in turn was bolted through a thick wooden baseplate to a 2.4 m high pedestal accessed by stairs. The clever connection design incorporated strips of leather to accommodate large flexures caused by the wind without cracking the glass.

Risk management was highlighted in the project plan. Ronalds had queried Wheatstone of the building: 'Is it insured against artificial lightning?'[C108] Fortunately Ronalds had a good appreciation of lightning conductor design and configured a lead strap in close electrical contact with the apparatus to run along the pedestal and down the exterior of the building to the drains. The strap had minimal joints and changes in direction, and he preferred any bends to be 'curvilinear'[C330] rather than right-angled and for joints be well soldered. He also included details of conductor design in his instructions to other observatories. Years later he inspected the tower of the Padua Observatory after it had been struck by lightning and noted the poor connections and many bends along the conducting wire.[J85]

The principal conductor could receive electric charge from the air by induction and/or absorption. If induction was to be used, a hollow copper ball could be placed at the tip of the rod. Ronalds always preferred absorption methods, as he made very clear to Wheatstone in his project plan, and he devised a series of experiments at the time to demonstrate their mode of behaviour and greater effectiveness. He noted in his diary in 1859 that a new friend, the renowned Italian physicist Francesco Zantedeschi, 'approved (highly) of my preference'.[J85] Wheatstone apparently held a different view and later publicised an alternative French design based on induction at the 1849 BAAS conference, citing several advantages over Ronalds' arrangement.

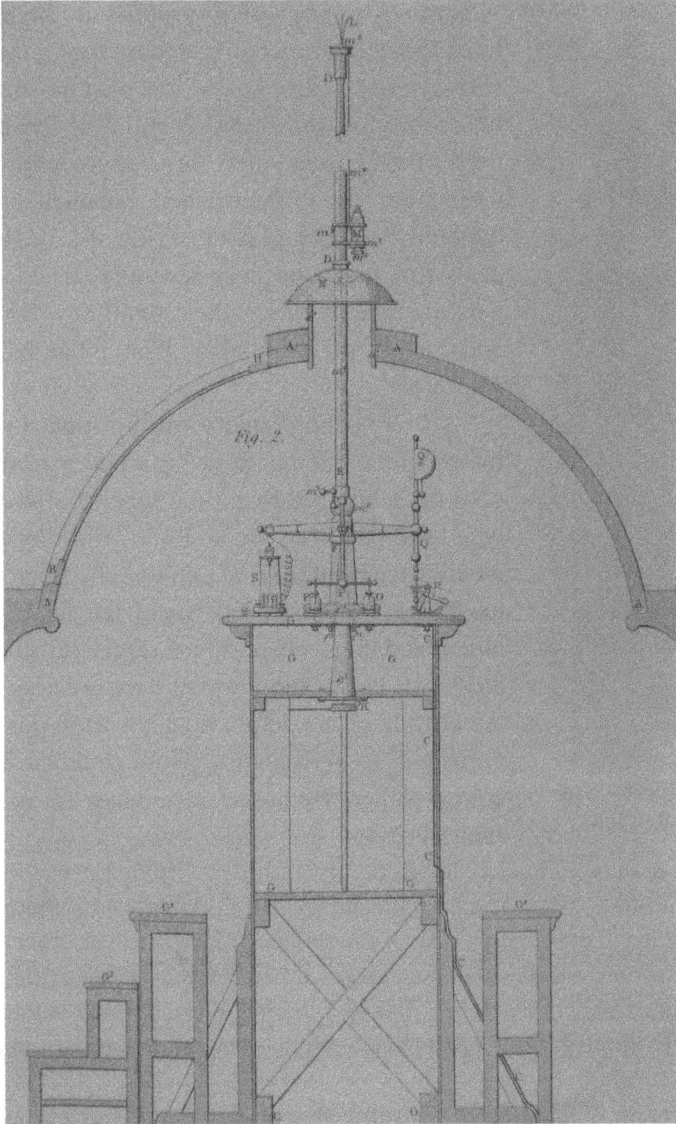

Fig. 14.2:   Principal conductor in the dome (1843). Ref. R12.

A later Director of the Kew Observatory, Dr Charles Chree FRS, explained the various absorption techniques in his entry on atmospheric electricity in the eleventh edition of *Encyclopædia Britannica* (1911):

> In the earliest *[method]* the conductor was represented by long metal wires…and left
> to pick up the air's potential. The addition of sharp points was a step in advance; but
> the method hardly became a quantitative one until the sharp points were replaced by
> a flame…or by a liquid jet breaking into drops.[A143]

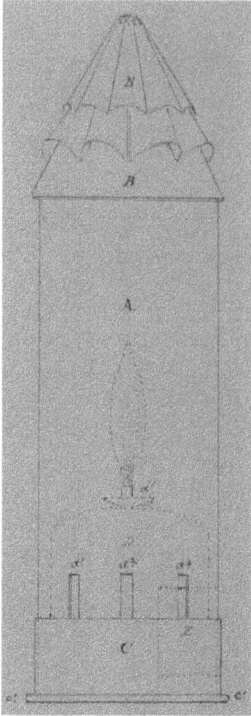

Fig. 14.3:   Collecting lantern (1844); Ronalds' original drawing. Ref. J62.

A burning fuse or a fluid dropper are both very effective electricity collectors because they generate a large surface area of droplets or ions that assist the conductor to acquire the same potential as the air. Ronalds used all of Chree's first three approaches at different times, but for his principal conductor at Kew he adopted a collecting lantern (Fig. 14.3). He honed the design of the lamp over several years to optimise its performance in all weather conditions. Both this lamp and that to the insulating glass pillar burnt olive or sperm oil.

The collecting lantern was positioned at the top of the principal conductor using a pulley at the tip and a cord descending through the tube to a small winch. It was lowered every six hours to trim the wick — this would have been the first atmospheric electricity apparatus to be commissioned where the lamp was always alight and thus observations could be made continuously. Access to the lowered lamp was via a ladder on the outside of the dome. Ronalds learnt quickly that the dome and its surrounding plinth could be dangerous in cold weather. He joked in a letter to Wheatstone in January 1844:

> The frost & snow…violently threatened my dismissal…Galloway had fixed a rope for me to hold by when I had to officiate for him in fixing the Lanthorn [lantern] & warned me of the slippery state of the Dome &c…I felt quite certain that something was wrong outside, and on getting out discovered an Icicle uniting the insulated copper cap with the dome…in endeavouring to reach which & forgetting the rope & the warning, my foot slid and a struggle for life & limb of some seconds ensued.[C131]

We have here an illustration of his somewhat absent-minded nature when focused on a technical problem. He highlighted the risk in his first BAAS report as a warning to other observers and a handrail was later added to ease the operation. He also documented the safest method of removing and reinstating the conductor when repairs were required.

Ronalds and Galloway had quickly gone to work to set up the observatory. The principal conductor was manufactured by John Newman, with whom Ronalds had been dealing with his drawing instruments, while the building modifications were completed by local workmen. In June 1843 he was able

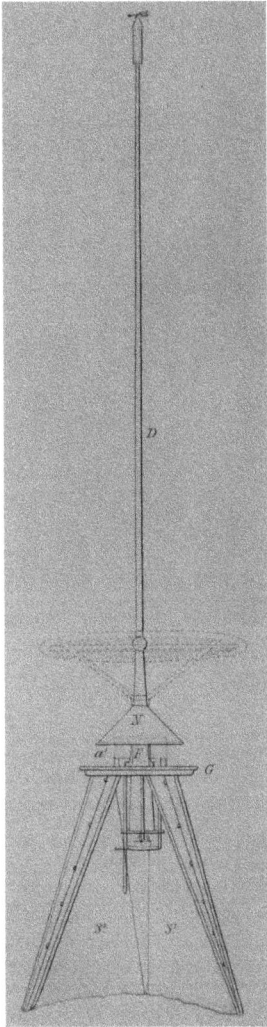

Fig. 14.4: Simplified atmospheric electricity apparatus (1850); supported on tripod with solid triangular legs. Dotted lines represent earlier pluvio-electrometer. Ref. R21.

to report their accomplishments to Wheatstone, which he did in a carefully edited letter:

> I hope that You will be happy very glad to hear of my our compleat success so far as in the erection of the rod and obtaining strong signs of the periodical electricity from it…I…shall be most should feel much gratified in shaking hands with you on the occasion when you can could spare time to come.[C112]

Wheatstone's colleague John Daniell* sent his congratulations as soon as he heard: 'I am quite delighted to find that you are once more actively employed in the Cause of Meteorology.'[C115]

Round-house apparatus

Ronalds realised that his apparatus, while robust, was also heavy and required specific conditions for successful installation. A particular problem was the need for it to be positioned at the top of the observatory. Architects did not generally share his zeal for cutting holes and making other such modifications to the elegant pinnacle of their observatory building.

He hoped to use a pagoda in the observatory grounds to house a more cost-effective and palatable configuration, but access was denied by the Commissioners of Woods and Forests. In 1850 he was able to develop and build an alternative. It retained his overall concept of a conductor with a collecting lantern and a warmed glass insulator, but these were supported on a sturdy tripod table of the type he had invented in the 1820s (Sec. 10.4). Sailcloth was wrapped round the legs to shield the heating lamp from wind in an outdoor set-up (Fig. 14.4). For greater protection a simple 'round-house'[R21] could be used (Fig. 14.5), just large enough to cover the apparatus, an observer and, if desired, self-registering equipment. The idea was that the arrangement could be positioned on the roof of a building or on open land. The copper rod was 3.7 m long.

The primary conductor remained in place at Kew until the mid-1850s, when its position was taken by a telescope. The Kew Committee's 1856 report advised

Fig. 14.5: Round-house atmospheric electricity apparatus (1850); equipment supported on tripod with base ties. Ref. R21.

that 'An apparatus of smaller size, but on the same plan, has been erected on the side of the dome'.[56] The new set-up was the round-house arrangement. The previous year, this smaller version of the apparatus had made the trip to Paris for the *Exposition Universelle*. The Kew Committee provided it to the Special Loan Collection in South Kensington (Chapter 12) in 1876. Sabine, as Chairman of the Committee at that time, would have been the driving force behind the significant array of Ronalds' inventions in the collection and thus deserves the credit for their survival. Today the glass insulator is still bolted to Ronalds' tripod and the 3.7 m long conductor and collecting lantern lie nearby in the Science Museum store, although they have been on display in the intervening years.

A description of the Kew Observatory in the 1897 *Record of the Royal Society* noted that the glass cupboards contained instruments 'of considerable historic

Fig. 14.6: Ronalds' atmospheric electricity apparatus made for the Madrid Observatory (1852), showing the suite of electrometers; positive print of a negative taken by Henneman at Newman's premises. © British Library Board, Talbot Photo 1 (349).

A Bennet gold-leaf electroscope, *B* Three registering or night electrometers around clock by which electrical contact is provided sequentially, *C* Henley electrometer, *D* Pair of Volta electrometers with eye pieces, *E* Spark measurer.

Writing near the bottom reads:

'Invented by F. Ronalds Esqr and Manufactured by J. Newman 122 Regent St London'

interest, such as Ronalds' apparatus for examining atmospheric electric potential'.[A139] It is not known to which parts this is referring although it confirms that more than one of the equipment sets he made were retained well beyond his departure.

14.2 Electrometers

The atmospheric electricity was measured by a series of electrometers (Fig. 14.6). Instruments were connected to four orthogonal metal bars projecting horizontally from the base of the principal conductor such that their active parts were insulated uniformly by the glass pillar below. Each instrument was optimised to measure specific electrical attributes and the set-up also allowed ready calibration and comparison of devices. Ronalds documented eight types of electrometer in his plan to Wheatstone, most of which he had first customised in his early years of electrical studies. Many of his subsequent innovations at Kew are mentioned

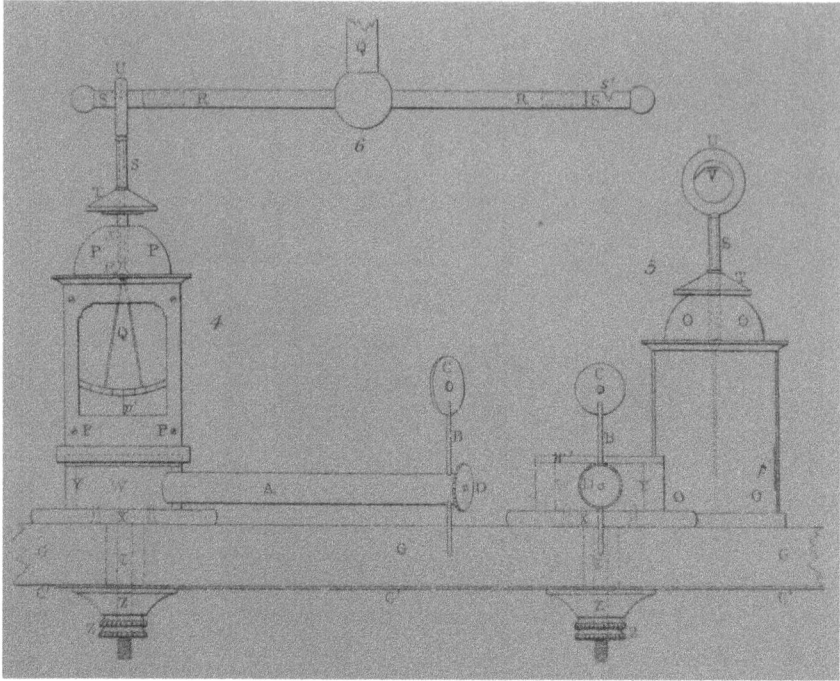

Fig. 14.7:   Improved Volta straw electrometers (1843). Ref. R12.

below, but the evolution of his electrograph is discussed in the next chapter on self-registration.

Volta straw electrometers

The work-horse of the atmospheric electricity measurement array was a pair of so-called Volta's electrometers (Fig. 14.7). Each comprised two thin straws and an arced scale to indicate the degrees of divergence when they were charged. Ronalds had made modifications to Volta's configuration around 1814, as mentioned in his Ideas Book, and he continued to hone the design for Kew. A friend of Volta's who visited the observatory told Ronalds that Volta 'would be much pleased'[R12] if he could see the improvements.

Ronalds' instrument is contained in a brass case 5 cm square with flat glass windows front and rear. The former glass bottle had the disadvantage of visually distorting the readings from the scale. The inclusion of significant metal also acted as a Faraday cage. A magnifying eye piece to assist in estimating fractions of a degree on the semi-translucent ivory scale was attached 30 cm away by a horizontal rod — this separation also helped reduce influences of the human body on the electricity.

Other design improvements involved the method of connection. The straws were suspended by hooks of fine copper wire through small holes at a precise spacing in the flattened end of a brass rod. The rod itself passed through a glass tube known as Singer's* insulation (Sec. 6.4) so that it was isolated from the case. The overall electrometer was hung from a steel knife-edge on the orthogonal bar. With this arrangement the electrometer could be readily attached and removed from the apparatus without breaking the conductor's insulation and receiving a shock. The second electrometer was made exactly five times less sensitive than the base instrument by adding weight to the straws: their hollow cores were filled with wire extending from their supporting hooks.

The devices came to be known as 'lantern electrometers' from their style. When Ronalds ordered these lantern electrometers for others, the cases were made by Newman, John Cary or Patrick Adie. He always completed the instruments however and provided the calibrated straws, which as he noted to Wheatstone: 'I believe no shop in London can do ready made'.[C108] He had also developed equipment to make perfectly straight wire indicators for electrometers, and to cut and fix gold leaves in the Bennet electroscope without puckering, as he outlined in his 1817 paper on atmospheric electricity;[R6] he brought both to Kew.

At least four of his straw electrometers survive. A linked pair with their eye-pieces formed part of the atmospheric electricity displays at both the Paris Exposition in 1855 and the 1876 Special Loan Collection and they are still attached to his tripod at the Science Museum. Others were acquired by the museum from the Met Office in 1957 and again in 1980 when the observatory closed. The latter is of an early style housed in a glass bottle and was possibly deployed in the initial Kew observations in 1843. The 1957 donation has an appearance of newness that suggests that the instrument's use at Kew extended beyond Ronalds' atmospheric electricity device. Another Ronalds lantern electrometer is catalogued at the Oxford Museum of the History of Science but has not been located in recent years.

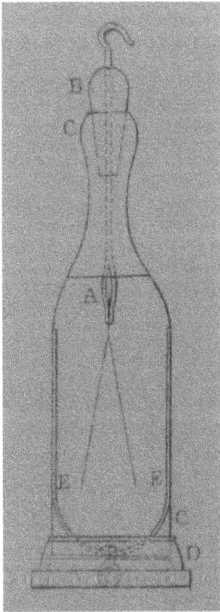

Fig. 14.8: Improved Bennet gold-leaf electroscope (1843). Ref. R12.

Bennet gold-leaf electroscope

The second indispensable instrument for electrical observation was Bennet's gold-leaf electroscope (Fig. 14.8). Being very sensitive, it was used to detect weak or changing signs. The leaves are suspended from specially configured forceps in Ronalds' design, and are housed in a glass bottle into which a long and wide brass strip is screwed. The leaves are therefore discharged when they diverge sufficiently to touch the strip and the charge is dissipated to earth rather than accumulating on the bottle. He further enhanced the arrangement in 1845 by storing the bottle in a glass receiver that contained the desiccant calcium chloride in a shallow annular trough. The electroscope thus remained dry and clean for use at any time.

The electroscope would have been on display at the Paris Exhibition, where Moigno* described its design as 'perfect and all-encompassing'.[A93] The Science Museum today holds three of Ronalds' gold-leaf electroscopes, acquired in 1876 and 1980. Their designs differ, with just one having the inner brass strip screwed to the metallic base as described and illustrated by him in 1844.

Henley electrometer

William Henley FRS invented an alternative electrometer arrangement in 1770. Ronalds' version was used to measure high charges, as he designed it to have one-fiftieth of the sensitivity of the baseline Volta electrometer. It comprised a pendulum with a pith-ball at its end, which swung between a pair of parallel semi-circular ivory plates marked in degrees (Fig. 14.9). Both the pendulum and the vertical support rod received the charge and the angle of rise of the former indicated the

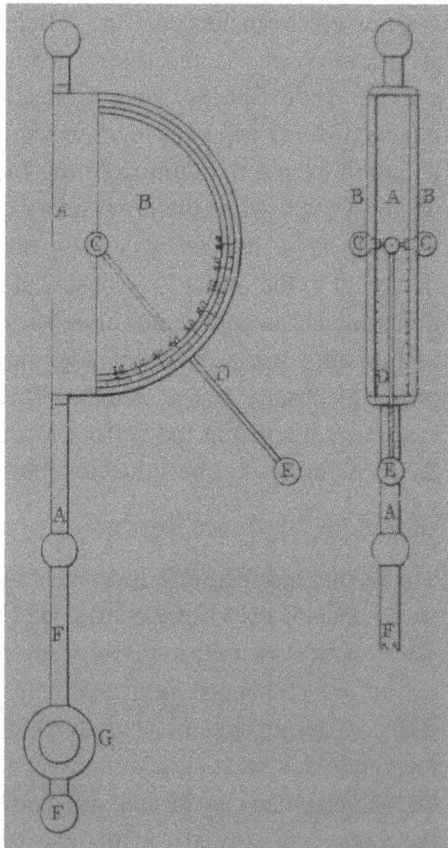

Fig. 14.9:   Improved Henley electrometer (1843). Ref. R12.

intensity of the electricity. The version of the device now in the atmospheric electricity array in the Science Museum is not of Ronalds' design.[B94]

Coulomb torsion balance

Ronalds' modified Coulomb Torsion Balance was his favourite electrometer. As he wrote in his 1842 project plan regarding different means of resisting electric repulsion to measure the charge: 'Is not Torsion better than Gravitation?'[C108]

The original instrument developed by Charles Coulomb in 1784 comprised a bottle containing a horizontal cross-arm suspended from a thread. At the top of the thread was a twistable and graduated torsion clamp. One end of the cross-arm held a gilded ball and the other a paper disc acting as a counterweight and damper. A second ball on a vertical rod, carrying the charge to be measured, was lowered into the vessel and, on touching the gilded ball, the cross-arm twisted away through repulsion. The angular separation of the balls could be read on a circumferential scale to give an indicator of the strength of the charge or, alternatively, the torsion required to bring the two balls back together could be measured. The device had both advantages and several disadvantages — it was awkward to operate, unsteady in use and readings were modified by induction and humid weather.

Ronalds mentioned his proposed alterations to the device in his project plan. According to Wheatstone's 1843 BAAS report, the modified instrument was in action by the middle of that year. He denoted it as having a range as great as that provided by Volta 1 and 2 and Henley combined, 'while it possesses the sensibility of the most delicate of Volta's straw electrometers'.[S1] Ronalds wrote the next year of his 'strong hopes that our principal use of all these electrometers will be that of comparing them with one torsion electrometer'.[R12] The configuration of the instrument was described in detail in his 1844–45 Kew Observatory Journal. Only a very brief précis of his report was printed in the BAAS proceedings, so his progress with this important electrometer is little known. A sketch and very brief summary of subsequent improvements also survive. In 1853 he denoted the instrument to Professor Henry Noad as: 'a little improvement on Sir W. Harris's improvement of the Coulomb electrometer'.[C346]

Ronalds' initial aim was to avoid the 'tedious manipulation'[R13] characteristic of the instrument. His configuration was also 'more easily read, far more comprehensive in its scale, and altogether, much more correctly and conveniently used'.[J61] His first adjustment was to suspend the arm by two parallel fibres rather than one (Fig. 14.10). The restoring force induced by rotation is thus due to changing inclination of the fibres rather than twisting of the fibre itself. This modification had already been introduced by Sir William Snow Harris in the 1830s. Ronalds also hung a weight below the fibres to keep them taut and housed them in a narrow tube made of brass rather than glass. Gold or platinum wires were substituted for the

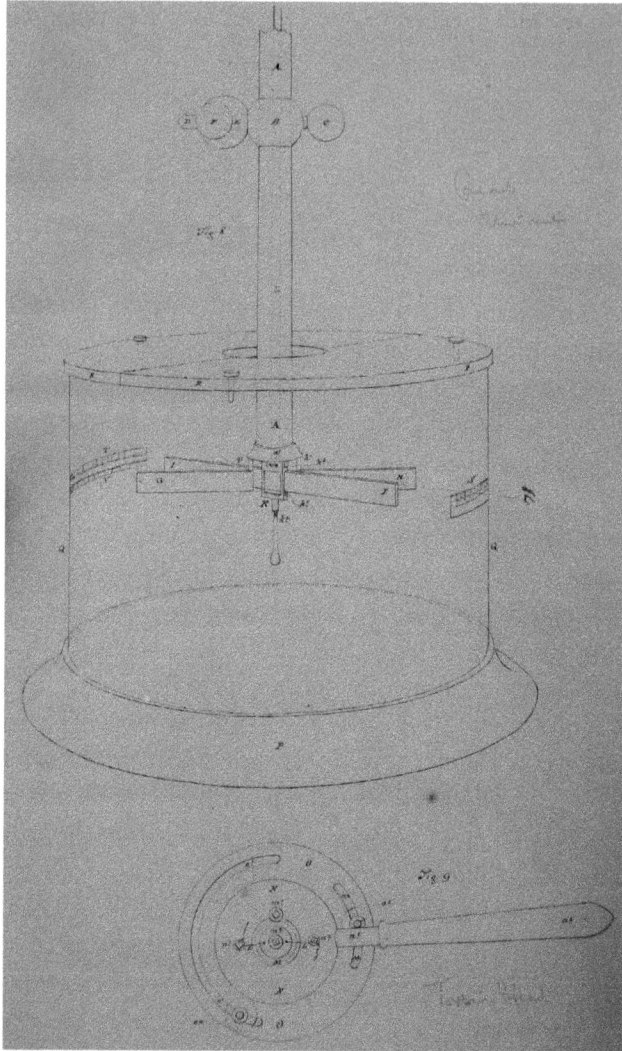

Fig. 14.10:   Modified Coulomb torsion balance — model 1 (1844); Ronalds' original drawing. Ref. J62.

silk fibre. His next alteration was to modify the insulated cross-arm to a thin metal needle and replace the vertical rod with a second horizontal needle in the same plane and fixed to the brass tube. Both needles were then able to be electrified directly from an arm of the principal conductor. In an early model, the instrument's glass container was warmed from below by a small lamp to keep the air dry. Finally, he pasted tinfoil on the glass 'according to Professor Faraday',[J62] although he planned to replace the glass container with one made of thin sheet brass.[J62]

At least two scales were placed around the cylinder to read the angle of divergence. The first was graduated to match the readings of the baseline Volta electrometer and another coincided with its less sensitive pair. To measure the higher charges, a handle on the torsion clamp was turned a fixed amount to introduce greater deflection into the wires and position the needle on the new scale. Further scales could also be added.

By 1845 the device was listed in Newman's *Catalogue of Philosophical Instruments* as 'RONALD'S improved torsion electrometer'.[B53] It is not known how many of the devices were sold.

Ronalds reported his idea of using the torsion balance as the basis of photographic registration of atmospheric electricity in 1846. The torsion balance was similar in concept to the magnetometer (see Sec. 15.6) in having a suspended horizontal bar and it was natural that similar approaches would be contemplated for measuring the rotations of both instruments. Indeed, when he described his torsion balance to the Astronomer Royal George Airy* in 1843, Airy suggested attaching a light mirror to the needle to indicate the angle of rotation, as he did with his magnets.[J65]

The revered physicist Lord Kelvin, perhaps best known today by the eponymous unit of absolute temperature, later turned his attention to the Coulomb torsion balance. He described his so-called divided-ring electrometer in 1856 and, as his ideas continued to evolve, first published his quadrant electrometer in 1867.[A112,A113] Torsion suspension became the most important electrometer configuration in the later 19th century, and Kelvin's instrument became the accepted standard for the emerging electrical engineering industry. It was not replaced by modern electronic instruments until the 1940s.

Ronalds' torsion balance may be regarded as an intermediate step in the evolution of this form of electrometer. Kelvin would have seen both his torsion balance and his magnetographs when he visited Kew in August 1850[J84] and it seems plausible that he was influenced by them, even if subconsciously. Similarities between Ronalds' and Kelvin's instruments include: a simple suspended needle; a fixed piece in the plane of the needle to replace the inserted vertical rod; automation of the charging of these metal elements through wire connections; employment of a Faraday cage; and the concept of photo-registration. A key difference was that Kelvin's electrometer indicated both the sign and magnitude of the charge. Ronalds believed the sign of the electricity to be unimportant in routine atmospheric electricity observations, although he did introduce a means of showing it in his photo-electrograph at Airy's request (Sec. 15.3), like Kelvin by creating a symmetric electric field. A later Kew Observatory Director, Dr Francis Whipple, described this last instrument as 'the forerunner of the quadrant electrometer'.[K14]

<u>Spark measurer</u>

Ronalds included what he called a 'safety valve' in his atmospheric electricity apparatus, which allowed the principal conductor to discharge across the gap between two

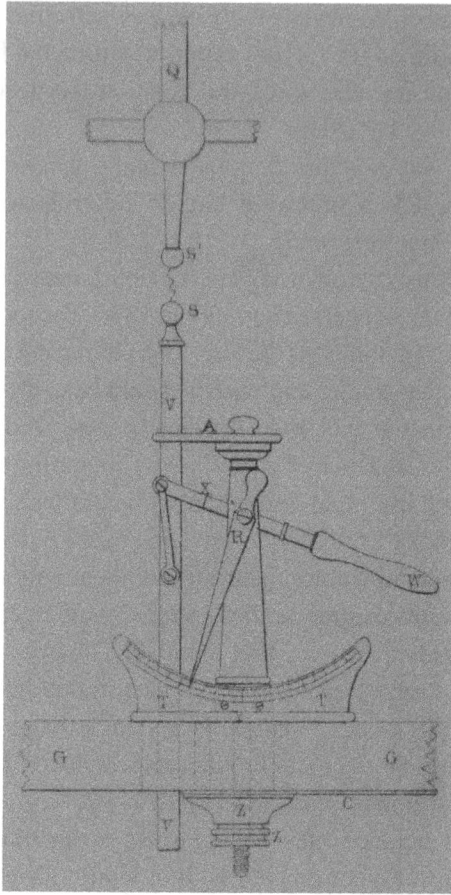

Fig. 14.11:   Spark measurer (1843). Ref. R12.

balls. The device was also an accurate 'Spark measurer'. The separation of the balls was adjusted by a glass lever, with the magnitude of the gap shown by an index on a precise multiplying scale (Fig. 14.11). High electrical activity could thus be quantified in terms of the frequency of sparks at a particular separation or the maximum length of the spark. He noted that there was close correlation between measured spark length and the degrees of intensity or tension on the Henley electrometer.

The instrument was more accurate and easier to use than the discharging electrometer invented by Timothy Lane in 1766. After display at the Paris International Exposition in 1855, it entered the Science Museum through the 1876 Special Loan Collection but has not been seen by the author.

Charge distinguisher

Ronalds brought a Charge Distinguisher of Beccaria's design to Kew in 1843 to enable the sign of the electric charge on the conductor to be confirmed. He also

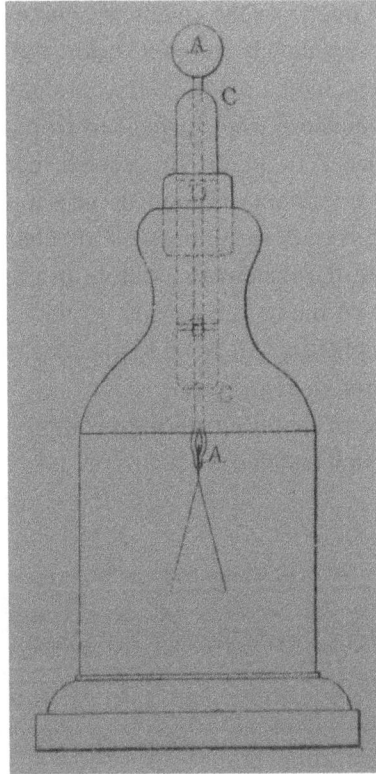

Fig. 14.12:   Charge distinguisher (1843). Ref. R12.

briefly employed dry piles for the same purpose but these were soon superseded by a new design (Fig. 14.12). He used as its basis a narrow Leyden jar with a pair of gold leaves suspended from the brass tube forming its inner lining. The whole was placed inside an outer bottle. Each morning the distinguisher was given a negative charge, which it retained for the 24-hour period. The resulting collapse or further divergence of the leaves when it was brought close to the principal conductor indicated the sign of the charge held. With this approach, unlike others, there was no risk of altering the electricity of the conductor through contact. One of Ronalds' electroscopes that entered the Science Museum from the Met Office in 1957 appears to be of this type.

Registering or night electrometers

We saw in Sec. 6.4 that one of the ways he took electrical potential readings while away from his apparatus was with a clockwork mechanism that made electrical contact at particular times with each of a series of electrometers. He returned to this approach in 1844 to enable observations during the night. He first made experiments to find the most suitable electrometer configuration — one that was sufficiently sensitive to drain little charge from the conductor, and then able to retain it so that the charge decayed only very slowly and in a steady and measurable way.

He determined the best indicators to be fine straightened wires with gum arabic at their ends. The indicators were suspended from a wire protruding from a vertical glass rod and placed in an airtight receiver containing calcium chloride (Fig. 14.13). The other important component was a spring-loaded device that extended outside the receiver. A revolving arm attached to the arbour of a timepiece and connected with the conductor (visible in Fig. 14.6) was brought into light electrical contact with the device for just a short time — this was achieved without breaking the airtight seal. An arced scale aided the reading of the resulting degrees of divergence through an eyepiece.

He began using three of these registering or night electrometers consistently from 1 January 1845, which enabled 24-hours-per-day observation. One survives in the Science Museum.

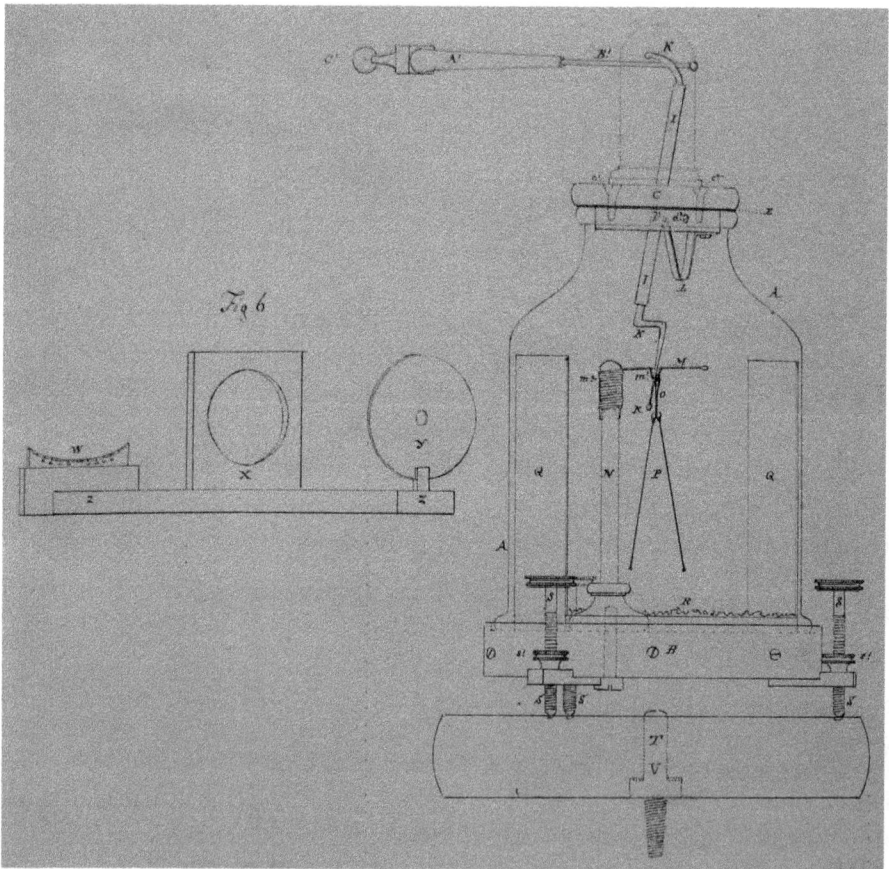

Fig. 14.13:   Registering or night electrometer (1844); Ronalds' original drawing. Ref. J62.

In designing the instrument he had observed the rate of decay of charge under different conditions, understanding that air was a conductor. As he noted, his desire was 'to prevent as much as possible dissipation'.[R12] As early as 1823 he had mused regarding atmospheric electricity: 'The *conducting power of air*, relatively to its temperature, to its humidity, and to its rarity, are all circumstances which should be taken into consideration, and their influences examined.'[R7] He also mentioned the value of studying dissipation and conductivity in his 1842 project plan, but was unable to complete and document his work. Charge dissipation experiments in different atmospheric conditions were formalised late in the century by physicists including Linss, Elster and Geitel and were then able to be explained in terms of the ionised particles in the air. Measurement of air conductivity commenced at Kew in 1909 and some years later its importance to atmospheric electrical potential was confirmed when an inverse relationship between the two was determined.[K13,K15]

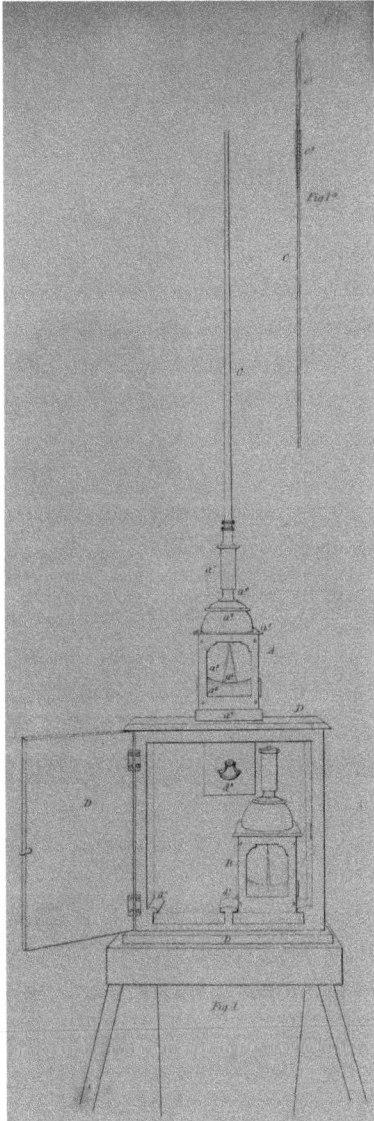

Fig. 14.14: Portable electrometers (1844). Ref. R21.

Pluvio-electrometer

Ronalds developed what he called a pluvio-electrometer in 1843 to measure the electricity of rain. He had observed that 'a hard shower of rain, &c. as frequently robs our conductor of large doses of electricity as that it brings them'.[R12] The device comprised a shallow copper dish a metre in diameter that was mounted horizontally on the usual heated glass insulator. It was supported on a tripod on the roof terrace and connected to electrometers inside the observatory to measure the electrical state of the dish in comparison with the principal conductor. The apparatus was little used and later morphed into one of his outdoor atmospheric electricity collectors (Fig. 14.4). Study of the electric charge carried by raindrops was re-instigated around the turn of the century by Elster and Geitel. In the mid-1930s, a more sophisticated apparatus was developed at Kew with an insulated funnel connected to an electrometer and incorporating photo-recording.[K15]

Portable electrometers

A new Portable Electrometer arrangement was configured in 1844 for use remote from the observatory (Fig. 14.14). The Volta electrometers were made more robust in their details, including being secured in a mahogany case for transport and by allowing a tube to be inserted through their base to protect the straws when not in use. A tapered conductor one metre long could be screwed upon the cap. Stored in a draw in the case was a collecting candle comprising cotton threads steeped in sulphur, which was held in place at the tip by a helical copper wire. The rod itself fitted into a walking stick.

Ronalds described and illustrated his transportable electrometer set-up again, and in more detail, in his 1851 BAAS report. He gave his motivation as being: 'several eminent meteorologists having thought that these instruments would...

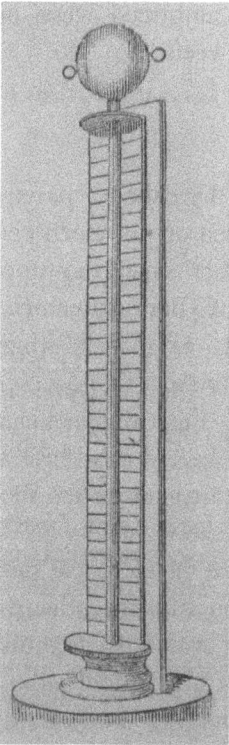

afford better approximate results in observations, on mountains, &c., that the portable instruments which have been usually employed'.[R21] He might also have been prompted by Wheatstone's very complimentary report two conferences earlier of an alternative portable electrometer based on induction, as mentioned earlier, or the observation made at the Great Exhibition that year that simple atmospheric electricity instruments were lacking.

Air thermometer for electricity

We end this section with mention of another electrometer Ronalds developed, although he did not use it with the atmospheric electricity apparatus. His 'Air-Thermometer, In Electricity'[A56] was introduced in the 1838 British Cyclopædia of the Arts and Sciences, edited by Charles Partington; Partington also lectured at various London institutions, which is how Ronalds would have known him. The device was additionally described in Hebert's Engineer's and Mechanic's Encyclopædia (1848).[A74] The Cyclopædia advises of it: 'There are several forms of this apparatus described by the early electricians, neither of which are at all equal either on the score of utility or simplicity to the accompanying instrument by Mr. Ronalds.'[A56]

Fig. 14.15: Air thermometer for electricity. Ref. A56.

At the top in Fig. 14.15 is a bulb containing a gas, to which opposing wires and rings are attached. When they are connected with a static electricity source, the shock depresses the fluid in the graduated thermometer tube. The apparatus could be used with air in the bulb to quantify the magnitude of an electric discharge through the expansion produced or, alternatively, the properties of different gases could be studied. Such a device for measuring discharge through its heating effect was first published by Ebenezer Kinnersley in 1763. A later version was developed by Snow Harris in 1827.

14.3 Meteorological Equipment

Ronalds' long interest and experience in meteorology is exemplified by the suite of instruments he took on his Grand Tour and the descriptions in his 1823 book of different weather conditions he observed there. In his 1842 project plan he listed eight parameters to be measured simultaneously with the atmospheric electricity: temperature, humidity, atmospheric pressure, precipitation, evaporation, wind direction and speed, and magnetic field. Instruments to measure several of these were again of his own invention and were brought from home or developed for the observatory. He also stipulated observations by eye from the roof terrace, including 'the forms of clouds &c, Color & Transparency of the air, fall of dew &c'.[C108] He had highlighted back in 1817 that the omission of these simultaneous measurements in atmospheric electricity studies to date were to be 'regretted'.[R6]

Photography was soon deployed to assist the observations. He used his camera to create an improved register of cloud types in 1850 while, as early as 1845, he was able to observe the minute-by-minute interrelationship between solar irradiation and atmospheric electric potential with his photo-electrograph (Sec. 15.3). He hoped to extend these investigations to humidity, barometric pressure and other parameters, all recorded photographically, but was unable to conduct the detailed work. The correlation of each of them with atmospheric electricity was later studied into the 20th century.[A137]

The quality of the instruments he did develop and deploy at Kew was recognised; Airy's deputy Robert Main at Greenwich wrote of Ronalds in 1851 that 'the science of atmospheric electricity, not to speak of other branches of meteorology, is mainly indebted to him for its present state of advancement.'[A84]

Rain and vapour gauge

Ronalds had developed a Rain and Vapour Gauge prior to his involvement at the Kew Observatory. Its novelty was that it measured net precipitation — the

Fig. 14.16:   Rain and vapour gauge — on land (*c*.1842). Science Museum 1876–0797.

combined effects of rainfall, temperature, humidity, solar irradiation and wind. Made of zinc, it comprised a vessel of 30 cm diameter, partially filled with water and open to precipitation and evaporation (Fig. 14.16). This was in communication via a pipe with a second covered vessel. A light dish floating on the water surface of the second vessel hung from a pulley with a silken thread and counter-weight, and its movement adjusted a magnifying index divided into water level intervals of 0.2 mm.[J58] The water level was noted at sunset each day before adding or removing water to return the dial to zero.

He noted in his first BAAS report in 1844 that the instrument worked 'with great delicacy and fidelity'.[R12] He had learnt by now that undue modesty created space for others to assume the credit. It also risked reduced Association support and delayed take-up of his inventions by other observatories.

After 13 years of use, an identical model was commissioned from Adie for the Kew Observatory's display at the 1855 *Exposition Universelle*. The Kew Committee offered this second gauge to the 1876 Special Loan Collection and it remains in the Science Museum collection today. A brief description of the device is included in *Knight's American Mechanical Dictionary*[A126] published in 1882.

At his home, he was able to install the gauge in the lawn but at Kew it was positioned on the roof terrace on one of his tripods. He mentioned as early as 1844 that it 'would be much more properly situated if the cylinders…were sunk into the neighbouring earth…It should perhaps be made to float upon such a *[water]* surface in a little boat'.[R12] This would have the advantage of more closely monitoring

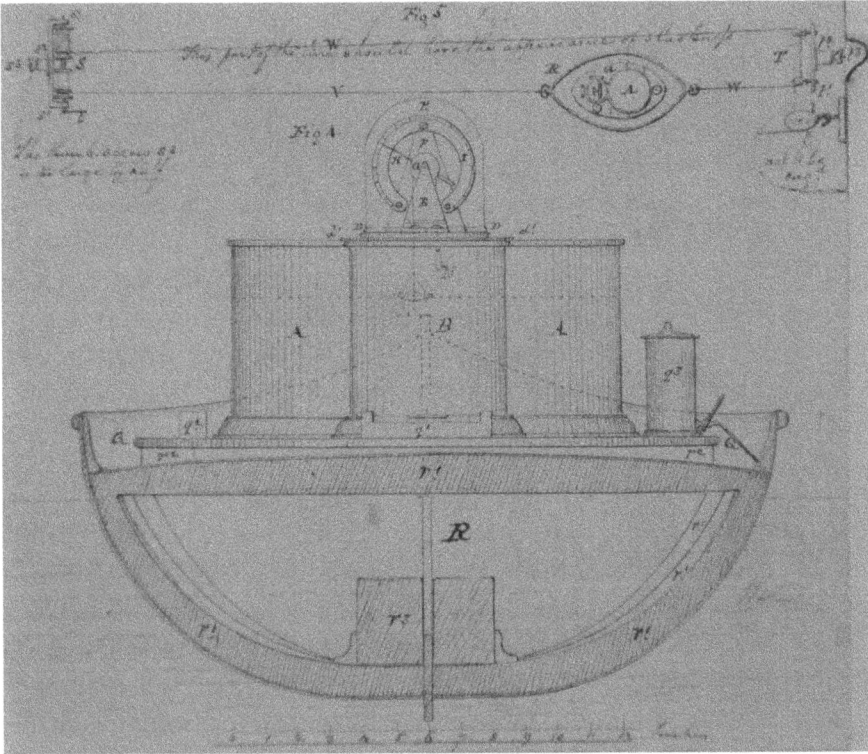

Fig. 14.17:   Rain and vapour gauge — on water (*c*.1854). IET 1.9.1.

true evaporation from a water source. As already mentioned, the grounds surrounding the observatory building were not able to be used at the time but he was
able to document his ideas when he retired. He conceived the gauge as being
installed on a small copper hull complete with keel, bulkheads and ballast water,
and with spirit levels to facilitate trimming (Fig. 14.17). The boat would be
moored on the water with a windlass such that it could be readily brought to the
bank to make daily readings.

Like other meteorological parameters, evaporation is still monitored closely
today and has been found to have altered over recent decades as part of climate
variability.

<u>Wind vane and spring anemometer</u>

As Ronalds said, his wind vane was 'rather more convenient and accurate than a
common weather-cock'.[R12] In addition to the vane and its counterpoise, he included
a fixed index that was positioned near eye level directly outside the window of the

observatory dome (Fig. 14.18). A light tin hoop that moved with the vane in the wind was marked with the compass letters and in this way the index pointed to the correct wind direction. His friend Dr John Lee* wrote to him in 1847: 'A few days ago I received a parcel containing a beautiful Wind Vane, which was packed with the greatest care and judgement, and I was inclined to think that it could only have come from you.'[C201] It was a gift for Lee's observatory, because he had admired it.

Ronalds soon tried a modification to the vane to create a spring anemometer that would also provide a simple visual measure of the wind pressure. A slider on a spiral spring was added to the top of the vane, enabling it to rise under the action of wind on a set of flyers. It did not work as well as the balance anemometer below and was discarded.[R12]

Balance anemometer

Ronalds had written in 1844 that he was 'driven to the necessity of inventing' an anemometer, 'for the lightest zephyr is as important, at least, as the stiffest breeze to electrical meteorology',[R12] and existing instruments did not achieve that range. Wind has a significant influence on the conductivity of the air. His instrument comprised a light deal board one foot (30 cm) square, and a sheltered scale-dish, each counterpoised on the arms of a cruciform that rotated on 'nicely-turned brass pivots'[R12] (Fig. 14.19).

Fig. 14.18: Wind vane (1843). Ref. R12.

Attached to the balustrade of the roof terrace, the board was oriented manually into the wind with the aid of a small vane, and weights were placed in the dish until balance was restored. The cruciform rotated with a weight of less than five grains but could also carry more than 4.5 kg (14,000 times as much). Galloway was the fabricator. The device was locked in position when not in use.

The instrument was listed in the Catalogue of the 1876 Special Loan Collection with a detailed description, and remains in the Science Museum. It was also mentioned in various meteorology books even in the 20th century — for example, Sir John Moore's 1910 *Meteorology*.[B113]

Ronalds noted in his account of the device that it could be modified readily so that it would face automatically into the wind and be observed without going outside. The first improvement was achieved in mid-1845 by the addition of wings. The 'somewhat rude and wooden'[J62] model was replaced at this time with a more solid and sensitive one in brass. Similar anemometers were probably made for several other observatories, including that at Stonyhurst.

In 1850, he met with Follett Osler, who kindly arranged for Kew to receive one of his self-registering anemometers. Its plate was also one foot square and was

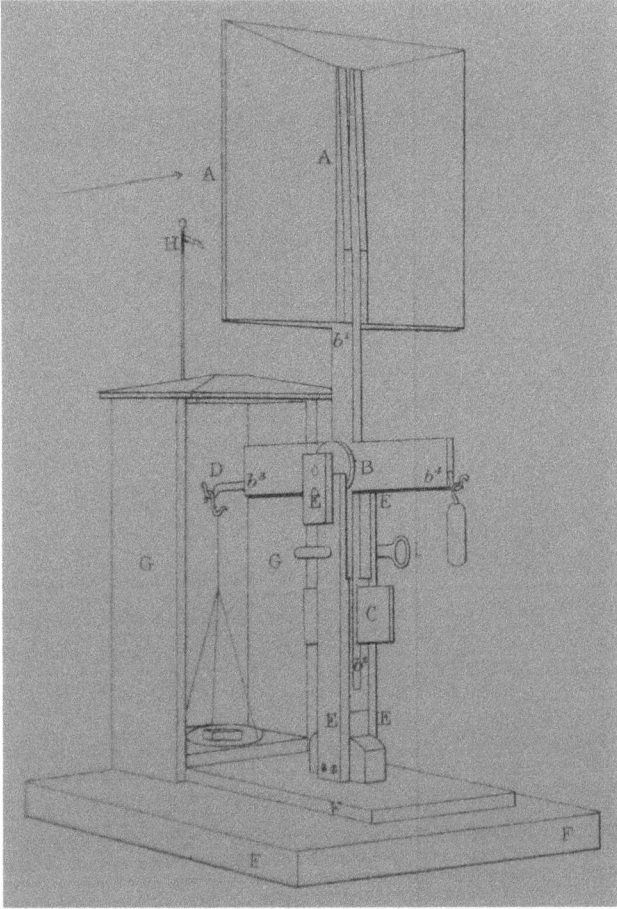

Fig. 14.19:   Balance anemometer — model 1 (1843). Ref. R12.

oriented into the wind by a large vane. The wind pressure was resisted by a spring, with its movement communicated by a wire running down the supporting tube to a pencil in the room below. The variations in wind pressure as well as direction were marked on paper moved by clockwork. Ronalds made mention the next year of further enhancements he was planning. The year after that — 1852 — he pre-scribed an instrument for the Madrid Observatory, noting to its Director: 'Your Ostler's [sic] Anemometer is an improvement upon all that have been previously made.'[C335] This last anemometer (Fig. 14.20) would have incorporated features from both inventors.

Instrument screen

A substantial range of meteorological instruments was collected at Kew over time, both conventional and novel, with Ronalds' devices supplemented by other

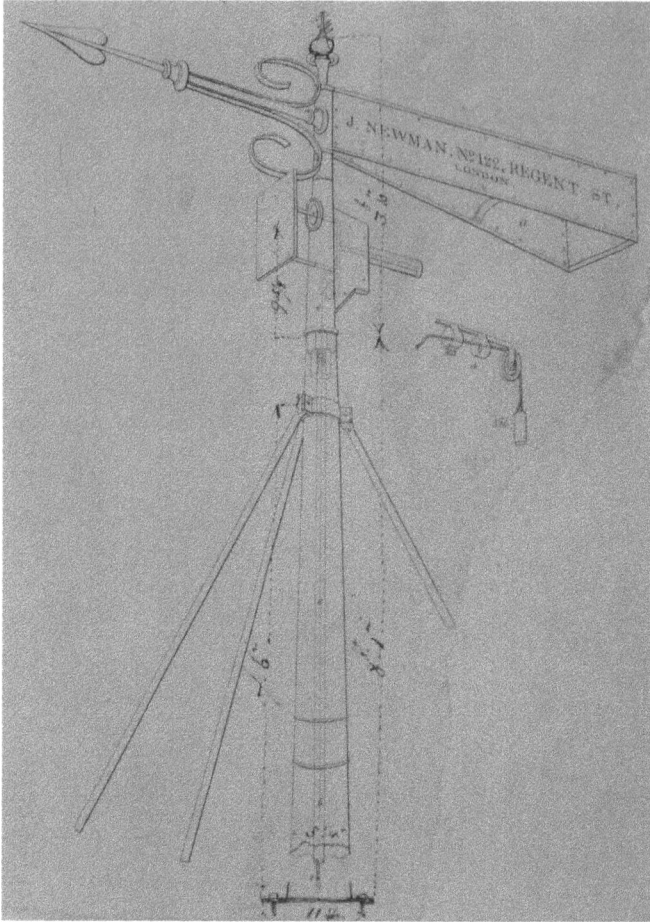

Fig. 14.20:   Osler self-recording anemometer (1852). IET 1.9.1.

donations and purchases. He developed a wooden triangular-shaped screen in 1849 to house part of the array, which was attached to the balustrade at the northern entrance to the observatory. The instruments were fastened to rungs of adjustable separation or set on a shelf (Fig. 14.21). Having a roof and slanted wall, and being able to rotate, allowed the instruments to be at least partially protected from sun, wind and rain, although he noted that 'the difficult problem of always protecting them from all these influences simultaneously…and of preserving a sufficiently free circulation of air, has, I fear, still to be solved.'[R21]

He acknowledged in his design the new thermometer stand recently erected at Greenwich. This is still referred to today as the Glaisher* Stand. It was superseded by the still-standard louvred Stevenson Screen later in the 19th century, which was

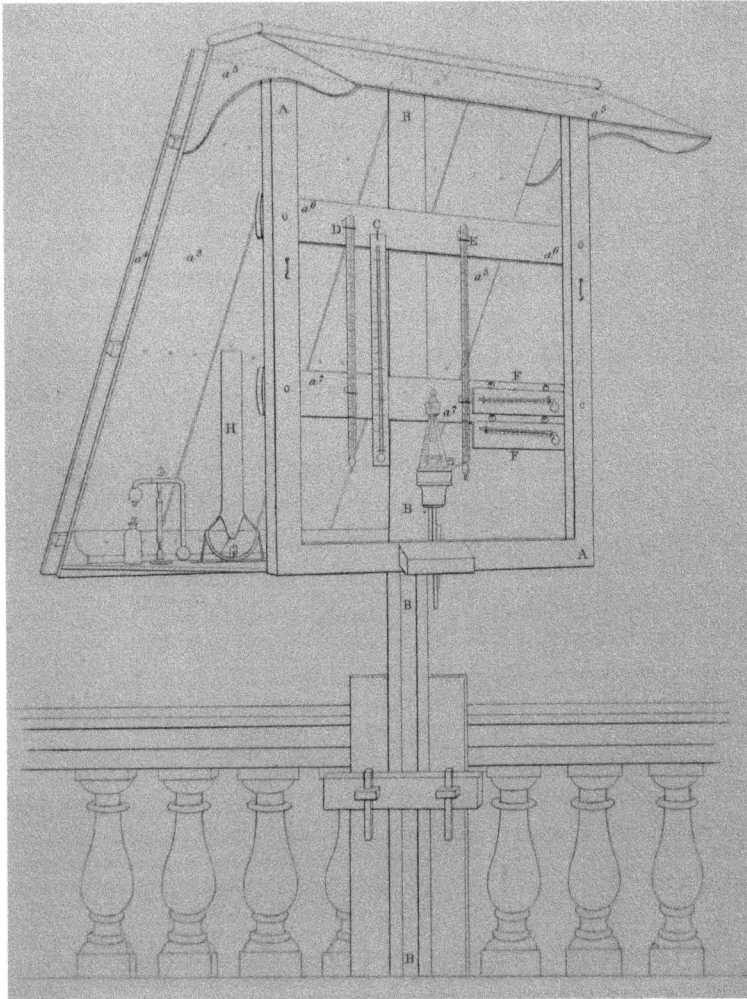

Fig. 14.21:   Instrument screen (1849). Ref. R21.

designed by author Robert Louis Stevenson's father. This last resembled a screen designed by Ronalds' former assistant John Welsh in 1854.

Hygrometers

Humidity measurements were made both by comparing the readings of wet and dry bulb thermometers and with alternative designs. Horace de Saussure's mechanical hygrometer was exhibited at the 1876 Special Loan Collection with the label: 'Eight-Haired Saussure's Hygrometer...formerly the property of Mr. Francis Ronalds, and used by him at the Kew Observatory'.[B94] The instrument remains at the Science Museum.

Kew obtained a Regnault Hygrometer in 1850, a device recently invented by renowned physicist Henri-Victor Regnault FRS, where the dew point is determined by cooling the wet thermometer bulb. The cooling process involves bubbling air through a surrounding cylinder containing ether to accelerate its vaporisation. An aspirator is used to create the air bubbles, with the suction created by allowing water to flow slowly out of a connected vessel.

Ronalds soon found several practical difficulties in the use of the device: the cylinder was prone to leakage at the connection between its glass and silver sections; there was ether wastage in charging it for each observation; the aspirator also required frequent refilling (and there was no plumbing nearby to supply water); unlike the well-known Daniell hygrometer, there was no clear dew point demarcation to assist readings; and it was additionally hard to read at low temperatures.[R21] These were just the sort of avoidable problems that Ronalds disliked and he quickly devised an alternative configuration. In addition to redesigning the hygrometer itself, he developed a new aspirator. It now comprised two rectangular vessels that could revolve around a fixed horizontal tube (Fig 14.22). The vessel

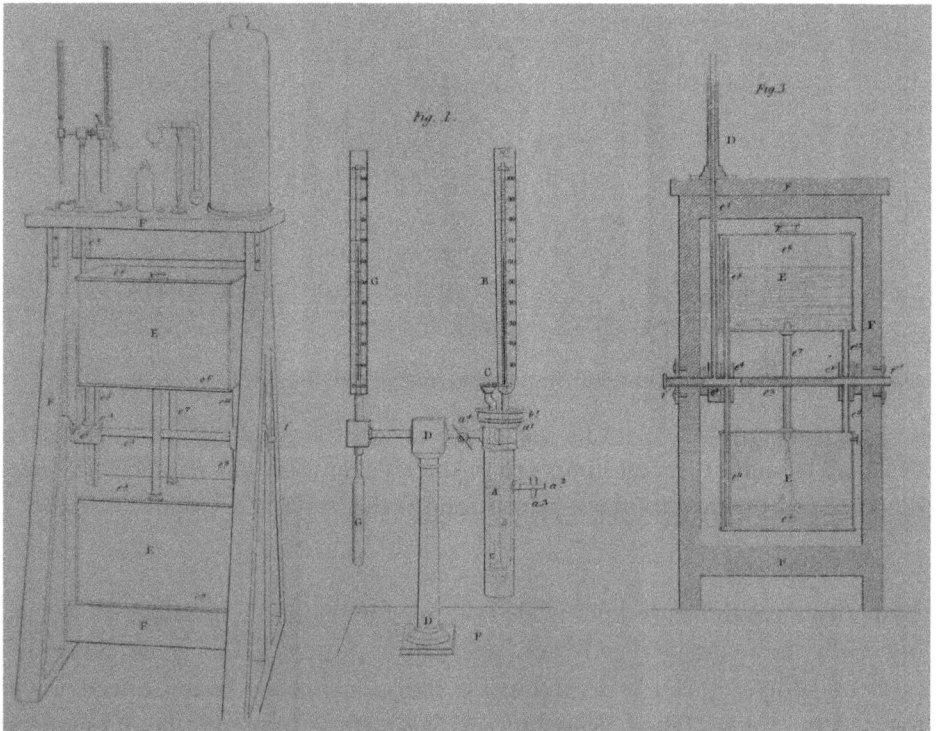

Fig. 14.22:   Improved Regnault hygrometer and aspirator (1851). Ref. R24.

in the upper position was in communication with the atmosphere via a pipe passing through the hygrometer stand and into the ether cylinder. Water draining slowly from the initially filled upper vessel to the empty one below created a vacuum that generated the air bubbles through the ether. The filling lower vessel in turn displaced air into the atmosphere. Turning the two vessels through 180° on their bearings restarted the process. When not in use a stopper prevented access to the air and thus avoided loss of ether.

Kew Committee members tested the new apparatus and advised that the alterations were 'decided improvements'.[J83] Clear zones with and without condensation formed readily on the polished tube, even when using lower-quality ether or naphtha, and thus operating costs could be reduced. Knowles Middleton later described the instrument as 'imposing' in his 1969 text *Meteorological Instruments*.[B154]

The machine was first exhibited at the 1851 BAAS conference in Ipswich. A Dr Ansel immediately asked for one for the new St Mary's Hospital in Paddington and others were in progress, including for Prof Forbes in Edinburgh; they were made by Cary and Newman.

Regnault soon inspected the new configuration at Kew.[J84] He recommended separating the observer from the instrument, to avoid the warming effects of exhalation, and using two telescopes to read the dry and wet thermometers. Ronalds quickly developed a suitable set-up, which he described in his French publication *Descriptions de quelques Instruments Météorologiques et Magnétiques*. The aspirator was now connected to the thermometer by a long, flexible metal tube, with condensed ether collecting in a trap at its lowest point (Fig. 14.23). A single telescope was mounted on top of the aspirator in a brass frame, and knurled screws in elongated holes enabled it to be quickly swivelled to pre-set positions to read either of the thermometers or to monitor the formation of dew.

At least two of the new devices were made in 1851 — one for Kew and another for Professor Dixon of Trinity College Dublin. It was the earlier and more compact version however that was exhibited at the Paris Exposition in 1855, with a position having been made for it in Welsh's new instrument screen.[S6] It is recorded in the Science Museum's 1922 Catalogue that Ronalds' hygrometer 'was employed for many years at Kew'.[B136]

Storm (or observer's) clock

In addition to measuring the basic meteorological parameters, Ronalds developed instruments for more specialised applications. He completed his Storm Clock (also called an observer's clock) in early 1845. Its basis was an inclined writing

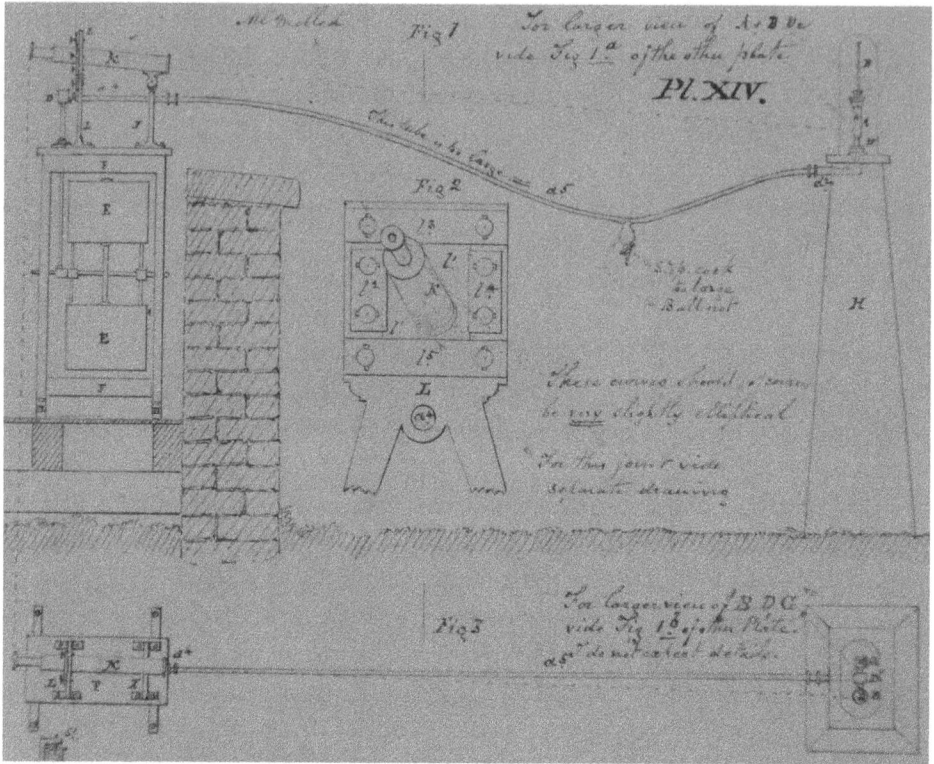

Fig. 14.23:   Improved Regnault hygrometer and aspirator — remote observation (1851). IET 1.9.1.

board on which paper was pinned (Fig. 14.24). A small brass index travelled at a chosen speed down the paper (typically 1.5 cm per minute),[J62] carried by a gut-line from the barrel of a clock. The apparatus was started and stopped by pressing a handle and the index plate was returned to the starting position by another weighted line wound in the contrary direction. Observations were jotted down at the current point of the index, thus avoiding the need to also be checking and recording the time. The times of particular phenomena were subsequently determined with the aid of a scale ruler.

He soon noted that he 'did not expect so much regularity & real utility as we obtain from this affair'.[J62] It was deployed primarily to record the changing indications of the instruments during storms or the rate of re-electrifying after the conductor had been earthed. Several authors highlighted it in early write-ups of the observatory,[K3,K5] and it is likely that one was also deployed at Greenwich. Ronalds acknowledged Airy's input in devising the concept, who had loaned him a clock for the purpose.

Fig. 14.24:   Storm (or observer's) clock (1844). Ref. R20.

George Symons FRS referenced Ronalds' 'useful'[A132] storm clock in describing in the 1890 *Proceedings of the Royal Society* his new apparatus for monitoring meteorological parameters during thunderstorms. In this device the paper travelled at 3 cm a minute, again driven by a clock, and phenomena of interest were graphed with pens. The term 'storm clock' gradually gained currency. It is included for example in Robert Morris Pierce's 1911 *Dictionary of Aviation* with the definition: 'a clock-like apparatus for recording or indicating storms: a meteorograf, especially the one devised and named by Francis Ronalds.'[B130]

Airborne instrument platform

As early as 1819, when Ronalds was musing after his ascents of Vesuvius about the variation of atmospheric parameters with height, he wrote: 'Balloons might be sent up with registering thermometers and then altitudes taken geometrically.'[J18]

He and William Birt conducted experiments at Kew in August 1847 in a similar vein. They deployed a kite rather than a balloon with the goal of creating an instrument platform for upper air observations. Ronalds developed the method of holding the platform in a fixed position. The hexagonal kite was first raised using the usual rope opposite the tail. Cords attached to each side were then pulled downwards and the three ropes tied to stakes in a large equilateral triangle. The arrangement gave the kite great lifting power. He wrote:

> The place of the kite did not seem to vary so much as one foot in any direction, and it really appears...probable that a very large kite or kites might be employed in this kind of manner *often* and very cheaply as a substitute for a captive balloon in meteorological inquiries...An anemometer, a thermometer, an hygrometer, &c. of some registering kinds, &c., might be hauled up and lowered at pleasure (like a flag)...by means of a line passing through a little block attached to the kite.[R16]

Again the focus was on unmanned ascent with automatic recording.

On the day, they studied the wind pressure, comparing the results with the balance anemometer.[C197] Ronalds hoped to continue the kite experiments, to explore 'their real utility in meteorology',[R19] but there were insufficient resources. He was able to publish the experiment in the *Philosophical Magazine* however; it was his first paper in that journal for over 30 years, which was now edited by Sir David Brewster*. The article was reprinted in the *Year-Book of Facts in Science and Art*.

Apart from their work, kites and self-registration were generally neglected at that time in favour of manned balloon flights. Welsh conducted four balloon ascents in 1852[2] with funding provided to Kew by the Royal Society and observed the variation of air temperature and other parameters with height. James Glaisher of the Greenwich Observatory conducted similar ascents in the 1860s, including one where he fell unconscious at very high altitude — luckily his 'aeronaut'[A176] was able to bring the balloon down safely. The advantages of kites over balloons in cost, control and exposure of instruments were rediscovered at the end of the century, and Ronalds' paper was printed again in the November 1896 *Monthly Weather Review*.[R16] Kites were used for regular self-registering weather observation by meteorological bureaus in Europe and America as well as at Kew from then until the Second World War. Whipple described the set-up once more in 1937, noting it as: 'One invention of Ronalds' which has been neglected, but which might be fruitful even now'.[K14]

Ronalds had suggested in his and Birt's paper that their configuration might have other applications. Marcel Maillot devised a kite in the 1880s with the goal of raising a person into the air. Its stabilising technique was 'remarkably similar'[A159] to Ronalds' according to Clive Hart's *Kites: An Historical Survey* (1967).

---

[2] *Illustrated London News* 4/9/1852 192.

14.4 Observations and Results

<u>Atmospheric electrical potential</u>

Several European observatories monitored atmospheric electricity at this time. Adolphe Quetelet made a daily observation at the Royal Observatory in Brussels around noon, and in specific periods also took numerous measurements through the day. Dr John Lamont FRS made observations at the Royal Observatory of Munich in 1850–51; both used the induction instrument that Wheatstone promoted.

Observations started formally at Kew on 1 July 1843, a year prior to Quetelet's commencement, and included the full set of meteorological as well as atmospheric electricity measurements. Initially readings were made at sunrise, 9 a.m., 3 p.m., sunset and approximately hourly through each day to find the morning and afternoon minima and maxima — except on Sunday afternoon, which Galloway was able to have to himself. Ronalds appeared to come to Kew to Sundays, perhaps to give Galloway a longer break. During storms and other interesting phenomena, many more observations of the rapidly changing electric potential were recorded using the storm clock.

By the beginning of January 1845, the registering or night electrometers were also in regular use, giving additional readings through the night at midnight, 2 a.m. and 4 a.m. Galloway then took two-hourly readings at even hours from 6 a.m. to 10 p.m. to complete the daily cycle, plus sunrise and sunset. He recorded all the observations in a Journal, along with detailed comments on the weather conditions. The original Journals survive at the Met Office Archive and short sections have been published in the BAAS reports and elsewhere.[K1]

The quality of Ronalds' apparatus and Galloway's recording enabled them to demonstrate that electricity was always present in the atmosphere — the primary conductor was constantly electrified. Ronalds noted as early as 1844 that already 'the electrical part of the journal…is more complete and accurate than any such hitherto recorded'.[R12] It was the first systematic, precise, comprehensive, continuous and extended study of the phenomena. The importance of the dataset was quickly recognised by his peers. Glaisher advised:

> I cannot but regret that so valuable a journal should not be reduced and made available at once. I wish the British Association would find funds for this purpose, I would not mind superintending the Reductions gratuitously…There is one thing however I should be very glad to have from you and that is a short report upon the Electricity of the past Quarter at the end of each quarter — no one else in the country can do this besides yourself.[C208]

Ronalds was eager to study the results himself, but could not: 'it has been quite out of my power to devote the time and attention to the subject which it eminently deserves.'[R19] We saw in the last chapter that BAAS funding was provided in mid-1848 for Birt to undertake a 'reduction' and analysis of the data. His results were presented the same year that Wheatstone described the alternative induction instrument and its observations. Birt's comprehensive paper — running to 87 pages — delineated the annual variation and demonstrated the double daily cycle of atmospheric electricity that was usual at the time in inhabited areas. In seeking to find a characteristic curve for serene weather, he discarded the high intensities that generally occurred during fogs. The morning maximum then almost disappeared and a much more even daily curve resulted. Birt posed the question: 'Is the diurnal march of atmospheric electricity…a single progression? In other words, does the electrical tension of dry air present a curve having simply an ascending and descending branch…?'[A77] This is just what Ronalds had observed with the dry sirocco wind in Sicily thirty years earlier and is now known to be the case in clean air. Birt's daily curves using the full and abridged data may be compared in Fig. 6.9.

Herschel commented to Sabine on the work:

> I have seldom read a more interesting report than Mr Birt's of the Kew Observations… In a word I think this report makes the subject of Atmospheric Electricity very much less puzzling than I always used to think it.[C260]

Auguste de la Rive wrote on the results in his *Traité d'Électricité* (1858):

> The observations which, on account of the very long time (several years) during which they have been continued, as also by the precision and care with which they have been directed, appear to us calculated to inspire the greatest confidence, are those of Mr. Ronalds at Kew, collected and discussed by Mr. Birt, those of M. Quetelet at Brussels, and those of M. Lamont at Munich.
>
> The Kew observations…have been to the number of 15,170 for a period of five years; of this number 14,515 are of positive electricity, and 655 of negative. The observations of positive electricity have furnished the elements of the determination of the ordinates of diurnal and annual curves of atmospheric electricity, principally during the three years 1845, 1846, and 1847, to which 10,176 observations have contributed.[B75]

Antoine Becquerel FRS (another French physicist and author) and his son used similar words in their 1858 text *Résumé de l'Histoire de l'Électricité*.[B76] As late as 1937, Whipple noted:

> Perhaps the paper has not received as much attention as it deserves. It includes a thorough analysis of the diurnal variation of electric tension and also an interesting discussion of the readings of the electrometers during thunderstorms.[K14]

Birt went on to publish further papers on the effect of water vapour on atmospheric electricity[A78] but with no mention of Ronalds.

Regular observations ceased in August 1848, after five full years, when the decision was made to close Kew and Galloway was dismissed. Ronalds' hope that they would recommence with Birt's arrival at Kew was rewarded by just three weeks of observations. His third assistant Welsh made measurements regularly over the period September 1850 to March 1851, and then much more sporadically for another two months. The next halt was to be temporary according to the Kew Committee minutes, so that a formal trial of Ronalds' three photo-magnetographs could begin in earnest. After the trial, however, the Kew Committee had new priorities for Welsh in thermometer calibration and balloon ascents. Ronalds strove to streamline the observations by establishing photo-registration but resources to operate his new machine were not available. It finally became clear to him that he would not be able to continue his atmospheric electricity studies at Kew. He wrote to Noad in 1853 of 'an accident having sometime since occured [sic] to the collecting lantern, which has not been repaired', his photo-electrograph being 'now rather ill conditioned' while, with his favourite Coulomb torsion electrometer, 'the parts…are I fear scattered.'[C346] Ronalds was very disillusioned.

The Kew Committee reinstated regular meteorological observations in 1854, just after Ronalds had left, but they only very briefly included atmospheric electricity. Electric potential measurements began again in the early 1860s when Kelvin installed his newly built apparatus. He had developed forms with a burning fuse collector, and also a dropper, which constituted a fine stream of water projected into the air. The latter was adopted at Kew, positioned at a window of the observatory and combined with photo-registration using his divided-ring electrometer. Readings stopped after a few years but, from 1873, became essentially continuous using improved versions of the instrument. The apparatus became the standard for atmospheric electric potential measurement internationally into the 20th century. Its and later observations have been compared with Ronalds' early work and found to be qualitatively similar, but with differences that can be understood.[K13] As Andrew Drummond wrote in a review article on atmospheric electricity in 1944: 'Ronalds' observations…clearly indicate that there has been little or no change in the nature of the daily and seasonal variations during the hundred years under review.'[K15] Observations continued at Kew until it closed in 1980, and the results spanning nearly 140 years form the longest atmospheric electricity data series in the world. Numerous aspects of the modern understanding of atmospheric electricity have been elucidated from these measurements.

Air–earth currents or 'frequency' observations

Right from the beginning, Ronalds wished to measure not only the magnitude of atmospheric electricity but also the rate at which the principal conductor received its charge. He had done this as early as 1810 at Highbury and then Hammersmith.[R6] In 1814 he defined the frequency at which the gold leaves of the electrometer diverge and then collapse when striking the grounded lining of the electrometer bottle as the 'quantity' as distinct from the 'intensity' of electricity (Sec. 6.2). A resumption of such work was included in his first report to the BAAS, where he noted it as:

> that great desideratum in atmospheric electricity, a means of noting the dynamic effects which are perhaps coincident, if not identical, with the property...called... "frequency," a property of great importance possibly considered in relation with the various opinions and theories which...are entertained concerning the natural agency of atmospheric electricity, in...the magnetism of the earth, the aurora, &c.[R12]

Wheatstone contributed a galvanometer made by Gourjon to Ronalds' atmospheric electricity equipment for the purpose. It gave indications during 'violent rain'[R21] but according to Ronalds was 'not to be depended upon as to measures'.[R21]

He reverted to his early ideas and built a new frequency apparatus in late 1843. It comprised two identical external conductor rods in proximity, each having a collecting lantern, a warmed glass insulator and a Volta electrometer supported from its orthogonal bar. One conductor was discharged momentarily and the rate at which its reading returned to the value exhibited on the other electrometer was noted. The arrangement is seen in Fig. 14.1. He soon observed that the ratio of the intensity of the electricity to its quantity varied widely with the weather, with high quantities (or currents) occurring in fogs and heavy dews.[R12] He would have been very interested to see if there were variations during magnetic disturbances. Others seemed to support the research: when Faraday visited Kew on 15 November 1850, he 'viewed the Electrical apparatus and held a very interesting & instructive conference on the subject of Frequency'.[J84]

Ronalds had little luck however in prosecuting the experiments to the extent he wished. He raised the matter again in his 1849 and 1850 BAAS reports: 'Their results may form a link in the chain of phænomena connecting the static with the dynamic electricity of the atmosphere; for it is only when frequency is great that *galvanometers* manifest a current.'[R20] Both Birt's and Welsh's observations in 1850–51 were focused in particular on frequency but were of insufficient duration to reach any clear conclusion. Ronalds developed a means of photo-recording the observations but the equipment could not be completed. He resorted to encouraging (begging) other observatories for which he supplied atmospheric electricity apparatus to conduct the dynamic studies and had some success at Greenwich and Bombay. As late as 1869, at age 81, he took the opportunity of receiving the latest volume of Greenwich observations to remind the Director Airy of the value of recommencing frequency observations.[C395]

In the 20th century, the manifestations of atmospheric electricity in both potential gradient and current, and their interrelationships, came to be accepted in terms of a global atmospheric electric circuit. Observations of air–earth currents recommenced at Kew in 1909 and continued until the observatory closed. The approach adopted was qualitatively similar to Ronalds' in timing the rate of change in potential of a conductor after it was earthed.[K13,A183] The air–earth current density is now understood to flow vertically downwards in fine weather with amplitude of order 10^{-12} ampere per square metre.[A178]

Global observations of atmospheric electricity and geomagnetism

An aspect of atmospheric electricity that was of considerable interest to Ronalds was its relationship with geomagnetism. It was natural after the electromagnetic discoveries of Oersted, Faraday and others that scientists would contemplate links between electricity and magnetism at global scale.

Sabine's Magnetic Crusade had already delineated a great deal with respect the spatial and temporal variation of geomagnetism. It was known that magnetic forces were larger near the poles than at the equator and how they varied on daily and annual cycles. Ronalds believed there were similar variations in atmospheric electric potential and he wished to explore them further: as early as 1823 he expressed disbelief in the low charges reported on Captain William Parry's 1819 expedition to the Arctic (where Sabine was science officer), suspecting faulty apparatus to be the cause.[R7] As soon as he completed his first photo-magnetograph in March 1846, Ronalds advised Sabine 'of the marked coincidence between the usual minimum of electrical charge & the first extreme westerly position of the needle & of certain other concurrents *[sic]*...My interest in this view of the subject has been greatly increased by Dr Faradays late discoveries on Solar or atmospheric magnetism.'[C168] He proposed 'to institute comparative self registered observations of this Electrometer *[his improved Coulomb torsion balance]* with those of the Declination magnet'.[J63]

The next step would be to extend this internationally by adding atmospheric electrical apparatus to the geomagnetic observatories at key locations around the world. The write-up of his presentation at the 1847 BAAS conference shows that he addressed the need there with some passion:

> Mr. Ronalds...concluded by expressing a strong desire that we should possess strictly comparable observations of atmospheric electricity, magnetism, and the *aurora* made at Alten or Hammerfest (in Finmark), at Bombay, and in this country; for, he said, it may be safely affirmed, that notwithstanding all that has been written about the relationship of the aurora with electric and magnetic phænomena, the regions, or apparent regions of the aurora have never been subjected to electric observation in a manner at all approaching to accuracy, or comparability with obser-vations made here or elsewhere. Neither has an electrometer at Bombay ever yet been compared with one in England...We have surely been deducing conclusions on this very curious subject without sufficient *materiel*.[R17]

The article was reprinted in the *Literary Gazette* and the *Year-Book of Facts*. He raised the subject of 'the establishment of electrical observatories…in distant parts of the globe'[R20] again three conferences later. We will see that he quickly succeeded in India, but that comprehensive Arctic observations were still many years off.

In 1852 Sabine published the critical connection between terrestrial magnetism and the sun. German astronomer Heinrich Schwabe had already documented that sunspot activity followed an approximately decadal cycle, with a minimum falling in the year he published — 1843. Sabine found that the mean effects of magnetic disturbances were likewise at a minimum in 1843 and a maximum in 1848.[A81] Lamont had also noticed a decennial periodicity in geomagnetism. Meanwhile Birt's analysis of the Kew observations indicated that there was a significant increase in atmospheric electricity potential in 1847 in comparison with 1845–46. Ronalds would have been aware at a qualitative level of other changes since then, but Birt had not analysed and published those results.

Another related phenomenon that created great interest was the *Aurora Borealis* or Northern Lights. The proceedings of the 1833 BAAS conference, for example, advised that 'a public discussion took place on the phenomena and theory of the Aurora Borealis'[S1] while, at the 1849 BAAS, a committee was appointed 'to consider the best mode of promoting'[S1] its observation. Eyewitness accounts of nights when the lights were particularly dramatic were published in newspapers and Airy devoted ten pages of the Greenwich *Magnetical and Meteorological Observations* (1847) to qualitative descriptions of it.[S4] The spectacular nature of the aurora made it all the more frustrating that so little was understood about its cause. As Ronalds wrote in 1847: 'We know almost nothing about the influence of the Aurora on Tensional Electricity & very little about its influence on Magneto-electricity'.[J69]

Yet another occurrence was now also able to be observed for the first time. A network of overhead telegraph lines was emerging. In 1847 and 1848, railway engineers found that the needles on their telegraph instruments deflected sometimes even when they were disconnected from their batteries. The deflections were particularly strong during a brilliant aurora. Ronalds' friend Charles Walker* observed the effects on the South Eastern Railway[A105] while William Henry Barlow FRS, resident engineer on the Midland railway, took systematic measurements of the currents at Derby.[A76] Ronalds retained data from Derby that he was given by the Superintendent of the Great Western Railway's Telegraph Office. He was advised immediately when on 9 August 1848 equipment was damaged simultaneously at the interconnected Paddington and Slough stations by electrical activity: 'About 1.50 during the Storm there was an Explosion in the Office similar to a Gun being let off. I found on inspecting the Instrument the Coils were Fused'.[C220]

Edward Highton, telegraph engineer to the London & North Western Railway Company, documented disturbances to his equipment during an aurora borealis on 17 November 1848. Carlo Matteucci (whom we met in Sec. 13.3) observed similar effects on the Pisa to Florence telegraph on the same day.[B78]

When Ronalds travelled to the Birmingham BAAS conference in 1849, he arranged to visit the resident engineer for the London to Birmingham Railway, where a telegraph system had been installed two years previously.[C263] He made mention at the conference of the recently observed anomalies on the telegraph lines and the value of the new infrastructure for observing atmospheric electrical phenomena at scale. He would also have wished to assist in developing solutions for what was emerging as a major hazard for the telegraph companies. He used the general terms 'atmospheric electro-magnetism' and 'magneto-electricity' for what is now known as geomagnetically-induced currents and considered them as another facet of atmospheric electricity.

The year 1847–48 was clearly an important one. It was the year of the first observed peak of both sunspot activity and magnetic disturbances in their just-discovered cycles. It was also the first time that it was possible to experience their effects on long telegraph wires. Sadly it also marked the end of Ronalds' five years of observations of atmospheric electricity. No wonder he had started to become disillusioned.

His wish was for these complex interactions to be explored in a holistic inter-national study and he was uniquely positioned to shape the undertaking. First of all, he had the vision and knowledge to pull the disparate threads together in an improved understanding of electro-geomagnetic behaviour. He had established the best equipment in the world for measuring atmospheric electricity potential and observed air–earth currents using his frequency apparatus. He had also developed the most accurate magnetographs — he thus had the capability to photo-record both geomagnetic and electrical effects simultaneously. From Kew he was able to distribute these quality instruments around the world. As Faraday advised Oxford physician Henry Acland in October 1850:

> in respect of observations of the Electric & Magnetic condition of the atmosphere…
> Mr. Ronalds of the Kew Observatory would be able & I have no doubt happy to tell
> you all that can be communicated from one to another & could assist you much [C314]

Ronalds experimented in 1851 with taking photos of the sun through a small tel-escope.[J84] At a larger scale, he had access to the telegraph system with its invalu-able signals of geomagnetically-induced currents. He would also have received updates from Sabine as to emerging trends in the magnetic data he was receiving and analysing from the Colonial observatories.

In fact he was unable to progress his vision. The other ingredient he needed, but did not get, was Sabine's powerful backing to bring both Kew observations and a coordinated global network of atmospheric electrical measurement to fruition — Sabine in fact hindered rather than promoted the goal. Sabine meanwhile was able to continue and enhance his system of magnetic observations (in part with Ronalds' photo-magnetographs) to find the geomagnetic relationships for which he is renowned. Ronalds made a veiled criticism of Sabine in 1852, after he was thanked by the Director of the Madrid Observatory for supervising the manufacture of atmospheric electricity apparatus for them:

> I am amply repaid by the pleasure of witnessing and being in any degree instrumental in…judicious promotion of enquiry on subjects…which I have often regretted the neglect of by those who are best enabled to patronise them effectually.[C331]

Surely adding insult to injury in Ronalds' eyes, Faraday communicated a report from Quetelet in the 1851 *Philosophical Magazine*. Quetelet had observed low atmospheric electricity levels in 1849, and opined:

> Nothing is more fitted to make one appreciate the deficiency which still exists in our relative knowledge of the electricity of the air, than the doubt which surrounds the anomalies I have observed during the first part of the year 1849. I have been unable to find any observations which may serve to control my own.[A79]

Quetelet noted that the decrease coincided with an outbreak of cholera. Captain John Lefroy* commented on a similar relationship between geomagnetism and cholera in Toronto at the same time[A149] — they were echoing the belief of Dr Thomas Southwood Smith and others mentioned in Chapter 9 that diseases were spread by bad air. Ronalds gave a clear enough answer to an enquiry at this time from the Secretary of the Meteorological Society of London: 'observations at Kew have never exhibited any direct coincidence between electrical peculiarities of the Atmosphere and the prevalence of Cholera.'[C273] He must have been beside himself. He would have now suspected an alternative reason with a periodic basis for the 'anomalies' but was prevented from pursuing his ideas. The paper is one of nineteen publications by Quetelet in the Ronalds Library.

Ronalds' ambitious goals perhaps seemed to be within reach later that year. His vertical force magnetograph at Toronto captured a considerable magnetic disturbance associated with an aurora on 3 September 1851, and telegraph lines in New England were disrupted correspondingly. The Kew magnetographs also traced the event and a similar one at the end of the month.[J84] He had succeeded in setting up his improved photo-electrograph in May,[J81,J84] but it was not in regular service, and he tried as well at this time to provide atmospheric electrical equipment to Toronto. Comprehensive combined electrical and geomagnetic observations were not to be, however, in his tenure.

He might have had one last attempt two years later. When he started his long sojourn to the Continent in October 1853, he stopped along the way for about five weeks to pursue an idea concerning the telegraph; he mentioned abstrusely to Manuel Johnson* that it was:

my wish to attempt an improvement upon telegraphic communication by visiting Folkestone, Dover &c.[C347]

Folkestone and Dover were on the South Eastern Railway line and were the site of some of Walker's telegraphy trials. Johnson replied:

I am very glad you are turning your mind to your old pursuit and this time I hope you will not allow others to run away with the fruits of your labour.[C348]

Two months earlier, Ronalds had told Noad regarding his atmospheric electricity work:

I have a project for registering the Frequency also in embryo[C346]

Nothing more is known about these ideas. It seems more likely however that his stay at Folkestone would have concerned geomagnetically-induced currents, rather than be a return to innovation in telegraphy itself after a break of approaching forty years.

The world had to wait for the next sunspot peak in 1859 to receive another sequence of clues. Ronalds was still in Europe, but would have followed proceedings with interest. On 1 September a solar flare was observed for the first time, now known as the Carrington Event after one of the observers. Richard Carrington visited Kew a day or two later and it was found that the event corresponded with an abnormality on the magnetographs and preceded the most dramatic geomagnetic storm in recent history.[A104] Aurorae were seen almost all over the world. Walker again reported major disturbances on the telegraph lines after a number of years of quiet: telegraph operators received shocks, and fires started in several of their machines. Recent reanalysis of atmospheric electricity readings made in Melbourne suggests that the electric potential was also enhanced.[A192] Airy at Greenwich responded by attempting to collect further first-hand accounts from telegraph clerks. With Walker's help he went on to build experimental lines along the nearby railway to investigate the correlation between geomagnetic forces and earth currents using photo-registration.

Not all scientists accepted the couplings. As late as 1892, Kelvin used his Royal Society Presidential Address to give his opinion: 'the supposed connexion between magnetic storms and sun-spots is unreal, and…the seeming agreement between the periods has been a mere coincidence.'[A136] Years earlier he had 'attributed the cause of the currents observed by Mr. Barlow *[on the telegraph at Derby in 1847]* to chemical action on the wires'.[3] He did use his 1874 Presidential Address at the Society of Telegraph Engineers, however, to call on telegraph operators to make further measurements on the lines to help delineate the phenomena.[A119] It would be the mid-20th century before the physics of the linkages was understood. Electric currents induced by geomagnetic storms remain a risk today to telecommunications

---

[3] *Literary Gazette* 24/7/1847 536.

systems, power transmission grids and pipelines, and is managed with warning systems and protection mechanisms.

Ronalds was right to suspect that the interplay between solar activity, geomagnetism and the aurora would extend to electricity, but we know today that these interrelationships are complex. Net effects on the global atmospheric electricity circuit would not have been easily discernible. Positive correlations between sunspot numbers and atmospheric electric potential at Kew and elsewhere were published in the 1920s, but these were dismissed a decade later when additional data became available.[K13] It is now understood that cosmic ray modulations, coronal mass ejections and other solar events can affect the potential gradient and air–earth current.[A178,A192] Interesting periodic and irregular variations in the potential might have been apparent near the poles, if the opportunity to observe them had arisen in Ronalds' time.[A178] Had he remained in the UK, he would have had more luck studying geomagnetically-induced currents using the telegraph infrastructure or perhaps his Exploring Wires if they could have been constructed. Overall, he seems to have been years ahead of others in his big-picture view of the phenomena and how to delineate them.

14.5 Dissemination

We retrace our steps a little now and review the scientific community's perceptions of Ronalds' atmospheric electrical equipment as well as his labours to establish it in other observatories. He, his apparatus and his observations had quickly gained a strong reputation. It was only a short time before flattering accounts of his work appeared in standard texts, and scientific luminaries recommended widespread adoption of his methods. The publicity was a direct result of his affiliation with the BAAS and was in stark contrast with previous decades of his career where he had worked in relative obscurity. Indeed the original publication of his atmospheric electricity apparatus and results nearly 30 years earlier had received little notice. He did not now seek the personal kudos — he did use the BAAS forum however, as we have seen, to try to prosecute his goals for global observations.

A major international conference was held in conjunction with the BAAS's 1845 meeting to discuss future directions for the network of magnetic and meteorological observatories. Herschel's report at the conclusion of the meeting recommended: 'That instruments for the observation of atmospheric electricity on the principle of the apparatus at Kew should be employed in the observatories'.[S1] Herschel had written to Wheatstone before the conference about 'his' equipment: 'your new form of observation in atmospheric electricity might be mentioned by you with much advantage'.[C148] The misconception discussed in the previous chapter that Kew's atmospheric electricity apparatus was due to Wheatstone was in full throes at this time.

Despite Ronalds' energetic efforts, take-up was in fact sporadic — the topic was scientifically challenging, expensive, time-consuming and involved some personal risk; furthermore, as already mentioned, Sabine did not encourage its adoption in the observatories under his care. Over time his equipment was supplied to at least ten institutions other than Kew and his electrometers would have been used by more. The apparatus was advertised for sale by Newman by 1845 and Ronalds was probably not aware of some of the purchases made. Close to home, equipment was installed at Greenwich, Dublin, Oxford, possibly Norwich,[C302] and perhaps several other regional stations. The Royal Agricultural Society also expressed interest in obtaining the apparatus. Destinations on the Continent included Pisa (see previous chapter), Madrid and Marseilles. Further afield were India (Bombay, Trivandrum and possibly Madras), the Arctic and Toronto (Table 14.1). The latter destinations nicely complemented Kew's location at latitude 51°. His ultimate goal was not able to be achieved however — coordination and comparison of the various results to build up a global picture of atmospheric electricity did not take place during his years in meteorology.

Table 14.1: Distribution of Ronalds' Atmospheric Electricity Apparatus.

Year	Full electrical laboratory	Round-house	Simple (portable) set-up	Electrometers only
1843	Kew[†]	—	—	—
1844	Greenwich	—	Kew	Dublin
1845	—	—	Pisa?	—
1846	—	—	—	—
1847	Bombay	—	—	—
1848	—	—	—	—
1849	—	—	—	Toronto
1850	—	Kew	—	—
1851	—	Trivandrum	Arctic expedition	—
1852	Madrid	Norwich??	—	—
1853	—	—	—	—
1854	—	—	—	—
1855	—	—	—	—
1856	—	—	—	—
1857	—	—	Marseilles	—
1858	—	—	Oxford[†] Stonyhurst?	—

[†] Included photographic registration

Some of the earliest visitors to Kew had arrived straight after Wheatstone's slightly misleading announcement of the atmospheric electricity apparatus at the 1843 BAAS conference. The announcer was not going to miss it; he advised Ronalds:

> Prof. Daniell and the Astronomer Royal will pay a visit to the Observatory on Thursday next at twelve o'clock. I will be there to meet them.[C121]

Daniell followed up with a second visit the next year, after Ronalds and Galloway had acquired significant data. This time he contacted Ronalds:

> I propose taking a ride over from Norwood on Saturday next with my girls to pay a visit to your Observatory at Kew. I shall be vexed not to have the pleasure of seeing you…But pray do not put yourself to any inconvenience…I think that the subject of Atmospheric Electricity is becoming every day more interesting & your labours more important.[C135]

He was preparing the third edition of his renowned text *Elements of Meteorology* and wished to expand the section on atmospheric electricity: 'I am full of admiration of the beauty of your Journal.'[C143] Sadly he died two months after writing those words, during a meeting of the Royal Society's Council. The book was published posthumously by his editors with his description of Ronalds' apparatus but no commentary on the atmospheric electricity observations at Kew:

> A new mode of insulation was described by Mr. Francis Ronalds, in…1817, which has lately been applied with the most perfect success to atmospheric conductors…at Kew,…a model of the most perfect apparatus which our present knowledge of the subject will perhaps allow of our constructing.[K2]

Walker was another early visitor. He gave full details of the apparatus in an addendum to his 1845 translation of Ludwig Kaemtz's *Complete Course of Meteorology*[K1] and was able to include detailed tabular observations made with the atmospheric electricity and meteorological equipment. Astronomer Dr John Drew first visited the observatory in 1851. He noted in his book *Practical Meteorology* (1860) that a description of the apparatus at Kew 'will be all that can be required to give a general view of what is doing in the best observatories throughout the British Empire'.[K5] Drew included the round-house apparatus and Ronalds' frequency observations in his account. After Noad visited in 1853, he gave a detailed description in his standard text *Manual of Electricity* (1859),[B77] down to drawings of the various electrometers and comment on Birt's analysis. Snow Harris also included key results in his *Rudimentary Electricity* (1853).[B64] *Modern Meteorology*, published in 1879 under the auspices of the Meteorological Society, advised: 'Years ago, when Francis Ronalds was director of Kew Observatory, the upper portion of that building was fitted up with such a collection of electrometers as had never been established before, and has never been equalled since'.[B101]

The centenary of Ronalds' effort was commemorated in 1944 in a review article on atmospheric electricity. The author Drummond started: 'One hundred years ago last summer the first systematic measurements to be made in this country of the electrical potential gradient…commenced at Kew Observatory under the direction of Francis Ronalds' and went on to describe 'what then must have seemed a very ambitious electrical programme'.[K15]

Greenwich

Ronalds' first customer was Airy. The original atmospheric electricity apparatus at Greenwich had been designed with Wheatstone's guidance in the early 1840s as part of the establishment of the 'Magnetical and Meteorological Department' there. It did not work well, being insufficiently insulated. As we have seen, it did not take long for Airy to hear of the new apparatus just set up at the rival observatory at Kew and he wrote to Wheatstone in September 1843:

> I should like much to see your electric instruments at Kew, which I understand to be in a very active state. Would you tell me how and at what hours I can see them?[C119]

Wheatstone replied:

> The fixed apparatus established there is perfect, constantly retaining its charge even in the worst weather we have had, and showing with beautiful regularity the hourly changes in serene weather.[C120]

There was no mention of Ronalds.

After his initial visit to Kew, Airy wrote to Faraday:

> A few days ago I went to the Kew Observatory to see the Electrical Apparatus (for Atmospheric electricity) mounted by Mr Wheatstone and Mr Ronalds. I was delighted with every part of it, and shall endeavour as far as circumstances admit to imitate it here.[C123]

Faraday wrote by return:

> I am delighted to have your testimony about the electrical apparatus though I could not doubt that all would be done that was possible considering who had it in charge[C124]

to which Airy replied:

> I am acting entirely in concert with (or rather under the direction) of Mr Wheatstone and Mr Ronalds, and am expecting them daily to come to Greenwich in order to prescribe for me in electrical matters[C125]

Airy soon realised that he should be dealing with Ronalds directly. Nonetheless for some reason he gave equal credit to the two men in his address to his Board of Visitors in mid-1844:

> In the autumn of last year, I had an opportunity of examining the beautiful arrangements of the Atmospheric Electrometer at the Kew Observatory, which have been made under the superintendence of Professor Wheatstone and Francis Ronalds, Esq. It was impossible to see these without perceiving that considerable improvements might be made in our own, by following the same plan[S3]

Ronalds showed his usual generosity in assisting in the construction of the apparatus at Greenwich. He made various visits, sent detailed technical letters and drawings to Airy, supervised Newman in designing and expediting the work, and completed and tested the electrometers himself. He also assisted in troubleshooting when the equipment was found to be less successful than that at Kew, providing comparative data, conducting experiments to help verify the problem and loaning scientific articles and his own instruments.

Anyone other than Ronalds would have regretted his efforts. As soon as the apparatus was operational, a suggestion was made to the General Committee of the BAAS:

> That it be recommended to the Council to consider whether the Electrical Experiments at the Kew Observatory should not be discontinued...it was stated that the discontinuance had no reference to pecuniary considerations, but arose from the fact, that similar observations were now being made at the Observatory at Greenwich, under the superintendence of Prof. Airy.[4]

In spite of Airy's actions, Ronalds continued to assist the atmospheric electricity efforts at Greenwich for as long as he remained at Kew.

The technical challenge at Greenwich was that Airy wished to emulate Ronalds' design by improving his existing arrangement, rather than starting afresh. Ronalds later explained the difficulties to the Director of the Madrid Observatory, who was consulting him on the erection of atmospheric electricity apparatus there:

> The Astronomer Royal applied to me for advice &c before the collocation of the Greenwich electrical Apparatus but (wishing to avoid some inconveniences) he adopted my suggestions in a partial, and very defective manner; consequently the printed Greenwich Observations of Electricity display very long periods of total deficiency of Signs...soon discovered (by himself) to be due principally to the Dissipation of Electricity from the long wire which connects the collecting Lanthorn &c at the summit of his Mast with the Electrometers &c below[C343]

Airy kept Ronalds updated on progress in the modifications. A particular focus was to provide improved insulation through the incorporation of a warmed glass funnel, which was made 'as nearly as the workman could manage'[C127] like Ronalds'. Airy admitted in his 1844 Visitors Report that, even with the new arrangements, 'the indications are much less constant than those at the Kew Observatory.'[S3] He told Ronalds: 'I really believe that the trees or some other local circumstance prevent our apparatus from receiving the same atmospheric charges which yours receives.'[C134] The next year he noted that 'The Galvanometer and the Induction Ball have been nearly useless.'[S3] He continued to persevere and late that year advised Ronalds: 'I really think that our Electric apparatus is getting into a feasible state. We have indications usually at all times, day and night. This is owing simply to better management of our lamps and to varnishing our glass.'[C150]

---

[4] *Athenæum* 28/6/1845 639.

Several years later, in 1848, Airy's assistant Glaisher published atmospheric electricity observations indicating that the conductor had negligible charge for nearly three months.[A73] The apparatus at Kew was electrified throughout the period. Ronalds wrote to him immediately advising of the difference and sending his Journal data. Airy replied:

> Mr Glaisher has shewn me your letter of March 7. In regard to the absence of indication of the electrical instruments during a part of this winter, I believe that their representation of the state of the atmospheric electricity is correct.[C204]

Ronalds clearly disagreed and could not resist making brief mention of the discrepancy in his *Epitome* publication that year.[R18] In 1851, Airy's First Assistant Main admitted in an article on Greenwich:

> the electrical apparatus at Greenwich was copied from that at Kew, with such modifications as were necessary, and…Mr. Ronalds gave his willing and able assistance in every part of it…The action of the Kew apparatus is, however, much more continuous and perfect than that at Greenwich, owing, probably, to the length of the conducting wire used with the latter.[A84]

Ronalds' preference right from the beginning had been to position equipment of his design in the Octagon room at Greenwich. He trialled his portable electrometers on the Octagon roof in 1845, showing them to perform better than Airy's recently revised arrangement. Airy floated to his Board of Visitors that year a proposal 'to mount an electrical apparatus, nearly similar to that of the Kew Observatory, on the top of the Octagon-room',[S3] but it did not go ahead. When Ronalds developed his simpler 'round-house' observatory in 1850 he immediately offered to build one for the Octagon Room. We saw in the previous chapter that Airy responded to this proposal with an attempt to purloin the primary Kew apparatus itself, which was quickly rejected by the Kew Committee. He then reverted to Ronalds' offer:

> I beg leave gratefully to acknowledge this offer, and to ask if you could within a few days furnish me with estimate of the probable expence…The objects to be registered are the kind and degree of electricity. Probably this information would guide you sufficiently.[C319]

Ronalds wrote to Airy giving a detailed description of the round-house and its apparatus, right down to the screws to be used in building the outer structure. He also offered to make further electrical trials on the Octagon roof prior to commencing construction. He had a skeleton built of the observatory in wood at a scale of 1:12 to assist in its explanation and made models himself of the internal equipment. The quality was such that he advised Airy concerning its little tripod stand:

> you will find even this little deal affair remarkably steady.[C328]

It did not take long for Airy to dismiss the whole idea:

> I have examined your very pretty model of an Electric House…I am obliged at this time to decide whether such a thing shall or shall not be attempted next year: and, on consideration, I have determined not to propose it.[C329]

Airy's autobiography entry for the next year includes:

> A wire for the collection of atmospheric electricity is now stretched from a chimney on…the leads of the Octagon Room to the Electrometer pole…There appears to be no doubt that a greater amount of electricity is collected by this apparatus than by that formerly in use.[B115]

The Greenwich equipment lasted 25 years longer than that at Kew — it was dismantled only in 1880. The 1877 *Results of Magnetical and Meteorological Observations*, for example, described the apparatus including the conductor and collecting lantern, and right down to: 'Volta's Electrometers, denoted by Nos. 1 and 2; a Henley's Electrometer; a Ronald's Spark Measurer…originally constructed under the superintendence of the late Sir Francis Ronalds'.[S4] Earlier reports had also made mention of his Bennet gold-leaf electrometer.[S4,B63] Despite its longevity, Airy was never able to overcome the arrangement's inherent shortfalls. *Modern Meteorology* (1879) for example advises: 'At the Royal Observatory, Greenwich, attempts have been made for many years to observe atmospheric electricity, but they have been very unsuccessful'.[B101] Greenwich did not collect comprehensive 24-hours-per-day data either. Detailed measurements were reported on selected days but generally only qualitative morning and afternoon observations were recorded.

Ireland

Another early visitor to Kew was Revd Professor Humphrey Lloyd FRS, Professor of Natural and Experimental Philosophy at Trinity College Dublin. Ronalds arranged to have electrometers made for him and completed them himself in the usual way. It was Lloyd who the next year made the original recommendation at the 1845 BAAS conference, echoed by Herschel, that Ronalds' atmospheric electricity equipment should be deployed in all the magnetic and meteorological observatories now set up.[S1] Romney Robinson* also visited Kew several times and in 1852 expressed interest in acquiring the equipment for the Armagh Observatory. Ronalds made some initial plans and, four years later, mentioned subtly to him: 'I frequently ponder on the idea which you kindly mentioned of erecting an Electrical Apparatus in your Observatory.'[C365] It is probable that it did not go ahead.

Toronto

The first chance to supply a Colonial observatory came in 1846 when Sabine brought his subordinate Lefroy to Kew. Lefroy was Director of the Toronto Observatory and was in London on honeymoon. It was the start of a considerable effort by Ronalds in supplying magnetic and electrical instrumentation to Canada, although it started slowly. Lefroy contacted Ronalds two months after the visit, asking 'whether you have taken any steps towards procuring for the Observatory at Toronto a set of your Electrical apparatus, as established at Kew. I think that when Colonel Sabine and myself were at Kew, we came to the understanding that you should order a duplicate set.'[C178] 'In reply', wrote Ronalds, 'I must say that I had not the least idea of your wishing to have an apparatus...I should be glad now to know as precisely as possible your wishes and will attend to them diligently and with the greatest pleasure.'[C179] To which Lefroy responded: 'I can only thank you for your kind offer to superintend the preparation hereafter of an Electrical apparatus, but cannot venture at the moment to specify what we want at Toronto.'[C180]

Lefroy's interest increased over the next few years, and in 1849 he ordered Ronalds' electrometers from Newman to attend a primitive collection device he was building. Sabine was dismissive — he wanted Lefroy to concentrate on geomagnetism: 'he has quite enough before him during the remainder of his directorship'.[C265] Ronalds was in a difficult position — he wanted to help Lefroy develop high quality electrical equipment, yet not did not care to incur Sabine's ire. He apparently steered a good path: Sabine referred to his reply to Lefroy as 'very sensible'[C269] while Lefroy called it 'very kind and instructive'.[C272]

Lefroy continued to seek Ronalds' advice and assistance and, when he visited Kew again in 1851, Ronalds offered personally to partially defray the expenses of providing better apparatus to Toronto.[J84] The episode culminated in Lefroy's formal request (at Ronalds' suggestion) to be given equipment now disused at Kew, as discussed Sec. 13.4. The 1853 Kew Committee Report advised that 'the Committee cannot recommend that any portion of it should be withdrawn from the Observatory, more particularly as Mr. Newman could supply a more perfect apparatus under the superintendence of Mr. Ronalds at a comparatively trifling cost.'[S6] It is unlikely that such equipment was made and sent to Toronto: there were budget limitations at the observatory, which had just been transferred to the Provincial Government, and Ronalds had other commitments at Oxford.

As we have seen, the Committee's resolution was the final straw in Ronalds' decision to retire from Kew. It was particularly annoying for him because he now knew that the only way he might achieve his electrical goals was through other observatories. Toronto was highly advantageous both in terms of its strong aurora and geomagnetic activity and by already being equipped with his photomagnetographs for comparative studies.

Interestingly, it was the value the Kew Committee apparently placed in his earlier work that enabled the atmospheric electricity apparatus to be preserved and later presented to the Science Museum. Ronalds himself had no concept of historical worth. He was a practical person and his objective was simply to facilitate the global study of atmospheric electricity.

India

The first actual order for apparatus for a British Colony came in late 1846. Observatories had been established in India by the East India Company and it was that organisation's Director William Sykes (rather than Sabine) who instigated the purchase. Newman contacted Ronalds when the request came through: 'Having had many applications and received an order for an Electrical apparatus similar to that at Kew…may I be permitted to ask if there is any alteration necessary to be made in any part (from your experience)'.[C189] Ronalds lent Newman 12 pages of his working drawings. The equipment was for the Colaba Observatory in Bombay where geomagnetic and meteorological measurements had commenced a few years earlier. Its equatorial location (latitude 19°) presented an exciting opportunity for Ronalds to broaden his observing network. Observatory Superintendent Charles Montriou sent him the first results from the new apparatus in 1850 and Montriou's colleague Edward Francis Fergusson visited Kew the next year for instruction on the use of the equipment.

An observatory had also been established by the Rajah at Trivandrum, which is close to the tip of India at latitude 8.5°. Its inaugural director John Caldecott first visited Kew in early 1847. His replacement John Allan Broun spent time at Kew in 1851 learning about Ronalds' instruments while a round-house version of the apparatus was built for him by Adie. There is mention in the Ronalds Archive of the proposal to fabricate a round-house for the East India Company's Madras observatory as well but it is unlikely that this last request was completed. With its east coast location, it would have given with the other two observatories good overall coverage of the Subcontinent.

Stonyhurst

In 1848, Father Alfred Weld, Director of the Stonyhurst Observatory near Blackburn, contacted Ronalds:

> I am very happy to announce to you that we are going to take up the electricity &
> I shall be obliged to you if you will send me a plan for the tower &c…I should like
> the tower to be neat & rather pretty (perhaps castellated at the top &c)[C215]

Weld's account of the aurora that year is in the Ronalds Library. Correspondence ensued for some months, with Ronalds preparing plans (complete with a suitable castellated tower) and providing detailed assistance, until:

> According to your advice, I consulted an architect in the neighbourhood, and got an
> estimate from him as to the expenses of the building which he proposed, and I am
> sorry to say that it would be something much greater than…the college could afford
> to give me…I shall keep the plans and all the instruction which you have so kindly

given me...I am distressed that I have put you to so much trouble about it, to no purpose. At the time when I wrote to you I thought I saw my way so clearly as to be able to say without doubt that we should carry the design through.[C255]

Seven years later, the two caught up in Paris. Ronalds recorded that Weld 'was about to prosecute the proposal formerly discussed...of establishing registering Instruments at Stone...but without building a tower separate from the Observatory.'[J85] Whether he was talking about atmospheric electricity and/or other meteorological or magnetic phenomena, Weld was still a little premature. He applied to the Kew Committee for 'Magnetical and Electrical apparatus'[S6] in 1858 and began the next year to assist Sabine with magnetic observations, but it was a successor who finally oversaw the installation of photo-magnetographs and a barograph in 1866. The following year Stonyhurst became one of the stations in the Met Office's observing network and acquired a complete suite of self-registering instruments.

It seems quite likely that Weld did obtain atmospheric electricity apparatus of Ronalds' type, as several English observatories sought the equipment in 1858 and we know that Adie made at least one set that year.

Arctic

There was understandable interest in establishing an observatory in the proximity of magnetic north as part of the Magnetic Crusade. Simple meteorological observations had been reported from Alten (near the top of what is now Norway) for some years and Ronalds was one of several who expressed strong desire for their continuation and extension in the later 1840s. Alten's high latitude (70°) and close proximity to the band of greatest aurora frequency were ideal for achieving greater understanding of geomagnetic and electrical phenomena. He offered electrical equipment of his own and from Kew for the observatory[R18] and Birt applied to become Observer there, but the proposal did not go ahead.

Ronalds had another unfulfilled opportunity in 1851 when he provided his portable electrometers to Lieutenant John Powles Cheyne, who was to participate in a search for Sir John Franklin's missing expedition to the Northwest Passage. Cheyne was on one of five ships under the command of Sir Edward Belcher. Suitable thermometers were also made at Kew by Welsh for the voyage. A year later Cheyne updated Ronalds:

> I cannot begin until we are frozen up in our Winter Quarters as we are constantly at work...and have no means besides of elevating the Instrument, for we are blocked up amongst the Ice about 20 miles from land, and are likely to be in this situation for some time — we are at present in company with thirteen whalers, all jammed up tight by the Ice, there were fifteen of them at first, but two have been smashed up. One of them is at present lying a wreck close to us. By the return of these whalers, our letters go home.[C334]

The expedition's ships were later abandoned, although the crew survived.

The criticality of observations at high latitudes in understanding electro-geomagnetic phenomena was recognised in organising the first International Polar Year in 1882. It was not until the late 1920s however that a permanent observatory was established at Tromsø, just near Alten, to study the aurora. Photo-recording instruments were deployed to monitor terrestrial magnetism, earth currents, atmospheric electric potential, vertical currents, and their interrelationships in an initiative funded by Rockefeller's International Education Board.[B140] Observations continue today.

Madrid

Considerable interactions with the Royal Observatory in Madrid commenced in mid-1848, with a visit to Kew from Dr Manuel Rico y Sinobas, who wished to establish meteorological and magnetic work there. He returned three years later with his colleague Juan Chavarri, then Director of the facility. The observatory was at latitude 40° with a rather high elevation of 670 m above sea-level.

Ronalds agreed to superintend the construction of full atmospheric electricity apparatus, together with a self-registering anemometer and the provision of standard thermometers and a barometer. All were made by Newman. On his return to Madrid, Chavarri provided drawings of the observatory and Ronalds designed structural alterations to the stone domed lantern atop the building where the electrical equipment and anemometer would be housed. Under pressure from Ronalds, Chavarri was able to convince the architect to make the necessary alterations. Ronalds also provided considerable detail on set-up and operation in his letters. He additionally advised on the design and manufacture of magnetic apparatus as mentioned further in Sec. 15.9.

Two photographs survive of Ronalds' atmospheric electricity apparatus with its glass insulating pillar and suite of electrometers. One is a negative and is held at the British Library and the other a faded positive in the Ronalds Archive. Certainly the negative is a studio shot as the principal conductor rests on a cloth and is not bolted into place. It was taken using waxed-paper, a process developed around 1851, and Dr Larry Schaaf, an expert in 19th century photography, believes it to be the work of Nicolaas Henneman.[A174] Henneman had a studio on an upper floor of Newman's shop. The positive print appears to show the upper portion of the pedestal that Ronalds designed and Newman built for the Madrid dome application. The photographs must therefore be of the apparatus just about to be shipped to Spain on its completion in mid-1852. Figure 14.6 is a print of the negative.

Sinobas, who was now Observatory Director, provided an update to Ronalds in April 1855. The electrical apparatus had been in action all the previous year and sparks 'of a dazzling brightness'[C360] had been observed with the spark measurer. Ronalds' reply encouraged him to include frequency observations in his work. He continued:

> I would, if I had now strength & health sufficient, make a journey to Madrid for the express purpose of conferring with you on the subject of printing *[the observations]* & of trying to induce you and your Government to institute barometric thermometric & other observations of the kind now in progress at the Radcliffe observatory, Oxford.[C367]

He was hoping that Sinobas might be able to institute the detailed photographic correlations between atmospheric electricity and other meteorological parameters that Ronalds was not able to complete.

Oxford

Ronalds had long tried to convince Manuel Johnson, Radcliffe Observer at Oxford, to erect atmospheric electricity apparatus. He used Sinobas' success to tempt him and, as we have just seen, did the reverse with Sinobas regarding photo-registration equipment at Oxford. Telling Johnson about Sinobas' work he hinted: 'I am quite sure that a regular course of Electrical Observations made in that latitude and high locality would afford some very interesting points of comparison with those which I still hope that you are about to procure.'[C361]

Two years earlier, he had sent his model of the round-house observatory down to Oxford, where it remained until he requested it to be sent to Paris for the 1855 Exhibition. He then had it returned to Oxford. In 1856, Johnson updated him: 'The other day I proposed the erection of an Electrical Observatory at the top of our Tower. I did not succeed in my point but it was only a question of time.'[C366] By 1858, the Radcliffe Observatory report made mention of a simple electrograph, which included a pointed copper rod and Volta electrometer. The collector was not successful and in mid-1861 'the single spike was replaced by a bunch of copper points'.[S5] From August that year, days with 'Active electricity'[S5] were recorded in the annual report. In the 1930s, Ronalds' 'atmospheric electrometer' and a 'Ronalds Lantern Pattern Gold Leaf Electroscope' were gifted to the Museum of the History of Science at Oxford from the observatory. He had succeeded in twisting Johnson's arm to procure atmospheric electrical equipment but would have been disappointed that it did not incorporate a lantern collector.

Other British Regional Observing Stations

Captain Henry James FRS, Director-general of the Ordnance Survey, had started visiting the Kew Observatory in 1850 and soon began to play an important role in the development of the UK's regional meteorological stations. He wrote to Ronalds in late 1857 seeking atmospheric electricity apparatus that would 'yield results comparable with those obtained at Kew'.[C268] Ronalds' friend James Peacock arranged for copies of his reports to be forwarded to the Captain and recommended he visit Newman. It was very shortly after this that the Oxford and Stonyhurst Observatories sought to procure atmospheric electrical equipment and it is possible that other regional stations did so too. The Ronalds Library contains James' book on meteorology.

France

Ronalds took the opportunity of his extended stay in France to coax further Continental observers to embark on atmospheric electricity studies. In December 1855 he met Benjamin Valz, the Director of the Observatory of Marseilles: 'He invited me to visit him at Marseilles, and I said I should be very glad to do so, if I could persuade him to erect an Electro-Atmosp[l] Apparatus…Said that if he did anything in the way of self registration I would remain some weeks at Marseilles perhaps.'[J85]

Ronalds did not reach Marseilles until mid-April 1857 but, once there, worked quickly. It was agreed that the round-house electrical apparatus would be installed under his supervision on the highest flat roof of the observatory. He learnt immediately from scientific instrument-maker Augustin Santi that the local glass was made with soda rather than potash and was 'totally unfit for electrical insulation'.[J85] Ronalds arranged for the glass to be tested by the Dean of the Science Faculty, Professor Auguste Morren. In the meantime he reverted to his portable electrometer apparatus with two Volta electrometers and a rod complete with collecting lantern. Valz being '<u>very</u> deaf'[J85] made the job more difficult but, by the time Ronalds left at the beginning of May, he had supervised erection of the alternative apparatus and noted that the more sensitive electrometer showed a divergence of 15° in initial readings. The copy of his booklet *Météorologiques et Magnétiques* that he presented to Morren survives in the library of the *Aix-Marseille Université*.

Ronalds met Count Théodore du Moncel in Paris in 1856, who was a prominent author in electricity. His diary notes:

> I explained my Barograph, and some other Instruments, particularly the Electrograph, which seemed to interest him greatly, and he said that he would erect such kind of apparatus at his house at Cherbourg, which is extremely well situated for Observations of Atmospheric Electricity. I insisted upon the expediency of making <u>Frequency</u>-Observations, and offered to assist him by speaking to *[Ignazio]* Porro *[Paris-based instrument-maker]* about the construction and cost of the Apparatus.[J85]

Ronalds must have been sounding like a broken record by this time. Nothing more is known about this last opportunity. Du Moncel did present Ronalds with a number of his 32 works in the Ronalds Library and also made mention of one of Ronalds' electrical ideas in a book he published the next year (as outlined in Sec. 16.1).

14.6 Conclusion

Ronalds took stock of his atmospheric electricity and meteorology achievements at Kew — modestly as always — in his 1860 autobiographical letter:

My first care…was to insulate, by a modification of the means which I had devised in 1817…a very high Voltaic conductor, & to provide it with improved & new Electrometers &c…It may be said that this apparatus was <u>constantly</u> electrified during Eight Years with a few <u>very small</u> intervals of transition from a positive to a negative state, and a few days for repairs &c, and that the results of 15.170 Observations made in 5 of these years consecutively, taken in conjunction with the Observ[ns] of M Quetelet of the Brussels observatory, & with those of M[r] Lamont of the Munich observ[y] &c, have contributed notably to the advancement of our atmospherico electrical science…Had not <u>mean</u> molestations intervened in 1851–2 I think that I should have been enabled to elucidate <u>a little</u> the subjects of atmospheric electro-magnetism & the Aurora, by its & by other means.

Soon after its completion I was consulted as to the adoption of similar arrangements at the Greenwich observatory. In consequence of certain local & other inconveniencies, my methods of observ[n] &c were here disadvantageously modified; the Conductor was therefore not <u>constantly</u> charged, but improvements were subsequently made & I trust that the observ[ns] recorded in the Greenwich tables will not be found useless.

A complete appar[s] & instructions were sent to the Madrid R[l] Observ[y] & others to that of Bombay &c. The director of the former made good use of his instruments immediately & has recently sent me an academical memoir (to Kew) on the subject which I have not seen but which doubtless contains valuable matter for his building and the <u>region</u> of it are eminently adapted to such researches.

I will not dwell on various other Instrum[ts] described in my first & in some other Reports…excepting one, relative to a proposed occasional substitute for the "Captive Balloon" in certain meteorol[l] Observ[ns]…I attached 3 light cords to a Kite & to 3 little stakes in the earth in an equilateral triangle. By this means it was retained at a nearly constant elevation, whereas even small increments of the wind's force produce great depressions of the Captive Balloon. This rough exper[t] was described in the Phil Mag [C375]

Ronalds was able to take his long-term interest in atmospheric electricity and meteorology to new heights in his years at Kew. The success of his electro-meteorological observatory gave him his first taste of widespread scientific renown, occurring as it did at the same time as the commercialisation of the electric telegraph. The quality of the instruments he designed and built was quickly recognised and several of his electrometers appear to have been used for many years in various applications. Other experimental set-ups like his pluvio-electrometer, frequency apparatus and electrical dissipation equipment were reinvented decades later for more detailed studies, with his torsion balance being an important step towards Kelvin's industry-standard quadrant electrometer. Ronalds simultaneously developed a number of novel instruments for meteorological measurement. Some of these were exhibited at the second International Exposition in Paris and survive today. Others, like his storm clock and airborne instrument platform, again influenced subsequent devices employed in the 20th century.

The complete apparatus produced the first comprehensive, long-term and accurate dataset of electrical and associated meteorological parameters and its analysis the most detailed picture of the phenomena to date. It was the precursor to much more extensive studies that continued until the Kew Observatory closed in 1980 and revealed many attributes of the atmospheric electrical circuit. Similar apparatus was established through Ronalds' personal efforts in various locations around the world. It is little appreciated that his actions were not driven simply by kindness, but an earnest desire to be able to delineate the geographical distribution and periodic and irregular variations of atmospheric electricity globally, as well as the electrical effects of geomagnetism. His understanding of a subject that still holds mysteries for atmospheric astrophysicists today was remarkably advanced for the period and the multifaceted manner in which he tackled its challenges has been underrepresented in science history.

Chapter 15

PHOTOGRAPHIC RECORDING INSTRUMENTS FOR METEOROLOGY AND GEOMAGNETISM

Ronalds' instruments and associated photography and reproduction innovations to record continuously the variations of physical parameters, and their take-up and long-term use around the world. The rigged competition set up by Greenwich Observatory and how it was finally settled by the Prime Minister.

We have seen that when the Kew Observatory was established, a large international science investigation was underway in geomagnetism and meteorology. The development of self-registering instruments to ease the load on observers and improve data consistency was therefore a key goal. It was mentioned in the initial prospectus for Kew and the observatory's first grant from the British Association (BAAS) was to build such an apparatus.

That grant was made to Charles Wheatstone*. Kew's inaugural annual report at the 1843 BAAS meeting included a description of his 'just completed'[S1] meteorological recorder and advised that he would present a 'record of the daily working of the instrument'[S1] at the next annual conference. No further report appeared. It is noted in the Ronalds Library Catalogue[R28] that the machine was never installed at Kew, correcting what was stated in various publications. Multhauf later wrote in his review of self-registration devices: 'Wheatstone's instrument left a very ephemeral record in the meteorological literature, and appears to have been defective or out of fashion with its time'. [A154] It was now over to Ronalds.

Ronalds' initial priority on starting at Kew had been to build his electro-meteorological observatory. He was also a pioneer however in the field of recording scientific observations. Section 6.4 outlines how 30 years earlier, in the period 1813–15, he had developed three different ways to register atmospheric electricity: a series of electrometers engaged sequentially using a clock; a pendulum linked to a ratchet mechanism and dial; and the rotating resin plate.

More recently he had been experimenting with the inverse operation of his tracing instruments following a suggestion by Lord Stanhope* (Sec. 10.8). This period coincided with the beginnings of photography in 1839. Ronalds immediately began to consider how these instruments might also be adapted for scientific photography by modifying the magic lantern with its light source, concave mirror and lenses, and he described his initial proposals for registering the movements of magnets, the thermometer and barometer to friends and colleagues in 1840–41. Details of these early ideas survive in his 1845–46 Kew Observatory Journal[J63] and his Memorial written for the Admiralty.[J73] It did not take him long at Kew to pull all these threads together in a comprehensive suite of recording instruments.

With private subscribers and the BAAS having funded him to develop atmospheric electricity apparatus at Kew, it was natural that he would first apply his self-registration ideas to that parameter and, to ensure rapid results, he would start with concepts in which he had most experience. He first updated his rotating resin plate device. The central spindle around which a string wound in a spiral was replaced with a balanced lever supported from one of the orthogonal bars on the principal conductor in Sec. 14.1. It rested lightly on the coated disc that revolved under the action of clockwork (Fig. 15.1), transferring the charge in the conductor rod directly to the recorder.

George Airy* at the Greenwich Observatory was very interested in self-registration and was 'delighted'[C123] with Ronalds' resin plate system when he saw it in September 1843. He approached Faraday* about improvements to the concept, who replied:

> nobody can be more competent or so competent as Wheatstone & Ronalds to work out practically the desired result; and…it is unnecessary & undesirable to think of it whilst in their hands.[C124]

Airy persisted and early the next year Faraday wrote:

> I do not see the way to a perfect register of the different conditions of the atmosphere in relation to electricity but fancy one must work the way out by degrees, and at present the point & revolving cylinder is the mode of registering I should first work at.[C132]

Faraday went on to design and build an experimental apparatus, although Airy quickly found fault with it. Airy then continued using Ronalds' equipment until at least late 1846.

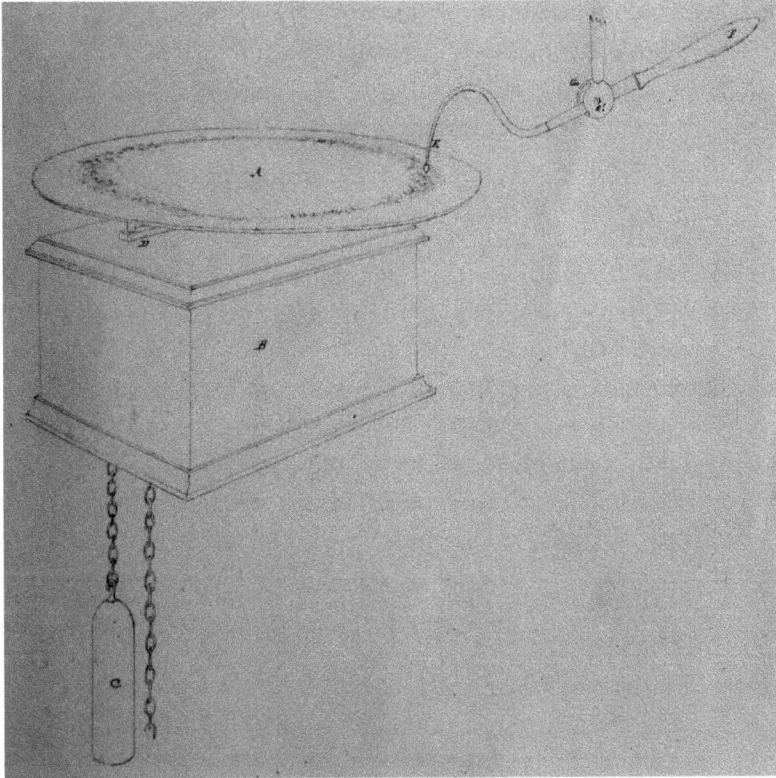

, Fig. 15.1: Improved atmospheric electricity recording apparatus with rotating resin plate — model 2 (1843); Ronalds' original drawing. Ref. J61.

Ronalds himself was never happy with his apparatus because of the 'tædium and difficulties'[R12] of making the resin coating. He had told Airy in 1843: 'I could never procure an approximation to an accurate figure, on a plate which had not been constructed with scrupulous attention to uniformity of thickness and compactness of material'.[C126] It was for this reason that Airy contacted Faraday. Ronalds instead turned to another aspect of his earlier work and built a series of registering or night electrometers (Sec. 14.2) to preserve readings taken at different times during the night. These were in daily use at Kew from January 1845 to give 24-hour observation and were found to be more accurate than the resin plate approach. He also kept Airy informed of progress in this form of self-registration.

He then found time and resources to trial his ideas in photo-recording. By October 1844 he had explained his scheme in detail to others and in April 1845[C375] he was able to build the basis of a cheap and simple photo-electrograph. He went on to design and construct equipment tailored to eight different phenomena, including temperature, humidity and atmospheric pressure. Three different machines were used at that time to delineate terrestrial magnetism — they measured individually the compass direction of the field (or its declination) and its horizontal and vertical force components. He also made a self-recording anemometer (Sec. 14.3) and offered advice on techniques to register rainfall. He quickly foresaw the application of photography to yet wider scientific pursuits, but lack of resources prevented these ideas being developed during his tenure at Kew. His progress was followed in the international literature, with the journal *Fortschritte der Physik* for example publishing six summaries of his reports and publications on self-registration in the period 1846–51.

Limited materials meant that his photo-recording machines did not run regularly at Kew: for a while he had only one suitable timepiece and a single lens tube. After each apparatus was developed and proved accurate, it was partially disassembled to enable a new device to be trialled. Equipment he built for observatories like Toronto and Oxford, however, was used routinely. Other manufacturers then took his concepts and supplied numerous locations around the world. In total more than 84 machines were made (Table 15.1).

Creation of the original instruments required substantial innovation in the camera itself, its ancillaries and in photographic processing; these were of course highly interdependent. We start by outlining Ronalds' camera before describing some of his considerable efforts in photographic manipulation. The remainder of the chapter is devoted to details of each of his instruments, with their manufacture and use, and the assistance he gave to other observers who expressed interest in procuring the equipment. Comparisons are made throughout with the technologies developed by his competitor Dr Charles Brooke*, and full details of their contest are included in Sec. 15.10.

15.1 The Camera

Ronalds' photographic apparatus was in some ways a very early movie camera, although he described it as 'a peculiar adaptation of the *lucernal microscope*'.[R14] The latter device had been invented by George Adams (whom we met in Sec. 10.5) in the late 18th century; in it a condensing lens focused light through a specimen slide that was viewed through a magnifying lens. It was also somewhat akin to a magic lantern and in this sense Ronalds' camera reflected his earlier ideas for projecting a drawing onto a wall.

The heart of the camera was a rectangular, horizontally aligned wooden box (Fig. 15.2). Using the language of his microscope analogy, at the 'object end' (the left in the engraving) was a light source. Facing it was a glass plate or condensing lens. In later models the outer cover to the glass was connected to the clock

Table 15.1:   Distribution of Ronalds' Photo-recording Machines.

Parameter	Ronalds' superintendence		Updated design		Kew-pattern			Approx. total
	Year	Location	Year	Location	Years	No.	Modifications	
Meteorology:								
Electrograph	1845	Kew	1858	Oxford				3
	1850	Kew						
Barograph	1845	Kew	1864	Lisbon	1867–	19	Slight	27
	1847	Kew	1864	St Petersburg				
	1854	Oxford	1865	Coimbra				
	1855	Paris	1866	Stonyhurst				
Thermograph	1845	Kew						1
Thermo/Hygrograph	1853	Oxford	1862	Lisbon	1867–	19	Significant	23
	1855	Paris?	1865	Coimbra				
Geomagnetism:								
Declination Magnetograph	1846	Kew						
	1851	Kew						
	1852	Madrid†						
Horizontal Force Magnetograph	1849	Toronto			1858–	21	Major	30
	1850	Kew						
	1852	Madrid†						
Vertical Force Magnetograph	1850	Toronto						
	1851	Kew						
	1852	Madrid†						
							Grand Total	84

† No photographic registration.

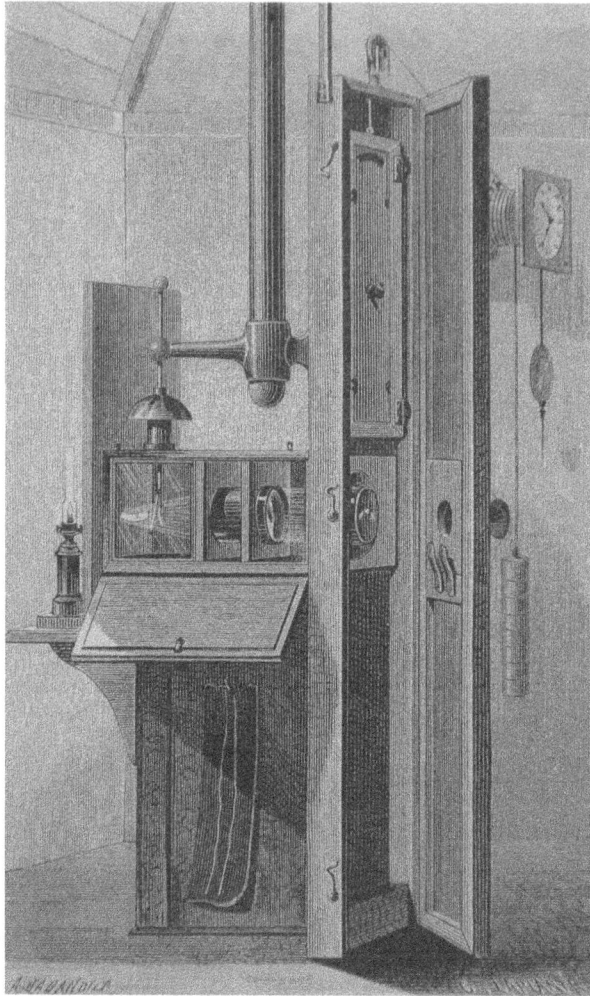

Fig. 15.2:   Idealisation of the photo-electrograph — Ronalds' first photo-recording machine. Ref. B91.

mechanism driving the camera such that starting the clock slid open the cover and allowed light into the camera, and vice versa when the clock was stopped. It was effectively a combination of the lens cap and shutter of today's camera. There was also a slit of suitable form at this end to shape the light. Next was the object to be photographed which, uniquely, was inserted inside the camera box. Closer to the 'eye end' was a tube containing achromatic doublets. An achromatic lens limited chromatic aberration caused by different wavelengths (or colours) of the light spectrum refracting through the lens at slightly different angles and therefore converging at different focal lengths; the concept had been patented by the Dollond family. The position of the tube could be adjusted readily on its support to focus the image precisely on to the photographic surface. The lenses also gave the desired magnification and condensed the light. They were followed by a diaphragm with another adjustable slit, shaped and sized to enhance sharpness and give suitable light exposure. This was a primitive version of the mechanical aperture diaphragm in a more modern camera.

A special feature of the camera was that the photographic surface moved continuously so that the variations over time of the object being photographed could be recorded. This was achieved via a frame carrying either a Daguerreotype plate, or photographic paper sandwiched between two glass plates, reflecting the two forms of photography that had just been invented. The frame was positioned by springs and drawn by a pulley on the clock so that it travelled slowly on rollers along a brass ruler inside a long darkened case. A sliding cover to the frame protected the sensitised surface from light when not in the case but allowed progressive exposure in the camera. A small closable microscope was inserted into the case so that it could be confirmed with the assistance of a fine linear scale that the image was correctly positioned and also in focus. Ronalds saw this as sufficiently valuable to explain and illustrate to professional photographer John Egerton, who 'most gladly'[C169] adopted it. Indeed, in many 20th century cameras the viewfinder showed the object to be photographed and not the image being taken of it and Ronalds' functionality has become the norm only in the digital age.

He had contemplated alternatives to the long case, including a revolving cylinder around which photographic paper was wrapped, and a hexagonal drum that remained stationary for a short period when each face was aligned with the camera lens. Both were rejected because of their lack of precision and flexibility.

He needed to juggle a range of factors in optimising his camera design. The amount of light to which any spot on the photographic surface was exposed depended on its intensity, properties of the lenses (including their aperture and focal length), the width of the slit in the diaphragm and the speed of travel of the frame. Greater magnification of the image was at the expense of brightness and distinctness. The photographic process needed to be tailored to these parameters to ensure a clear picture would be created. The nature of the object being photographed was also critical — the more rapid its variation, the faster the frame needed to traverse to capture the fluctuations and the quicker the chemical reaction must take place. The length of recording was generally set at 12 hours which, together with the frame speed, determined the required lengths of the photographic surface and the darkened case. All of these parameters were open for adjustment in perfecting the configuration for each application and, conversely, an innovation in any component meant a redesign of the system.

The light source was clearly a key consideration — a brighter light allowed a less-sensitive photography process and its spectral properties were also important.

A lamp needed to be able to burn optimally without trimming for as long as required. Lamplight was in addition an expensive item — various cost estimates made by Ronalds over the years concerning different inventions always itemised candles, oil and coal separately. His choice of glass or a condensing lens at the camera's object end was dictated by the light source. He tried to use natural light whenever possible to reduce costs, by positioning the camera near a window. At night he used a lamp and maximised the light through a condensing lens.

He experimented with a variety of artificial light sources and modified the lamps to maximise intensity with minimum maintenance. He initially used an oil lamp 'of new construction'[R21] but based on that invented in 1780 by Aimé Argand. The lamp was held in a guide so that it could easily be replaced in its optimum position. By 1849, he had moved to a Polyflame Lamp inspired by an invention by Sir Benjamin Thompson FRS (Count Rumford) in 1812.[R19] Ronalds modified Rumford's design, 'after many vain attempts'[R19] choosing three flat wicks raised and lowered by rackwork. He recommended cutting each wick 'with very sharp bent sithers [sic]'[J75] so as to be 'a little hollow in the middle'[J74] and settled on a tall square chimney made of copper with a narrow glass plate opposite the brightest part of flame.

He also trialled various types of oil, including spermaceti (from whales) and the renowned Lucca olive oil. In early 1850 he sought his nephew Dr Edmund's advice on possible additives to the oil that might give the light a blue or green tinge, but he could not recommend any chemical that would not harm the brilliancy of the flame. By the end of the period Ronalds was able to use coal gas. He helped configure the supply system for his machines at Oxford, together with lamps having several gas jets at slightly different heights and a glass chimney.[J85] It was in 1856, after he had retired, that reticulated gaslight was installed at Kew.

Before long, Greenwich also began to develop photographic registration equipment. The inventor was Brooke, who was supported by Airy. His general approach, shown in Fig. 15.3, differed from Ronalds' in various respects. Brooke adopted cylindrical lenses placed transverse to the light beam rather than an achromatic variety arranged longitudinally. With this arrangement the light converged in just one dimension to change its aspect ratio, for example from a pencil to a spot. Perhaps the most obvious difference in the two arrangements was that Brooke used only photographic paper, so he was able to avoid Ronalds' frame and long case and wrap the paper around a drum. The drum comprised two 'French shades'[A72] of slightly different diameters between which the paper was held; Brooke made mention of the effects of variations in size of the glass shades and their imperfect cylindricality in reducing the accuracy of his photos.[A72]

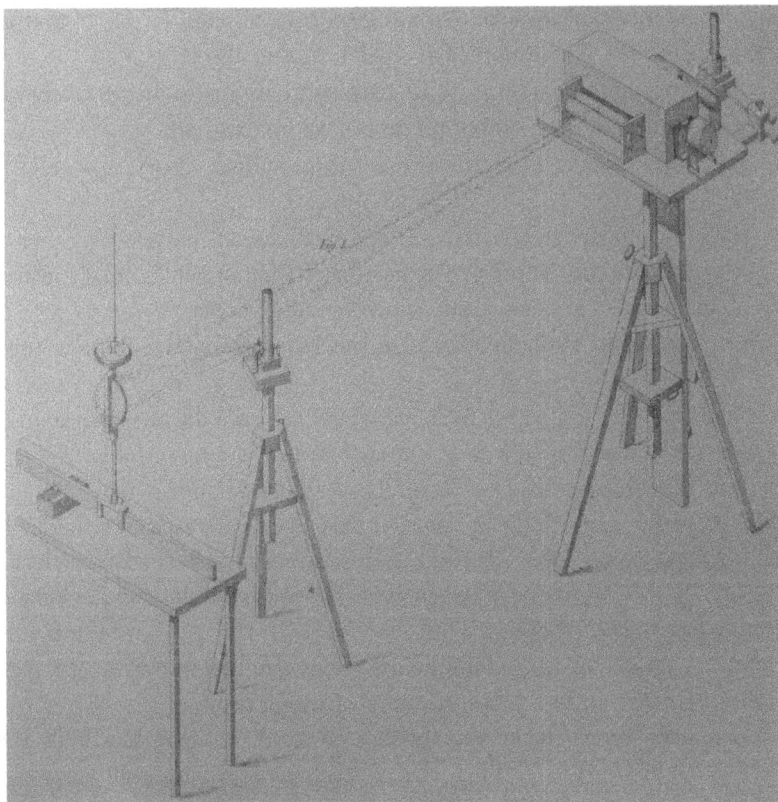

Fig. 15.3:   Configuration of Brooke's declination magnetograph. Ref. A72.

A further variance, in particular for the magnetographs, was that Brooke's photography system was designed as additional elements to the existing equipment at Greenwich. Ronalds in contrast needed to start from scratch, with his first steps being to design and find money to procure the components needed to observe the attribute of interest. These were then integrated with the camera in a self-contained apparatus.

Brooke's fuel of choice had the trade-name Camphine and was a purified turpentine spirit patented in 1839. It gave a very brilliant white light that facilitated photography greatly. It was also expensive, had limited availability and caused safety concerns. Edward Sabine* had advised Sir John Herschel*, even before Brooke was awarded the Admiralty's prize for his instruments:

In considering the relative merits of Mr Brooke's & Mr Ronalds' modes of Automatic registering, & without meaning to imply a general preference of one to the other, there is one point which makes Mr Ronalds far more useful for <u>Colonial</u> Observatories — namely that the Sun during the day & an Argand lamp (fed with oil) during night suffices. M[r] Brooke requires <u>Camphine</u> throughout the 24 hours...but there are few of the Colonies in which it can be had at all.[C219]

This was an important consideration for Sabine given his focus on observations in remote regions to build up a global picture of geomagnetism.

Captain John Lefroy*, Director of the Toronto Observatory, advised Sabine in June 1850:

> I have but a few minutes in which to report a serious accident, very nearly resulting in the destruction of this building. The barometer lamp of Mr. Brooke's instrument set fire to the massive wooden stand supporting the Cylinder[A149]

Sabine must have told Herschel because the latter advised Sabine a few weeks later:

> As to Camphine I have a horror of it and would not have a Camphine lamp in my house in common use for the world — ever since the burning of *[missing]*[C293]

Late that year Sabine advised Ronalds of a conversation with Airy:

> M[r] Airy & I have had a long & amicable talk upon the respective merits of the 2 kinds of Photometric Ins[ts] — He was greatly surprised at the difference of cost of illumination & said he wondered whether M[r] Brooke could do with daylight — we shall bring him round.[C318]

There is no evidence that Airy changed his views. He had in fact suggested the use of gaslight to Brooke in late 1848, to which Brooke replied:

> I have carefully compared the illuminating power of ordinary Gas-light and am satisfied that it does not exceed that of a good oil lamp, which...will leave no trace of rapid & considerable changes *[with his instruments]*[C244]

Overall, it may be said that Brooke's configuration was simpler than Ronalds', but also less sensitive and accurate. We have already seen that precision was all-important for Ronalds — he mentioned to Manuel Johnson* at Oxford concerning his new barograph, for example, that he was 'Feeling anxious that an Instrument destined to occupy an important position should be scrupulously exact (if possible)'.[C336] At the time, impartial observers were unwilling to record their views on the relative merits of the two competing approaches — it was only those affiliated with one or other of them who did so. Airy's assistant James Glaisher* was perhaps an exception in his letters to Ronalds quoted in Sec. 15.10. By 1897 all the key characters were deceased and Sir Philip Hartog, the author of Ronalds' entry in the first *Dictionary of National Biography*, was willing to write that 'Brooke... began his research at a somewhat later date' and his method was 'somewhat inferior in its optical arrangements'.[M6]

We turn now to the alternatives Ronalds trialled and the procedures he developed for creating a photograph with his camera, and then reproducing it and using it for scientific purposes. His efforts are put into context by a very brief introduction to the early history of photography.

15.2 Photography: Media, Processing, Reproduction and Documentation

The invention of photography will always be associated particularly with two men: Louis Daguerre in France and Henry Fox Talbot FRS in England. Daguerre's success was announced in January 1839 in Paris. The core of the Daguerreotype was a plate coated with silver, polished to a high gloss and fumed with iodine to form a silver iodide coating. After exposure, the latent image was 'developed' to visibility by mercury fumes and then 'fixed' by removing the remaining silver iodide so that it did not continue to react to the light.

Fox Talbot was working on an alternative process at the time. He first displayed his work at the Royal Institution a couple of weeks after Daguerre's announcement, and the Calotype (also known as the Talbotype) was introduced to the public in 1841. Its basis was paper soaked in iodised silver. Developing produced a 'negative' with the dark and light areas reversed, from which a 'positive' print could be made by re-exposing through the negative on to treated paper. The terms 'positive' and 'negative' were first coined by Herschel. Herschel's discovery that sodium thiosulphate would dissolve silver salts created a very effective fixer.

Both methods were patented in England and the first portrait studios were set up in London in 1841. Purchasing rights to Daguerre's process, and later assisted by Richard Nicklin*, Richard Beard opened up at the Royal Polytechnic Institution (see Chapter 12). An early patron was Crabb Robinson*, who recorded of his portrait: 'The first taken, was so bad that an offer was made to have it taken again'.[1] Beard defended his rights vigorously in suits against Egerton and also Antoine Claudet, who studied under Daguerre and established his own studio at the rival science exhibition called the Adelaide Gallery. A later competitor was Claudet's erstwhile employee Thomas Richard Williams. Talbot meanwhile had licensed Henry Collen, a portrait painter, who set up in Portman Square. Talbot's former assistants Nicolaas Henneman and Thomas Augustine Malone later established a studio in Regent Street.

Daguerre's and Talbot's inventions created huge excitement as their myriad artistic and scientific opportunities began to emerge. With many additional contributors, the new concepts saw rapid technical advance for many years: there was no knowing where the next breakthrough would occur or which of the two methods, or others about to be discovered, would prove best for any particular application. As the Juries Report of the 1851 Great Exhibition noted: 'Whether the followers of Talbot will ever obtain a pre-eminence over those of Daguerre, or *vice*

---

[1] Ref. J13 18/10/1841.

versâ, is a question for time to solve'.[B61] Innovation extended beyond the photographic process itself to related aspects including photographic paper and camera lenses. Costs could also change rapidly. It was with this in mind that Ronalds designed his camera to accept both media and also put considerable personal effort into both to achieve the improvements he desired. In doing so, he worked with all the UK-based people mentioned so far, and more.

Early alternatives to the original processes involved the incorporation of organic substances. In one, an albumen (egg white) mixture constituted the base for a sensitised coating on a glass plate. The 1851 Juries Report included the recipe and other details. A new concoction called collodion was announced soon afterwards. It in turn was replaced by gelatine in the 1880s, which continued in use in film-based negatives in the 20th century.

Ronalds' earliest recorded interaction with the new photographic community was in April 1844 and it related to his old rotating resin plates. He took one of his powdered plates to Collen's studio to be photographed 'in the manner usual for portraits'[R18] by the camera obscura. He immediately posted the photo to Airy — from the studio — noting it was 'a very rough first attempt (compleated about 5 minutes since)' and envisaging 'graphic reports of our Kew Observations'.[C133] This was perhaps the first articulation of how photography could revolutionise scientific reporting. Airy acknowledged receipt but did not pass comment on the photo. The one-hour curve of the electricity of dew remains in the Greenwich Observatory Archive at Cambridge.

Ronalds displayed and discussed further examples at the BAAS conference that year (Fig. 15.4). For presentation, the photograph was glued to paper marked as a clock face and with corresponding electrometer readings written around the circumference. He also explained his vision in more detail: he hoped to be able to take photos regularly and distribute copies of the 'pictorial register of atmospheric electricity'[R12] among interested meteorologists. To do so, the price per image would need to reduce and he began negotiating. In mid-October 1844 he noted to Collen: 'But five shillings per diem for chemicality &c would not do. Would it?'[C138] Two months later he wrote: 'I think that the result of our contest was a fixature [sic] of the amount at £2-6-0. Could we not save the trouble & expense of fixing the Image — would it not remain without fixing long enough to read merely.'[C142]

Ronalds had foreshadowed in his 1844 BAAS report that he was already working on improved self-registration methods. He continued to interact with Collen as he developed his continuously recording camera and applied it to the electrometer, barometer and thermometer. There is some hint that the machines were announced in August 1845,[C328] but the first technical write-up was in February 1846 in the *Philosophical Magazine* — and the author was Collen rather

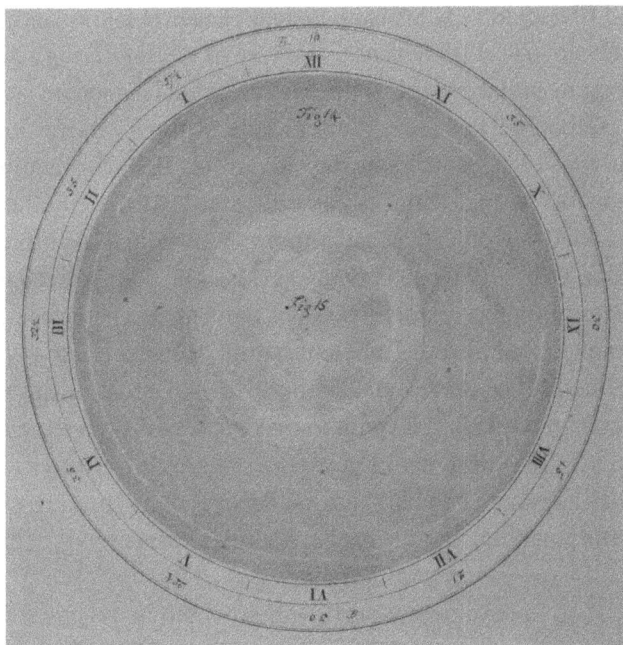

Fig. 15.4:   Collen's photos of two atmospheric electricity recordings on resin plates, created over short and long timescales (1844). Ref. J61.

than Ronalds. Ronalds was aware of the forthcoming article, having provided a negative for it, but was not impressed when Collen suggested that the machines were 'made by us conjointly'.[A68] The Ronalds Library Catalogue notes of the paper: 'Collen claims a share in my inventions unjustly. — F.R.'[R28] Here is another example of Ronalds' sensitivity. He did not wish to hog the limelight, but the originality of his ideas was pivotal to him.

It was fortunate in some ways for him that Collen did prepare the paper — Ronalds would have delayed writing until he was perfectly satisfied with all aspects of the apparatus and, in doing so, would have lost publishing priority to Brooke. He reduced his interactions with Collen at this time, although was gentlemanly enough to acknowledge the contributions he felt he did make in his next BAAS presentation: 'Mʳ Collen's prepared paper has been found to be the best... *[and]* Mʳ Ronalds derived*[?]* at first some valuable information from that Gentleman relative to photographic manipulation.'[J67]

Their early progress was written up in *Fortschritte der Physik*:

The practical application of the chemical action of light to the self-registration of instruments has made significant progress in 1846, and it can no longer be doubted that this registration method will soon be able with advantage to replace actual observation in many cases.

Messrs COLLEN and RONALDS have used the camera obscura to record the details of various meteorological instruments, with very nice results being achieved.[A75]

Ronalds began collaborating with Malone and Henneman as soon as they opened their Talbotype business 'Sun Picture Rooms' above John Newman's premises in August 1847. Ronalds had a suite of challenging requirements for his photographs, just as he had with his camera, and these were quite different from those being tackled by others. Most important of course was precision — the images needed to be as sharp and delicate as possible, with high contrast. This was dependent not only on the relative speed of the sensitised surface and the object being photographed, but also the inherent nature of the surface. Paper produced a fuzzier image than a Daguerreotype because of its fibrous texture. Irregularities could be reduced by purifying the pulp to make a very fine paper and using special finishing processes but these could not reduce the light-sensitivity of the paper's subsequent coating. Another challenge with paper was shrinkage and distortion caused by wetting and drying during the chemical processing, which could alter the scale of the image in some ill-defined way.

He also wished to develop a workflow that would be practical for an observatory conducting continuous daily photographic registration. The processing could not be too complex, time-consuming or expensive. As he explained to Airy in late 1846: 'I hope to be able to reduce the Knack-ery of the paper process to a simple, certain, & easy machine-ical operation.'[C188] Materials must be able to be purchased in bulk, prepared in advance without deterioration and be tolerant to delays at any point in the procedure caused by other priorities in the observatory. The images themselves needed to be able to be copied readily for distribution and to retain their features over time as part of an ongoing record. Finally, consumables and operator time could not be too costly as multiple daily photographs would be required. He was pushing photographic suppliers to the limit. A note survives from this time in the Ronalds Archive: 'Mr Talbot thinks that the conditions are not compatible[?] with economy. There would be no objection to give double the market price if it answered all the conditions required. He has the subject under his own consideration.'[J72]

Malone and Ronalds trialled different papers, a variety of sizing methods to improve the finish, and a range of chemical treatments both before and after exposure to the light. They had an early breakthrough with the processing, with Malone writing:

> it has kept white in the shaded portions — I am glad to find it so…This power of keeping the latent impression is valuable as it can be brought out at leisure… Mr Talbot is in Town but fears he will not have time to go to Kew [C200]

They were able to produce good images 30 hours or more after preparation.[R18] Unfortunately the new paper was 'very dirty'.[C202] Results remained disappointing, with Malone later updating Ronalds:

> I am just now very nervous about our new paper. We have received a few sheets unsized, I have not yet been successful in sizing them myself, therefore have given orders to have some sized at the Mill. It is very tough & I think very close in texture. It has not been bleached at the Mill. I remained there until the day before the last stage viz the drying which has been carelessly managed…This is vexing, so much care having been taken in the first part of the process. Still I hope we shall find it good in every other respect…Failure is impossible — nevertheless we may have some trouble to get it right.[C218]

They were still at work after 18 months, with Malone advising Talbot:

> Mr Ronalds of Kew has been with me in the <u>evening</u> experimenting for his registration plan [C247]

The trials were funded from Ronalds' Royal Society grant. Ronalds acknowledged all the assistance in his 1848 and 1949 BAAS reports, to which Malone expressed his appreciation for 'the honour you have done me'.[C286] Samples of several of the photographic papers used are retained in the Ronalds Archive, including from the Chafford Mill run by the Turner family, Whatman's paper, as well as Talbot's.[J71]

Despite 'much time and pains',[R18] Ronalds' exacting standards could not be reached, so he turned to Daguerre's method in early 1849 and began to interact with Beard and Williams. He then needed to optimise his workflow anew. Daguerreotypes had inherent advantages over paper in their sensitivity to light and image sharpness. Disadvantages included the cost of the silvered plates and the time and labour of processing. The most critical problem he needed to solve, however, was a means of copying the impressions and then reusing the plates, for it was too expensive to preserve them.

He trialled numerous approaches to copying. The first success was due to J. E. Wood, who was at the time engraving Ronalds' drawings for his 1849 BAAS report. The photograph on the plate was traced with a drypoint needle such that it could produce multiple prints in a printing machine but without damaging the silver coating. Sabine was pleased with the result:

> if I understand you rightly the plate will do again when polished, as its surface is neither etched nor ground. If this be so we may have our plates more thickly silvered by & bye, which will be cheaper in the end — Can you obtain half a dozen more copies from Mr Wood. I should like to send one to Herschel, one to Lord N. & one to Lord Rosse, also one to Lefroy, to stimulate him. If Kew continues we may get up either a Bifilar or Declinometer, or both for the winter, and have a good disturbance engraved in this way as a supplement to your report [C262]

Fig. 15.5: Early print of trace of horizontal force magnetogram made using gelatine sheet (16/7/1850). Ref. J64.

This was a time when Kew was threatened with closure, and yet Ronalds was still working at full pace. Lords Northampton and Rosse, Past- and current Presidents of the Royal Society, were important stakeholders.

Ronalds' idea the next year was even better. Herschel had just trialled gelatine for photography, and it occurred to Ronalds to use the transparent and strong gelatine sheet as tracing paper. It formed a ridge where the edges of the Daguerreotype curve were traced with an etching needle that showed up when the lines were inked and then printed in the manner of a copperplate (Fig. 15.5). He noted that 'The experiment succeeded on the first trial.'[R20] Further work proved banknote paper to be the best for the print, being soft enough to conform to the engraved line, but also tough and having little adherence to the gelatine in the press. He soon started to import gelatine supplies from the Continent. A case, still containing several leaves, is retained at the Met Office Archive and other engraved sheets are in the Ronalds Archive. He passed some to Herschel for his photographic trials.[C282]

He advised in his 1850 BAAS report in Edinburgh that 'printed engravings can be produced in any required quantities with the greatest facility and expedition'.[J64] The write-up in the *Athenæum* went on to note:

> Sir D. BREWSTER *[Conference President and respected photographer]* wished to suggest to Mr. Ronalds that by taking a negative impression of the positive photographic curve, copies might in a much simpler manner be multiplied to any extent[2]

Ronalds was not so sure. He had already worked with Malone on using the Talbotype method to copy Daguerreotypes. The process was longwinded and the copies not sufficiently sharp. He summarised his views to Sabine:

> Some recent conversations which I have had with Ross, of Edinboro', and Sir David Brewster & the Astronomer Royal there convince me that as yet the gelatin traces & printings are more efficient for copying in every respect than the common calotype or the albuminous-type — but we must live & learn.[C302]

Others agreed. After trying it in Toronto, Lefroy wrote to Sabine:

> The very beautiful discovery of Mr. Ronalds that traces may be printed from the etching on gelatine paper, removes a great difficulty.[A149]

---

[2]*Athenæum* 10/8/1850 839.

When a very large number of copies was required, a print of the gelatine tracing was made on lithographer's paper and then transferred to stone. Ronalds included a lithograph in his 1851 BAAS report, although he stressed: 'It is not pretended that the lithograph is equal, in sharpness and accuracy, to the original impression from the gelatine.'[R21] In this period he was falling back on his previous experience in engraving, lithography and printing (Sec. 10.7) and he would have used his presses for the trials. Corresponding methods were adopted in 1869 when his early vision at last became reality and curves from the machines at the Met Office's observatories began to be reproduced for publication.

Ronalds' Daguerreotype plates were a highly atypical size of 33 cm × 9 cm and he approached numerous suppliers to find the best and most cost-effective way of making them. Stanton Brothers' price for this size was 17 shillings. Egerton charged £3.5s.0d per dozen.[C322] William Henshaw offered silvered copperplates at 2/6 each, or three shillings 'with more silver on that will take more impressions…I have put them in at the lowest possible price supposing you will want large quantities of them.'[C266] Ronalds also enquired after 'M[r] *[William]* Radcliffe's *[sic]* price for The copper plates planished & finished'.[C263] It seems that he was contemplating electroplating the copperplates himself. He later advised: 'Plates made by electrotyping are by far the best.'[R24] The silvered plates were also too long to be either prepared for exposure or developed readily at commercial facilities so he set up a full processing laboratory at Kew for his photographic experiments. Examples of the accessories he designed and made formed part of the overall package when he sent his first magnetograph to Lefroy in Toronto in 1849.

The full photographic workflow he documented for Lefroy is retained at IET and in the Met Office Archive. It had been honed 'from consultations and operations performed in the presence of several of the best photographists in London'[J75] as well as his own experience. Once again, quality and reliability were paramount. As he noted in his French publication *Météorologiques et Magnétiques*: 'Careful preparation is critical, a failure is irreparable as the observations of a previous day…cannot be repeated, and gives an unfortunate gap in the register.'[R24] He admitted that his approach needed 'plenty of "elbow grease"'[J75] while Lefroy called it 'laborious'[C310] in his feedback to him.

The first and perhaps most difficult step was to remove the previous image and render the silvered plate perfectly clean. Ronalds designed polishing boards that were fixed to the table and held the plate firmly in place for polishing. He noted that they 'seem to be improvements upon the frames employed by Daguerreotypists usually'.[R21] He also made a series of polishing buffs, each comprising a curved piece of wood covered with flannel and thick plush velvet or another suitable material and provided with a handle.[R20]

The first buff was saturated with a cream of rottenstone and olive oil — in Lefroy's words, 'that detestable oil and rotten stone'.[C295] Rottenstone is a fine

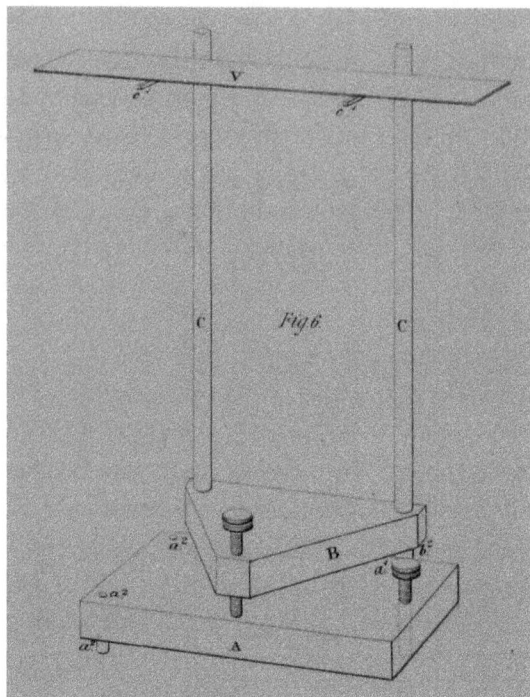

Fig. 15.6:   Burning off and fixing stand (1849). Ref. R20.

powdered porous rock that acts as a polishing abrasive. Final traces of the mixture were removed from the plate with cotton wool. 'Buff N°. 2'[J75] of linen or silk and a second polishing board were used with charcoal powder and the plate was rubbed until a black polish was produced. The process was then repeated with a third buff.

The next step took the plate to the Burning Off and Fixing Stand (Fig. 15.6). The stand Ronalds developed, although very simple in concept, epitomises him as a scientist. He designed it at a time when he was building two challenging magnetographs for Toronto, mastering a new mode of photography, facing additional pressures like the observatory's closure and challenging staff issues, and lamenting the halt of his atmospheric electricity observations — yet he took the time to devise a convenient mechanism for ensuring the plate would be held horizontal. His idea was to separate the actions of levelling about each of the two axes of the Daguerreotype plate. This was achieved by giving the stand two platforms, both having a pair of fixed legs (in perpendicular planes) and a third leg comprising an adjustable screw. As he noted, it 'allows of much more rapid adjustment for horizontality than the usual stand having three adjusting screws.'[R20] On the stand, any dirt remaining on the plate was burnt off. A cotton tuft soaked with ether or

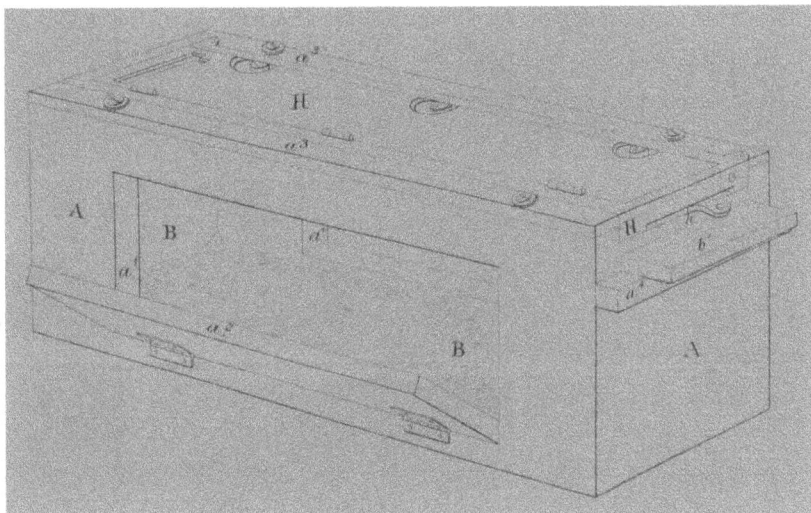

Fig. 15.7:   Coating box (1849). Ref. R20.

spirit of wine was lit and the flame applied under the plate until it was almost red hot and had a slight cloudiness.

The fourth buff was then used to polish it again: 'the whole black polish should be as perfect as you can possibly make it.'[J75] He added the advice: 'Do not take snuff during the cleaning & polishing work, or blow, or spit upon the plate.'[J75] A final polish was done with a fifth buff of white silk and a little charcoal immediately before the coating process and the buffs were cleaned and stored in a dry place.

After insertion of the plate in its sliding frame, the frame was placed in the top of a coating box with its interior glass vessel containing iodine crystals. The frame's sliding cover was then removed via a handle to expose the plate to iodine fumes. A mirror on the open door allowed the changing colour of the plate to be monitored by reflection (Fig. 15.7). The process was then repeated in another coating box to apply the accelerator bromide and distilled water before returning to the iodine box. Once the plate was ready, the protective cover was replaced and the plate transferred to the camera.

After exposure, the covered plate was taken to the darkroom and inserted in a mercury box, where the image was developed with fumes generated by heating mercury with a spirit lamp. The image was then fixed by plunging the plate into a sodium thiosulphate bath, washed in distilled water and dried. Further fixing using *sel d'or* (gold hyposulphite) could also be performed using the burning off and fixing stand.

Then to tabulation. The image was studied and its readings recorded — it was these numbers that formed the basis of subsequent analysis to understand the physical phenomenon under investigation. Ronalds noted how he 'hit upon an

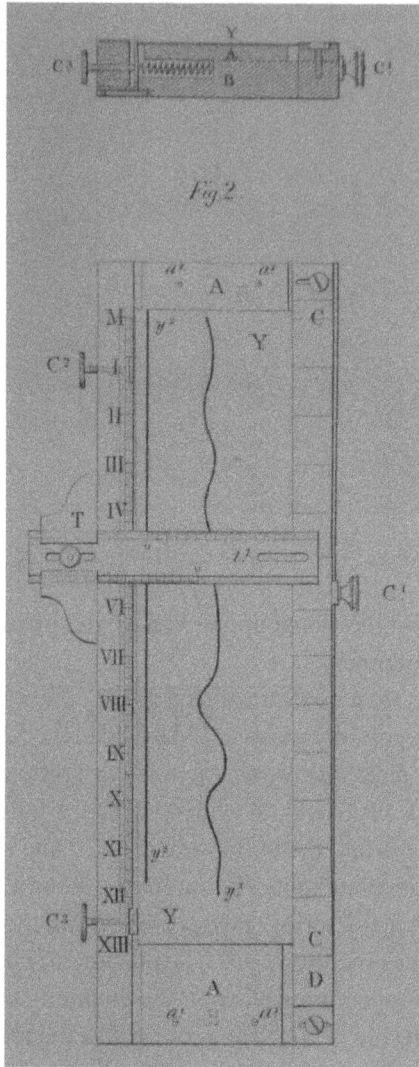

Fig. 15.8:   Ordinate board for linear readings (1849). Ref. R20.

obvious, but very useful addition'[R19] — an ordinate board to facilitate measurement of the separation between the curve and the zero line (Fig. 15.8). The Daguerreotype plate was held in a mahogany case, marked along its length in five minute intervals at a scale of 25 mm to an hour to match the movement of the plate. A tee-square slid up and down the plate, with scales engraved in 0.4 mm and 0.5 mm wide divisions (sixtieths and fiftieths of an inch) along each edge. Ordinates were read with the aid of a compound magnifying lens, typically at half-hourly intervals and at points of interest like local maxima and minima, giving up

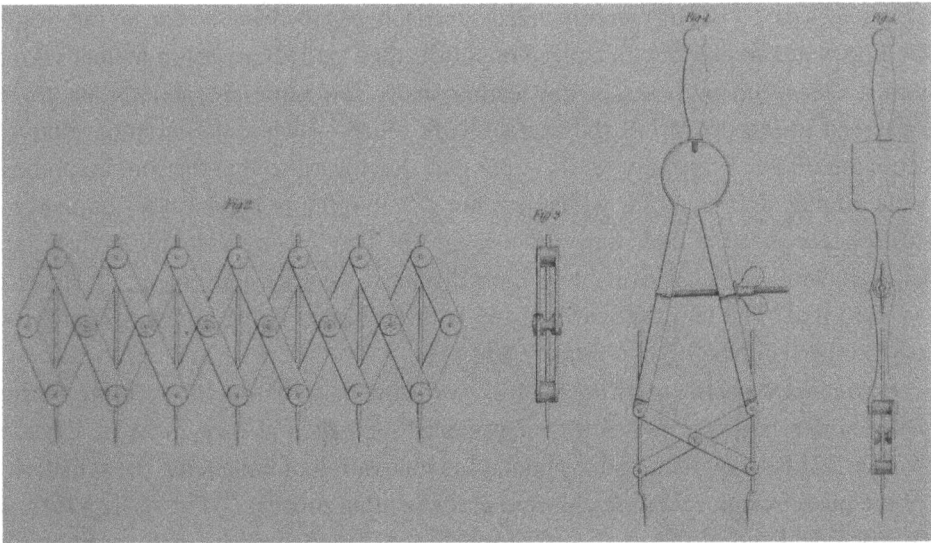

Fig. 15.9:   Two dividing instruments for splitting a trace into a number of equal intervals (1850). Ref. R20.

to 75 measurements in a 12-hour period. The ordinate board formed part of the Kew Observatory display at the 1855 *Exposition Universelle* in Paris.

To assist in the analysis stage, he also created dividers that could section the photograph into a number of equal intervals (Fig. 15.9). Its lattice form was similar in concept to his fire escape ladder described in Sec. 12.13. The pointed vertical members were fixed to the lower joints of the lattice but slid through the upper ones. With the points not being able to be brought very close together, a second instrument was created for subdividing a small interval. The first instrument could then transfer these subdivisions across the other intervals. He took a prototype to Holtzapffel* in October 1850, thinking that the instruments might also be of use for draftsmen and engravers; Holtzapffel charged 30 shillings for making it.[J84]

After analysis of the photograph, the next step was to clamp the plate together with a gelatine sheet on a tracing board to reproduce the image;[R21] copies were subsequently printed for filing and distribution. Then it was back to the beginning — cleaning and polishing the plates to remove the image and winding the clock in readiness for the next half-day of observations with each machine. Experience suggested that the plate could be cleaned and reused around thirty times before needing to be re-silvered.

Other parts of the set-up also needed constant attention. Lefroy complained to Ronalds about the maintenance requirements of the lamp, and he in return paraphrased Lefroy in emphasising lamp-cleaning to be "'detestable" but

<u>very necessary</u>'.[A149] His recommended approach was to clean the burner interior in boiling water and soda with a piece of soft leather 'and afterwards a feather taking care to leave no particles of the leather in it. The same precautions are to be observed in respect of the sliding rackwork piece — when the lamp is removed from the oil get & turn it with its upper part downward so that the foul incombustible oil may drain out of it whilst it is hot'.[J74] 'In order to prevent very unpleasant spilling &c of oil a Table carrying a <u>large</u> Tin Tray having a double bottom one pierced with holes all over at ½ inch distances <u>is very convenient</u>.'[J74] He also advised that 'The lamp should be lit ½ an hour before being applied; in order to ensure the flame being in a proper state.'[J74]

Ronalds was able to refine his full workflow in an observatory setting during a six-month trial of three of his instruments conducted at Kew in 1851. Claudet had visited Kew by then and explained his innovation of enhancing the sensitivity of the plate by incorporating chlorine in the coating process.[J84] Punch-lists survive in Ronalds' hand of the key preparatory steps for the workflow that were hung in the Chemical Room and the rooms housing the cameras.[J70]

Sabine proudly gave Herschel a detailed update on the overall process in early 1850:

> [Ronalds] now manages to make the silvered plates for w[h] Beard originally charged us 11 shillings for 3[s], or with an extra quantity of silver for 3[s]6[d]…I enclose a small piece of gelatine tracing paper used…for taking off the trace from the silver plate…a numerical record is kept daily…& for this purpose M[r] Ronalds has invented a scale w[h] enables the reading to be taken with great rapidity & extraordinary accuracy from the plate.[C278]

Lefroy struggled with the exacting nature of Ronalds' Daguerreotype workflow and returned for a time to paper processes with which he was familiar. Ronalds' view of the sample he was sent was that:

> The outline, as viewed by the naked eye, seemed tolerably well defined…but when examined under the same magnifying power as that used in tabulating the curves received upon silver,…exhibited a little fringe…and was quite inadequate to measurements, equally minute, with those of lines received upon a silver surface.[R21]

So that was no good. Ronalds also explored the new albumen and collodion processes in 1851–52 but found they offered no advantage for his application. His former assistant John Welsh conducted further trials with collodion at Sabine's request in 1854. As Ronalds had underlined:

> the Kew magnetographs were always *equally applicable* to either the Talbotype or the Daguerreotype process at the pleasure of the observer or photographist. They can be as easily applied to *any* new photographic process I believe.[R21]

By the time Ronalds' barograph for Oxford was ready in late 1853, there had been further advances in photography. Ronalds was now on the Continent but Radcliffe Observer Johnson kept him abreast of progress. He started with Talbot's approach, although he prepared his own paper because Henneman charged the 'exorbitant' price of '£2.2 for 100 sheets'.[C354]

He was then fortunate in procuring the services of young William Crookes for a year. Crookes went on to become a chemist of note, later being knighted and becoming President of the Royal Society. Crookes chose the newly developed waxed-paper technique, where the paper was saturated initially with wax before chemical sensitisation. The wax gave the paper high reactivity to light and produced an image with increased sharpness and contrast. It also prevented shrinkage during the various chemical operations. Reducing the workflow to continuous 24/7 practice was aided by the paper retaining its sensitivity for a week, having a relatively slow development time, and being suitable in a range of weather conditions. The negatives were readily archived and multiple prints could be made for distribution. In all, it was the breakthrough that photo-recording machines needed. Crookes' Handbook of the process is in the Ronalds Library. Success at Oxford meant the same practice was later adopted for Ronalds' instruments at Kew and elsewhere until replaced by gelatinised paper in the 1880s.[K11]

Ronalds summarised his views and experience on photographic processes and equipment in early 1854 following a request for advice:

> The Daguerreotype method was latterly employed by me in preference to the Talbottype *[sic]* (or paper process) in consequence of the greater sharpness and precision of the image...— I find here *[Paris]* that the paper method has greatly improved in regard to distinctness.
>
> ...Perhaps at a very rough estimate £60 might be taken as the mean cost for each Instrument...the whole time of an intelligent and zealous Observer who is a good Photographist would be absolutely required for the proper manipulation &c of the Instruments [C356]

Brooke and Airy seemed be less fussy than Ronalds about photographic method. As early as October 1846, Glaisher informed Ronalds: '*[Brooke]* tells me that he can give me receipts *[recipes]* for preparing the paper — which will not fail.'[C181] The method might have been straightforward, but it did not meet Ronalds' quest for the best. The paper needed to be sensitised just before use and remained damp throughout the exposure, after which early development was required. The resulting image was not particularly crisp. Greenwich continued to use this same process for many years.

The next sections provide more detailed descriptions of each of Ronalds' registration instruments and their take-up.

15.3 Photo-Electrograph

We have seen that Ronalds' first photo-recording machine — in April 1845 — was for registering atmospheric electricity. He had contemplated two different

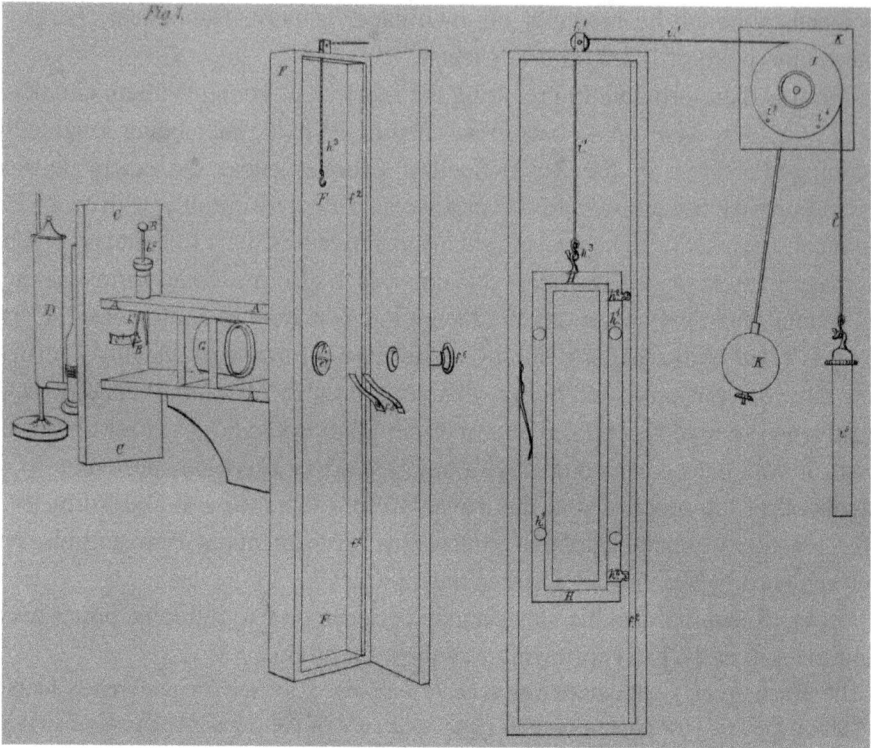

Fig. 15.10:   Photo-electrograph (1845). Ref. R15.

electrometers for the purpose: his modified Coulomb torsion balance, discussed in Sec. 14.2, and his Volta electrometer. Only the latter approach was published (Fig. 15.10). He chose the dimensions of 7.5 cm square and 40 cm long for the horizontal camera box, which was built by his assistant John Galloway. The photographic paper in its frame travelled vertically in a case 76 cm long at 2.5 cm per hour when pulled by a 10 cm wheel on the clock (which Ronalds had turned on his lathe). The lens tube was provided by Andrew Ross.

The straws of the Volta electrometer were charged by the conductor rod and protruded through the top of the camera box. Only their tips were visible through the slits at each end, which were arced to match their rotation. The photograph thus depicted a pair of curved lines with their separation varying according to the electric potential. Ronalds later designed an arced ordinate board (Fig. 15.11) to enable the angular divergence of the straws to be read very accurately at precise times along the photograph.

In the sample negative included in Collen's 1846 paper (Fig. 15.12), the electrometer lines have the whiteness of the photographic paper as they did not receive any light, while the darkest bands occur when the sun is strongest. Ronalds noted with interest in his 1846 BAAS report that these periods of

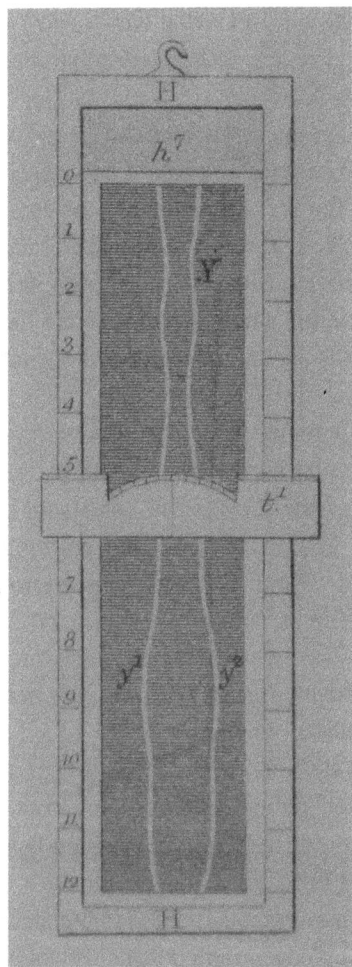

Fig. 15.11: Ordinate board for angular readings from photo-electrograph (*c.*1849). Ref. R24.

brighter weather coincided with perturbations in the atmospheric electricity reading. Images made as early as July 1845 were later displayed at the Royal Society.

He sent a detailed description of his apparatus to Airy, who was less than impressed:

> quality or sign of electricity is in my judgment more important than quantity *[Airy means intensity, see Sec. 6.2]*, and therefore mere records of diverging straws would not do.[C152]

This frustrated Ronalds, who explained by return:

> Under the impression that a conclusion arrived at by every observer of the periodical electricity of serene weather is correct viz that its character is positive, & that the exceptions to this law are extreemly rare and generally (if not always) accompanied by an easily distinguished feature I still hope that such records may serve as a good means of facilitating observation[C153]

Airy replied:

> I assure you that I do not set the valuation of your "Voltaic straw" result so low as you suppose. Only that I think the noting of the changes of quality to be exceedingly important: and that I expect the same thing may be managed by your ingenuity[C154]

Within a few weeks, Ronalds' 'ingenuity' had risen to Airy's challenge. He inserted a third electrometer index into the camera that moved through attraction and repulsion between two oppositely charged Leyden jars (Fig. 15.13). It thus confirmed that the atmospheric electricity was positive or indicated the much more unusual negative electrical state. He also placed a thin central bar across the slit to provide a fixed centreline. An 1845 example of this form of electrogram is retained in the Greenwich Archive (Fig. 15.14).

Gaston Tissandier's 1873 book *La Photographie*[B91] gives the clearest general picture of the photo-electrograph (Fig. 15.1), together with a detailed description. Even publications intended for 'the Working Man' like the first issue of William Gibbs' *Decorator's Assistant* (1847)[R14] included a summary of the instrument. Photography in all its guises had captured the public imagination and Ronalds' electrograph was seen as a pioneering machine in the new science. The framework

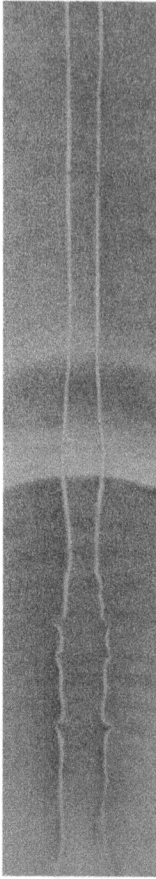

Fig. 15.12: Engraving of early photographic negative from photo-electrograph. Ref. A56.

of the machine was still carefully preserved at Kew in 1937, according to Superintendent Francis Whipple,[K14] but has since been lost.

Being Ronalds' first photographic machine, he strove to return to it to introduce improvements suggested by experience. He wished to facilitate the recommencement of electric potential observations at Kew through automation, and he was also eager to adapt the concept to record air–earth currents. He completed and trialled a new instrument for the first application by 1851, but the second instrument does not seem to have been finished.

The next apparatus was built for the Radcliffe Observatory at Oxford. Johnson's 1858 annual report describes it as 'an *Electrograph*, erected by Mr. Adie, consisting chiefly of a long copper rod, terminating in a sharp spike and pointing upwards above the tower, and giving its indications photographically by connexion with a straw *Electrometer*'.[S5] Electrograms dated up to 1874 survive at Oxford, suggesting that it was in use over a significant period. It is possible that other machines were built contemporaneously with Oxford's for regional meteorological stations such as Stonyhurst. Starting in the 1860s, however, photo-electrographs would have been built with Kelvin's modified Coulomb torsion balance as their basis, as outlined in the previous chapter.

Brooke did not apply his photographic method to the electrograph. Nonetheless Greenwich never trialled Ronalds' apparatus, even though atmospheric electricity continued to be observed there for 35 years using instruments of his design.

15.4 Barograph

Ronalds applied his camera to the registration of temperature and atmospheric pressure in August 1845. He sent sample photos to Airy the next month but they have not been seen at the Greenwich Archive. It was in this period that the future of Kew was under review for the first time, so he was both short of funds and bearing considerable uncertainty.

He was able to obtain a U-shaped siphon barometer to record atmospheric pressure, the open end of which was inserted vertically into the camera box from below. The narrow-slitted diaphragm at the object end of the camera was now immediately adjacent to and parallel to the barometer. Its width was adjustable to

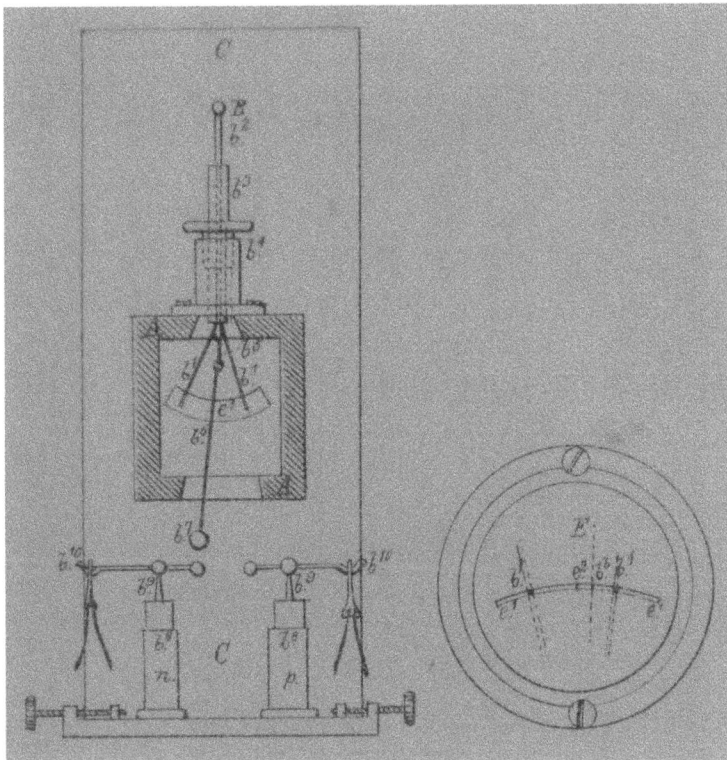

Fig. 15.13: Photo-electrograph incorporating third index and centreline (1845). Ref. R15.

suit the lighting and its position also minimised reflection from, and refraction through, the glass tube. The changing height of the mercury was captured on horizontally moving photographic paper (Fig. 15.15). A dark undulating band was generated in the positive print where the mercury blocked the light, with its height representing the atmospheric pressure at any time. He believed the accuracy matched that of the usual readings taken by eye. Initially the upper part of the mercury was photographed directly but soon a light, black pith-ball of almost the internal diameter of the tube was inserted atop the meniscus to create a sharper upper boundary. This was omitted from his subsequent machines and he was pleased to note that the meniscus was then 'totally unencumbered by any ball, piston, plug, float or machinery interfering with the *free motion* of the mercury in the clean tube'.[R18] As Met Office Director Robert Scott noted in *Elementary Meteorology*, published late in century:

> 'all instruments which are provided with a mechanical contrivance for registration must necessarily be more or less sluggish in their action as compared with the simple instrument, for the mercury has not only to move in accordance with the changes of pressure, but to do the mechanical work of moving the float.'[B109]

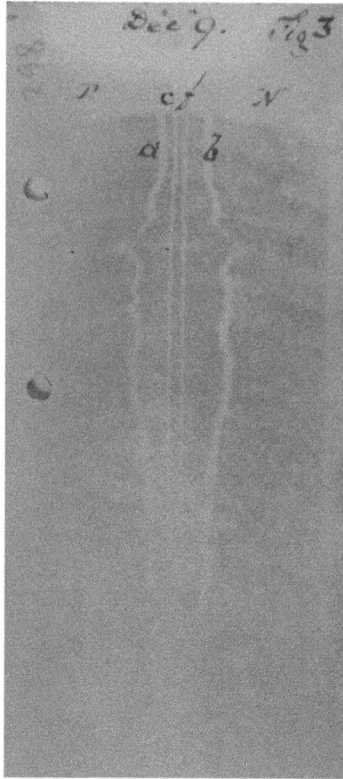

Fig. 15.14:   Electrograph negative with third index and centreline (9/12/1845). Cambridge Archive RGO 6/701 298.

Fig. 15.15:   Section of positive print of mercury from Oxford thermograph (28/5/1855). IET 1.9.1.

The photo-barograph concept had initially been trialled as early as 1839 by Cornish instrument-maker Thomas Jordan.[A55] His was simpler than Ronalds' arrangement with no slit diaphragm plate or lens to shape, condense and focus the light and sharpen the photograph. It thus captured a shadow rather than an image of the mercury. Collen noted in his paper:

> The projection of shadows on photographic paper...was at once objected to by Mr. Ronalds, whose knowledge of the delicacy required in observing and registering the various instruments at the Observatory, made him fully aware of the necessity of obtaining as perfect definition as the best optical arrangement would produce[A68]

Ronalds soon configured a more sophisticated barograph, and by late 1846 it was under construction. Siphon barometers had several disadvantages, including a tendency for the mercury to stick in the open limb. He wished to use a high quality barometer incorporating a cistern of diameter much larger than the tube itself to maximise accuracy. He was also anxious to incorporate a method of temperature compensation to account for the varying density of the mercury. Herschel had noted in his 1844 annual report to the BAAS on the coordinated magnetic and meteorological observations that 'a barometer which shall register its readings *corrected* for temperature would be of the utmost value'[S1] and 'earnestly' recommended that a machine be developed. Such an instrument would reduce manual post-processing of the observations.

We know that Ronalds had contemplated temperature-compensating mechanisms for years, as illustrated by his extensive experiments on the battery-operated clock in 1815 (Sec. 6.3). His Ideas Book mentions 'Hardy's Pendulum applied to El. *[electric]* Column, to Thermometer & Hygrometer'[J14] in 1822. Hardy had developed an accurate clock for the Greenwich Observatory where the pendulum bob was a tube of mercury. Lengthening of the metal pendulum rod with increasing temperature was counteracted by the mercury rising so that the overall centre of gravity of the pendulum, and hence its period, did not alter. A related idea was the gridiron pendulum developed by John Harrison. This device had several pairs of rods made of different materials. They were connected on either side of the clock pendulum such that the pairs extended either up or down under increasing temperature and the complete pendulum's effective length did not change.

Adapting these ideas to the barograph, Ronalds deployed a vertical zinc rod on each side of the barometer tube (Figs. 15.16 and 15.17). Their bases were pinned to each end of a lever with its fulcrum positioned at one-third of the separation. At the top, one rod was fixed in place and the other attached to a second lever, half as long and again able to rotate about its third point. This last pin position was adjustable to enable fine-tuning of the instrument. The barometer itself hung from the other end of this upper lever by a silk skein. By this mechanism, an extension of each of the zinc rods by δ caused the barometer to descend

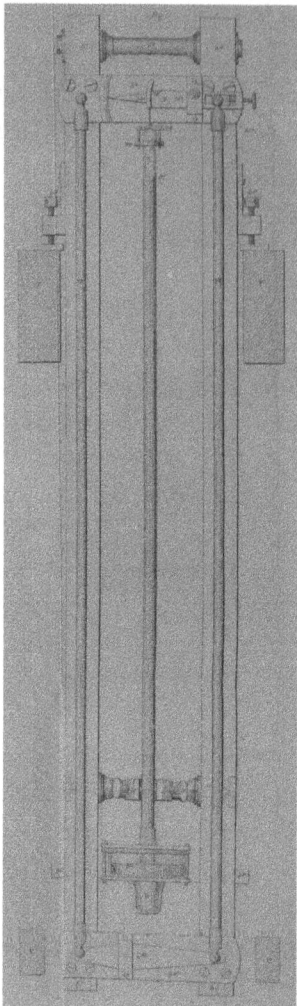

Fig. 15.16: Temperature-compensation mechanism for barograph (1847). Ref. R21.

by 6δ. The coefficient of thermal expansion of mercury is almost exactly twice that of zinc but the apparatus was configured for linear expansion of the zinc and volumetric expansion of the mercury in the tube. The surface of the mercury at a particular atmospheric pressure was thus always the same distance from the zero of the scale, independent of temperature. Extending beyond the tip of the upper lever was an index to a curved scale to show the extent of the barometer's vertical movement.

Ronalds advised Airy of his work in January 1847, who replied: 'I shall be curious to see your "barometer-self-corrected-for-temperature" arrangement.'[C190] Ronalds duly sent a description and drawings and again Airy was dismissive:

> the machine is rather complicated…The gridiron pendulum is not often to be trusted…
> the exactness of the correction at every moment is essential: if it ever goes wrong, the result is totally lost…there is another consideration. The laws of expansion are not accurately known. I have found that, for brass and iron, the expansion for each 1° of the mercurial thermometer is greater at high temperatures than at low ones.[C191]

Ronalds then sought Herschel's advice. He was more positive — luckily — or the apparatus would probably never have been published:

> I am glad to see the subject taken up efficiently. Your plan appears ingenious and I do not see why it sh[d] not answer.[C199]

Ronalds did not publish the barograph until 1851 'because time and opportunity have scarcely permitted an examination of its qualifications'.[R21] He was perhaps being even more cautious than usual after Airy's reaction. Welsh had now been able to complete a detailed comparison with a standard barometer and the results were found to be 'very satisfactory'.[R21] The quality of the barograph was then quickly recognised more widely. Sabine wrote to Ronalds almost immediately:

> I think of asking you to make one of your Barometers for Toronto. If the American Experiments go on it will be most important that they should have so admirable a means of keeping a continuous record[C320]

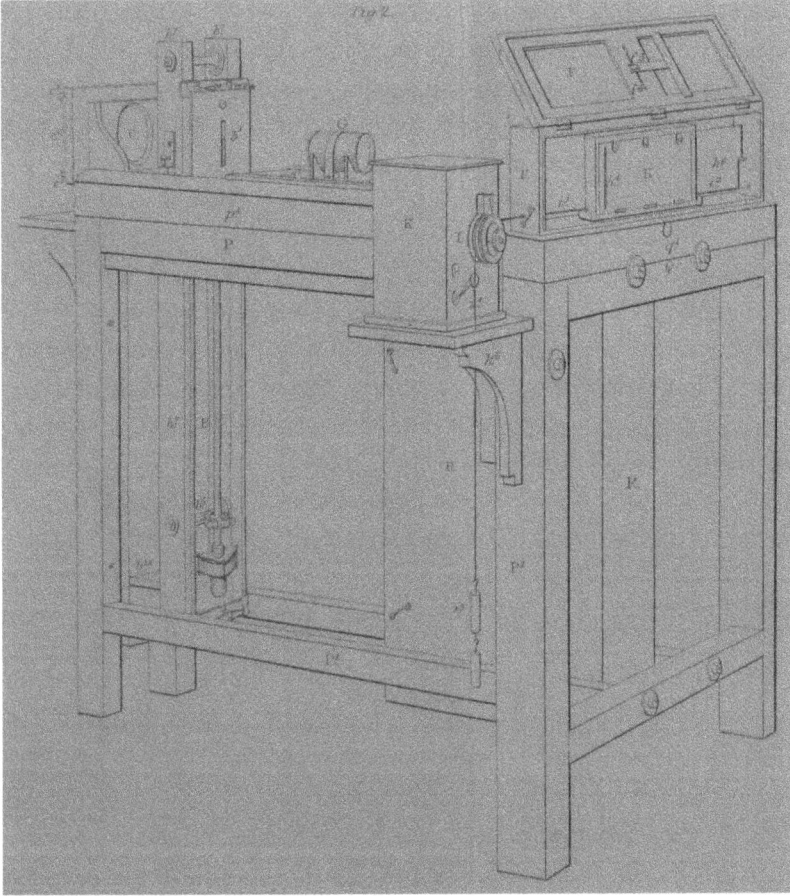

Fig. 15.17: Overall arrangement of barograph — model 2 (1847). Ref. R24.

In fact Toronto was not supplied until 1874, but this machine was then in almost constant use until 1940.[B150] The design also featured in various meteorological texts. A 20th century monograph on barometers by Knowles Middleton advised: 'This excellent instrument worked very well indeed'[B150] and, as well as the compensation system, singled out the 'excellent projection system'[B150] for the image.

Ronalds' third barograph was built for Johnson at Oxford and set in action in 1854. Details of its manufacture and cost are given below. Differences from the second model were confined essentially to the supporting elements to increase the machine's stability and durability for extended use. They included resting the machine on a strong iron bed and substituting glass for wood in the verticals of the outer frame that carried the levers. He also added a rod to ease friction on the pivots by acting as a partial counterpoise to the barometer.[R24] When Johnson had mastered its use he informed Ronalds that it 'is in my opinion as perfect as any Barometer and during the last 18 months I don't think we have lost 24 hours record.'[C366] Johnson published several papers based on the results of the machine. He reported to the 1855 BAAS conference for example that there is a momentary spike in barometric pressure when a thunderclap occurs.[A99]

Ronalds' original temperature-compensated barograph was taken to the *Exposition Universelle* in Paris in 1855 as part of the Kew Observatory display. Sample barograms made by both it and the Oxford machine were also exhibited. Forwarding his photographs to Ronalds, Johnson noted modestly: 'A Parisian Photographer will smile at our rude effort but as we are Meteorologists I think it better not to do anything that could interfere with the truthfulness of the copy.'[C363]

The display was the catalyst for French instrument-makers to begin manufacture and distribution of the instruments. Tissandier devoted several pages to the barograph in *La Photographie*, starting: 'Mr. Ronalds, and afterwards Mr. Salleron, have adopted the ingenious arrangement which we shall describe.'[K7] Jules Salleron was a Paris-based instrument-maker. His machine included the temperature-compensation mechanism as well as the long case carrying the sliding Daguerreotype plate, but also incorporated a thermometer and hygrometer. He supplied one to the Lisbon Observatory in 1864[A191] and there is record of another machine for *l'Observatoire du Dépôt de la Marine* around 1860.[A107] Ronalds ended the description of his barograph and thermograph in his 1855 book *Météorologiques et Magnétiques* with advice of additional instruments being made: 'They have been well received by numerous renowned physicists...A barograph and thermograph of this type will be built for the Imperial Observatory of Paris.'[R24] The new Director of the Paris Observatory, Urbain le Verrier, had a keen interest in meteorology. He outlined his needs for photo-recording machines among a range of equipment in his first Observatory report that year.[B71]

While the Oxford barograph was in continuous use for over 25 years, that at Kew was intended as a proof-of-concept — there were no resources for regular recording. That changed with the advent of the Meteorological Department (Sec. 13.5). Scott wrote in his history of the Kew Observatory in 1885:

> In 1862, at the suggestion of Admiral Fitz Roy, who agreed on the part of the Meteorological Department of the Board of Trade to bear the expense incurred, the Barograph designed by Mr. Ronalds was fitted up, and has been from that date kept in constant operation.[K9]

These early barograms are retained at the Met Office Archive.

Results of the two English machines were soon being compared. In a paper read at the Royal Society, sudden changes in atmospheric pressure during various storms in 1863 were found to occur at Oxford about 50 minutes earlier than at Kew, giving early insights into the progression of storms.[A111] Several other copies of the apparatus were also made at this time, including for St Petersburg, Coimbra and Stonyhurst, as the machine's reputation continued to grow.[K17,A191]

Scott's assertion is not quite correct, however. The original barograph was in daily use for several years, but in 1867 — 20 years after its initial invention — an updated model was developed. The principal modification was that the zinc rods and levers were used not to lower the barometer but to move a long index ending in a shutter.[B85] This generated a curved baseline on the photograph, with the temperature-corrected atmospheric pressure reading being given by the distance between the mercury surface and the edge of the shutter. The new Kew instrument was still in use up to the observatory's closure in 1980.[A161,J87]

Kew Observatory records indicate that these new 'Kew-pattern' instruments were supplied to 19 British and overseas observatories over the next two decades.[K9,K17] They were made by several English manufacturers, including Patrick Adie, R. & J. Beck and James Hicks, who advertised it for £68 with a detailed drawing in his *Catalogue of Standard, Self-Recording and other Meteorological Instruments* (c.1870).[B102]

It seems therefore that at least 27 barographs were made of Ronalds' general design, with quite possibly others on the Continent and in the USA. Its extended worldwide use and resulting scientific and social impact in helping to understand the weather make it arguably his most effective invention.

The Kew Committee of the Royal Society provided Ronalds' first temperature-compensated barograph to the 1876 Special Loan Collection (Chapter 12). The exhibition also included one of the Kew-pattern machines, noting it to be 'the improved form of Ronald's barograph'.[B94] Both are now in the Science Museum. The earlier model was acquired in 1876 and was on display again in the 1980s, while the updated version was received in 1926 after many years of use. Ronalds' Radcliffe Observatory instrument was donated to the Oxford Museum of the History of Science, where it also remains today, along with sample barograms.

Turning briefly to his competitor, Brooke developed his barograph in 1846. He deployed a syphon barometer with a mechanical arrangement to magnify the changing mercury level. A float was inserted on top of the mercury to which a stem was attached that passed through friction rollers. It interacted with two pivoting wooden arms at right-angles that converted the vertical movement of the mercury and stem into horizontal motion of a slit at the far end of the 76 cm long arm. Light shining through the slit created the photograph.[A72] Brooke did not develop a second model and Airy never adopted a temperature-compensated instrument; Greenwich thus retained an observing system distinct from other UK and Colonial stations throughout the century.

It should be mentioned here too that photography was not the only way of retaining a record of the movements of mercury in a barometer, although it was the means used predominantly across the British Empire in the 19th century. Electromagnetic and mechanical means (such as that used on Wheatstone's 1843 instrument) later became popular, particularly in Europe and the USA.

15.5 Thermograph and Hygrograph

Ronalds records:

> In 1841 I proposed to Sir James South[*] a method of registering the variations of the Thermometer photographically by placing the instrument behind a screen through which a narrow slit should be cut, and by receiving the shadow of the mercury passing through the slit upon photographic paper fixed in a frame and made to move horizontally by a time piece. A rough drawing of the scheme was made and shewn to several persons [163]

The drawing was later in the possession of Romney Robinson* at the Armagh Observatory. Ronalds' ideas continued to evolve and came to fruition in August 1845. He chose at that time a thermometer with a broad, flat bore; the portion with varying mercury level was inserted into his camera while the bulb was outside (and remote from any lamp). In other respects the arrangement was comparable with his first barograph (Fig. 15.18).

Also in 1845, alternative, simpler schemes were published by Scottish inventor Mungo Ponton and Irish physicist Henry Hennessy.[A67,B153] Unlike Ronalds' model, they had no optical system to create an image of the varying mercury level. Brooke's variant developed in 1846 also photographed the mercury shadow.

Ronalds noted that the thermograph was readily extendable to also perform the function of a hygrometer by including both wet and dry bulb thermometers. It is unlikely that he was able to build one at Kew but he configured a combined

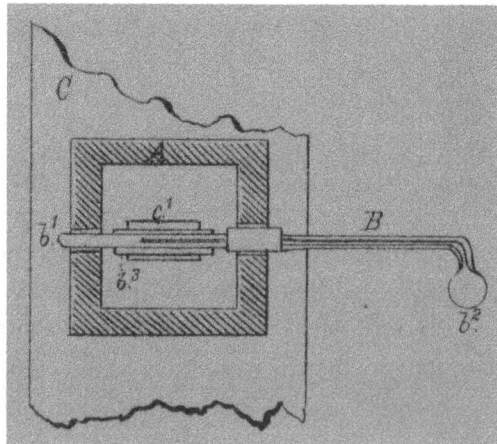

Fig. 15.18: Thermograph — model 1 (1845); section through camera box showing thermometer and slitted diaphragm. Ref. R15.

thermograph and hygrograph in 1853 for Johnson at the Radcliffe Observatory. He altered the set-up on the basis of his experience, in particular replacing the focusing lens tube with condensing lenses between the thermometers and the photographic surface. He was also able to arrange for the bulbs to project through a window to be in the open air.

Crookes made some alterations to the apparatus when he was commissioning it for use. Ronalds kept a summary of a letter he sent Johnson in 1855: 'Acknowledged receipt of his sketch of Mr Crook's modification of my Thermograph. Said it differs from my original mode only in the number and position of Lenses...My figures sharper.'[J85] Johnson included a copy of one of Crookes' thermograms, as well as a barogram, in his 1854 annual report.[S5] The instrument was altered again, probably in the late 1850s, when an air bubble was inserted near the top of the mercury column. Its movement was then recorded on the negative as a thin dark curve rather than the previous light band of variable height. The instrument's components were arranged such that the lines of both dry and wet bulb thermometers were captured on the same paper, from which the relative humidity could be computed readily.

The same approach was adopted when the Kew-pattern thermo/hygrograph was developed in 1867. It was still in service at Kew in 1966, using an electric lamp rather than a gas flame.[B153] Notwithstanding the differences from his original model, the example of the 'new' machine in the Science Museum is catalogued as 'Ronalds' photographic thermograph, 1867'. It was presented by the Met Office in 1926 and in the 1960s was displayed as it would be set up at an observatory. One of these machines had also been exhibited in the 1876 Special Loan Collection.

Distribution of the Kew-pattern thermo/hygrograph mirrored that of the barograph. First to receive them were the new Met Office's official regional observing stations: Stonyhurst and Falmouth in England, Glasgow and Aberdeen in Scotland and, in Ireland, Armagh and Valentia (at the landfall of the just completed Atlantic telegraph). Oxford later became a 'reporting station'[A184] and in 1880 its original machines were replaced with those of the Kew-pattern to ensure complete compatibility — these continued in use until 1924. Overseas observatories were issued with the instruments in the approximate order: Bombay; Mauritius; Melbourne; Sydney; Toronto; Brussels; Shanghai; Adelaide; Hong Kong; Tokyo; and Jakarta.[K9,K17] Hicks advertised the thermo/hygrograph for sale at £82.[B102]

15.6 Declination Magnetograph

Ronalds had also long contemplated the photo-recording of geomagnetic variations:
> In the Autumn of 1840 we *[Ronalds and his friend Samuel Sharpe*]* held a conversation on the subject of self registering...the variation of the Declination magnet...

by the employment of an image projected from a concave mirror, attached to the magnet, upon photographic paper moved by clock work[163]

Similarly, to a sketch dated 12 November 1841, of an instrument that projected a perspective picture onto the wall (Fig. 10.22), he added the note:

Registration, magnetic Variations[?], concave mirror on machine, Image [?] upon a photogenic paper on a Cylinder wound by Clockwork

We will see that these two brief descriptions are very similar in concept to the magnetograph developed some years later by Brooke.

Ronalds started serious work on the registration of the earth's magnetic field in late 1845, as soon as he had completed his first electrograph, thermograph and barograph. Not only was this a high priority for Sabine, but Ronalds also had increasing interest in the possible interrelationships between geomagnetism and atmospheric electricity. It was a new area of endeavour for him however and he had to start at the beginning in creating his machines.

He began with the declination magnetograph. The essence of a declinometer is a magnet suspended parallel to the magnetic meridian, rotations of which indicate the instantaneous horizontal orientation of the magnetic field at a particular place. Glaisher arranged for him to borrow a rectangular magnet 61 cm long from Greenwich, where the magnetometers were also this size. The silk skein was 2.7 m long and attached to a stirrup that held the magnet. It was supported at the top by a graduated circular plate that could be rotated in plan to adjust the orientation of the magnet or measure the torsional stiffness of the arrangement. The concept has strong resemblance, except in scale, to the Coulomb torsion balance described in Sec. 14.2. Ronalds configured the overall apparatus to be supported in a large frame, incorporating two strong pillars that formerly held the King's telescope (Fig. 15.19).

Oscillations caused by magnetic disturbances were damped by a metallic hoop around the magnet. In accordance with Faraday's law, the changing magnetic field sets up a current in the hoop; the field induced by this current tends to oppose the magnet's field and thus diminish the vibration. The damper on the declinometer was made of mahogany that had been electroplated with a copper coating, but Ronalds' later dampers were able to be made of pure copper. The temperature in the magnet box was monitored with a thermometer — not just the current in the hoop but also the heat of the lamp risked undesirable temperature effects.

For registration, he adopted a counterbalanced right-angled brass index attached to the underside of the stirrup. It translated the lateral rotations of the magnet into index movements in a vertical plane close to the lamp end of the camera. The camera box itself was longer than his earlier model to allow the motions to be magnified significantly in the image. This required a new optical arrangement and he purchased a Voigtländer lens from Egerton for £10 as part of

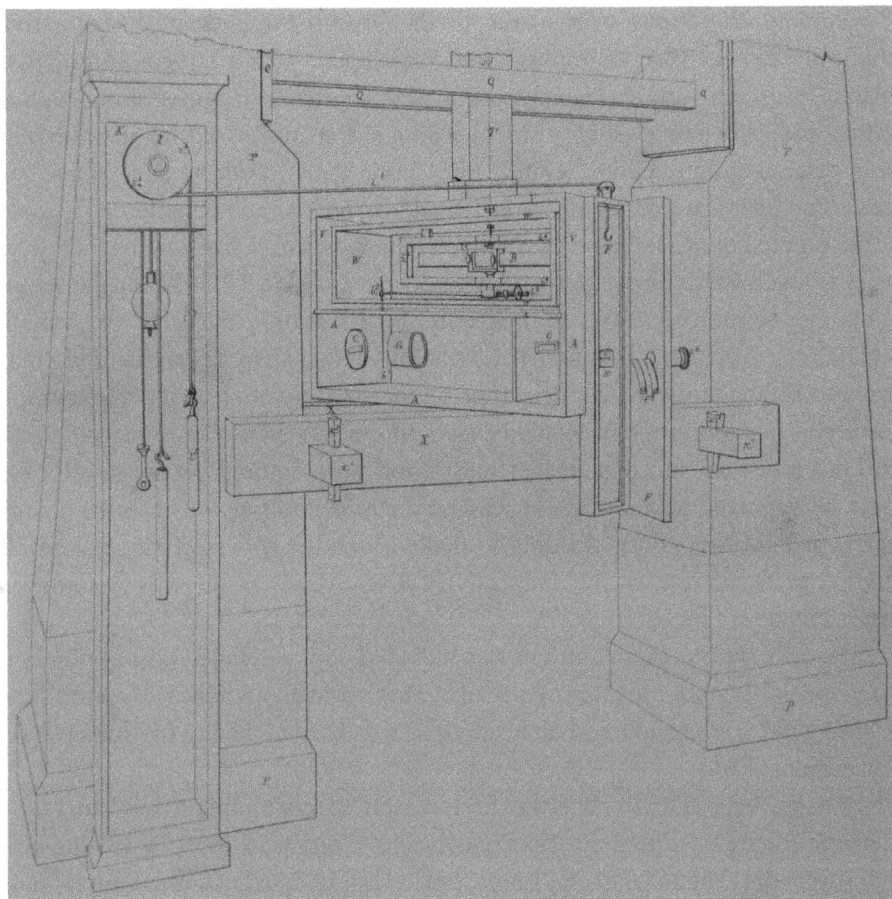

Fig. 15.19:   Declination magnetograph — model 1 (1846). Ref. R15.

Fig. 15.20:   Early negative from declination magnetograph (20 June 1846). Cambridge Archive RGO 6/676 405.

his first Royal Society grant. He was able to tell Sabine in March 1846: 'I get a very sharp & strong image'.[C165] He also sent numerous magnetograms to Airy for his review; a six and a half hour example survives in the Greenwich Archive (Fig. 15.20).

Ongoing alterations were made to the instrument until it was effectively rebuilt in 1851. In 1847 he replaced the Voigtländer with a £6 lens from Ross to enlarge the scale further. He adopted a new approach to registration, with the vertical indicator replaced by a shield with a slit, as discussed for the horizontal force magnetograph below. Perhaps most importantly, both he and Sabine were keen to compare the performance of smaller magnets, and Ronalds also wished to trial pointed ones. Sabine egged him on in late 1850: 'What size magnet do you mean to take? Will you <u>condescend</u> to six inches *[15 cm]*?'[C320] A smaller magnet would be more responsive to magnetic perturbations, but also to draughts, although the latter might be assisted by aerodynamic shaping. It required greater mechanical and optical magnification of its rotations and added photographic sensitivity to capture them. Ronalds used 30 cm and 38 cm long magnets in his later machines but felt that further innovation was required before smaller ones could be deployed accurately. He was very pleased that he was able to organise repairs to the observatory building to make it wind- and watertight, 'which rendered a great service to the magnet'.[C213] The overall result was a significant improvement in precision.

The declination magnetograph was included with a detailed description in the 1876 Special Loan Collection. Parts of it survive in the Science Museum today, including the original counterbalanced vertical index attached to the stirrup and its replacement shield.

Ronalds was actually not the first person to contemplate how to register magnetic forces — a method was conceived by Jordan in 1839.[A55] He suggested both suspending the magnet and supporting it on a knife edge to increase its stability for photography. An index extending longitudinally from the magnet generated a moving shadow on photographic paper placed immediately below. As with Jordan's barograph, there were no optical elements.

Brooke's first application of self-registration was for the declinometer. He produced photographs in early 1846 and his approach was deployed at Greenwich from 1848. The Greenwich magnets had, until then, been read by eye with the aid of either a mirror or lens attached to the magnet. Brooke replaced these with a concave mirror 13 cm in diameter that both reflected and focused the light from the lamp onto the photographic paper 3.8 m away.[A72,S4] The arrangement is seen in Fig. 15.3. The mirror had a complex elliptical shape with two quite different focal lengths to accommodate the positions of the lamp and the photograph. Repeatability of results was also dependent on accurate repositioning of the lamp when trimmed and refilled as its precise location affected the angle of reflection onto the paper. Numerous early examples of his curves have been published.[A72]

Ronalds disliked the mirror arrangement, believing it added weight to the magnet, caused loss of light, brought errors to the reading and offered less scope for fine-tuning the machine. It was for these reasons that he discarded his similar ideas from 1840–41.

15.7 Horizontal Force (or Bifilar) Magnetograph

The success of Ronalds' declination magnetograph prompted Sabine to request a horizontal force (HF) magnetograph for Lefroy at the Toronto Observatory. As suggested by the name, its purpose is to delineate variations in the horizontal component of geomagnetic field strength. The magnet is hung horizontally at right angles to magnetic north using a bifilar suspension of two silk threads; these create a torque when twisted that balances the magnet's tendency to rotate to the magnetic meridian. The varying angle of the magnet can be related to the changing magnetic force through calibration of the instrument's torsion stiffness. The configuration has clear similarities to the declinometer and altering the mode of suspension enables this adaptability.

Sabine believed that Ronalds would be able to develop an instrument capable of recording the large magnetic storm responses seen at Toronto while also giving information on small daily variations. He would then be able to extend his research on the periodic laws of magnetic disturbances, described in Sec. 14.4. Colleagues were doubtful. Airy was dismissive even of the need to monitor large disturbances, giving the opinion that it was the gradual (and more easily recorded) changes that were of interest. Lefroy shared his view.[C292] Geomagnetism expert Humphry Lloyd's advice was that 'the same instrument cannot be employed both for the regular diurnal changes, & the disturbances… for if it gives the former with sufficient precision, the ranges for the latter would be excessive.'[C232]

It was the perfect challenge for Ronalds — to take photo-recording to a new level of precision. The machine needed to be sufficiently stable in action that all minute movements were due to magnetic rather than other causes. The photographic surface must be sensitive enough to capture large, sudden movements and the image so sharp that small perturbations could also be measured accurately. It was at this time that he moved to the Daguerreotype method.

He naturally wished to ensure that his machine was scientifically 'perfect' and reasonably straightforward to use before sending it so far away as America. The inherent technical difficulties of a first-of-a-kind configuration with little experience to fall back upon were exacerbated by him not being able to trial the instrument adequately — the UK had somewhat smaller magnetic storms as well as different weather conditions for photographic processing. Another problem with the instrument being designed for extreme disturbances was that most of the time there was negligible activity to be seen:

In reference to the scale, I will just remark that I followed orders, and greatly to my dissatisfaction (as a showman) have been obliged to exhibit straight lines in lieu of curves, but pleased to find that I had executed my task satisfactorily to our excellent friends.[A149]

It is also interesting to recall from Sec. 13.3 that Sabine's tough request coincided with both the BAAS decision to close Kew and the award of the self-registration prize to Brooke. It all added up to a challenging situation for him.

With the machine being intended for long-term use and its stability being critical, he ensured that all key elements were linked together to form rigid frames. Brass was used extensively in the camera and magnet attachments; these were bolted to marble slabs that were in turn supported by massive masonry pillars on strong foundations. Wood was used only for covering and protecting the apparatus. Relative movements in the apparatus were thus minimised as were expansion effects under varying weather conditions.

Ronalds consulted Lloyd on the machine's form, who had designed the magnetometers (without photo-recording) used at the Colonies in the Magnetic Crusade. He chose not to follow much of the advice he received but, in a series of letters, Lloyd came to understand his preferences and acknowledge their value.

One suggestion that he did adopt with thanks was to substitute a slit shield for the counterbalanced brass index. Ronalds configured it as two very thin semicircular plates positioned to overlap slightly (Fig. 15.21 and also Fig. 15.22). The lower one was fixed in position and had a small vertical slit at its upper edge, creating a baseline on the image. The upper plate moved with the magnet and had a narrow vertical slit at its lower edge. It generated a wavy line that represented the changing force on the magnet.

He developed a simple Magnet Arc Amplitudes apparatus to determine the scale of the photograph. It employed two small glass rulers, each with a scale marked in 0.5 mm divisions. One was built into the sliding photograph frame, with the other placed at the position of the slitted diaphragm. The two scales were able to be compared through the microscope in the long case to determine the precise magnification generated by the lenses. The process also confirmed that there was negligible distortion of the scale across the breadth of the image. Together with the geometry of the slit shield arrangement, the angular rotation of the magnet corresponding to any ordinate of the magnetogram could then be computed.[R21,A83]

The overall instrument was complete in mid-1849. He also made a scale model of the machine to send to Toronto, with Sabine's assistant Younghusband noting of it: 'It is so nicely made that I could not fail to have a perfectly clear conception of the working of the whole apparatus.'[C258]

Ronalds was able to build a second very similar camera the next year — for use at Kew — with equipment from Woolwich. This allowed improvements to be

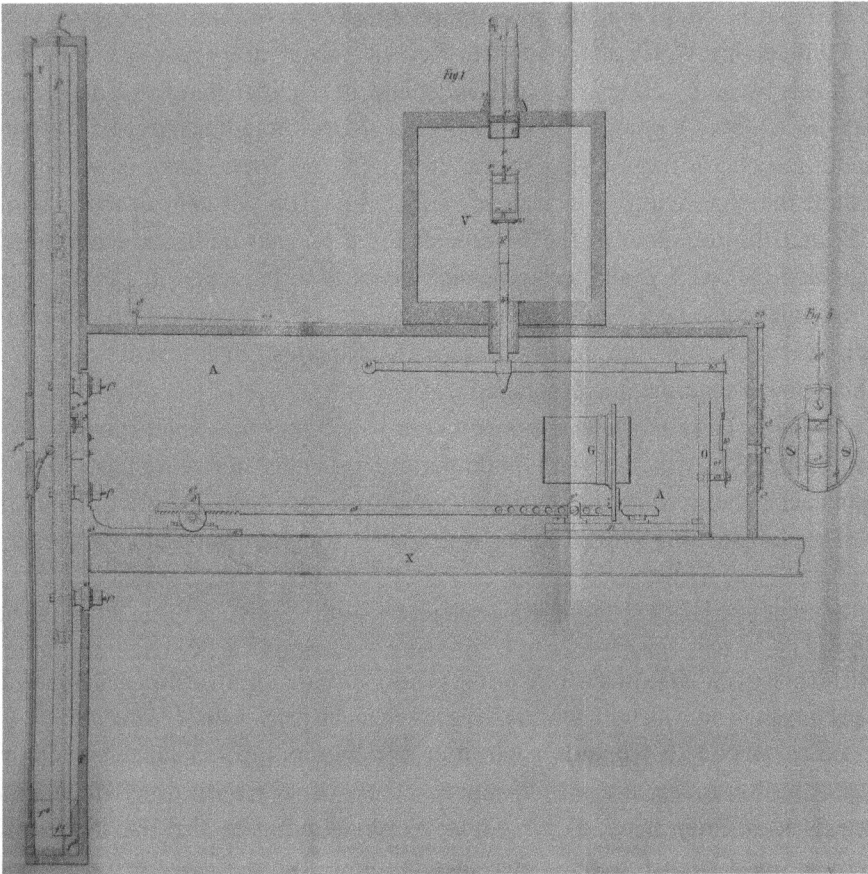

Fig. 15.21: Horizontal force magnetograph (1849); longitudinal section through camera box showing slit shield arrangement and rackwork to adjust position of lens tube. Ref. R19.

investigated, while also facilitating support for Lefroy and enabling comparison with the Toronto results. Interesting data survives regarding its performance. The optical magnification of the lens group was determined using the magnet arc amplitudes apparatus to be 3.46 times.[A83] The speed of the photographic plate and the usual width of the adjacent aperture were such that each point on the plate was exposed to the light for 75 seconds.[R21] Ronalds noted in the Kew diary that this was significantly less than half that of Brooke's instrument.[J84] It was practical considerations in the configuration of the machine as much as the photographic sensitivity that was the determinant — he had procured specimens with as little as 20 second exposure times.[R19]

With the aid of his ordinate board, he estimated that he could measure the separation between the edge of the curve and that of its baseline to an accuracy of 0.05 mm although, as he told Sabine, 'My sight is not strong'.[C302] Brooke had advised that he could read his photographic curves to 0.2 mm.[A72] A test was conducted involving Welsh and visitors to Kew like observatory manager John Broun and it was found by comparing their independent readings that Ronalds' precision could be achieved routinely.[J84,3] Lefroy concurred, although he advised: 'I find the minute inspection most trying to the eye sight'.[C310] For a curve extending over 8 cm of the plate width, the accuracy equated to 0.06 per cent of the maximum value and the rotations of the horizontal force magnet could be determined to 13 seconds of arc. Even higher precision might have been possible by increasing the magnifying power of the board's eyepiece: 'M[r] Wheatstone was of opinion after careful examination that the trace on M[r] Ronalds' plate would bear a higher optical power than the one employed'.[C298]

The polyflame lamp on the instrument was found to use less than a pint of olive oil per day, which cost £1.7s.0d per month. Other operating costs including chemicals, gelatine and wear of the silvered plates totalled 13 shillings a month.[J84]

The comparisons with Brooke's instruments were due in part to a request from Sabine:

> The point which I most desire to learn at present is the minimum reading that can be relied on in your apparatus & in D[r] Brooke's — This will depend not only on the definition of the edge of the trace on the plate or on the paper, but also on the general stability; a point which I know you have taken much pains with.[C299]

Lefroy wrote to Ronalds soon after commissioning his machine of 'every inflection…very insignificant movement…those of very short time which are frequently beautifully marked',[C310] while Welsh reported to the BAAS that 'The instrument has been found capable of recording, in a perfectly distinct manner, almost all the magnetic changes which occur, and with a delicacy of scale quite sufficient to represent even the most minute movement.'[A83] Ronalds' second-generation machine was a success. It continued in use at Toronto for many years, although the photographic registration apparatus was removed in 1860.[B82]

Kew's instrument was displayed in Paris in 1855[S6] and also the 1876 Special Loan Collection. It was again exhibited at the Science Museum in the 1980s but is currently in storage.

15.8 Vertical Force (or Balance) Magnetograph

Ronalds next turned his mind to the vertical force (VF) or balance magnetograph, which measures changes in the vertical component of the geomagnetic force. The magnet in this case is fitted with a steel knife-edge on each side that rests on horizontal agate to enable rotation in a vertical plane (Fig. 15.22). A screw is

---

[3]*Athenæum* 12/7/1851 748.

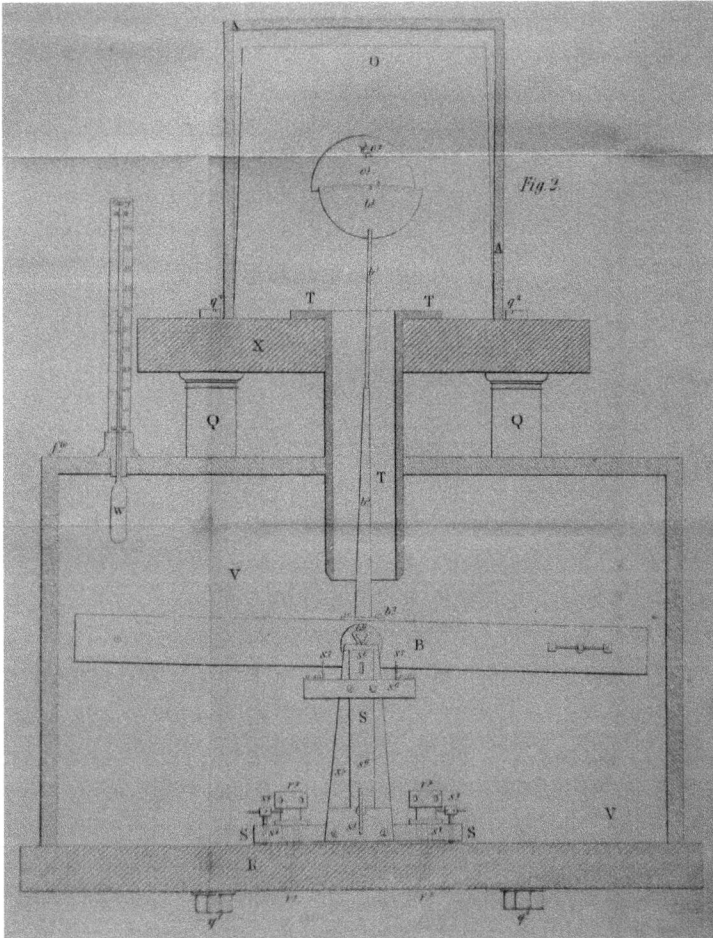

Fig. 15.22:   Vertical force magnetograph (1850). Ref. R20.

provided at each end of the magnet so that its centre of mass can be moved longitudinally and vertically. The mean magnetic couple is thus balanced by gravity while the sensitivity to disturbance can also be adjusted. He finished his first vertical force magnetograph for the Toronto Observatory in March 1850 and a second for Kew in early 1851. Both had 38 cm long magnets and their oscillations were magnified nearly four times by lenses especially made by Ross.

Sabine eagerly anticipated the suite of magnetic instruments finally being complete. He updated Herschel in February 1850:

> Lloyd whose judgement in such matters is very good greatly prefers Mr Ronalds' arrangements to those of Mr Brooke. I however await the reports from the Toronto Observatory where both will be tried together.[C278]

When Faraday visited Kew later that year, he suggested that Ronalds trial a more substantial copper damper for the horizontal force instrument, which he implemented immediately. Sabine again updated Herschel:

> M[r] Ronalds has one of his Magnetographs (the Bifilar) at work, and the other two will very shortly be ready also. He is naturally very anxious to shew them to you… We also, (the Committee for visiting the Kew Observatory), are extremely desirous that you should see the Instrument & its performance…Faraday was there yesterday, & was greatly pleased…Faraday gives the Bakerian Lecture at the R.S. on the 28 Nov[r] on his theory of the Diurnal magnetic variation.[C316]

Faraday in fact chose to talk on another topic that evening, the possible connection between electricity and gravity, and it is unlikely that he made mention of Kew or Ronalds' work.

The vertical force magnetograph did not live up to its promise. The support mechanism of the magnet was challenging to perfect — Lloyd's model did not work well[A172] and Airy wrote in 1863 of the 'fretting motion' and occasional 'dislocations'[J86] of the Greenwich instrument. The problem manifested itself at Kew as a gradual change in the mean orientation of the magnet. The knife-edges were found to rust slightly over time and the agate plates could move a little out of alignment.[R21] Ronalds returned the support system to the manufacturer Henry Barrow, who reshaped the knife edges and levelled the plates anew, but the benefits were only temporary. The Toronto machine also demonstrated problems and Lefroy brought the magnet to London when he visited in 1851. Barrow again repaired damage to the knife-edge and tightened a loose screw. On his return, Lefroy continued to conduct detailed trials with the instrument for several years but its readings were not published in the observatory's reports.

Airy suggested the possibility of supporting the magnet by 'springs'[C327] or slender steel strips rather than knife edges and asked Ronalds in 1851 to explore the practicality of the idea. Ronalds agreed that 'The Knife edge is an abomination'[C327] and set about developing an alternative. He sent a model of a magnetometer with his own configuration of spring suspension to Airy the next year for perusal. Airy replied in early 1853:

> I am utterly ashamed of the long delay on my part in acknowledging the receipt…of the spring suspension of Vertical Force Magnet…I wish much that you could set it up for trial with photography.[C337]

This did not happen as Ronalds was now beginning his exit from Kew.

Ronalds continued to ponder the problem and four years later floated another idea to Lloyd:

> If you were to suspend a magnet by a pair of wires (or threads) and to adjust its centre of gravity properly could you not obtain both the horizontal and vertical force? and could you not make it describe on a fixed photograph spherical surface, the kind of figure presenting nearly the daily resultant?[C368]

He acknowledged that he had received inspiration for the support concept from a short book by Professor Elie Wartmann entitled *Mémoire sur deux Balances a Réflexion* (1841),[B45] which described novel approaches to precision weighing. He might also have remembered a comment Sabine made in 1850:

> we want to obtain from the registered curves of H.F. and V.F., the equivalent variations of the Dip and Total Force, which must be <u>calculated</u> until some ingenious man like yourself shall have devised machinery to trace curves of dip & total force from those of H.F. and V.F.[C307]

When Ronalds arrived in Geneva some months later he befriended Wartmann, who offered to explore the concept.[J85]

15.9 Ronalds' Contributions to the Early Take-Up of the Instruments

In supplying his photo-recording machines to other observatories, Ronalds demonstrated the same level of customer service as he showed for his drawing instruments in Sec. 10.6 and his atmospheric electricity apparatus in Sec. 14.5.

Toronto

The first overseas market for photo-magnetographs was the Toronto Observatory. Director Lefroy summarised the episode in his autobiography:

> *[In mid-1846]* Sabine…wished me to take the opportunity of learning it *[photo-recording]* of Mr. Charles Brooke…It was altogether premature; Brooke had not half mastered it himself, and seemed to have little idea of instruction. It was not for more than a year after this that they got it to work at Greenwich Observatory…To make matters worse, there was a rival method developed by Mr. F. Ronalds…they made my life a burden for two years.[B114]

Letters written at the time have a slightly different tone, indicating that Lefroy valued the opportunity of gaining this experience. When Sabine suggested another post for him in 1850, he replied:

> No one would, I think, willingly bring into operation a system so full of promise and of interest as the new photographic methods of magnetical registration, to let another reap the fruits of it.[A149]

Sabine had suggested to both Ronalds and Brooke in 1846 that they might prepare photo-magnetic equipment for Toronto. He was equally supportive of both men, especially in the early years — his goal was good self-registering equipment for his observatories and he was keen to start an impartial trial of the two systems. As he later said to Herschel, perhaps a little tongue-in-cheek: 'I employ both at Toronto because Camphine is there obtainable, & because I want a good practical report as to their comparative working — To expect this for Greenwich would be to expect too much from an <u>Astronomical</u> observatory'.[C245] For Brooke, Sabine's request was straightforward — he simply sent his existing photographic apparatus for Lefroy to mount on the declination magnetometer there. Ronalds' stand-alone machine, having just become operational, was far from 'perfect'; he replied to Sabine's request in early April 1846: 'before I could esteem apparatus sufficiently exact for regular observation much expensive kind of work must be done'.[C168] The specification evolved over time as a result into the two much more specialised machines for observing large disturbances.

Once the horizontal force magnetograph was finally designed, Ronalds organised the various suppliers and manufacturers. Ross provided the lenses, Edward Dent the timepiece and Newman made the slit shields. Nicklin manufactured some of the brassware and smaller accessories and a local carpenter the wooden outer boxes. As is often the case, final completion and testing took longer than expected and Ronalds came under increasing pressure from Woolwich. Younghusband wrote kindly: 'I am sorry you are so hurried.'[C261] The magnetograph finally set sail on the *Pearl* in August 1849. In addition to the machine itself and its supporting slabs, the package included the model and drawings of it; scissors to shape the wicks of the lamps; around ten silvered Daguerreotype plates; all the processing equipment and chemicals; a full equipment list; detailed instructions for unpacking, erection and use including 'Precautions';[J75] and sample photographs. Cholera broke out on board and the captain died. The vessel pulled in to a Spanish port where it was plundered. It then set sail again, but turned back partway across the Atlantic because the St Lawrence waterway would soon be frozen. Once back at Plymouth, the authorities refused access to the cargo for fear of involvement in insurance claims. The boat would start off once more for Canada in the spring. Even if Ronalds' equipment had survived Spain, there was great fear that it would now be damaged severely from the effects of time, dampness and motion. Parts of the machine would be rusted or broken, chemicals would be spoiled and the Daguerreotype plates tarnished. It was a worrying period for him given all his efforts in perfecting the apparatus — and this would only have been exacerbated by Birt's behaviour at the observatory at this time (Sec. 13.3).

Fortunately the vertical force magnetograph had an uneventful journey on the *Lady Elgin*. Lefroy was able to advise Sabine of the arrival of both ships in Montreal in May 1850 and immediately sought out 'the Instruments about which I have heard from Mr. Ronalds, and thought so much, due to his anxiety'.[A149] Sabine updated Ronalds: 'I have been delighted to find that your instruments have arrived…with so little damage, which is most creditable to the packing'.[C299] Lefroy was keen that Ronalds come to Toronto to help in establishing the machines, but that did not come to pass.

Lefroy reported back regularly to Ronalds and Sabine however on his progress in mastering the instruments in a remote location. One of the early problems was: 'I have already had to dispossess some spiders'.[A149] It was later found that a film of moisture had formed between the lenses that prevented good photographs at night.[C313] He sent sample magnetograms and gelatine prints from an early period, on which Ronalds provided advice and ongoing encouragement with typical modesty: 'I may safely say that you have succeeded better than I did by far at first.'[A149] Lefroy was able before long to put the horizontal force magnetograph into full operation.

Madrid

In 1852, Ronalds became involved in the design and fabrication of meteorological and magnetic equipment for the Madrid Observatory. As well as his full atmospheric electricity apparatus, the Director Juan Chavarri had ordered magnetometers of the Greenwich configuration, but apparently without photo-registration. Ronalds preferred instruments more akin to Lloyd's and his own and stepped in to oversee the revised design, construction and set-up. He chose magnets of 30 cm length and consulted Lloyd on other aspects of the new arrangements. Unfortunately Barrow took advantage of the changing situation to inflate his manufacturing charges considerably, which caused difficulty for Chavarri and embarrassment for Ronalds. After detailed review and negotiation, Ronalds was able to bring the price back to close to the original guesstimate that had been made. The magnetometers arrived at Madrid in September 1852.

Oxford

A nice illustration of Ronalds' earnestness in promoting the adoption of his instruments is provided in his interactions with Johnson at Oxford. Johnson first visited him at Kew to seek his assistance on photo-registration in February 1852, the month before his mother died. He juggled his family commitments and Kew business with this exciting opportunity for much of the next two years, and he and Johnson continued to interact until at least 1856.

Ronalds agreed to supply a barograph and a thermo/hygrograph, together with an ordinate board. He first designed improvements to the barograph, after which he superintended its manufacture and assembly, including the structural alterations required at the Radcliffe Observatory and the provision of gas to the instruments. He additionally assumed financial management of the work — he had learnt his lesson with the apparatus for Madrid in not taking on technical supervision without also being informed of costs.

The first instrument-maker he chose, Edward Dent, fell terminally ill and there was no progress for some months. He moved key parts of the work to Newman, but he was very busy. Ronalds then admitted to Johnson in early 1853:

> On a former occasion I could fairly attribute delay in the progress of work…to other persons rather than myself. I must now confess that I am almost the sole delinquent, and will not waste your patience with excuses which I might perhaps urge, founded on private engagements, calamities &c.[C336]

Newman's contribution included the barometer and its temperature-compensation mechanism, for which he charged £20, with 70 per cent being labour.[J85] He also made the clock and the iron support frame for the machine. Ronalds sourced other components from a range of providers. He selected an achromatic lens group from Ross and a condensing lens from Dollond, 'after trial'.[J85] Holtzapffel fabricated a second sliding photograph frame and case, the first by another manufacturer having being 'badly made'.[J85] Ronalds also worked with various brass-founders, smiths, cabinet-makers, and masons in London and Oxford.

He had the usual share of issues in coordinating the suppliers. His diary noted in May 1853 that he had 'complained of <u>excessive</u> delay & <u>annoyance</u> on the part of Ross & Newman'.[J85] He also realised that this might be his last chance to create a barograph and he wanted it to be a model of precision. Meanwhile Johnson was starting to feel pressure from his trustees to see results. He tried to find a middle ground, writing to Ronalds in mid-1853:

> I know how much trouble the thing has given you and I know how anxious you are to have it as perfect as possible but I believe the best way will be to get it done… quickly…in such a way…that it will be effective and improve upon it by degrees.[C344]

Newman finally completed the compensation mechanism in July. Activity then centred at Oxford, where Ronalds spent more than two months.

With the thermo/hydrograph, Newman supplied the thermometers and the clock, and Dollond the lenses, but the machine was built largely by local craftsmen in Oxford.

Ronalds later tallied up his effort on the job from his diary entries. He had made eight journeys from London to Oxford in a 12-month period from October 1852 and had a total of 184 visits to the Radcliffe Observatory and to 'Artificers &c in London & Oxford'[J85] concerning the work. Of these, there were 77 trips to see Newman at either his Regent Street or Camden Town premises, 13 to lens suppliers, and 69 to various other craftsmen. He also spent 30 days working on drawings of the equipment and its set-up.[J85]

In Oxford, he interacted with various new and old friends, as well as the manufacturers. Johnson passed on their regards after he had left:

> I have had a great many enquiries after you from Professor *[John]* Phillips*[*]* & *[Charles]* Daubeny *[Professor of Chemistry]*[C348]

Only this morning I have had <u>tender</u> enquiries after you from *[local manufacturer]* Old Taylor[C357]

Ronalds' final departure from Oxford in late 1853 had been rather sudden, as it was from Kew (Sec. 13.5). In Paris he remained most anxious to hear about progress and, according to Johnson, sent 'much good advice'[C349] by return. Johnson forwarded his first photographic results from the barograph in November: 'Though <u>very</u> imperfect you will perceive that all is right as far as the Instrument is concerned and it only requires a little experience on the part of the Photographer.'[C349] He followed up a few days later: 'I have been much pleased with the distinctness of the boundary line of the Mercury. It looks almost as if one had drawn a fine line along but it is quite the <u>honest</u> work of the light.'[C350] Several of these early photographs are retained in the Ronalds Archive.

Johnson greatly appreciated Ronalds' generosity in establishing the machines for him. His 1854 annual report, where the results are first described, notes:

> These Instruments are of Mr. Ronald's design, and I must express my great obligation to that gentleman for the care with which, at a great sacrifice of time and personal convenience, he superintended not only their construction, but also their erection.[55]

Johnson admitted even two years later to Ronalds: 'I often want you to consult.'[C366] He was also naturally proud of his achievements: 'I hope some day you will come and look at your work again.'[C362] That almost certainly did not happen due to Ronalds' extended stay on the Continent.

<u>France</u>

We have seen that Ronalds' barograph and magnetograph and their photographic curves were exhibited at the *Exposition Universelle* in Paris in 1855. His instruments were positioned close to Brooke's and so the complaint he made in his autobiographical letter is certainly possible:

> The number (or Ticket) of my Barograph was exchanged for a number tending to shew that it was an invention of my rival Mᵣ Brooks, & curves, produced at the Radcliffe Observatory Oxford by my Instrument, were shown, by M Le Verrier, at the French Institut, & taken by many for Mᵣ Brooks's curves. With the aid of the Abbé Moigno this <u>Mis</u>-take was detected & rectified.[C375]

Ronalds' instruments created quite a stir, featuring in a heated debate at the *Académie des sciences* on how best to equip Colonial observatories. The story played out in Moigno's* *Cosmos*, starting in late 1855:

> His Excellency the Minister of War had requested the advice of the Academy concerning the establishment of meteorological observatories in Algeria...After three years of silence, the commission *[appointed by the Academy]* made its report...which...raised a very lively discussion...Speaking of instruments that record observations automatically using photography, M. Pouillet said they have so

far given promises, not actual observations that could be relied upon; M. Le Verrier protested earnestly against this assertion…M. Pouillet tried to defend his report with its system of hourly observations with three separate employees, but with no new or compelling argument. Field-Marshal Vaillant, Minister of War, echoed M. Le Verrier and stated the views of the commission were not acceptable to the administration… MM. Becquerel and Mathieu took a turn in the defence of the committee. M. Despretz shared M. Pouillet's opinion in believing photographic recording apparatus cannot give the actual air temperature, which enables us to say in passing that his objection may not be justified. M. Regnault…added his assertions to those of MM. Pouillet and Despretz…M. Le Verrier could barely keep his composure… Voices started popping up everywhere, asking that the discussion be adjourned to another meeting…The proposal, adopted unanimously, ended the debate. We will leave our reflections for another time.[A92]

The nub of the debate was also picked up in an article 'The Weather and its Prognostics' in the 1856 *North British Review*:

M. Le Verrier maintained that the report of the Commission failed in its object…he referred to the observations made at Kew and Oxford with Mr. Ronalds' instruments as a formal refutation of it. Field-Marshal Count Vaillant took the same view of the subject as M. Le Verrier…Prince Charles Bonaparte…spoke in high terms of the automatic system of registration [A98]

Ronalds' work was presented formally at the *Académie des sciences* in April 1856 by physics professor César-Mansuète Despretz.[A95] Detailed and illustrated descriptions of the barograph and thermograph, as well as Ronalds' recommended processes for preparing Daguerreotype and Talbotype photographs, appeared in *Cosmos* the next month.

A final mention in *Cosmos* at this time was in relation to a pendulum, described by instrument-maker Ignazio Porro:

The elements of the recording device to be applied here follow M. Ronalds' recording instruments for meteorology and magnetism. Components key to all recording instruments of this kind needed in science today, including the electrical and mechanical devices for more sophisticated astronomical observations, can be seen at present in the workshops of the Institute technomatique, with the consent of the famous author.[A96]

The *Institut technomatique et Optique* was Porro's business premises.

One of the reasons Ronalds was so eager to share his methods with French instrument-makers and scientists was that he had learnt that le Verrier's request to Newman to build a barograph of his design for the Paris Observatory had been declined.[J85] It is not clear how many photo-recording instruments were built by which manufacturers in this period. Renowned optician and photographer Jules Duboscq was recommended to build the machine for the Paris Observatory. By late in the century French observatories at Lyon, Nantes, and Perpignan also had photo-recording machines in action that were not supplied by Kew.[K11]

Interest was expressed to Ronalds from two different quarters in Montpellier. Dr Charles Frédéric Martins, a professor at the University of Montpellier and author of the *Annuaire Météorologique de la France*, had visited Kew in 1850 and 'was much interested with the Photo Barometrograph'.[C302] When Ronalds stayed in the city seven years later, he was taken by Professor de Mouchy to inspect the old observatory with a view to determining its suitability for geomagnetic studies. Ronalds offered advice and introduced him by letter to Lloyd. The instruments could not be afforded however at that time. He wished to kick-start the observations by arranging for the Kew Committee to present his old equipment to the Montpellier Academy, but realised after his experience regarding Lefroy and his electrical apparatus (Sec. 13.4) that it would not do so.[C368]

Other Opportunities

Ronalds received enquiries from numerous other observatories around the world at different times. Edward Fergusson, the incoming head of the Bombay Observatory, spent time at both Greenwich and Kew in 1851 'with a view to be able to judge of the two modes of photographic registry; & in the expectation that the Kew method will be better suited to India than that followed at Greenwich'.[C323] Emilio Faà di Bruno, later an admiral in the Kingdom of Sardinia, also visited Kew in 1851 on behalf of the *Accademia delle Scienze di Torino*. He went on to build a simple thermograph and barograph based on Ronalds' papers and with assistance from Ronalds himself. Di Bruno ended one letter to him: 'Be so kind to write me as oft you can; but use light paper *[to save postage costs]*.'[C341]

The Professor of Natural Philosophy and Astronomy at Dartmouth College in New Hampshire, Ira Young, inspected the barograph in 1853; he was visiting the UK and Europe to purchase equipment to outfit a new observatory at the College.[J85] Ronalds directed him to Newman. A final surviving expression of interest was for the Brera Observatory in 1861 when Ronalds met Luigi Magrini, Professor of Physics at the University of Milan.[J85]

15.10 The Prize for Self-registering Instruments and the Competition between Greenwich and Kew

The key observatory that would not accept Ronalds' assistance with photo-recording instruments was Greenwich. In fact, for much of the time he was at Kew, he was in the very uncomfortable position of being in a competition with Brooke and Airy. The rivalry became a central element in the early history of the machines — especially for Ronalds. It was instigated when a £500 prize was offered to encourage their development, which Airy embraced as a chance for him to show the superiority of Greenwich over Kew.

Another of Airy's various foes was Charles Babbage FRS, pioneer of the computer. Doron Swade's 2003 PhD thesis on their controversy described Airy as 'an influential behind-the-scenes advisor who allowed a personal grudge to bias his counsel to government'.[B186] The same words might summarise Ronalds' experience regarding the self-registering prize, although Airy's rancour was not really against Ronalds personally but focused on the observatory he was running. He also looked down on practitioners like Ronalds and Sabine in comparison with Cambridge mathematicians like himself and Herschel,[K18,A169] and he might have remembered that Ronalds had assisted Babbage's close friend Sir James South with his large equatorial mount, as described in Sec. 12.10.

We have seen that Ronalds and Airy started interacting on self-registration matters in 1843. The enthusiasm with which Ronalds aided Airy is exemplified by his sending his first photographic curve to Airy in April 1844, direct from Collen's studio. It was 18 months after this that the prize was established. The BAAS Council described a few years later what had happened:

> the subject of self-registering instruments was discussed at the meeting of the British Association at Cambridge, in 1845, upon the application for a grant of money...to enable Mr. Ronalds to complete an apparatus for that purpose at Kew...a recommendation was made on that occasion by the Association to Government...of the expediency of encouraging, by specific pecuniary rewards, the improvement of self-recording magnetical and meteorological apparatus.[S1]

Ronalds also recorded this conversation in his 1845–46 Journal. The BAAS saw that the vital importance of photographic recording across observational science should enable them to attract alternative funding sources to the challenge.

Airy gave a slightly different perspective in his autobiography:

> the transactions in this year were most important...for the subsequent introduction of self-registering systems, in which I had so large a share...I had remarked the distress which the continuous two-hourly observations through the night produced to my Assistants...I therefore proposed 'That it is highly desirable to encourage by specific pecuniary reward the improvement of self-recording magnetical and meteorological apparatus...' which was adopted. In October the Admiralty expressed their willingness to grant a reward up to £500. Mr Charles Brooke had written to me proposing a plan on Sept. 23rd, and he sent me his first register on Nov. 24th.[B115]

It is interesting that Airy was so precise in these dates in an otherwise high-level book. He also carefully preserved all his correspondence on the matter in a labelled file — he was proud of what he achieved. Other relevant documents that enable the story to be pieced together are at the Royal Society and in the Ronalds Archive.

The Admiralty advised Airy on 23 October that they would provide the prize but apparently asked that their role not be publicised. Airy immediately advised Brooke of the Admiralty's decision,[C149] and also Herschel in his role as BAAS

President. Ronalds had outlined his photo-electrograph to him back in April, yet Airy's next letter to him a week later made no mention of the prize. Indeed he requested the old apparatus developed in 1815 to record atmospheric electricity: 'I wish to ask if you could lend me for a short time your self registering apparatus with the chronometer movement and the sealing-wax plates, and a few plates in a state fit for use: that I might try it here.'[C150]

Ronalds explained his new equipment using the Volta straw electrometer in detail in his reply and advised that he had now also completed the photographic thermograph and barograph:

it may at length be safely said that the promise which I made to you, at Greenwich… is…in a sure train of fulfilment. That promise might be stated in your own terms viz "The arrangement of an efficient self registering apparatus, preserving its records in a permanent form…"…

Now, my dear Sir, what is to be done? This photographic method is really…much superior to the oldest method (of the resinous plates &c)[C151]

Airy dismissed the camera concept and reiterated his request for the resinous plates. Ronalds expressed his 'very great disappointment' but advised: 'the resinous plates shall be ready.'[C153]

It was at this point, early November 1845, that Airy introduced Glaisher to Ronalds and stepped back from the Kew relationship himself:

M[r] Glaisher will attend you at the Kew Observatory on Monday morning…He is well acquainted with the subject generally, and perfectly acquainted with our wants and capabilities, and I shall be much obliged by you giving him the fullest information in your power on this matter of self-registration.[C154]

Glaisher provided a detailed written report to Airy on his return, advising Ronalds' photography to be 'by far the best method'[C155] of self-recording. He added:

I feel it but right to say that every facility was given me to make myself acquainted with all M[r] Ronalds is doing and he has done, in the most generous manner possible and I am sure the service will benefit much from many of his valuable remarks.[C157]

Ronalds' next visit to Greenwich a few weeks later was abortive. Airy refused to see him even though Glaisher and others were not at their posts, explaining: 'interruptions…fall on me personally with extreme oppression'.[C159] Ronalds expressed his regret at the intrusion and went on in his reply to explain the intended purposes of his visit:

I wished, in the first place, to report progress on the photographic registration of electrometers exhibiting the kind in addition to the tension of their charges; my experiment having been undertaken in pursuance of your suggestion, or desideratum…

The second object of my visit was to petition for permission to examine, more minutely, your magnetic apparatus with a view to experiments for registration of a like kind upon some magnets at Kew.[C160]

The letter included a description of the latest form of his photo-electrograph that showed the sign as well as the magnitude of the electric charge, with drawings and a sample electrogram. Perhaps Airy felt a little guilty because he apologised: 'I am sorry that in haste I expressed myself so ill in my last letter'.[C161] Feelings of guilt or not, he made no acknowledgement of the instrument and its results, nor Ronalds' request. He had just seen Brooke's first photographic result. Glaisher did not seem to understand the game-plan however because he had already told Ronalds on seeing his photographs: 'I feel certain as you do that ultimately we must adopt the photographic process — but, it must be when you have surmounted the difficulties and we must adopt Ronalds' Process'.[C158]

Ronalds himself had no idea about Brooke's rival plan. Airy outlined in his autobiography some of what was going on:

> I was constantly engaged with Mr Charles Brooke in the preparation and mounting of the self-registering instruments, and the chemical arrangements for their use, to the end of the year. With Mr Ronalds I was similarly engaged: but I had the greatest difficulty in transacting business with him, from his unpractical habits.[B115]

The word Airy chose to describe Ronalds is an interesting one given that this book is full of his practical inventions. One of his justifications can perhaps be understood in terms of the very different modes of operation of the two men. Airy's son wrote of his father: 'The ruling feature of his character was undoubtedly Order'.[B115] Ronalds in contrast was typical of many highly creative people in being rather disorganised and forgetful about little details. He was also inclined to write as he thought whereas Airy's letters were very succinct and to the point. Airy later gave other explanations of the word's meaning, as described below. Regarding the substance of Airy's remark, there is little evidence of him interacting with Ronalds in 1846 but certainly Glaisher remained very supportive of Ronalds' photographic efforts. Again, Airy explained three years later how he distanced himself from Ronalds at this time. Ronalds continued to try to update him, particularly with progress on his declination magnetograph, but replies were increasingly delayed and evasive.

Brooke submitted his first paper, describing his declination magnetograph, to the Royal Society in June 1846. There must have been an exchange with Glaisher because Ronalds wrote to him: 'I am very much obliged by your very satisfactory letter...which shall be preserved as a flattering testimonial of your wish to do equal justice'.[C171] He followed up by providing photos to Greenwich to enable comparison with the declination magnet measurements taken there. Glaisher was impressed:

> whether your magnet moves as ours moves or not is quite immaterial, your purpose is to be able to register all the motions of your magnet...The beautiful agreement in your results with those at Greenwich is nevertheless highly satisfactory...As I conceive you have overcome all the difficulties in the self-registration of the different positions of magnets[C174]

Airy also admitted: 'Your images are pretty sharp.'[C170]

Brooke and Ronalds both presented their work in Wheatstone's Session at the Southampton BAAS conference in September 1846, as described in Sec. 13.3. Ronalds wrote a despondent letter to Glaisher after the event, to which Glaisher sent a motivating and helpful response:

> You know with what reverence I feel for the discovery, and the adaptation of Photography to the self-registration of so variable yet so weak a force as is Magnetism. Your Photographs are certainly the best I have seen — they more closely represent a motion corresponding in all its particulars with our obs^d motions, than Mr Brooke does — they are more sharply defined — &c &c. The better definition is the important part...At the Obs^y the best will be chosen — here will be a fair field and no favour...and whatever may be our decision after trial, will carry weight with it. Now here you stand well — your photographs are the best — and having done so much, leave not a stone unturned to complete that so well begun by having your apparatus fixed here...I expect Mr Airy home the latter end of this week...Let your photographs be here — send them to Mr Airy for inspection while he has those of Mr B. so that he may compare, and decide himself...I can tell you this you stand well with him. One great part in your favour is — that you are a better Mechanician than Mr Brooke — but Mr B. is indefatigable.[C184]

Ronalds did as Glaisher suggested. He wrote to Airy immediately:

> I now undertake to say under a feeling of great (I hope not arrogant) confidence that if my...apparatus for photographic registration were properly applied to your magnet as now suspended at Greenwich...it would produce Photographs more perfect as to sharpness &c than any of my electro-photographs...and I think that I can now as confidently assure you of my ability to place Mr Glaisher...in a position to procure the same thing. One of my motives for these few statements is sufficiently obvious to you, I dare say, but let me assure you that mercenary motives are not, & never have been, my only ones, in these affairs.[C185]

He delivered the letter and a book of photographs to Greenwich the next evening. Glaisher passed them to Airy with a cover note suggesting it was pure accident that he was there to receive the parcel. He added: 'He is anxious to have his method tried here.'[C186]

Airy had an excuse for not getting back to Ronalds regarding the comparison of the two sets of photographic records. Ronalds had quoted Glaisher's phrase about the 'beautiful agreement' of the Kew and Greenwich results in his presentation at Southampton and it was published as part of a summary of the paper in the *Athenæum*[4] and the *Literary Gazette*. Airy wrote instead:

> It is the rule in all establishments public and private that nothing issues without the express sanction of the head of the establishment — in this case, of myself...It is absolutely necessary to maintain this rule. Therefore in future would you apply to me

---

[4]*Athenæum* 26/9/1846 1000-1.

for the sanction for publication…in order that I may have the matter properly looked to, and that such a statement may be sent to you as I can properly subscribe.[C187]

Meanwhile, helpful Sabine nudged Ronalds to publish his work as Brooke had done. Ronalds was reticent — his apparatus was not yet perfect:

Let me reply to your last question Viz Should I not present some account of my apparatus to the.Royal Society in return for their donation of £50? It is one which has often been put by myself & the answer has been "Is it yet worthy the attention of the R.S?"[C177]

Pragmatic and opportunistic Sabine replied:

I think there can be no doubt that it would be both right & politic to make a provisional report to the R.S. of what is in progress by the aid of their donation… With such a report I think we might endeavour to obtain another £50 this next year. Without it, it could not be attempted.[C182]

Ronalds submitted his paper to the Royal Society on 18 November 1846 and read it to the meeting on 21 January 1847; the interval allowed review and was entirely typical. Brooke submitted a supplement to his paper five days after Ronalds but it was presented by Airy just three days later on the 26 November. Airy had presented Brooke's first paper only two days after it was received. Speed and precedence seemed to be important for Greenwich. All three were published in the same volume of the *Philosophical Transactions*. Brooke's follow-up paper described his barograph and thermograph while Ronalds' paper covered all four of his instruments, including the electrograph. Charles Weld made mention of the 'two very remarkable Papers'[B57] in his *History of the Royal Society*.

After these events, Ronalds altered his relationship with Greenwich — the earlier friendliness was gone. He did resolve to try to maintain a business link in spite of Airy's evasiveness and criticisms. He genuinely wanted the observatory to have the best equipment and continued to strive to have his approaches trialled there, but it was not to be. He also now appreciated Airy's powerful position in the photo-registration competition and had finally developed something of competitive streak. He could not have realised how much the odds were against him. Glaisher was quite upset at this new attitude: 'not having heard at all from you lately makes me fearful that you are not well…I write this to ask you how you are and to tell you how glad I should be to hear from you again.'[C194]

In 1847, when Ronalds was again in need of development funds, he broached the matter of the prize more overtly in another letter to Airy:

I have also <u>long</u> and before any other person been <u>successfully</u> employed in procuring efficient & ~~perhaps~~ the best self registering system for Meteorolog[1] & <u>Magnetic</u> Instruments; in which pursuit you have lent me some valuable assistance, and I trust that you will not now refuse me a due share of Encouragement in this not unimportant object. Whatever sum of <u>money</u> I may receive shall be spent upon it (and an account kept). I do hope therefore that I shall not be compelled to encroach upon a small patrimony *[his inheritance]* more than I already have; and which I cannot afford to do with common prudence…Excusez *[excuse me]*[C193]

Airy fully appreciated the purport of the letter: he filed it with the note: 'Letter from M[r] Ronalds, and answer…about reward for self-registering Mag[l] and Meteor[l] Apparatus.' For a man of pithy communication, his reply (when it finally came nearly three months later) was remarkably abstruse:

> I ought long ago to have thanked you for the communication…But there were some remarks which I could not answer (and cannot now) relating to money matters…I have then merely to say that I have not at present and am scarcely likely to have any share in money managements except those which relate directly to the Observatory [C196]

Airy had already written to Brooke regarding how he would apply for the prize for him.[C192]

Airy submitted a detailed application on Brooke's behalf in August 1848, immediately after the decision had been made at the BAAS conference to close Kew (Sec. 13.3). Ronalds was mentioned at the end, but simply to discount any claim he might make:

> A correspondence on the same subject was commenced with M[r] Francis Ronalds of Chiswick, but it came to a close without any formal termination*[?]*, and the mere circumstance of the distance of M[r] Ronalds' residence would probably had made it impossible for him to mount apparatus at Greenwich. So far as the interests of the Royal Observatory are concerned, there is no competition with M[r] Brooke.[C221]

(The distance from Chiswick had not concerned Airy when Ronalds was helping to install atmospheric electricity apparatus at Greenwich in the period 1843–46, as described in Sec. 14.5.) Airy recorded his success in his autobiography:

> Being satisfied with the general efficiency of the system arranged by Mr Brooke for our photographic records (of magnetical observations) I wrote to the Admiralty in his favour, and on Aug. 25th the Admiralty ordered the payment of £500 to him.[B115]

How many Lords does it take to change a prize? In this case it was four. In addition to the First Lord of the Admiralty (the Earl of Auckland), the Marquess of Northampton became heavily involved in his roles as BAAS President and immediate Past-president of the Royal Society, as well as the incoming President Lord Rosse and the Prime Minister Lord John Russell. Admiral Sir Francis Beaufort FRS (who created the Beaufort scale for wind force and was responsible administratively for Greenwich) also offered his assistance. The prime-movers however were Sabine and Herschel.

The process started when Sabine heard what had transpired. He was in Ireland but quickly informed Ronalds and Herschel.[C226] To Ronalds he first broke the news, before suggesting what Ronalds might do: 'do you not mean to put in your claim for reward also?…Were I in your case, I incline to think that I should at once address a memorial to Lord Auckland', and then providing some ideas as to what a submission might contain. He finished: 'I beg that you will have the kindness[?] to consider this letter as private.'[C225]

Herschel soon contacted Airy:

I cannot help thinking it a pity that Ronald who has worked hard and so far as I have been able to judge successfully on his part should not have come in for a share of the reward. Do you think it likely that the Adm^y would now attend to his application?[C228]

He received the following reply:

I have no doubt that they [the Admiralty] would be extremely unwilling to make another grant [one can sense Airy's glee]…

The footing upon which long ago I put the grounds of claim to Brooke was that it must be for services in the way of self-registration rendered to the Observatory…I had at first some correspondence with Ronalds, but we could not get on together — I mean that his distance from Greenwich and other inconveniences made it impossible for us to work together, and the matter quietly dropped.[C229]

So that explains the evasive behaviour.

Herschel began to suspect underhand behaviour and responded to Airy:

Now as I gather…that Mr Ronalds researches were mentioned by you to the Admiralty…And as…you offer me copies of your correspondence with the Admiralty on the subject I should be very glad to have them…I may place them in Lord N's hands he having called on me…to join with him…in representing Mr Ronalds claims to the Admiralty which I feel fully disposed to do[C234]

He also addressed Airy's rather territorial attitude towards the award:

There is nothing that I am aware of officially before the B.A. to connect either yourself or the Kew[?] observatory with the matter.[C236]

Airy retorted in a letter headed 'Confidential':

I stipulate absolutely that they [his submission package] shall not be printed…

When you come to look at the matter, you will find it necessary to understand the practical view which I take of it. I never contemplated assisting to reward an abstract invention à la [Charles] Babbagienne which was not made really serviceable to us: I believe it would have been difficult in any case for Ronalds to have set up instruments actually at work here, but when Brooke had applied first and had the instruments to a certain degree put into his hands it became almost impossible to trial[?] with another person.[C237]

Airy continued to try and justify his actions for many years, although some of what he said was less than truthful. Sabine later relayed to Ronalds the argument Airy put forward at a Royal Society meeting in 1850:

it came out that Mr Airy was not aware of there having been any definite proposal for a self-registering magnetometer before he brought forward his proposition for a Government reward at Cambridge. I told him it was my belief (speaking only at the moment from recollection), that you had been engaged upon your plan some time before that…Mr Airy said he was quite sure Mr Brooke had not thought of it before the proposal of the grant, to which I assented, but said you had.[C290]

In 1863, Airy described events to his Board of Visitors as follows:

the transaction at Cambridge had excited two able men to turn their thoughts to the construction of photographic self-registering instruments. Mr. Charles Brooke wrote to me on 1845, September 23, and Mr. Francis Ronalds on 1846, February 16. Much correspondence followed, with many trials, chemical and mechanical; till on 1846, November 11, I determined on adopting Mr. Brooke's form.[J86]

The correspondence dated 16 February 1846 was Ronalds' request for assistance in obtaining a skein of silk to suspend the magnet Glaisher had loaned him, but Airy is forgetting the various letters and photographic specimens Ronalds sent in 1844 and 1845.

As late as 1871, Airy advised the Royal Society that 'Self-registering instruments were introduced by me… ten years later, the Kew Committee established self-registration at Kew'.[B100]

Returning to 1848, Ronalds followed Sabine's suggestion and prepared an eight page 'Memorial'[J73] explaining the timeline of development of his instruments. Sabine commented: 'I think your memorial very well drawn up, and very much to the purpose — I have made in pencil all the suggestions that occurred to me — they are however very trifling compared with the whole body of the memorial which I could not improve.'[C230] The Admiralty did not agree, and simply suggested by reply that Ronalds approach the Royal Society.

He was next advised to send the document to Lord Northampton, who was at a loss what to do. Sabine and Herschel continued to plot a way through the problem, in Herschel's words: 'to facilitate & smooth the Adm^y's rectification of the uncomfortable state into which (doubtless from want of due consideration) they have suffered this matter to get'.[C240] They then synchronised letters to Lord Northampton with their proposed approach. Herschel's started:

I was very sorry to hear that the Admiralty had granted the whole sum of £500 to Mr Brooke…without any reference having been made to the Societies with whom the recommendation originated[C235]

while Sabine continued:

As the case stands at present a substantial injustice has been done to Mr Ronalds, which cannot fail to be much canvassed; and a very unfavourable impression will be produced if his claims should be denied a fair consideration[C233]

Sabine continued to update Ronalds on what was happening over the ensuing months:

I took Lord N's & Sir J. H's letters to the Admiralty yesterday. It was fully admitted that they had acted hastily & inconsiderately upon an ex poste statement & an interview was recommended between Lords Northampton & Auckland, as the readiest & smoothest path towards a satisfactory settlement.[C238]

Northampton and Herschel next sent Ronalds' memorial to Auckland. Sabine gave him a copy of their detailed cover letter, noting:

> One thing you may be sure of this time that it will not be replied to quite so unceremoniously as last time...the Adm^y can hardly help for their own sakes retracing their steps as far as circumstances will now enable them to do so.[C241]

A little later:

> The Adm^y have replied that having given the whole £500 to Mr B. under the advice of the Ast^r Royal, they have no more money. Lord N...expressed his belief that the matter could not rest so — & suggested that in his opinion the next move should be a communication into Lord J. Russell [C242]

Sabine also endeavoured to keep Ronalds' spirits up:

> I am sure it will give you pleasure to know that your collection of photographic traces was taken to the B^d of Visitors of the Royal Observatory *[Greenwich]* on Friday last by Sir J. Herschel...and that they were extremely admired.[C243]

Sabine's empathy is the more remarkable given that he was very ill at this time.

Northampton, Rosse and Herschel then planned a visit to Prime Minister Russell. Herschel and Sabine reasoned that £250 would be better than nothing, the former even following Airy's lead in finding a plausible justification:

> it appears to me a handsome enough recognition when accompanied with an understanding that the two parties stand on equal footing as regards merit of invention and that the difference in the sums received refers to work specially done for Greenwich by one of them.[C240]

They knew that it was not the money itself that was of importance to Ronalds, but recognition of the originality and validity of his ideas. Any award he would receive would be spent on equipment for Kew.

It was late April 1849 before Northampton was able to advise Sabine: 'I have received a letter from Ld Russell promising the Two Hundred & Fifty Pounds to Mr Ronalds, which I am sure you will be very glad to hear.'[C251] When Sabine thanked Herschel for his efforts, the latter replied:

> I am sufficiently rewarded for any little trouble I have had in the matter by the idea of having aided in serving so meritorious a person as Mr R. from suffering injustice and putting right a matter which might otherwise have been regarded as mismanaged.[C253]

He was more politically correct in reply to Ronalds' note of appreciation:

> I will hope the handsome conduct of Gov^t in the affair will produce a good effect in shewing that Scientific invention is duly appreciated at head quarters.
> Allow me also to hope that you will continue to believe me
>> With the highest esteem
>> Very faithfully yours [C256]

The episode ended with the formal announcement of Ronalds' and Brooke's awards at the next BAAS conference.

Brooke retained firm belief in his achievements. He wrote to Sabine at this time:

> I dare say you will look upon a paternacious adherence to the pre-eminence of one's own child, as a pardonable at least if not an amiable weakness...
>
> I am happy to say that I have completely succeeded, & shall be happy...to shew you some Photographs equal in distinctness, & far superior in cleanliness of the paper, to any you have yet seen [C259]

15.11 Evolution to Continuous Geomagnetic Recording at Kew Observatory

It was after these events that Sabine became more proactive in Kew affairs. The main motive for focusing more of his busy day in that direction was that he was losing his Colonial observatories — funding of coordinated magnetic observations had ceased and Toronto, the most advanced operation, would soon as be handed over to the Provincial Government.[A147] Kew could become his new observatory. He also had all the competitive spirit that Ronalds lacked, and was now relishing a fight for supremacy between Kew and Greenwich.

Sabine soon conspired a way for Kew to start regular registration of magnetic phenomena. The first step was to conduct a long-term trial with Ronalds' three new machines. He wrote in October 1850:

> What do you say to the following suggestion: — I will suppose that by the beginning of the month of December or thereabouts, you will have the three Magnetographs in working condition. Why should you not then ask the R.S. to give you...a sufficient sum to work those three Inst[s] for one twelvemonth; the records...to be presented at the close of the year to the R.S., with such comments on the working as you may think proper to add [C312]

The idea was of great interest to Ronalds in testing the value of his instruments and perhaps at last being able to unite geomagnetic and atmospheric electricity measurement in a rigorous way. He was hoping however that beforehand he could finish rebuilding his first instrument (the declination magnetograph) and improve his horizontal and vertical force magnetographs on the basis of experience. Sabine had no time to waste — it was a typical example of Ronalds' quest for scientific perfection versus Sabine's shrewd political positioning. A compromise was reached. Sabine even scoped out the form of the application to the Royal Society to speed up the process.[C321]

The trial was conducted from 1 April to 1 October 1851. Ronalds' assistants Welsh and Nicklin conducted his full Daguerreotype photography process, including the tabulation and gelatine copies, and Welsh provided a formal report on the equipment's performance to him, which was published in the BAAS proceedings.[A83]

Sabine knew that the trial was sailing close to the wind in terms of the agreed segregation of roles between Kew and Greenwich (Chapter 13), and furthermore Airy was the incoming BAAS President, so he came out guns blazing at the next conference. His speech detailing numerous advantages of Ronalds' machines and processes was printed in the proceedings and in the *Athenæum*. The latter also relayed Airy's response:

> The ASTRONOMER ROYAL said that, while he admitted that at Kew the photo-graphic method of registering these or similar observations was first introduced successfully, and that the methods there used were essentially the same as those of Mr. Brook, which had been adopted at the Royal Observatory, Greenwich, he could not help thinking that there was one important distinction. In Mr. Brook's method, a sheet of photographic paper…recorded the working of the instrument; and these original records, carefully dated, were kept…he had strong doubts that any person could transfer by hand all the minute fluctuations of the instrument[3]

In fact Brooke's curves were traced in ink so they would survive fading of the photographic image. Nonetheless it was a nice admission by Airy of the prece-dence of Ronalds' work over Brooke's — if indeed it was reported correctly.

The trial was successful, with the BAAS Council reporting the next year that 'Mr. Ronalds' adaptation of photography to record the magnetic variations is an effective and practically useful invention, supplying to those who may desire it the means of making and preserving a continuous registry of the phænomena.'[S1]

Ronalds began to reduce his commitment to Kew at this time, and Sabine and the Kew Committee focused their attention on other scientific goals. It was 1855 when Sabine asked Welsh to conduct a second trial with the three magnetographs, supported by a £250 Royal Society grant. As reported at the 1857 BAAS conference:

> Since the time when the magnetic self-recording instruments…were constructed under the direction of Mr. Ronalds, very considerable improvements have been made in the art of Photography, and the six months' trial…has led in several other respects to suggestions for improvements which could not but be expected…in instruments of so novel a kind, while at the same time the…trial…has placed beyond doubt the sufficiency of…self-recording instruments, for the examination of the solar mag-netic variations.[S6]

This generous recognition of Ronalds' prior work was typical of Sabine — and it also helped argue Kew's superiority over Greenwich. In fact the new magnetographs designed by Welsh in 1857 were influenced by both Ronalds' and Brooke's earlier models as well as subsequent developments. Perhaps the most obvious change was the scale, with the magnets now being just 13 cm long.[A103] Like Ronalds' earlier machines, the instruments were supported on stone pillars rather than the wooden tripods used at Greenwich.[S4] Gas was the light source, having been installed for the purpose. The confidence gained in using waxed-paper in Ronalds' machines at Oxford negated the need for flexibility in photography method and so a drum could be inserted with confidence. Ronalds' two semicircular slitted diaphragm plates, one stationary and one moving with the magnet, were replaced by two semicircular mirrors performing the same task. They varied from Brooke's arrangement in being exactly planar, with the addition of lenses to focus the light, rather than his concave mirror that provided both these functions but had no provision for a baseline. Brooke later acknowledged the superiority of the new approach — his image was less distinct due to aberration at oblique angles of the light and the plane mirror was also a lighter weight on the magnet.[B95] The curves were tabulated by Sabine and his team at Woolwich using an updated version of Ronalds' ordinate board that Sabine believed enabled measurement to 0.03 mm accuracy.

The Kew Committee exhibited the new magnetographs in the next world fair — the 1862 International Exhibition in London — where they won a medal. Just about everyone was a judge in the Philosophical Instruments Class: Brewster, Sabine, Gassiot, Wheatstone, Glaisher and Brooke.

The *Encyclopedia of Geomagnetism and Paleomagnetism* (2007) advises that 'The instrumental developments by Brooke and Ronalds established the standard technique employed for magnetic observatory recording worldwide for more than a century.'[B190] By 1900, the combined apparatus had been sent to 20 other observatories.[A138] Sabine's initial justification for the new instruments had been to re-establish self-recording for the next magnetic peak expected in 1858. Continuous magnetic recording with these instruments actually continued at Kew until 1925; they were then receiving increasing interference from electric tramways. The Kew machines were transferred to a remote observatory established in Scotland and observations continue there today. Slow variations of the magnetic field remain critical for navigation, while magnetic survey tools used in the minerals industry rely on accurate reference data, and magnetic storms pose risks for modern infrastructure. The increasing shift of the north magnetic dip pole in recent years has attracted popular interest.

15.12 Conclusion

Self-registration instruments have been placed among Ronalds' most important scientific achievements for many years. As early as 1847, his and Brooke's work was deemed sufficiently important to be included in one of the first summaries of the history and progress of photography, published in the *North British Review*.[A71] Nearly two decades later, when John Phillips was BAAS President in 1865, he made mention of 'the inventions of Ronalds and his successors...measuring and comparing contemporaneous phenomena…over large parts of the globe'.[S1] The ability scientists now had to photo-record meteorological and magnetic parameters was still regarded as a crowning achievement of the BAAS and Ronalds was its instigator. Similarly, his entry in the 1897 *Dictionary of National Biography* advises that 'Ronalds's invention was of extreme importance to meteorologists and physicists…it is employed in all first-rate observatories, and has been used in many physical investigations.'[M6]

Ronalds was the first to develop and document a successful 'movie camera', complete with optical system, to capture the continuous modulation of natural phenomena using photography. His initial model was built in April 1845 and original photographic curves from that year survive. He went on to apply the concept to a range of meteorological and magnetic parameters while also achieving a step-change in precision. In doing so he collaborated with leading practitioners in enhancing the quality and flexibility of photography and developed a valuable means of reproducing Daguerreotype images. He then personally supervised supply of equipment to observatories in Toronto, Madrid and Oxford as well as Kew, and probably also Paris — he certainly assisted instrument-makers on the Continent in their manufacture. Perhaps 23 machines to his design were made in that period. His instruments were subsequently used in justification of developing a 'Met Office' and, after being updated to varying degrees, many more of the machines were sent around Britain and the world in later decades. All of this was achieved in spite of Airy at the Greenwich Observatory, who actively undermined Ronalds' work in favour of a rival plan of Brooke's.

Of the various photographic instruments, Ronalds' temperature-compensated barograph is perhaps the most noteworthy. It played a key role in the development of systematic weather observation and prediction in the UK — with all its attendant economic importance and public interest. The near-perfection of the original machine meant that few modifications were required prior to its widespread international adoption, and the instrument was still used routinely in some observatories 130 years after it was invented. Few inventions can claim that sort of track record.

We will now explore what Ronalds did during his well-earned 'retirement' from Kew activities.

Chapter 16

LAST YEARS AND LEGACY

Ronalds' extended sojourn on the Continent, his last years in southern England and his final innovations (including the Ronalds Library). A wrap-up of what he accomplished in his life, and the broader acknowledgement over time of several of these achievements.

Ronalds lived for another 20 years after he had left the Kew Observatory. He spent nearly half of this time on the Continent, while his last years were passed in south east England. Two threads run through these periods that continued well beyond his death — the Ronalds Library, and increasing recognition of his achievements — and friends and family had close involvement in both. The major contributors were his baby sister Maria Carter and her family, but they were assisted by nephew Dr Edmund, cousin Thomas Field Gibson, nieces Julia Ronalds and Sarah née Martineau, and her husband Charles Flower. Colleagues including Edward Sabine* and Latimer Clark* and family friend Mary Sharpe* also lent a hand. Ronalds himself had little interest in personal renown and would be less recognised today without the efforts of these influential people. A brief analysis of what he accomplished through his long life concludes the book.

16.1 Travels 1853–62

An approximate itinerary of Ronalds' post-retirement travels may be gleaned from surviving letters and diary entries, and is included in Fig. 8.1. He had reached Paris by mid-December 1853. He stayed three and a half years in France, mostly in Paris but also in the south, and spent even longer in northern Italy. In between, there was a year in Switzerland. He headed home again through France and probably arrived around the beginning of 1863, giving a total of nine years away (Table 16.1). Being absent for so long, he was unable to say final goodbyes to his

Table 16.1: Approximate Itinerary of Retirement Travels.

	Years	Dates	Place	Months	Visited Before	Contacts Name	Papers Collected
France	3½	Dec 53–Jun 56	Paris	30	Yes	Galignani	5
						Moigno	
		Jul 56	Bordeaux			Abria	7
			Toulouse			Petit	16
			Carcassonne				
		Apr 57	Marseilles		Yes	Valz	
						Morren	
		May 57	Montpellier			De Mouchy	
		June 57	Lyon		Yes		
S	1	Aug 57–Aug 58	Geneva	12	Yes	Wartmann	35
Italy	4	Sep 58	Turin		Yes		
			Milan		Yes		
		Oct 58	Verona	3	Yes	Zamboni	16
		Dec 58–Mar 59	Venice		Yes	Vaughan	
		Mar 59	Milan		Yes		
		Mar 59–Jun 60	Padua	15	Yes	Zantedeschi	92
						Orsolato	1
		Jun–Jul 60	Venice	2	Yes		
		Aug 60	Brescia		Yes		
		Aug 60–Jun 61	Milan	10	Yes	Williams	47
						Magrini	
						Volta	74
			Trips to Como				
		Jun 61–Jan 62	Bellagio	7			
		Feb–Aug 62	Milan	6	Yes	Brugnatelli	19
		Jun 62	*Trip to Pavia*		Yes	Cossa	2
F		Sep 62	St Michel		Yes		
S			Geneva		Yes		
F			Paris?		Yes		

brother Charles, his favourite Uncle Charles, and cousins Betsey Ronalds and her sister, but four of his UK-based brothers and sisters lived to see his return.

He had several reasons for going abroad. One immediate benefit was that he regained the freedom he had lost at Kew. He quickly returned to the life of his Grand Tour many years ago, pursuing the interests he wished to at the time. His lack of a firm plan for extended travelling is made clear when he queried John Gassiot* on his arrival in Paris: 'Has Mr Welsh[*]…sent, as promised, a box containing my Slide Rest, and some of my other Instruments which I packed to go to Holzapffell's[*]…for repair'.[C351] He was anticipating being away from Kew for only a matter of weeks.

The clearest objective for his 'retirement' was to enhance his electrical library, as discussed in detail in Sec. 16.3. In the course of doing so he was drawn to places he had visited 40 years earlier and was able to meet eminent scientists or their descendants. He also had broad scientific goals — to provide encouragement and assistance to observers to continue the investigations he had started at Kew and to pursue various other ideas in the back of his mind that he had been unable to tackle there. The wording of his note to Manuel Johnson* when he first arrived in Paris suggests that it was one of these ideas that prompted the sudden trip, and the library was a collateral benefit: 'I am taking the opportunity of being here to improve my electrical Catalogue, and Library'.[C353]

A telegraphic connection between the observatories in Greenwich and Paris — along Charles Walker's* South Eastern Railway and across the Channel — was being established at this time to transmit time signals for research purposes. The Paris Observatory Director had just died; he was François Arago, who had carefully studied the relationship between the aurora and variations of the magnetic needle. It is possible that Ronalds' trip across to Paris from Folkestone and Dover concerned the telegraph link and perhaps his desire to incorporate the study of geomagnetically-induced currents (see Sec. 14.4). Other scientific instigations for the trip can also be surmised.

Once in Paris, he settled into the local life much as he had done in Naples years ago (Sec. 8.1). He would have enjoyed the slower pace and being easily able to partake of the opera and other cultural and social activities after many years of single-minded focus at Kew. One of his close associates at this time was Moigno*. He would also have interacted with numerous other local scientists, like Professor Antoine Becquerel and his son who have 55 of their publications in Ronalds' library. His postal address was Anthony and William Galignani's business in Rue Vivienne, which was a bookshop, reading room and club for English residents and visitors. He arranged for his good friend James Peacock of Hammersmith to handle his mail at home and to assist in monetary affairs. He later bequeathed

Peacock £300 in acknowledgement of his kindness and, when Peacock predeceased him, transferred the gift to Peacock's unmarried sister.

He had spent a year in Paris when Gassiot told him that his instruments would be included in Kew's display at the *Exposition Universelle*. He would have chosen at this time to stay for the exhibition and then later to publish the booklet on his meteorological inventions in the display. He also drafted part of *A Walk through the Universal Exhibition of 1855*,[R22] published by Galignani to assist English-speaking attendees. In it he outlined Great Britain's 'philosophical instruments' display, including his own machines as well as others' devices. His booklet, the steady stream of visitors, and the interest generated by his exhibits, would have occupied much of 1855. In 1856, he was assisting Porro in building his photo-recording equipment for French observatories. With all this, he had remained in Paris very much longer than he intended and he could see ongoing opportunities there. There was now certainly no reason to return to Kew. There were other goals he wished to achieve, however, and his advancing age would have been playing on his mind, so he made the decision to begin his travels.

Ronalds headed south around July 1856. His first draw was probably Dr Joseph Abria, Professor of Physics at Bordeaux, who published regular meteorological observations and also had interests in electricity and geomagnetism. He then travelled to Toulouse to visit astronomer Dr Frédéric Petit; he was Director of the observatory there, and had a small geomagnetic laboratory in his garden. Ronalds acquired seven of Abria's papers and 16 of Petit's for his Library. It is not known if he helped develop observatory equipment for these gentlemen during that nine month period, but he did at his next major stop, Marseilles, and he also provided advice the following month at Montpellier, as discussed in the two previous chapters.

He was in Switzerland by August 1857, having travelled via Lyon. It was his third journey to Geneva, which he made his base for a year. He became good friends with Elie Wartmann and his father Louis, with whom he shared wide interests across precision instrumentation, new applications for electricity, as well as astrophysical phenomena including the aurora, shooting stars and meteorites. He collected 35 of their papers on these topics. It was at this time that he promoted the new support mechanism for the troublesome vertical force magnetograph, as mentioned in Sec. 15.8.

Another subject they explored was the use of electricity for lighting. Sir Humphrey Davy had demonstrated in the first decade of the 19th century that, when two pieces of charcoal connected with a Volta battery were brought together and then separated, a spark of brilliant white light was produced between them. Reducing the idea to practical form took decades. One of several problems to be overcome was that the light was inconsistent — it started well, but the charcoal burnt quickly and unevenly, creating a cone at each tip that altered their separation constantly and made the arc flicker and then stop. Impurities in the charcoal exacerbated the intermittency. In addition to better quality carbon and special electrode shapes, numerous complex electro-mechanical regulators were developed around mid-century to feed the carbon automatically so that a suitable gap would be retained.[B119]

An understandable wish in deploying electric lighting was for multiple lamps to be linked to a single power source, to give more diffuse lighting over a wider area. The temperamental nature of the carbon made this a challenge as one lamp's behaviour would affect the others. Depending on the wiring arrangement, any lamp could break the circuit or alternatively swamp the other lights. Parisian instrument-maker Joseph Deleuil trialled various combinations, using lamps with a single battery and multiple batteries, but could retain uniform illumination for at most 15 minutes.

Moigno described Deleuil's work in *Cosmos* in early 1856, and then outlined Ronalds' proposed solution to the problem:

> Mr. Ronalds, before Mr. Deleuil, called our attention to the great advantage of using a multiple unit, a kind of electric candelabra…The carbons burn slower and last longer, one of the branches of the candlestick could shut down without anyone noticing, and the others continue to operate, with the provision of an additional mechanism to close off the failed branch of the circuit.[A94]

The article was paraphrased in Count du Moncel's *Exposé des Applications de l'Electricité*[A100] in 1857. It was Wartmann who alerted Ronalds to the new citation, as he had just published a paper on a similar topic.[A155] It appears that Ronalds' idea was not pursued, and the problem was not solved adequately until the late 1870s, after which electric street lighting began to be introduced.

The arc lamp also began to be trialled for lighthouses at this time. Ronalds advised Captain Tremblay of the French Navy, whom he had met through Moigno:

> the Light-House here *[in Geneva]*…not having succeeded satisfactorily, in consequence of the usual Intermittance *[sic]* (or flickering) &c, I…proposed…interrupting the communication of the battery with the charcoal apparatus, at given regular intervals of time, by means of clock-work…the light which would be emitted during the periods of communication, would be more uniform, and stronger…The part of the charcoal apparatus might remain as usual for regulating the distances &c, or perhaps might be modified by causing the Clock-work to remove the accumulated cone.[C372]

The concept was very simple but only applicable of course where regular pulses of light were required.

Again, there is no evidence that it was pursued. At age 70, he was still abreast of current technical challenges and coming up with practical solutions, but he had even less interest in commercialising them himself than he had shown 40 years earlier. He simply passed the thoughts on to others. English trials began in late 1858 under Faraday's* oversight, using a large steam-powered electric generator for the first time rather than a bank of batteries. By 1882 there were five electric lighthouses in the UK.[A155]

The family was despairing by now that they would ever see Ronalds again. Dr Edmund implored: 'I wish you would make up your mind to revisit Great Britain & spend a few weeks or months during the Summer with us.'[C370] Ronalds replied that he would be back in the UK before long, and sought his nephew's advice on places and people to visit in Germany regarding his library on the way home. In August 1858 he determined on a short side trip to Turin and left considerable luggage (presumably mainly books) in Geneva awaiting his return. The reality turned out to be four years in Italy and no German travels.

By December that year he was on the other side of Italy, in Venice, where he had earlier in his life spent several weeks in quarantine. One of his reasons for going there would have been to collect papers by Ambrogio Fusinieri, of which his library contains 42. Henry Vaughan became his closest friend in the city, an art collector and philanthropist whose collection of Turner watercolours is now displayed in the UK and Ireland in January each year. He also met the respected photographer Carlo Ponti and naturalist Giovanni Domenico Nardo. He mentioned to them an idea for diminishing the excessive blackness that occurred commonly in photos at that time — those parts of the negative could be painted in a suitable tint to intercept the required amount of light when the positive impression is made. He noted in his diary: 'They had (neither) heared [sic] of such a project, and thought it a fully worthy one for trial.'[J85]

Ronalds had contemplated a new telescope support back in Geneva, its purpose being a simple, cheap means of steadying a traveller's handheld telescope. He had several sets of these telescope forks made in Venice. A set comprised two wooden rods, each inserted firmly into a candlestick. Two small hardwood cylinders were attached at the upper end of each rod to create a fork and these supported the telescope near the object and eye ends. The apparatus was set up to view objects on the horizon but the inclination could then be adjusted by twisting a stick on its axis to cause that end of the telescope to ride up a little in its oblique fork. A local turner charged the equivalent of 8½ pence to make each apparatus. Vaughan found it 'very effective, convenient, and steady (increasing the power of Telescope considerably.)'[J85]

Ronalds modified the design while he was relaxing at Bellagio a couple of years later. The forked rods now fitted snugly into a hole near each end of a horizontal piece of hardwood. A central support on the linking horizontal fitted into a candlestick or, for outdoor use, a metal screw on the top of a post. The improved arrangement facilitated smooth horizontal panning around the horizon. Vertical adjustments were again made by twisting one of the forked rods, but now with the aid of a peg passing through it. He found it so convenient in watching activity on the lake below his hotel that he suggested:

> at Hotels, Inns, &c, situated on Sea-coasts, border of lakes, &c, &c, frequented by exploring travellers who often carry telescopes, the Apparatus…mounted in a very solid and handsome manner, and applicable to <u>any</u> ordinary portable telescope… would be found a very acceptable piece of furniture[J85]

Surrounded by gondolas in Venice, he had come up with another innovative concept in March 1859 that he called the 'Propelling Rudder'.[C381] It combined the propulsion and steering mechanisms of a boat in a single apparatus. Ronalds' interests in steam propulsion were introduced in Sec. 12.12 with mention of Fulton's early steamboat propelled by paddlewheels. The paddlewheel began to be superseded by a bladed propeller or screw after its first deployment in 1839. The propeller shaft was aligned down the longitudinal axis of the vessel, with steering provided by a separate rudder. Ronalds' idea was to position the propeller in a frame having an outer profile similar to a rudder and attached to a vertical shaft. A universal joint inserted in the drive shaft at the position of the vertical shaft transmitted spin to the propeller while it and its rudder frame could rotate in plan. The apparatus would be powered and operated by the gondolier.

He did not think any more about his idea until 1862, when *The Times* carried several detailed articles on the poor steering demonstrated by the new armoured frigate HMS *Defence*, particularly at slower speeds; the ship had become grounded in sea trials. He wrote to the newspaper explaining his original concept and its extension to steamships by replacing the universal joint with a much more substantial arrangement of three mitre wheels in a U-shape. He also addressed design details to withstand the large thrusts generated and would have had working models made except that he was very ill at the time, as discussed below. The letter was not printed. He described the concept to friends as 'a scheme for causing the Screw propeller to take any angular position, relative to the keel, in order that it may act in the additional capacity of a much more efficacious rudder than the common rudder: which common rudder acts only when the vessel is under weigh'.[J85] Steering was enhanced by the high flowrates past the rudder-like frame generated by the propeller. One of these friends noted in a letter to the *Athenæum* (that was published): 'Mr. Ronalds's plan was much approved by competent judges'.[1]

As is frequently the case, similar concepts were developed and honed over the ensuing decades. The experimental torpedo boat USS *Alarm* trialled a 'steering-propeller' patented by William Mallory in the 1880s. It was the mid-20th century

---

[1] *Athenæum* 20/9/1862 374–5.

when the azimuth thruster pod, called by its inventor the 'rudder propeller', began to be marketed and became a common feature on ships.

Ronalds and Vaughan took an excursion in March 1859 from Venice to Milan and Padua by train. In a diversion from science, Ronalds took Vaughan on 8 April to visit sculptor Rinaldo Rinaldi in Padua, who was making a 'colossal' likeness in marble of Ronalds' 'poor old friend and fellow traveller'[J85] Giovanni Belzoni — the man who had brought Ramesses II to London (see Sec. 8.2). Ronalds would have been proud to explain to Vaughan that he played a small part in enabling this commemoration in Belzoni's home town.

Manufacturer and inventor Sir Edward Thomason records how the story started in his autobiography:

> In 1819, Belzoni returned to England from his laborious researches in Egypt...It appeared to me that his fame was worth recording upon a medal. I, therefore, wrote to my friend Mr. J. Brockedon, the celebrated artist, to try to obtain for me a likeness of G. Belzoni [B56]

William Brockedon FRS wrote to Thomason in May 1821:

> I regret that circumstances have prevented my sending the model earlier; 'tis my first attempt, but it has the merit of being a very strong resemblance. All his friends are much pleased with it [B56]

It was Ronalds who knew the parties and brought them together. He held 'a dinner party in Queen Square'[J85] — the family home in 1821 — with Belzoni, Brockedon and Sir Frederick Henniker* at which Belzoni agreed to Brockedon creating his profile. The medal was cast a short time later. It was timely, as Belzoni died two years later in what is now Nigeria. The large medallion in progress at Padua was based on the same profile. Ronalds advised Rinaldi that 'the countenance was too aged' and the sculptor agreed to 'try to alter it a little'.[J85] It remains on display today in the *Palazzo della Ragione*.

Vaughan returned to Venice after that edification but Ronalds decided to linger in Padua, despite still keeping a hotel room in Venice. Vaughan sent him a care package, including his umbrella. He ended up staying over a year but it was not all of his own choosing. The Second Italian War of Independence commenced on 29 April 1859 with Austria's invasion of the region, and the French under Napoleon III quickly came to the aid of the Italians. Vaughan wrote the next week from Venice: 'I may report as fact that our Consul has telegraphed for a ship of war to come here to take away the Inglese *[English]*.'[C373] The large Battle of Solferino took place around 100 km from Padua in June, involving a quarter of a million men. Vaughan provided another update a couple of weeks later: 'The 'Gauls' had shewn their readiness for work, by appearing in the offing yesterday morning with a force of at least 40 vessels. The floating batteries looked like good sized houses.'[C374] Vaughan was using Ronalds' telescope forks in 'spying'[J85] on the ships. Over the next year, much of the Italian peninsula was unified as an independent state, although the region surrounding Padua and Venice remained under Austrian control until the Third War of Independence in 1866.

Ronalds kept himself occupied in various ways. He inspected the lightning conductor at the Padua Observatory just after it had been struck by lightning, as mentioned in Sec.14.1. He mused at the time that a safe gunpowder magazine might consist of a large metal container 'like our large Gasometers'[J85] that would act as a 'Franklin's Can'[J85] or Faraday Cage in shielding the interior from external charges.

He made drawings and pasteboard models of various new concepts. One was a 'Watch-Box...for convenience of packing',[J85] which he planned to show to local watch-makers. Another, which he arranged to have made, was a 'Sun Dial Meridian Alarum, founded on the application of an Electro-thermo magnet'[J85] and had the purpose of signalling solar noon. It appears to have comprised a means by which at midday the sunshine would hit a spot where it was focused through a lens to induce a thermoelectric current. The current, in combination with an electromagnet, could act on the minute hand of a clock to correct its time or be used to ring a bell. Connecting the device to a telegraph line would enable the time signal to be sent to other towns. This last application is an automated version of the time ball system described in Sec. 12.6 that was run from Greenwich in the 1850s.

He additionally spent time with surgeon Giuseppe Orsolato, who was editor of the local Academy of Science's magazine. They 'Conversed about the light emitted from the Tropaeolum *[nasturtium]* on the bursting of the pollen',[J85] with Ronalds acquiring his paper on the subject and suggesting an electrometer arrangement to observe whether the pollination involved electrical effects. He presented his booklet *Descriptions de quelques Instruments Météorologiques et Magnétiques* to the Academy.

It was June 1860 when he returned to Venice to pick up his belongings. From there he went once more to Milan — he was now at last homeward bound. He spent two years however in the Milan area. In his circle there was a Prince from the Lusignan family. The Prince's father had been the ruler of Armenia and was killed in 1829 supporting the Shah of Persia against Russia. Tsar Nicholas I had apparently confiscated the family's property and funds. The Prince was seeking a pension from Persia and wished Ronalds to edit a letter he was drafting in English to that country's Ambassador in London. He kept Ronalds informed on progress over the ensuing 18 months, including showing him 'Mr Downing Street's answer'[C383] to another letter written to the British Foreign Secretary.

Ronalds finally crossed the border into France in September 1862. He used the Mt Cenis pass, as he had first done in 1818. This time he arranged to inspect 'the machinery for forcing air into the end of the...tunnel'[J85] — a railway was under construction and air pumps were being used to ventilate the work gallery. This first major tunnel in the Alps was opened in 1871. He then returned to Geneva to gather up his goods and made his way to London.

16.2 Final Years

Once Ronalds had returned to the UK, he rented a house at Battle, close by Maria and Samuel Carter. He retrieved the rest of his belongings from the Pantechnicon and his orphaned niece Julia Ronalds came to live with him, to help sort through them and his many Continental acquisitions. It would have been a major job. Julia also did a little of his writing for him.

He retained an active interest in science, as exemplified by his letters to Airy at Greenwich (Sec. 14.4), and to the new Society of Telegraph Engineers (STE) in July 1872, as mentioned below. A document written in 1870 described him as being still 'in the possession of full mental vigour'.[J96] He had his mother's strong constitution.

He must have been quite ill in November–December 1866, however, because he suddenly wrote his will. It was a very short one — there was just one thing on his mind: 'I give and bequeath all my books papers manuscripts philosophical instruments and apparatus of every kind to Samuel Carter...and in the event of his dying before me I give and bequeath the same to my nephew Edmund Ronalds'.[W11]

When he had recovered, he thought a little further and itemised some additional wishes in four codicils. The final codicil was written three months before his death. He appointed Samuel and Dr Edmund as executors, and enlisted the assistance of two other nephews (James Montgomrey Jnr and Edmund's son Hugh) in investing in debentures to support Julia and also Edmund's unmarried daughters. These were the only family members in the UK who might need financial support. Henry Ronalds Jnr had explained the situation two decades earlier:

> Spent the day...in company with my cousins Julia Ellen and Eliza who were on a visit to their Aunt Emily all penniless girls brought up at least the two latter of them with every prospect of affluence but all their young hopes crushed in consequence of my cousin Edmunds speculation as a silk manufacturer...they are truly to be pitied as the chances of marriage are decidedly against governesses under the present constitution of Society in England.[J90]

Ellen and Eliza had since wed, one into the wealthy Greg cotton-spinning family, but their sisters Charlotte and Janet had not and were the recipients of Ronalds' bequests. He in addition repaid tenfold the vital £50 he had received from the Wollaston* Fund in 1846 to continue development of his photo-recording machines while the Kew Observatory was under threat of closure (Sec. 13.3).

Fig. 16.1:   Battle Cemetery in 2012; Ronalds' grave is just left of centre in the foreground.

His estate was valued for probate at 'under £3,000'. This was considerably less than his very wealthy brothers-in-law Peter Martineau and Samuel, his sister Emily and even Edmund (Table 4.2). He had not yet spent all of his inheritance and earnings on science — but most of it. He would have been pleased how it had panned out.

He was buried in the public cemetery in Battle. It is a simple grave with no headstone (Fig. 16.1), as would have been his final wishes, and is now in poor condition from exposure to the elements. The only writing appears to be in the centre of the slab:

SIR FRANCIS RONALDS K[T] F.R.S.

BORN IN LONDON

21ST FEBRUARY 1788

DIED AT BATTLE 8TH AUGUST 1873

The location of grave is mentioned in a 1951 article in *Nature* entitled 'Britain's Scientific Shrines'.[A151]

Samuel would have sent a death notice to the papers. Obituaries quickly appeared in many of the national newspapers, including *The Times*, *Pall Mall Gazette*, *Daily News*, *Morning Post*, *Glasgow Herald*, and of course the local *Hastings and St Leonards Observer*, as well as periodicals like the *Athenæum* (the *Literary Gazette* had ceased), *Nature*, the *Journal of the Society of Arts*, the new *Journal of the Society of Telegraph Engineers* and *Dodsley's Annual Register*.

That giving the best representation of Ronalds' personal traits was in the *Athenæum*, which started:

> SIR FRANCIS RONALDS must not be allowed to pass from amongst us without a few words in recognition of his labours, which were always useful, and which were never obtruded upon public attention

and ended:

> At the ripe age of eighty-five, on the 8th inst., this true lover of science died, an example of a good man, working always purely for the sake of truth, and never with any desire for fame.[D7]

The *Morning Post* was perhaps the first to suggest publicly that he should have received greater recognition in his life:

> The learned and scientific world will learn with regret, though scarcely with surprise, the news of the death of Sir Francis Ronalds, who was known not only as the first director of the Royal Observatory at Kew, but also as a man devoted through life to the advancement of electrical science and its many practical applications. He has for many years held a very high rank among scientific circles…The man…deserved some better and more substantial honour at the hands of the nation and its Government…he closed his long and laborious life, full of years if not of honours, on Friday last, in the 86th year of his age.[D3]

His passing was also noticed internationally. *Appletons' Journal*, published in New York, advised of the death of the 'eminent English scientist'[D11] and numerous Australian papers carried the death notice. The Australian relatives long remained very proud of him — the death notices of at least four of Alfred's older children made specific reference to their inventor uncle.

His life was commemorated at the STE's annual meeting:

> That well-known name — the father of Electric Telegraphy — Sir Francis Ronalds, has passed from amongst us; his name was associated with Telegraphy from so old a date that he was the one link which bound the present to the past.[D13]

Clark squeezed in a further tribute in his 1875 Presidential Address:

> Our sorrow for his loss is tempered by the remembrance that he lived to witness, to an extent perhaps never before vouchsafed to man, the wondrous success and development of that telegraphic system which he had done so much to perfect and to advance in his early life[T11]

Airy* also made mention of Ronalds' passing among others in his Presidential Address to the Royal Society:

> Sir F. Ronalds for his knowledge of electricity, his introduction (collaterally with others) of photographic self-registration, and his attempts at establishing a telegraph not by galvanism but by electricity[D14]

It was classic Airy. No obituary was published in the Royal Society Proceedings, but it is drawing a long bow to attribute this omission to Airy.

16.3 The Ronalds Library

Ronalds' electrical library has been mentioned in most chapters of this book and was the most constant goal over his long career. From the beginning it had two parts; the first was to create an exhaustive bibliography of all works on his subject, and the second to acquire a copy of those works of greater relevance and interest. He believed he had succeeded and that his library was 'approximately complete'[C396] to the time of his death. The theme itself broadened over time — once the connection between electricity and magnetism was confirmed, the latter was naturally added to the scope. Meteorological and astrophysical papers were also collected, given their possible interrelationships with atmospheric electricity.

He had always immersed himself in the literature of any area he embarked upon, making abstracts of his readings that ran to numerous volumes through his life. He collected books and papers on electrical science as soon as he began experimenting with electricity. He had formulated the concept of an electrical library by 1815 and within a year had documented designs in his Ideas Book of 'Bookshelves for Large Libraries'.[J14] He wrote to Samuel Carter in 1869 of 'my Catalogues formed between the years 1815 & the present time'.[C396] It was a lifetime's work.

Books entered the bibliography and collection through various routes. The standard approach was to visit renowned libraries to study their holdings and then seek copies of important works through local or international booksellers. Evidence survives of numerous instances where Ronalds diverted attention to such activities. Book-collecting was a focus of his first documented trip to Paris and Geneva in 1814 and the early and late stages of his Grand Tour, particularly in Florence and Strasbourg. He timed his return trip to Sicily so that he could attend a renowned book fair near Lyon in 1823. He also found BAAS meetings very useful: he perused the Bodleian Library's printed catalogues during the 1847 conference in Oxford and, in Edinburgh three years later, his diary notes: 'Went with M[r] Welsh to many booksellers in search of electrical & magnetic books (found but one) and to the College-Library.'[J80] He again spent considerable time at the Bodleian while staying in Oxford in 1853 to manage the assembly of his barograph, and ordered works from booksellers in Germany, Italy and France. These arrived after he had left and were duly delivered to Johnson, who kindly took care of them: 'I have got a heap of books for you which <u>have been paid for</u> so you need not trouble yourself about them we can arrange matters when you come back.'[C357] It is not clear how that happened, as Johnson had died before Ronalds returned to England.

In his first year in Paris, Ronalds spent much time at the Imperial National Library, now called the *Bibliothèque nationale de France*. Private libraries had been seized in the French revolution and a central inventory created.[A180] This major step in information management would have been of immense value to him. He continued his task throughout his European travels. A typical diary entry (in Brescia) reads: 'Visited the Biblioteca Comunale, and made some extracts. Went to Valentinieri's Magazzino, and bought some books.'[J85] He developed a design for a Pamphlet Guard in Paris to hold his growing collection and had more made in Padua.

In a second approach, he took advantage of book auctions organised by the executors of deceased scientists. He utilised Arago's estate sale in Paris in June 1854 — his library contains eight of Arago's works as well as twelve of his copies of others' papers and his portrait. Back in England, he benefited from the library sales of Faraday in 1868 and Dr Peter Roget FRS (of Thesaurus fame) in 1870. Where a person eminent in the subject had possessed and perhaps annotated the work is generally noted in the catalogue — Ronalds realised their inclusion added interest to the collection and also assisted in the preservation of the works.

A great advantage he exploited in assembling his collection was that he lived in an era when much progress was made in his field and he knew many of the contributors: the numbers of papers in the library authored by characters in his story are noted through this book. Personal contact was important in obtaining the works — he was not the only scientist who chose often to print works for private distribution rather than publish. A key activity on his Continental travels was to seek out leading academics in each town to discuss their work and acquire key papers. They also assisted him in identifying important works by other local scientists, past and present, and advising how to get them. He would have enjoyed these interactions immensely.

He was able to pursue this tactic with Abria in Bordeaux, Petit in Toulouse and the Wartmanns in Geneva. When he realised he would not be able to visit his Kew guest Carlo Matteucci in Pisa, he wrote to him from Milan: 'I send a list of those of your works on Electricity &c which I have been able to procure & another of some of those which I cannot yet meet with & which I fear I shall not obtain without your kind assistance.'[C376] He acquired a total of 50 of Matteucci's publications.

Ronalds always offered to pay for the books. As he told Matteucci:

> May I...assure you (on my honour) that this petition is not proffered with the slightest expectation of Gifts? — I devote a principal part of a little Civil list Pension to purchases of the above Kind[C376]

At the same time he was conscious that his funds were limited. He had written to Gassiot from Paris:

> being in this expensive Babylon, at work (in the Libraries, and amongst the Booksellers) endeavouring to complete the French portion of my electrical library, I find it necessary to use all prudent available means of going so; even to the extent of begging[C351]

Fortunately many scientists made a complimentary present of their work, as Ronalds had always done through his career. In other cases he showed his characteristic generosity in buying works as much (or more) to help the family, or as a way of paying his respects, as for their value to his collection.

He was able to spend considerable time with Professor Francesco Zantedeschi when he was holed up in Padua. On his second visit, Zantedeschi: 'had prepared for me a Packet of more than 50 of his works (and a Catalogue of them) which works (pamphlets &c) he gave me separately, stating various interesting particulars concerning them.'[185] Ronalds was able to present in return only his *Descriptions of An Electrical Telegraph* and *Descriptions de quelques Instruments Météorologiques et Magnétiques*, although they were the two publications of which he was most proud. The two men continued to meet regularly. One of the topics they discussed was the relative merits of absorption and induction methods of observing atmospheric electricity (Sec. 14.1). Another was whether Gian Domenico Romagnosi had discovered the relationship between electricity and magnetism 20 years before Oersted (Chapters 6 and 7). Zantedeschi argued that he had but Ronalds made clear in his library catalogue his view that he did not. They remained in contact after Ronalds left Padua and one of Zantedeschi's 92 papers in the Ronalds Library is dated as late as 1869. Ronalds also obtained a signed copy of his lithographed portrait.

A few of Ronalds' visits had overtones of commemoration or reminiscence — a walk down memory lane. He enjoyed seeing Giuseppe Zamboni's studio again in Verona after nearly 40 years (Sec. 6.3), where he was hosted by Zamboni's nephew and their close friend, the physicist Gaetano Spandri. He purchased papers written up till Zamboni's death in 1846 as well as two eulogies, his portrait and two of his more recent dry piles.

Ronalds had first visited Pavia, in search of Alessandro Volta, in 1818. It happened to be the day that Volta's close colleague Dr Luigi Brugnatelli, the Professor of Chemistry, died. Brugnatelli had invented electroplating in 1805 using Volta's battery, but ceased publishing the work when it was not well received. The technique was patented in 1840 in England and founded a major industry. Ronalds probably felt sympathy for the story, which had strong parallels with his telegraph. He returned to Pavia in mid-1862 and arranged to meet Brugnatelli's daughter-in-law. He purchased a number of books, which formed part of his collection of sixteen works, a portrait and six of Brugnatelli's copies of papers by other authors. He also acquired three of his son's papers. The key items are well annotated in the Catalogue, while his diary notes of the day:

> I well remembered my experimenting upon Brugnatelli's invention of Electro-Metallurgy in 1815, and that those operations of Brugnatelli were then well known to many Professors in England and France.

Brugnatelli has evidently been most unjustly deprived of his fair fame in the matter of Electro gilding &c.[J85]

He also spent time with Dr Alfonso Cossa in Pavia, whose dissertation was on the history of electrochemistry. Cossa presented him with two of his recent papers on the topic, and the catalogue includes the note: 'Shows clearly Brugnatelli's priority in invention, and complete practical application of electro-plating, &c.'[R28]

Ronalds did not always meet with success. He seemed to fail in seeking out the descendants of Count Prospero Balbo in Turin, who was Beccaria's patron. The catalogue contains the suggestion under Balbo's entry that 'Possibly some of Beccaria's valuable or curious MSS. left in his hands may be discoverable.'[R28]

His most fulfilling result in this mode of collecting concerned his hero Volta. A notice appeared in *The Times* in early 1860 advising the family's wish to sell his manuscripts and scientific articles:

> The failure of the crops and various other unfortunate circumstances have so burthened[?] the economical position of the sons of the immortal Alessandro Volta, of Como, that unless some prompt ASSISTANCE be rendered them their future prospect will be a very sad one...they make an appeal to all those in Europe and America who venerate the memory of the great chymist to come forward to assist this worthy family.[2]

Ronalds was able to reach Milan by August that year. One of his early activities was to travel up to Como to see the statue of Volta in the Piazza Volta and to visit his eldest son Count Zanino Volta. Zanino 'expressed a strong desire'[J85] for the Royal Society to purchase the manuscripts and instruments, and Ronalds sent off a letter seeking the Society's views. In the meantime he offered 20 francs for *De vi Attractiva Ignis Electrici* (1769) — as noted in his catalogue, this was Volta's 'first Electrical work, much esteemed.'[R28] He had little interest in his other publications, having already acquired them. Zanino agreed to search for the paper and then check with the estate's executors about its sale. In the end, he presented the book to Ronalds in appreciation of his assistance. Ronalds had been planning to leave the area at this time but, as was often the case, that did not happen. When the Royal Society declined to purchase the papers and equipment, citing lack of space, he decided to help Zanino himself.

He returned to Como from Milan several times. In March 1861 he spent a week 'rummaging amongst the numerous and most confused collection of... books, MSS, and instruments'[C378] and putting them in order. He also visited Volta's country house, where there were more papers. He finally confirmed as he suspected that the electrical books 'were almost exclusively Duplicates of some in my possession'.[J85] He was able to select 70 works, 25 of which had been

---

[2] *The Times* 26/1/1860 5b.

presented to Volta by others, for which he paid Zanino 200 francs. He presumably planned to give away some of his own copies to friends. The Ronalds Library holds a total of 74 works relating to Volta, including two portraits and numerous descriptions of his life and work written after his death. Ronalds was very pleased to record later that year that the Como Council voted 20,000 francs (equivalent to £8,000) to the family for the purchase of Volta's manuscripts and equipment.[J85]

Now aged 73, the exertions in Como proved detrimental to his health: 'I acquired a very bad lumbago, by much stooping & groping…(& I fear other damage) the Books being massed on the floors.'[J85] It seems he initially recovered well, but as summer approached: 'the almost unexampled heat and <u>dryness</u> in these parts…fevered and weakened me (and many others)'.[C378] He left Milan in June for the cooler lakes, staying at a newly opened luxury hotel in Bellagio. 'I intended to return shortly — Soon afterwards however a very severe indisposition…rendered me incapable of doing so'.[C382] Niece Sarah Flower provided a little more information when she visited the region three years later: 'Then to Bellagio, to the Gran Bretagne where the landlady had taken such good care of uncle Frank and how delighted she was to talk of the dear old gentleman.'[B151]

Ronalds remained in Bellagio for seven months. He then returned to Milan where he stayed with the Williams family, who continued to look after him for another seven months — Revd John Williams was the British chaplain in the city. It would have been either John or a family member who wrote to the *Athenæum* on his situation:

> It has been our privilege to see him very constantly during the past year, though most of it has been passed by Mr. Ronalds in great suffering from a severe illness contracted last summer, greatly in consequence of his unremitting labour in collecting a vast number of works treating on electricity.[1]

He was finally well enough to head homeward from Milan in August 1862.

He summarised his Continental travels in his 1860 autobiographical letter:

> In 1854–5 Worked hard in the Imperial Library of Paris on the Bibliography of Electricity Magnetism &c…(a labour which, at sundry times with long intervals, I had undertaken ever since 1815) and as hard in dealing with the 'Bouquinistes' *[second-hand booksellers]* (almost as difficult rogues as horse-dealers)…
>
> Since about the summer of 1856 I have wandered into many public & academic libraries & book shops of France and Italy; gathering a <u>little</u> fruit to be added, on my return to Dear old England, to the English German &c store; but I shall rot long before my fruit is ripe.[C375]

The result of his efforts was a library of around 2,000 books, together with 5,000 pamphlets and papers that were later bound into 250 further volumes (Fig. 16.2). His bibliography contains a total of 13,000 entries. The librarian who managed the collection after Ronalds' death, Alfred Frost, noted of the two:

Fig. 16.2:   Portion of the Ronalds Library in 2012; the books are now held in a strong-room.

> The catalogue…is believed to contain a record of nearly all the important books and papers bearing on the subject, published in any language, up to within a short time of its author's death…
>
> The collection is a very complete one, particularly in Italian, French, and German[A122]

The earliest documents listed are 5th and 6th century, while the oldest items in the library are 'scarce and curious'[R28] works dating from the 1550s. The most recent international works were published in 1872 — he was collecting actively until the end. The historical and current material is complemented by a considerable array of 18th-century works by revered scholars like Beccaria, Franklin and Priestley*, from whom Ronalds had learnt so much when he entered the field himself. Many of the books display his bookplate with the Clanranald crest or coat of arms, as discussed in Sec. 3.1.

Clark was also creating a library and the two exchanged notes regularly in Ronalds' last years. The newest title in Ronalds' library was authored by Clark and published just prior to Ronalds' death in August 1873. Clark lived a further 25 years and 'his correspondence and records give every indication that he was very anxious to make a finer collection than Ronalds.'[A142] Clark's library was purchased from his estate by the US benefactor who made that boast and it was described at the time as the 'second most important electrical and scientific library in the world'.[A199]

It is perhaps unsurprising, given all we have learnt about the man, that Ronalds developed an original method for cataloguing his library. Frost recorded that he 'adopted what is now known as the card system'.[A122] The point was emphasised in 'Progress of the Modern Card Catalog Principle' published in 1902 and again in 1969, where Minnie James advised that 'the first modern card catalog in actual use appears to have been that of Sir Francis Ronalds in 1820'.[A141] James and Frost would have been known to each other through the Library Association in London.

Ronalds used a separate card for each document of interest, giving the ability to insert additional works while keeping the chosen order and with room for information to be added to any card as it was discovered. He always included the sources of his information.[A122] The cards would have been stored together in some manner of his design that facilitated access and manipulation without risk of losing the sequence. The system had considerable advantages over the usual book form of catalogue.

He was not the first to think of the idea of cards or strips of paper. The *Bibliothèque nationale de France*, for example, had chosen playing cards to collect data on their new regional holdings. They were a temporary tool however to ease central collation before transcribing the information as a book.[B194] A couple of English bibliophiles had also put information on playing cards or paper slips in the late 18th century. Larissa Brookes nonetheless reiterates James' view on the important historical position held by Ronalds' catalogue in her 2009 thesis on early library cataloguing systems.[B194] Certainly, his system was highly novel for its time. Card catalogues only became popular in the later 19th century — hence Frost's phraseology — and remained at the core of most substantial libraries until the digital age.

Ronalds had long hoped that his library would be considered to be of sufficient quality to be accepted for safekeeping by a learned body. It was natural that he would contemplate the Royal Society for this role and, during his convalescence in Milan, he broached the idea subtly with the new President of the Society, Sabine:

> You well know that we have no really complete English history of Magnetism & Electricity, & I thought that incompetent as I might be (even if spared long enough) to fill up the void, I might be enabled…to contribute some, not unimportant materials to the work of a competent hand *[he was still hoping to update Priestley's electricity text, a task he began in 1817 (Sec. 6.6)]*.
>
> Now this is said by way of apology for a petition, namely, that you would oblige me by your opinion as to the disposal of the said Books MSS &c. My hope has been (almost) hitherto that I should have the opportunity of assembling them in England (with the rest) of making in some sort a selection, of having this suitably bound & catalogued (in a special manner) and of giving or bequeathing the whole in such

manner that it might be conducive to the above suggested purpose, or, at least, to facilitate the sound study of my (life long) favourite science. I would fain prevent the fruit of much of my labour from being thrust aside, as antiquated lumber; for I am sure that there is a great part of it which is not so. (Even Dr Priestley's old History is deficient in facts which are, even now, interesting & important.)[C379]

Sabine was noncommittal in his reply:

It gave me great pleasure to hear from you again, and Mrs Sabine & I rejoice in the prospect you give us of seeing you in London in the course of the summer *[Ronalds was being typically optimistic with his homecoming date]*. We will then converse over the good use you propose to make of the electrical & magnetic books you have collected…

You will find Kew in good working order[C380]

It is not entirely clear what took place after Ronalds' return to England. Our three key friends all had scope to influence the Royal Society's perspective, with Wheatstone being on the Library Committee and Airy replacing Sabine as President in 1871. It seems however that the Royal Society remained the preferred destination until late in Ronalds' life. Clark wrote to him in 1869:

Is it true that it is your intention to present your magnificent collection of books on Electricity to the Royal Society? If so, why not do it while you are yet living? One half of the world does not know that you are yet among us, & really some public recognition of your services to the cause of science ought to be made — Such an occasion would form a fitting opportunity of acknowledgement.[C394]

This of course was not Ronalds' style.

Another possible repository for the library appeared at the eleventh hour — the Society of Telegraph Engineers. During Ronalds' life, electricity had evolved from being simply a scientific novelty to a useful power deployed to run the world's telegraphs. The STE was established for the 'advancement of Electrical and Telegraphic Science'[A115] and the support of those working in the new field. He would have been very pleased to witness this further step towards the future profession of electrical engineering. He declined initially to join the Society, knowing that his advanced age and the distance from Battle would prevent him from attending any meetings, but accepted a second pressing invitation delivered by Clark.[C408] He contributed by submitting a letter for publication in the first volume of the STE's Journal[R27] — it concerned Volta's early proposal for an electric telegraph that he had come across when sorting through his papers in Como.

Ronalds' special status in the profession was emphasised by the membership. Frost wrote that he 'must always stand as the first of English Telegraph Engineers'.[M5] At the first general meeting held in February 1872, the inaugural President Dr William Siemens FRS (who was a younger brother of Werner in Sec. 7.3), made mention of Ronalds among other early practitioners. Clark, in seconding the motion of thanks for the Presidential address, underlined the sentiment:

> There is to me something peculiarly interesting in being present at the inauguration of what I trust will be a long-enduring and important Society. And it is a great pleasure to feel that this Society had been established while many of the oldest members of the art of telegraphy are still living...One — Sir Francis Ronalds, the pioneer of all — I would particularly mention on account of his venerable age, and the earnest and persevering efforts he made to introduce the art of Electric Telegraphy into the public service, at a time when the world was not yet ripe for it. I have always felt a peculiar interest in Sir Francis Ronald's career...and I rejoice to feel that he is a member of our Society.[A116]

Ronalds' library was well known at this time — the *Journal of the Society of Arts* called it 'the finest collection of electrical works in England'.[D10] Leaders of the fledgling Society were more than keen that it would find a future home with them. It would be not just be the dream 'foundation'[A119] for their library but also a huge boost to the stature of the STE itself. Representations were made to Ronalds and Samuel Carter of the benefits of such a course, led by Clark and Siemens. Certainly the collection would be cherished at the STE. What probably most swayed Ronalds was that the books would be available at the STE to the practitioners who needed them most. At the Royal Society, only that rare breed called Fellows would have access and there was also significant duplication with their existing holdings. Furthermore, Ronalds had seen from the Royal Society's response to the offer of Volta's materials that it could be quite indifferent to such collections — they had more than enough of value already.

The decision must have been a very tough one for him so late in his life. He had always envisaged the collection at the Royal Society and this alternative was risky, simply because it was so new. Many societies that are formed do not stand the test of time. Siemens' summary of the status in January 1874 was that Ronalds 'before his death had promised to give it his favourable consideration, if not absolutely to commit to our charge his library'.[A119]

Ronalds had wisely left the final decision with Samuel. Samuel watched and waited awhile and then used his renowned legal skills to negotiate a watertight trust agreement with the STE (Fig. 16.3). It contained a number of important stipulations. The first was that the collection would be termed the 'Ronalds Library'.[J97] The STE would fund the preparation and publication of the catalogue in book form to facilitate its dissemination as a bibliography of the discipline, noting which of the works were in the collection. The books and pamphlets themselves would be bound, with the words 'Ronalds Library' in gilt letters on their covers. The collection was to be insured for a minimum value of £5,000. Most importantly, the library was to be made readily accessible to all who wished to use

Fig. 16.3:   Original trust deed for Ronalds Library (1875). IET.

it for their electrical studies. Finally, the STE was to become incorporated within five years to help ensure its permanence. If these conditions were not satisfied through the life of the trust, the collection would transfer to the Royal Society. The trust deed was signed in June 1875.[197]

Samuel also included in the collection Ronalds' surviving business correspondence, science notes and unfinished manuscripts, as well as the relatively few personal papers that had been retained. It is unlikely that Ronalds would have welcomed this last addition, which exposed much about the nature of a deeply private man. Samuel clearly felt that the personal side was worthy of archiving and this biographer wholeheartedly agrees. Not everything was preserved. Ronalds gave away items of interest to friends and family; one example was his annotated copy of the first edition of Alfred's *The Fly-fisher's Entomology* that he presented to John Carter, and others were inventions like his perspective tracing instrument. Samuel kept the autobiographical letter that Ronalds had written to him from Padua, but fortunately it was later presented to the UCL by Edmund's granddaughter Ethel Ronalds. As with many families, most of the other pieces are now dispersed or lost.

The duration of the trust was stipulated to be until 20 years after all of Queen Victoria's grandchildren were deceased. This turned out to be 1976, and so the trust was in operation for just over a century. The young Society was not able to obtain a Royal Charter but became incorporated in accordance with the Companies Act. This proved sufficient for the Trustees. The STE's successor entity, the Institution of Electrical Engineers (IEE), was granted a Royal Charter in 1921.

Eminent trustees were chosen to oversee the Library — Kelvin, Siemens, Clark and Major Sir Francis Bolton, who was Honorary Secretary of the STE and Editor of the Journal. The President of the Institution of Civil Engineers, a much longer-established body, was ex officio member. The family was represented by Samuel, his son John Carter and son-in-law John Martineau Fletcher (both of whom were barristers), as well as Dr Edmund.

A new deed was drawn up in 1908. With Samuel and Dr Edmund now deceased, two of Ronalds' great-nephews joined John Carter and John Fletcher — they were engineer and amateur meteorologist (Hugh) Basil Ronalds and surgeon (Hugh) Ronald Carter, the eldest son of Hugh Carter the painter. The next generation was introduced in the 1922 deed through the engineer James Edward Montgomrey, Ronalds' great-great-nephew. John Carter was still there, providing continuity, along with Basil and Ronald. In 1949, Ronald and James were continuing trustees. The family monitored the success of the Ronalds Library for much of the long term of the trust.

Clark had been able to foreshadow the bequest at a meeting he chaired in late 1874: 'It will cost £200 or £300 to print the catalogue alone, and the books themselves are of enormous value…it is one of the most magnificent collections of electrical works that the world possesses. (Applause.)'[A118] A publication fund had been established at the onset of the Society in anticipation of just such a need — Ronalds had contributed £2 to it in 1872.

Clark would have been delighted that it fell to him to announce the agreement formally in his Presidential Address in 1875. *The Times* covered the story and continued to report progress on the Library for several years.[3] Samuel died in 1878, and Maria in June 1880, so sadly neither saw the work complete. The Library and Catalogue were finally able to be launched at a soirée on 10 November 1880, the delay being caused in part by a 'financial crisis'[B166] at the young Society. It was advised that the Library would be open daily (except Sunday) and four

---

[3] *The Times* 15/1/1875 9g; 24/12/1875 3d; 12/11/1880 7f; 1/1/1881 6b.

nights per week till 8 p.m. to facilitate access. The printed catalogue could be purchased for 16 shillings.

Accolades flowed for many years. Fahie, who wrote a glowing account of Ronalds' telegraph in 1884, was equally effusive about the Library, calling it 'A magnificent collection of books on electricity, magnetism, and their applications. The catalogue…is a monument of the concentrated and well-directed labour of its indefatigable author.'[T15] Prof Silvanus Thomson FRS, whose own book collection was subsequently bought by the IEE, spoke in 1900 of 'Sir Francis Ronalds, whose library is the glory of electrical engineers'.[A140] In the early decades of the 20th century, mention of the Ronalds Library was often prefixed by the adjective 'famous'.

Concerning the Catalogue, Sir William Preece, the 1880 President, described it as 'a book that every telegraph engineer should possess'[A124] while the STE annual report noted 'the praises so abundantly bestowed upon it by the English and the foreign press'.[4] The Catalogue is referenced in *Origins of Cyberspace* (2002) as 'the first bibliography pertaining to telecommunications, and probably the first published bibliography…on specialized subjects in physical science'.[B184] It is still held in numerous libraries around the world.

Ronalds would have been more interested to know whether it and the Library were of use to others, as they had been to him. Certainly an official history of the IEE in 1987 advised that the Library's 'practical value was limited'.[B166] One person who did quickly avail himself of the resources was Arthur Kennelly (1861–1939). His first job on leaving school at age 14 was at the STE, where 'he found the Ronalds' Electrical Library, and all his spare time was spent in studying electro-physics in the excellent collection of Ronalds' books.'[A148] He went on to become Professor of Electrical Engineering at Harvard and MIT. The Catalogue itself was employed in academic research. Clark for example highlighted its value in clarifying precedence in discovery[T11] and it was a major source for Dr Paul Mottelay's *Bibliographical History of Electricity & Magnetism* (1922).[B137] Booksellers of rare scientific works were another user group, and remain so, quoting from the Catalogue as a mark of quality for their books on offer.[B134] It has been republished several times in recent decades. It was reprinted by Maurizio Martino in the USA in 1990s and by Cambridge University Press in 2013, with the cover page noting that 'It remains an invaluable resource for students in the history of science.'[R28] Overall, Ronalds could be pleased with his final achievement.

We saw in Chapter 13 that, through Ronalds' actions, the Kew Observatory survived its early years and achieved longevity and international renown. His early influence aided similar outcomes at the STE. The Ronalds Library was both an

---

[4]*Electrician* 18/3/1887 417.

imperative for survival and a foundation for success — it gave gravitas, instigated the creation of a legal entity, and provided a scholarly base in training the new profession of electrical engineers. That was at the end of his life. He was also revered by his STE colleagues as the first of their discipline because 60 years earlier he was demonstrating pioneering engineering himself. Ronalds was thus not only the first electrical engineer but played a part in creating the profession as a whole; these contributions have now been largely forgotten.

16.4 Recognition and Commemoration

The Ronalds Library was not his first achievement to attract widespread acclaim. When he left the UK in 1853, he was well known in scientific circles through both his important position in the BAAS and the increasing notice of his early telegraph demonstration. The telegraph continued to unfold while he was away, and the time was also ripening for his photo-recording instruments to form the cornerstone of a new era of systematic meteorology. Many people in his position would have stayed in London to enjoy the reflected glory of these developments. Ronalds was the opposite and probably appreciated his relative anonymity in foreign climes. Clark went so far as to say in 1872 that 'he lived for many years the life of an exile in Italy'[A116] but that is an exaggeration. His extended absence did hinder further recognition of his earlier accomplishments — it was long enough for him to be largely forgotten and for nearly all those who had first-hand knowledge of his telegraph to pass away. The success of the Ronalds Library was a useful reminder of his lifelong contributions, and it was also a palatable one for him in that occurred largely after his death.

Royal Society

The first formal acknowledgement of the quality of Ronalds' work by his peers had been his election to the Royal Society on 1 February 1844, when he had been at the Kew Observatory for a little over a year. It was Charles Wheatstone* who organised the nomination; the candidate's 'qualifications' are recorded in his hand as being:

The Author of: various Papers on Electricity, Meteorology &c.

The Inventor or Improver of: several Electrical Meteorological & Mechanical Instruments.[J66]

Nice and vague, with no mention of the telegraph.

Fellows endorsing the admission could sign the form in one of two categories. Those who recommended him 'From Personal Knowledge',[J66] other than Wheatstone, were Faraday, William Brande*, Thomas Graham* and Brockedon, who was an inventor and travel writer as well as an artist. Signatures in the category of 'From General Knowledge'[J66] were Sabine, Edward Solly and Henry Moseley. The list is short in comparison with other nominations in the period, underlining his low profile at the time. Solly was a Dissenter and a lecturer in chemistry, including at the Royal Institution with Faraday. Another of his interests was the influence of electricity on vegetation and his paper on this topic is in the Ronalds Library. Moseley was a colleague of Wheatstone's, being Professor of Natural and Experimental Philosophy and Astronomy at King's College, London. It was thus also a tight circle and Ronalds was perhaps a little lucky to be admitted to the Society.

Monetary Awards

The next tributes to Ronalds' achievements were in the form of monetary rewards from successive governments. Prime Minister Lord Russell awarded him a £250 prize for his self-registering instruments in April 1849 (see Sec. 15.10) and, on 9 August 1852, Lord Derby's Government added him to the Civil List for 'his eminent discoveries in Electricity and Meteorology'.[5] He, like several others at the time, was granted £75 per annum. They were later described in an anonymous letter to the *Sussex Advertiser* as 'wretched pittances'.[6] A more usual amount was £300, which both Airy and Sir James South* had received for years and William Wordsworth was granted in 1842.

Pensions were a topic of interest at the time. The BAAS had formed a Parliamentary Committee in 1852, chaired by Lord Wrottesley, and one of its first considerations was the underrepresentation of scientists in the awards: 'in a country like this, which owes so much to Science, there should be…means of rewarding…cases of great merit, which have been brought to the notice of the Government by Scientific Societies'.[S1] It was the Royal Society who put Ronalds forward for consideration — specifically, the President Lord Rosse, Past-president Lord Northampton and Sabine as Vice President and Treasurer, whose idea it would have been. William Sykes also became involved in his role on the Kew Committee.[C333] The problem was that the approach came very late in the government's budget cycle: 'it was favourably entertained, but the funds were exhausted'.[S1] Influence must have been exerted because the small pension was found within a couple of months.

Sabine was also BAAS President that year and would have enjoyed announcing the award in his Address: 'The intimate association…of Mr. Ronalds for several years past with the Observatory of the British Association at Kew, must render this…selection peculiarly gratifying to our Members.'[S1] Ronalds had not left Kew at this stage, but was much occupied with his mother's estate following her death.

[5] *Daily News* 8/7/1853 5e.
[6] *Sussex Advertiser* 30/10/1866; IET 1.9.2.

Arranging some tribute at this difficult time for his many years of honorary service to the observatory and BAAS was a nice idea. Perhaps Sabine could also see, before Ronalds himself did, that it was becoming time for him to move on.

Distinguished Men of Science

Ronalds received a very special honour in the late 1850s, when he was selected for inclusion in a large engraving entitled 'The Distinguished Men of Science of Great Britain Living in the Years 1807–8'.[B80] The piece depicts a hypothetical conversation in the Library of the Royal Institution (Fig. 16.4). It was created by Scottish engraver William Walker and others, and accompanied by a book containing memoirs of each of the 51 men and an Introduction by Robert Hunt. The package thus contained the first formal likeness and the first biography of Ronalds.

What made it special was that he was a decade younger than anyone else and the only one still alive. As such, none of his close colleagues were there — not even luminaries of similar age like Faraday and Herschel*. Those mentioned in this story who were included are Wollaston*, Davy, Troughton, Telford, Huddart, Banks, Brunel, Rumford, Young, Kater, Watt and Thomas Thomson. Other familiar names in the picture are the father or uncle of Ronalds' associate — Herschel, Stanhope*, Rennie* and Dollond.

Ronalds' inclusion would have been due in part to Bennet Woodcroft FRS, who was in the process of reinventing the British Patent Office — he provided

Fig. 16.4: 'Distinguished Men of Science of Great Britain Living in the Years 1807–8' (*c*.1862) — Ronalds is seated second from right; engraving by William Walker and others. © National Portrait Gallery, London 1075a.

considerable assistance to the authors. Dr Edmund had written to Ronalds about him in 1858:

> Mr Woodcroft who…devotes his leisure to collecting likenesses of inventors & copies of their works applied to me some time ago for a likeness of you as inventor of the Electric telegraph & tells me that he has had your work upon the telegraph handsomely bound & put into the south Kensington museum. I sent him a copy of an old Photograph taken by Malone many years ago & which was much faded & he has since written urging me to get you to sit for another which he intends placing in his Gallery of likenesses of inventors…it appears to be the object of his life to fish out the real discoverers & appliers if scientific facts to useful purposes & have the merit given to those to whom it is due.[C371]

Ronalds almost certainly did not bother to sit for another photo as requested — Thomas Malone probably only got the first one under duress. It must have been taken when they were working together on self-registering machines (Sec. 15.2) and Ronalds was around 60. It captures what would have been a typical pose — he appears to be lacking the confidence to hold his head up and look straight into the camera (Fig. 16.5). His hunched posture would have been exacerbated by

Fig. 16.5:   Ronalds aged about 60; photograph by Malone and Henneman. © Royal Society IM/003876.

many years of close work and study. It is this faded photo that was the source for his likeness in the engraving.

Very soon after Ronalds had left Kew, Gassiot also acquired a copy of this photo from Henneman and Malone for his collection of portraits of Royal Society Fellows. He explained to Ronalds, who had made a self-deprecating remark about it:

> I got it to please myself, and I tell you candidly — I am very much pleased when I look at it — for it reminds me of one who at an early period of his life cast away all selfish ideas of £.s.d. & devoted himself to the pursuits of Science & who I hope will still live many years to enjoy trifling as the amount may be, still that *[pension]* which has been obtained for him through a President of the Royal Society, who is capable of forming a correct and just estimate of his Scientific attainments.
>
> I presume you will at any rate allow me to say that being a little Electrician myself — my selection of Portraits would scarcely have been complete without that of Francis Ronalds.[C352]

It is perhaps Gassiot's copy that is now in the Royal Society archives.

With the image sorted, focus shifted to Ronalds' memoir. Samuel was contacted and he wrote to Ronalds in early 1860 seeking information. Ronalds' reply from Padua is the document referred to through this book as his 'autobiographical letter,' from which numerous excerpts have been quoted. Although written in a light-hearted manner, it is also dense and largely unintelligible to anyone not associated intimately with the technical aspects of his career and the contributions of others in his fields. Yet he had spent considerable time on it, as he noted in his diary:

> I attempted 5 Times to draw up a list of my scientific labours and in a sixth attempt concluded a rough draught *[sic]* of a list trying to curtail it as much as seemed consistent but making it much too long & diffuse for Mr Carter's or Mr Walker's purpose. One of them must cut it down (the more the better) if either will take the pains. I cannot.
>
> …The list became an imperfect kind of narrative from about 1813 to this time of scientific matters which I have undertaken & published &c omitting very many things.[J85]

He still had little idea how to 'sell' his achievements. As a result, his entry in Walker's book was just one page in length, listing the telegraph and several other inventions from the letter and a few publications that Samuel would have given him.

In the second edition of the book, in 1864, his memoir is three times longer and written 'From particulars derived from authentic sources.'[M1] He had now returned to England and his family and friends must have worked hard with him and Walker to shape the article. It focuses primarily on the telegraph but also includes mention of other early electrical science, the tracing instruments and tripod and their use at Carnac, the Kew years and his electrical library.

Another biographer, Geoffrey Hubbard, is much kinder to Ronalds in his description penned in 1965:

> In the National Portrait Gallery there is an engraving of a meeting of the Royal Institution which shows him as a young man, a slim elegant figure among the solid engineers of the industrial revolution. They represent the age of steam; he was the advance party of electricity, which was ultimately to make the assured progressiveness of the steam age obsolete. He was also, fortunately, a lively and engaging writer.[B152]

The engraving and book were well received on publication. *The Times* denoted them as 'an appropriate monument of our greatest scientific epoch'. Ronalds' portion of the picture was explained in the article as follows:

> A group on the right depicts Crompton, inventor of the spinning mule; Tennant, the well-known chymist and bleacher, of Glasgow; Cartwright, inventor of the power-loom; and Ronalds, who in the year 1816 passed an electric message through a space of eight miles made in trenches dug in his garden at Hammersmith *[this is incorrect]*; and this group is listening to Charles Earl Stanhope, who is supposed to be describing his printing press and process of stereotype printing.[7]

The columnist in the *Athenæum* highlighted that Ronalds was the only one of the scientists still living but admitted he knew little about him. He invited Ronalds 'to communicate two of three columns of autobiography'[8] for publication. Williams responded to this call with the letter also cited earlier:

> I beg to say I fear you will get no response, his retiring modesty being no less than his talents…It will be interesting, I believe, to know that in the last years, in which Mr. Ronalds has seemed lost to his friends, he has still maintained his known industry, and been working for the advance of science to which he devoted his earlier life and strength…The pleasure we derived from Mr. Ronalds's society, as well as the regret with which we parted from him on his return to England last month, induces this answer to the queries of the *Athenæum*.[1]

Knighthood

The idea of commemorating pioneers of the telegraph through knighthood came to the fore in late 1866, as we saw in Sec. 7.6. Sabine reminded the Royal Society at this time of Ronalds' achievements in both photo-recording and telegraphy (Sec. 13.5), while Samuel worked with government where he was well known and respected from his years of shepherding railway bills through Parliament. Samuel convinced Ronalds that a letter to Prime Minister Derby would be appropriate and

---

[7] *The Times* 26/5/1862 6c.
[8] *Athenæum* 23/8/1862 247–8.

helped him draft a simple statement of his early work on the telegraph.[C386] Samuel followed up with a meeting with Derby's Private Secretary Viscount Barrington, after which Barrington advised him in December 1866:

> Lord Derby...is obliged to decline re-opening the question of Honors for the Invention of the Electric Telegraph, in consequence of the length of time that has passed since the name of Mr Ronalds has been connected with the Invention.[C390]

The 'catch-22' upset Ronalds, and with his family's support he replied to Derby:

> from the time when I first offered it to the Government fifty years since, to the present day, my name has never ceased, (as I venture to hope it never will cease) to be connected with the Invention of the Electric Telegraph.[C391]

This was the period when he was very ill and had just written his will.

An opportunity to acknowledge formally the significance of his early work appeared at this very moment in the shape of the 1867 *Exposition Universelle* in Paris. The overseers of the British display decided to honour the men who had revolutionised industry in Britain. Fifteen pivotal developments were depicted on large blinds hung in the windows of the British section of the Machine Gallery. Famous names were there like Richard Arkwright and his spinning machine, James Watt's steam engine and George Stephenson's 'Rocket' steam locomotive. There were two telegraphs, Cooke and Wheatstone's, and Ronalds'.[B86] It might have been mentioned to the Commissioners that crediting Ronalds' work in this way could help Britain prove international precedence in the telegraph's invention over those pesky foreign claimants like Morse, as discussed in Sec. 7.5. Two of these Commissioners were Thomas Field Gibson and Sabine.

Derby led a Conservative government. Samuel was elected to Parliament in 1868 as the Liberal member for his hometown of Coventry. It was only a very brief term of office, but after the Liberal party was returned to power late that year, he was able to approach closer friends. Wheatstone and Cooke having now both received their knighthoods, he persuaded Ronalds in early 1870 to put together a Memorial for Prime Minister William Gladstone. The seven page document touched on his work at Kew, the use of his photo-recording machines by the Met Office, and his library, as well as the telegraph. It even alluded to Morse's ongoing claim that first invention was in America: 'If what Mʳ Ronalds has done is entitled to recognition the date of his Pamphlet in 1823 would at once dispose of this question.'[J96]

Maria and Samuel had planned a lengthy vacation at this time and were concerned lest it would harm Ronalds' chances. Samuel sought an update through George Glyn, who was Parliamentary Secretary to the Treasury:

> As I am going abroad on Monday for three months, unless there be any chance that, by deferring my journey, I could forward the claims of my relative Mʳ Ronalds, I am sure you will forgive my asking you whether it is any use my staying.[C397]

He followed up a few days later:

I am off tomorrow for 2 or 3 months.

Send me one line this afternoon, <u>if you can</u>, that I may set the mind of my old friend at rest.[C398]

With no word, Samuel and Maria set off for the Continent. Just a week later, on 28 March, Prime Minister Gladstone wrote to Ronalds:

I have great pleasure in announcing to you that, in acknowledgment of your early and remarkable labours in telegraphic investigations, Her Majesty will be pleased to confer upon you the honour of Knighthood.[C399]

Ronalds' short reply is in the British Library.[C400]

Maria and Samuel were informed in Rome by Glyn, putting them in a bit of panic. Ronalds was now 82 years of age and it would not be easy for him to make the necessary arrangements. Samuel wrote:

My dear Sir Francis,

At last — better late than never…I wish I was in England to assist…I fear a Court dress will be necessary, & perhaps a Sword, or at all events, a Scabbard![C401]

He added: 'Tomorrow I shall deliver your packet to Sign. Volpicelli & get his reply.'[C401] Ronalds was still busy collecting for his library — it contains 39 of Professor Paolo Volpicelli's works. Maria followed up the next day with her congratulations:

This is my dear Frank a very great delight to both of us — that you should live to know that your real merit is recognised by your country — & I think my good[?] husband is proud of having helped to bring the truth to light.[C402]

The helpful advice in their letters came too late. A note written the day after Gladstone's advised that the investiture would take place at Windsor Castle in two days' time. It was suggested that he stay in London overnight so that he could take a special train that would be organised from Paddington to the Castle on the 31st. He would have spent the night before and after with a young family member, perhaps Maria and Samuel's daughter Jane and her husband John Fletcher. The press announced the news: 'Mr. Francis Ronolds arrived at the Castle, and was introduced to the presence of her Majesty by the Lord Chamberlain, and received the honour of Knighthood.'[9] (What he wore is not recorded.) Letters of congratulations started to arrive immediately. One of the first was from Thomas Field Gibson on 1 April:

My dear Sir Francis

I cannot resist the pleasure of writing such an address, and troubling you with my congratulations on this tardy but well meritted [sic] acknowledgment of your claim to an invention of which others have had all the glory and profit.[C403]

---

[9] *Standard* 1/4/1870 3g.

Ronalds appeared in the next editions of *Debrett's Illustrated Baronetage*[M3] and *Dod's Peerage*.[M4] The former added a relatively new accolade to the rather tired stories of the telegraph: 'A Barograph invented by him at Kew, was employed in Adm. Fitzroy's later observations, and is adopted at the Radcliffe Observatory, Oxford, and with slight modifications in all the new Govt. meteorological observatories.'[M3] Ronalds' other seminal invention was now beginning to make its name.

Likenesses

When Ronalds had first returned from the Continent, Maria and Samuel would have decided that a likeness or two needed to be taken of him for posterity. At this stage there was only one photo — the informal shot taken by Malone — and it was in poor condition.

Probably their first step was to arrange a studio photograph by Maull & Polyblank, who specialised in portraits of noted people. The partnership changed names in March 1865, which dates the image to 1863–64 — Ronalds was then about 75 years of age. He is depicted conducting experiments with a balance electrometer invented by John Cuthbertson (Fig. 16.6). The product is a *carte de visite*, the size of a calling card, and it was popular to collect portraits in this style in the 1860s. The cards were distributed and preserved carefully around the family. Rollo Appleyard used one in his 1930 biography of Ronalds, noting it to be 'a photograph lately brought to light by Mrs. T. C. Carter, widow of a nephew of Ronalds'.[M7] Friends also requested one — Clark wrote in 1866:

> do you possess a carte de visite? Or would you mind favoring me by having one taken <u>at my cost</u>, and forwarding me a copy of it. I am making a collection of every known name in connection with telegraphy which I shall bequeath to the <u>Institution of Civil Engineers</u> & I should much like to possess your carte de visite <u>and autograph</u>.[C387]

A drawing was also made around this time depicting Ronalds at work at his desk (Fig. 16.7). His perspective tracing instrument is on the right, in front of what might be his pamphlet holder or card catalogue. Part of his library is behind him and his favourite dog sits on an adjacent chair. The artist is unknown. It may possibly be Ronalds' friend John Williams in Milan. A reproduction is included in the 'Electricians' Album' at the IET created by Silvanus Thompson.

Samuel and Maria's son Hugh Carter was an artist of some note and he and his parents convinced Ronalds to sit for a large portrait in oils (depicted on the cover of this book). It was painted around 1867. A woodcut of the work was made to accompany a long article published in the *Illustrated London News*[M2] just after his knighthood (Fig. 16.8). This was another simple format for family members to

Fig. 16.6:   Ronalds with Cuthbertson's electrometer; photograph by Maull & Polyblank (*c*.1864). © Royal Society IM/Maull/003875.

collect. His only remaining paternal cousin, Robert, carefully pasted the print on to sturdy pulp-board, and it also features in Sarah Flower's scrapbook. A Daguerre-otype was taken at this time as well, similar to the painting but orientated towards the right side of his face; it is held by descendants of Ronalds' brother Hugh Ronalds in the USA.

Maria proudly hung the original painting in her home. It passed back to her son on her death, who donated it in 1897 to the National Portrait Gallery. An engraved tablet on the gilt frame explains that the subject was the 'Inventor in 1816 of the first working electric telegraph'; the family specified the wording and requested of the Gallery that the painting be hung at a height such that 'the tablet may be easily read'.[C417] It was still on display in the 1950s but is currently in storage. The artist made a copy of the painting for himself just before the presentation and this second version was acquired by the IEE in 1924 from his widow's estate.

Fig. 16.7: Portrait of Ronalds by JW? (*c.*1870); perspective tracing instrument of the first kind is seen on desk to far right. Thompson Library IET 3.B.2.64.

Fig. 16.8: Woodcut of Ronalds based on Hugh Carter's portrait. *Illustrated London News* 30/4/1870 464.

STE members had first sought the original portrait in 1872 but the family was not yet ready to part with it.[10]

Ronalds' youngest niece Julia Ronalds in Australia (cousin to the Julia residing with him) was also an artist. She later created another oil painting of him, probably based on the Maull & Polyblank photograph.

Once he had become Sir Francis, something more was required. It was Samuel who commissioned a life-sized bust in white marble. The statue was created by Edward Davis in 1871 and exhibited at the Royal Academy that year. Samuel also arranged for a number of plaster casts to be made. He presented one to the Royal Society,[C406,A134] but it is now lost. Sarah Flower noted in her journal concerning the date 1 April 1870:

> Aunt Emily had a letter telling us that Uncle Frank yesterday was knighted by the Queen as the inventor of the *first* electric telegraph that was made to speak in England. At last, in his old age, his labours are acknowledged and it makes his friends happy. My mother *[Mary Ann]* soon after gave us a bust of him on which his invention is recorded in the year *1816* long before Cooke and Wheatstone brought out their telegraph.[B151]

Presumably the casts were distributed quite widely around the proud family.

The bust was on display at a soirée organised by retiring STE President Clark in December 1875 at which the Ronalds Library was first shown to the membership. It was perhaps this event that prompted the commissioning of a second marble bust from Davis. It was presented by Siemens to the STE in December 1876 and resided for many years on the landing of the main staircase at the IET.

A marble bust of Ronalds sat for many years in the garden at Kelmscott House, where he had built his telegraph (Fig. 16.9). It features in a photo with Alfred's grandson Oscar Ronalds (Fig. 16.10), who visited the UK from Australia in 1926 and met with Carter descendants. Inscribed 'E. Davis 1871', it is the original sculpture and would have been gifted by the family in the early part of the 20th century. It has recently been brought inside the museum there.

Publications and plaques

Another key focus for family and friends was publication. Dr Edmund had looked out for opportunities to advise the community of Ronalds' achievements while he was overseas. In 1858, he arranged to provide information for Johann Poggendorff's forthcoming biographical dictionary, 'thinking that your name ought not to be omitted in a work of the kind'.[C370] An entry listing Ronalds' key publications was included in the first edition of the renowned *Biographisch-literarisches*

---

[10] *J. Soc. Telegraph Eng.* **II** 22/1/1873 15.

Fig. 16.9: Ronalds aged 83; bust by Edward Davis (1871) photographed in 2012. Kelmscott House Museum.

Handwörterbuch der exakten Wissenschaften.[A101] Ronalds used the dictionary himself in annotating his library collection.

Starting in 1866, there are records of Samuel's responses to articles that celebrated others as the inventor of the telegraph, in which he reminded readers of Ronalds' contribution. Charles Flower (who was a prominent brewer and benefactor) also contributed, as did Clark. After Samuel died, his sons continued to uphold Ronalds' memory. He was not the only family member they promoted — John Carter edited the 10th edition of Alfred's *The Fly-fisher's Entomology* in 1901 and Dr Edmund's son Tennent as well as John assisted in the 11th edition in 1913.

The birth of the STE created a setting where Ronalds' name was mentioned very frequently. He remained relatively unknown however outside the engineering profession. The family tried to address this by encouraging and assisting authors to document his accomplishments for a broader audience. The Carters' input is apparent in Frost's biography in the Ronalds Catalogue. Fahie approached

Fig. 16.10:   Alfred's grandson Oscar with bust of Ronalds (1926). Ronalds family papers.

Dr Edmund in 1882 when he was researching his book *A History of Electric Telegraphy*, but in this instance he could not help:

> I…am very sorry that I am not able to comply with your request for a copy of the original or reprint of Sir F Ronalds' pamphlet. I had several of the latter but have given them all away & now only possess one copy of the original, which as you will easily understand I am very reluctant to part with.[C411]

The Carters found another willing mouthpiece in W. G. Kemp; he was the publisher of *West London Sketcher and Theleme*, a magazine centred in the area where Ronalds and his family had resided for many years. Kemp prepared several illustrated articles in 1889 commemorating Ronalds' activities nearby, including his telegraph and at the Kew Observatory. The first was entitled 'Who Invented the First Electric Telegraph? A Popular Misunderstanding' and ended:

it would be a fitting occasion for Hammersmith to…grace the house *[where Ronalds had built his telegraph]* or park with some just memorial to one who should be publicly celebrated for his great services in the interest of science and of his fellow-men, and a tablet on the house where the discovery was made, would, if it could be arranged, from *[sic]* a pleasing memorial of the inventor's merits as a man of science and a philosopher.[T16]

The second article continued:

A considerable portion of Sir Francis Ronald's life was spent in Hammersmith, Kew, and Brentford, and these parts may fairly claim him as an eminent local celebrity, whose fame should be suitably recognised and published to the world.[K10]

With the groundwork laid, John Carter approached William Morris, who was living at Kelmscott House at this time. Twenty-five years later he shared what happened in a postscript to a letter in the *Spectator*: '[Morris] was much annoyed, and, walking rapidly up and down, said: "I have often doubted which has been the greatest curse to mankind — railways or telegraphs." He, however, eventually consented reluctantly to the request, and the tablet is there.'[11] The plaque was placed in 1891, organised and funded by the Carters. The tale appears in slightly modified form in Appleyard's biography, who received assistance from Ronalds' great-great-nephew James Edward Montgomrey. The story must have been passed down through the family. Another token of remembrance, in a similar vein, was the naming in the 1880s of the new street adjacent to the family home in Highbury as Ronalds Road; this was instigated by Mary Sharpe.

John Sime penned a biographical article on Ronalds in the magazine *Lightning* in 1892.[T17] It included quotes from his autobiographical letter (which was then still in the Carters' possession), a photo of Hugh Carter's portrait and a drawing of the new plaque (Fig. 16.11). Sime's article, like Frost's, was reprinted in pamphlet form by the family, deposited in the British Library and elsewhere and distributed to interested parties. Sketches were also published in several newspapers showing the plaque in position on the side of the coach house (Fig. 16.12). A new tablet was affixed in 1904 at the expense of the IEE when the building was remodelled;[B139] it is now at the centre top of the former coach house.

John Carter's published letters contain interesting first-hand insights into Ronalds' attributes. He noted in the *Spectator* that Ronalds only 'reluctantly allowed his friends to claim for him'[11] in his later years the honour of inventing an electric telegraph. In 1921 he wrote in *Nature*: 'there can be no doubt — I know he had none —' that the length of the telegraph could have been extended considerably from his demonstration; 'he is well known to have been a most cautious,

---

[11] *Spectator* 20/3/1915 16–17.

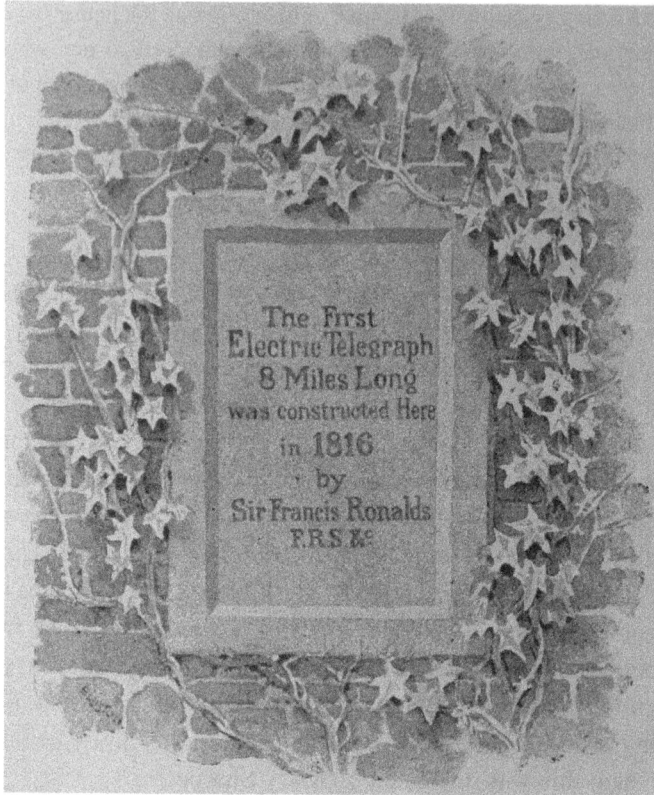

Fig. 16.11:   Original tablet at Kelmscott House (1891). Sime (1892).

prudent, and accomplished electrician.'[A144] The next year *Nature* commemorated Ronalds in a 'Calendar of Industrial Pioneers'.[A145]

16.5 Ronalds' Legacy

Ronalds was a prolific inventor for over half a century. His first idea to be documented was a new mode of electrical insulation around 1810; Singer made it his own and facilitated its wide adoption for over a century. The last known, in 1862, was a propelling rudder for ships, which has also seen extensive use since its reinvention in the 20th century. He is not associated in the literature with either of these technologies. He ended his autobiographical letter: 'There is a "Century of inventions" &c in Common-place <u>&c</u> not worthy of mention here'.[C375] His role in some of these is lost forever. (His allusion would be to *A Century of Inventions*, written by the Marquis of Worcester in 1663;[B29] it is a list of 100 of the Marquis' ideas, which were discounted by many as 'folly', but included the steam engine.)

Fig. 16.12:   Kelmscott House showing plaque in original position on left of former stable (*c.*1891). WU B5109.

A handful of Ronalds' many discoveries have had broad and lasting impact on the world, but his name is linked with just a few of them. Discussion has always centred on the telegraph, because his work was well documented and the outcomes of the technology itself were so far-reaching. The telegraph revolutionised global communication and instigated the practical use of electricity — few devices have been so important in helping to create modern living. It is right for his contribution to be remembered: various aspects of his original invention were borrowed by later scientists and practitioners and a direct line can be drawn from his work to today.

A more important invention in terms of his ongoing participation is the continuously recording camera for automatic registration of natural phenomena. He showed here that he was more than a scientist who could demonstrate an idea. He was also a developer (no pun intended), going on to build durable instruments and then driving their adoption and use internationally. His actual machines were in daily deployment in a leading observatory for 25 years and slightly modified versions for over a century. The whole community has profited from the results in terms of weather forecasting and improved knowledge of meteorological and geomagnetic variability. Despite these wide benefits, his contribution was, and is, little known outside the observatory network.

Also in this top tier may be placed the hinged tripod still found under the surveyor's theodolite and the slide-rest revered by Holtzapffel as a foundation of mass production. Ronalds participated in the extensive commercialisation of both but did not publish either, so his name remains unconnected with them. They also address more narrow domains than communication and the weather and do not capture the public imagination in the same way.

At a second level might be listed numerous novel devices that he created for personal use but which then saw wider distribution. He often made these items himself for friends and colleagues. Many of the meteorological instruments and electrometers invented at Kew fit into this category, as do his geometric pen, lathe accessories and various telescope supports. Demand was strong on occasion — in manufacturing and selling 300 of his patented tracing instruments he built a successful business. At other times he did not participate in commercialisation and simply documented the concept, for example with his semi-automatic electricity generator and his means of ensuring the electrostatic generator would work in all weather conditions. Again, these ideas had considerable take-up.

A further class of innovation occurred where Ronalds stopped after proving the concept to himself through prototyping and the device evolved in other hands over succeeding decades. Sometimes his original work was recognised, for example with his storm clock and airborne meteorological instrument platform. In other cases the instrument was invented anew. This subset includes the battery-operated clock, his fire finder for pinpointing the location of a fire, volute compasses for drawing a spiral and several instruments for measuring novel aspects of atmospheric electricity. Some of these inventions are still in use today.

The final group might be the myriad sensible things he developed for home and family, like his self-actuating fire alarm, table leg height adjusters, blinds that open from the top down, and toys for his nieces and nephews.

Ronalds was not only a creative inventor and a practical engineer, but also made important theoretical contributions to science. The standout is articulating signal retardation in long insulated wires due to induction, 20 years before Faraday. The mode of operation of the dry pile, including the delineation of voltage and current, and his early ideas on global atmospheric electricity, are similarly noteworthy. His explanation of the window of vision in fishing is still revered in the fly fishing world. He additionally generated surviving datasets of value. One comprises early survey results and drawings of the Carnac megaliths and another is five years of comprehensive atmospheric electricity measurements at Kew. These brought forward by two decades both the era of accurate visual recording and the longest time series of atmospheric electricity observation worldwide.

His inspiration was considerable. His extraordinary vision is first seen in his description of a new age of electricity and rapid communication. Soon after, he seemed to identify the need for a sulphur import industry in the industrial revolution. He developed a holistic fire risk management strategy in 1840 for his capital city that was decades ahead of its time. From there he went to Kew, where he shaped the mission for a world-leading observatory and nurtured its early development. Perhaps his biggest concept of all was the acquisition of a complete electrical library and its donation to help educate the initial generation of electrical engineers and thereby sustain their embryonic professional body. It has been argued in these pages that Ronalds himself was the first in this new cohort.

It is an impressive inventory and underlines the dilemma of why he is little known today. Much of the reason relates to his retiring nature. Sime noted in 1892:

> Ronalds was a man of great originality, whose modesty, amounting to diffidence, did much to hide his real merit from the world...He was content to do his share of the world's work quietly, modestly, and with a single eye to the promotion of scientific knowledge without regard to any personal rewards.[T17]

He did not promote his ideas through publication or lecture, nor did he join clubs or sit on committees. He would not have impressed in a gathering. In all, he was not seen as a leading scientist in his peer group and few records were created of his achievements. It is remarkable in fact that his long-term influence has been as significant as it is, and it is solely through the quality of his scientific output.

Another factor in his relative obscurity was that he worked in quite different domains over five decades. His meteorological activities were well understood in the BAAS years but his telegraph was then a piece of history. Only those with concurrent interests in turning inventions or telescopes would have recognised his past successes in such areas. By way of example, Sabine, who was his closest scientific colleague for nearly a decade, expressed surprise in 1848 when Ronalds mentioned that the tripod in Sabine's office was 'mine'.[C239] Not even his family appreciated the gamut of his portfolio — they worked hard to keep his name alive but their efforts focused on his telegraph. There was no upwelling in the scientific community to raise his stature through posthumous memorial or comprehensive biography. There seemed relatively little reason to remember him. There have been very many others like him, of course, and few stars remain bright through succeeding generations.

Ronalds' inherent lack of confidence was certainly exacerbated by two big knock-backs so early in his electrical career — Singer calling his clock a 'toy' and the Admiralty dismissing his telegraph as 'wholly unnecessary'. If there is a message for today in his story it is perhaps to spend a little time with the shy, bumbling person in the corner. Barrow and Singer found it easy to mock but much more would have been achieved if they had encouraged the sensitive young scientist. It is impossible to know where his quick-fire succession of electrical inventions and theoretical insights would have led next, but the world might well have had an electricity industry as well as many other 'modern' innovations many years earlier. I will let Frank have the last words, who made this point in his autobiographical letter:

> I will say that if the electric telegraph of 1816 had been fairly examined, an effective instrument might have been in the hands of the government & that after Dr Œrsted's…experiments *[in 1820]* an improved telegraph might have been in the said hands *[well before Cooke and Wheatstone's work in the late 1830s]*[C375]

and ended his 1871 reprint of *Description of an Electrical Telegraph*:

> I only cherish a hope that, as a pioneer…I may have contributed in some degree to the sum of human knowledge and happiness, and that my name may remain connected with an Invention which has conferred incalculable benefits on mankind.[R26]

APPENDIX
BRIEF BIOGRAPHIES OF RONALDS' ASSOCIATES AND FRIENDS

Professor Sir George Biddell AIRY FRS (1801–92)

George Airy was appointed Professor at Cambridge a few years after graduating there, and established a large telescope with an equatorial mount. He was then Astronomer Royal for 46 years — a role that included the Directorship of the Royal Observatory Greenwich under the auspices of the Admiralty. It was the highest office in civil science and he proactively extended his influence by advising broadly across government and industry.

Airy succeeded Sabine to be President of the Royal Society in 1871–73, having preceded him as President of the British Association for the Advancement of Science (BAAS) in 1851–52. He gives a nice illustration of his attitude in his autobiography:

> This year I was President of the British Association…Prince Albert was present…I was engaged to meet him at dinner, but when I found that the dinner day was one of the principal soirée days, I broke off the engagement.[B115]

His son wrote in the introduction to this autobiography that 'In debate and controversy he…was absolutely fearless',[B115] while his obituary noted he was 'a tough adversary'.[A135] One of his foes described him as 'adept in bringing his spite to bear on anyone who happens to offend him'.[A90] Ronalds was on the receiving end of some of this behaviour as inaugural Superintendent of the Kew Observatory, which Airy saw as a competitor to his own establishment. Nine of Airy's reports are in the Ronalds Library, mainly concerning equipment and results from the Greenwich Observatory.

Professor William Thomas BRANDE FRS (1788–1866)

William Brande was appointed to the Chair of Chemistry at the Royal Institution of Great Britain in 1813, where he worked with Faraday and Davy. The same age as Ronalds, the two probably met at Institution lectures. When Ronalds shared his work on the dry pile with him, he replied: 'I was anxious to see what you had done, as I shall notice the electric column in my next lecture'.[C7] Brande also witnessed Ronalds' telegraph demonstrations at Hammersmith. He was Editor of the *Quarterly Journal of Science, Literature, and the Arts*, where Ronalds published two of his papers, and Ronalds collected two of Brande's papers on electro-chemical phenomena for his library. Brande was one of Ronalds' nominators to the Royal Society.

Sir David BREWSTER FRS (1781–1868)

Sir David Brewster was a respected Scottish physicist, specialising in the areas of optics and photography. He served as Principal of St Andrews University and later Edinburgh University and wrote and edited numerous works on science.

Ronalds and Brewster interacted though much of their adult lives, with Brewster playing an ongoing role in publicising Ronalds' achievements. The *Edinburgh Philosophical Journal*, which Brewster established, picked up and published Ronalds' automated electric generator in 1823. Brewster made mention of Ronalds' atmospheric electricity work in his 1832 *Edinburgh Encyclopædia* and his photo-recording in the 1847 *North British Review*. He was a leading contributor to the 1842 and 1855 editions of the *Encyclopædia Britannica*, including writing the long article entitled 'Electricity' that described much of Ronalds' early electrical science and engineering.

Brewster was one of the founders of the BAAS, where he was President in 1850–51 and supported Ronalds' efforts at the Kew Observatory. He was also a juror at the 1851 Great Exhibition and its follow-up in Paris in 1855, at both of which he passed comment on Ronalds' instruments. There are nine of Brewster's works in the Ronalds Library, including the *Encyclopædia Britannica* entry, as well as several of the journals he edited and two portraits.

Dr Charles BROOKE FRS (1804–79)

Charles Brooke studied at Cambridge and then St Bartholomew's Hospital, becoming a surgeon. He invented several instruments and processes, both related and unrelated to medicine. Principal among these was photographic recording apparatus for the Greenwich Observatory under Airy's patronage — magnetic instruments of this type were also deployed in Toronto, Paris and the Smithsonian Institution. Brooke wrote later editions of Golding Bird's *Elements of Natural Philosophy*;[B83] the book mentions Ronalds in relation to his telegraph, but describes only Brooke's approach to self-registration, including his monetary prize and his medal at the Great Exhibition. It was probably Brooke who penned the obituary of his good friend Wheatstone in the *Proceedings of the Royal Society*, in which Ronalds' telegraph was described as a 'hopeless scheme'.[A121] Ronalds collected two of Brooke's photo-registration papers for his library.

(Josiah) Latimer CLARK FRS (1822–98)

Latimer Clark worked initially as a chemist before taking up civil and electrical engineering. He joined the Electric Telegraph Company in 1850, and maintained the association for 20 years. As Chief Engineer, he superintended the construction of many kilometres of telegraph line in the UK as well as subsea cables in the Atlantic Ocean and the Far East. He also filed numerous patents.

Clark had a strong interest in the development of the fledgling telegraphy industry and in preserving its early history. He took a leading part in founding the Society of Telegraph Engineers (STE) in 1871 and was President when the Ronalds Library was transferred to the Society in trust. He had first introduced himself to Ronalds in 1861, writing to him in Italy. Clark also built up his own library, which was later purchased by Schuyler Wheeler and presented to the American Institute of Electrical Engineers. The Introduction in the *Catalogue of the Wheeler Gift* begins:

> The philosopher…writing on hobbies of a high and inspiring nature, has two recent examples at hand to illustrate his theme, viz.: the example of Sir Francis Ronalds and that of J. Latimer Clark. These distinguished men were contemporaries and friends; both long-lived, moderately moneyed, and ardently fond of old electrical books.[B128]

Clark's and Ronalds' libraries each held 10 of the other's works. Other items in Clark's collection included several of Ronalds' letters and original drawings, his photographic portrait and a piece of his telegraph cable.

Professor John Frederic DANIELL FRS (1790–1845)

John Daniell was the inaugural Professor of Chemistry at King's College London and a 'mutual excellent friend'[C128] of Ronalds and Wheatstone. He and Ronalds shared interests in electricity, meteorology and mineralogy. Daniell was one of

Ronalds' early supporters and first visitors at the Kew Observatory, where Ronalds used his dew point hygrometer. He also invented the Daniell cell, which provided a reliable supply of electricity for the first time. The magnitude of the unit 'volt' was later designed to approximately equal the electromotive force created by one cell. Ronalds collected 11 of Daniell's publications, including the statement of facts written with Brunel on the relative contributions of Wheatstone and Cooke to their telegraph.

General Joseph-Antoine DE GOURBILLON and his Family

Prior to the French Revolution, Joseph de Gourbillon had been Secretary of the court of Marie Joséphine (the wife of the King's younger brother) and his mother and/or wife was her lady-in-waiting.[B19] Mme Marguerite de Gourbillon is known for having an intimate and long-lasting relationship with Marie Joséphine,[B192] with the countess writing frequent love letters to her. Marie Joséphine became Queen when her husband took the title of Louis XVIII, King of France, in exile in 1795.

Despite his powerful position, de Gourbillon's greatest interest was literature. His written works addressed topics as diverse as translations of Dante's letters and verse; English culture; the fate of House of Bourbon (as predicted in the Bible); comic opera; as well as his tour of Sicily. Ronalds met him in Naples in 1819; they travelled in southern Italy together for some months, precipitated in part by Ronalds' monetary problems, and it was a mixed experience.

Jean-André DE LUC FRS (1727–1817)

Jean-André de Luc was a Swiss geologist and meteorologist. The second half of his life was lived close to London, where he held the prestigious position of Reader to Queen Charlotte in George III's court. He is best known scientifically for his research on the dry pile he helped to invent. He also devised meteorological instruments, and Ronalds took a hygrometer of his design on his Grand Tour. The Ronalds Library contains eight of his works together with his letters to Ronalds.

Although they had a 60-year age difference, de Luc was an important support to Ronalds in his early years of electrical science. He provided little scientific input however; in preparing for Ronalds' first visit, de Luc warned: 'I shall be very happy to…convene with you for about an hour, for my head is not capable of holding long conversations.'[C4] Another time he admitted: 'my memory is now so deficient, that I sometimes am at a loss to find out what I mean, or the purpose for which I made such a remark'.[C11]

Professor Michael FARADAY FRS (1791–1867)

Michael Faraday was the first Fullerian Professor of Chemistry at the Royal Institution. His fundamental discoveries in electrochemistry and particularly electromagnetism underpin modern electrical engineering for the industrial world. The International System (SI) unit of capacitance (the farad) and a non-dimensionalised unit of charge (the faraday) are both named in his honour.

Faraday was of similar age to Ronalds and also a Dissenter; they would have first interacted at science lectures in London, perhaps as early as 1810. Faraday was later a nominator for Ronalds' election to the Royal Society. Ronalds collected eight of his works for his library as well as a portrait engraving. He additionally purchased around 27 of Faraday's copies of papers written by other scientists in an auction arranged the year after his death — these are so identified in the Catalogue to the Ronalds Library. Faraday also knew of his cousin Dr Henry Ronalds — they were expert witnesses on opposite sides of a court case in 1818 concerning the preparation of malt for beer.[B172]

John Peter GASSIOT FRS (1797–1877)

John Gassiot was a wealthy merchant. His company Martinez Gassiot was the largest shipper of port from Portugal to the UK and still exists today. He lived close to Ronalds' cousin George Field Jnr at Clapham Common and the two were friends.

A keen scientist, Gassiot was for many years Chairman of the Kew Committee, and he ensured the observatory's future in 1871 through a £10,000 donation. Ronalds acquired six of his papers, including Faraday's copy of that where he showed that it was chemical reaction rather than metal-to-metal contact that drove the action of a voltaic battery. Gassiot in turn collected an early photo of Ronalds for his collection of Fellows of the Royal Society.

James GLAISHER FRS (1809–1903)

James Glaisher worked as Airy's assistant for over 40 years, first at Cambridge and then as Superintendent of the new 'Magnetical and Meteorological Department' at Greenwich; his was effectively the first UK government meteorologist position. Their relationship was sometimes a little rocky: Airy exerted complete control while Glaisher's descendant and biographer advises that 'Glaisher was a man of uncompromising and fixed opinions'.[A176] Against Airy's wishes, Glaisher was extremely supportive to Ronalds in the early stages of his development of photo-registration. Ronalds and Glaisher developed a close friendship and Ronalds was one of his nominators to the Royal Society in 1849.

Glaisher later became President of the Royal Meteorological Society and was also President of the Royal Photographic Society for over 20 years. He is best remembered today for his daring balloon flights where he measured atmospheric conditions at altitude.

Professor Thomas GRAHAM FRS (1805–69)

Thomas Graham graduated from the University of Glasgow and went on to become Professor of Chemistry at the University of London. Crabb Robinson struggled with his science:

> I went to a Conversazione…Saw a solid matter proceed from the combin[n] of Gases, but I understood nothing in the lecture. D[r] Graham co[d] exhibit sights but attempted nothing in the way of explanation. It was a bad as possible.[1]

Graham later replaced Herschel in the prestigious role of Master of the Mint.

Graham was one of Ronalds' nominators to the Royal Society, the two having become good friends. He also supported the Kew Observatory in its initial subscription and in his membership of the 1846 and 1848 Review Committees. Ronalds aided his experiments on gases by collecting 'clean' air at Kew. Dr Edmund Ronalds' first academic position was in Graham's laboratory, as the Brentford cousins explained:

> Cousin Edmund…has I think most fortunately met with a situation exactly suited to him as assistant to a M[r] Graham the first Chemist in London which will occupy him from 11 O' clock to 5 every day and be the means of introducing him to become a popular man himself [C106]

There is just one of Graham's papers in the Ronalds Library.

Sir Frederick HENNIKER Bart (1793–1825) and his Family

Lord John Henniker-Major FRS (1752–1821) interacted with Ronalds on science matters, including witnessing his telegraph in operation and providing glass rods for use as electrical insulators.[C5] Watching the ballet in Paris, Ronalds later contemplated another application for the glass:

> The petticoat is a very usefull part of the Costume, were it not for this He's might come to be mistaken for She's. I have some idea of taking out a patent for ethereal petticoats, they shall be made of Spun Glass such as Lord Henniker sent me a specimen of some time ago from Bristol.[J17]

Lord Henniker's nephew Sir Frederick Henniker was a close friend of Ronalds — they would have enjoyed each other's intellect and wit. Both

---

[1]Ref. J13 25/4/1838.

conducted their Grand Tours at the same time, and they met up in Sicily although not, as hoped, in Egypt. Back in London, they worked together on their observations and Ronalds assisted with Henniker's book on his travels. Henniker then turned to politics and was preparing to stand in the 1826 election when he died suddenly at the age of 31.

Sir John Frederick William HERSCHEL Bart FRS (1792–1871)

Sir John Herschel was a mathematician, astronomer and chemist. He was elected to the Royal Society the year he graduated from Cambridge and was awarded a baronetcy for his astronomical survey of the southern hemisphere. Regarded as England's pre-eminent scientist, he was an important force in the establishment and early survival of the Kew Observatory, and Ronalds acknowledged his support in his autobiographical letter. They also interacted a little on photographic matters. Two of his works are in the Ronalds Library, one of which is his entry in the eighth edition of the *Encyclopædia Britannica* on Meteorology. He was buried at Westminster Abbey close to Sir Isaac Newton.

Charles HOLTZAPFFEL (1806–47) and his Family

After settling in London in 1792, John Jacob Holtzapffel (1768–1835) founded a mechanical and tool business specialising in treadle-powered lathes. His son Charles and grandson John Jacob Jnr (1836–97) helped the business continue into the 20th century.

Ronalds bought his first Holtzapffel lathe from John Snr, as well as other equipment, after he started to make scientific instruments for his experiments. The company also manufactured and sold several devices that he invented. Charles provided advice concerning the design of his perspective tracing instruments, and they also probably collaborated on curve-drawing apparatus. Business letters survive between them regarding his slide rest. It is likely that Charles was involved in a comprehensive book on turning that Ronalds commenced but never completed. Charles and later John Jnr went on to publish a five-volume set *Turning and Mechanical Manipulation*, which became highly influential texts in ornamental turning.

Manuel John JOHNSON FRS (1805–59) and his Family

Manuel Johnson was superintendent of the East India Company's observatory at St Helena, an island off the coast of South Africa, before spending 20 years as

Director of the Radcliffe Observatory at Oxford. He also served as President of the Royal Astronomical Society.

Ronalds worked closely with Johnson in the early 1850s when he developed photographic recording instruments for Oxford. He was invited to stay with Johnson and his wife Caroline: 'I shall be rejoiced if you would pitch your tent here…the only diffidence I have about it is that you will find us hum drum hosts.'[C339] Johnson advised after he departed: 'When you return you will find another inmate of our Family — whom my wife brought to light this morning. She is doing very well and always desires her best remembrances'.[C348] Ronalds was in fact on his way to the Continent, although they kept in touch — Johnson's photographs proved important in publicising the quality of Ronalds' equipment. Sadly Johnson died just a few years later, leaving Caroline and six children under the age of eight.

Dr John LEE FRS (1783–1866)

John Lee LLD, born John Fiott, inherited the beautiful Hartwell House near Aylesbury. It had earlier been let to King Louis XVIII during his exile in England after the French Revolution. Ronalds and others were invited to vacation at the property after several BAAS conferences. It was there in 1850 that the Royal Meteorological Society was formed, although Ronalds was unable to attend on that occasion. Lee and his wife Cecilia also visited Ronalds and his mother in Chiswick, and Ronalds' niece Sarah Flower stayed with the family in 1858.

Although Lee read mathematics at Cambridge, built an astronomical observatory at Hartwell and later became a barrister and QC, his and Ronalds' scientific interactions generally concerned meteorology and archaeology. Ronalds gave Lee a wind vane of his design, while Lee helped publicise his and Blair's work at Carnac. Lee was also the inaugural President of the Numismatic Society of London.

General Sir John Henry LEFROY FRS (1817–90)

After training at the Royal Military Academy at Woolwich, John Lefroy worked under Sabine in the Magnetic Crusade. Talented and ambitious, he took over from Lieutenant Charles Younghusband as Superintendent of the important Toronto Magnetic and Meteorological Observatory in British North America in late 1844. Eight years later, he managed the observatory's handover to the government of the Province of Canada. He subsequently served as Governor of Bermuda and then Administrator of Tasmania.

Lefroy corresponded regularly with Ronalds during his years in Toronto concerning equipment provided to the observatory. Lefroy's relationship with Sabine was at times strained and Ronalds risked being caught in the middle — particularly concerning the pursuit of atmospheric electricity observations, which were of great interest to Ronalds and Lefroy, but not to Sabine. Five of Lefroy's Canadian papers are in the Ronalds Library, including two on the aurora borealis, which was a related scientific interest they shared.

François-Napoléon-Marie MOIGNO (1804–84)

The Abbé Moigno was an opinionated French priest, scientist and author. He described Ronalds' telegraphic inventions in his 1849 book *Traité de Télégraphie Electrique* and was perhaps the first to denote him as a 'father' of the electric telegraph. Ronalds became good friends with him on moving to Paris in 1853, with Moigno penning various flattering articles on his inventions in his journal *Cosmos*. He also assisted with Ronalds' French booklet *Météorologiques et Magnétiques* and offered it for sale at the *Cosmos* offices. Five of his works are in the Ronalds Library, including both editions of his treatise on the telegraph.

John Bowyer NICHOLS (1779–1863) and his family

John Nichols, his father John Snr and son John Jnr enjoyed literary and antiquarian pursuits and together edited and printed the *Gentleman's Magazine* for nearly 80 years. The Nichols and the Ronalds families were friends over several generations.

John was born and raised in apartments above the family's printing business in Fleet Street. His father lived his last years in Highbury Place where he interacted with Ronalds' grandfather William Field Snr. John, his wife Eliza and their children later moved to Hammersmith and their activities there are documented in daughter Mary's diary.[133] With her interests in art, science and horticulture, Mary enjoyed her frequent calls on Ronalds and his mother in nearby Chiswick.

Richard B. NICKLIN (*c.*1809–54)

After employment at Richard Beard's Daguerreotype studio, Richard Nicklin resided at the Kew Observatory in the period 1849–52 where he assisted Ronalds and John Welsh. He built parts of Ronalds' second-generation magnetographs and performed the role of 'photographist'[R21] in a formal trial of the instruments. He also accompanied Welsh in his 1852 balloon ascents. In 1854 he was commissioned by the British War Office to document the Crimean War, so heralding the use of photography in war propaganda. He died later that year, and his work was lost, when his ship sank in the region.

Professor John PHILLIPS FRS (1800–74)

Geologist John Phillips held senior academic positions in York, London, Dublin and Oxford. He published the first global geological time scale, founded on fossil data, and helped to standardise terminology in the discipline. Ronalds collected two of his papers on the magnetic and electrical phenomena of rocks for his library. Crabb Robinson read several of Phillips' texts, commenting that one was 'really an instructive little book tho' elementary & for readers of the lowest class'.[2] Phillips was also one of the founders of the BAAS. He was the inaugural Assistant General Secretary, holding the position for 27 years, and became President in 1865–66. He was one of Ronalds' strongest supporters at Kew.

Revd Dr Joseph PRIESTLEY FRS (1733–1804)

In shaping science, religion and politics, Joseph Priestley would have been one of the people Ronalds most revered. The Ronalds Library contains the first five editions of his standard text *The History and Present State of Electricity* (1767) along with various other works on electricity and his portrait. Ronalds tried over many years to write an update of the book but it was not completed.

Revd Priestley's theology fully embraced exploration and questioning of the natural world, a philosophy that came to be associated with the Unitarian religion he helped found in England. He was also outspoken on various issues, including the laws that restricted the education and employment rights of Dissenters. While he was minister at the New Meeting in Birmingham (later the church of Alexander Blair's friend James Yates), the chapel, Priestley's home, and those of a number of his sympathisers were attacked and burnt in riots in July 1791. Priestley and his family migrated to the USA shortly afterwards. Ronalds' sister Mary Ann was the last baby he baptised before embarking for the New World.[B131]

Sir John RENNIE FRS (1794–1874) and his Family

The Rennies were a renowned engineering family. Sir John built the 'new' London Bridge adjacent to the Ronalds merchant business, which had been designed by his

---

[2]Ref. J13 16/3/1822.

father. Sir John's older brother George was a mechanical engineer, who focused in particular on the emerging steamship and railway sectors. All three were elected to the Royal Society.

Ronalds' and the Rennie brothers' paths crossed at various times. Ronalds met Sir John in Athens in 1820 and, in 1837, he sought advice from the brothers on his canal steamboat scheme. Both brothers subscribed to the prospectus for the Kew Observatory in 1842 and four years later Sir John credited Ronalds with the invention of the telegraph in his Presidential speech to the Institution of Civil Engineers.

Henry Crabb ROBINSON (1775–1867)

The son of a tanner, and a Unitarian, Henry Crabb Robinson wrote of himself: 'I am spoken of as a barrister & scholar well known in the literary world'.[3] Today, he is remembered principally for his voluminous diaries, which detail the activities and attributes of his circle. Brief excerpts included through these pages illustrate the highly interconnected nature of the upper middle class, and particularly the Dissenters. They also provide valuable glimpses into Ronalds' world outside his science. The diaries were bequeathed to Dr Williams's Library when Crabb Robinson died at age 91.

Crabb Robinson summed up Ronalds when they met in 1823:

At 3 I went to Mr P*[eter]* Martineau's with whom I dined & spent a very agreeable afternoon — Only Arthur Taylor *[printer and author]*, Miss Christie *[daughter of merchant John Christie]* & M^r & Miss Ronalds there. An uninterrupted convers^n on many subjects. R: has travelled — in Turkey &c. He is said to be a very expert Electrician — His Convers^n is not that of a thinking man.[4]

Ronalds got off relatively lightly — Crabb Robinson admitted his inability to understand science and described another engineer (the cousin of Hugh Ronalds' wife Kate) as: 'A dreadful bore. The very sight of him makes me uncomfortable — he talks of nothing but his mechanical schemes'.[5]

Revd Dr (John) Thomas Romney ROBINSON FRS (1792–1882)

Thomas Romney Robinson was Director of the important Armagh Astronomical Observatory in Northern Ireland. Ronalds would have interacted with him in the early 1830s in relation to Sir James South's equatorial mount and they remained

---

[3]Ref. J13 18/7/1834.
[4]Ref. J13 20/4/1823.
[5]Ref. J13 8/12/1828.

friends. Ronalds sought his advice on his early ideas for a self-registering thermometer and Romney Robinson's renowned anemometer was later installed at Kew. He was a strong supporter of Ronalds' atmospheric electricity work at the BAAS, where he served as President in 1849–50. The *Literary Gazette* said of him at that time: 'The action of the reverend orator is very odd, and consists in raising his left arm perpendicularly up and down, whilst the right is pinned to his side motionless, like the empty coat-sleeve of a wounded man.'[6] There are three of his papers in the Ronalds Library.

General Sir Edward SABINE FRS (1788–1883)

Edward Sabine obtained an officer's commission in the Royal Artillery at the age of 15. His interest in geomagnetism, sparked during his early tours of duty, was to be a lifelong passion. He became the first great bureaucrat of organised science — at the heart of the Magnetic Crusade — coordinating observers around the world to help unravel its mysteries. There are 25 of his papers on magnetism in the Ronalds Library, a number of them formerly belonging to Faraday.

Sabine was General Secretary of the BAAS for two decades (except when he was President). The role included responsibility for the Kew Observatory, where he provided invaluable support to Ronalds. Right at the beginning of their interactions he supported Ronalds' election to the Royal Society, having already served as its Vice-President. His ten-year Presidency of the Society followed over a decade as its Treasurer and he continued in office as Chairman of the Kew Committee until his death at the age of 94. In these positions he had a prolonged and eventually successful tussle with Airy for the supremacy of Kew over Greenwich in meteorology and magnetism.

Samuel SHARPE (1799–1881) and his Family

The Sharpes of Highbury were good friends of the Ronalds family. Residing in Highbury Place, Samuel Sharpe was a banker by profession, but is remembered as an Egyptologist, for his translations of the Bible and (as a Unitarian) for his strong support of UCL and Dr Williams's Library. Crabb Robinson described him as 'a sensible man with a little pretension'.[7] Ronalds asked Samuel's advice on his and Blair's manuscript concerning ancient monuments and also on his early ideas for

---

[6] *Literary Gazette* 15/9/1849 672–4.
[7] Ref. J13 16/11/1840.

photo-recording of meteorological parameters. Samuel viewed the Kew Observatory with him in 1848.

Samuel's brother William leased the Ronalds' former family home at 1 Highbury Terrace for many years. He was an attorney, in partnership with Ronalds' relation and friend Edwin Wilkins Field. It was almost certainly William's daughter Mary C. Sharpe who suggested in the 1880s that the access-way adjacent to her home be named Ronalds Road to commemorate Ronalds.

George John SINGER (1786–1817)

George Singer lived in London's West End and delivered lectures on electrical science; according to his *Dictionary of National Biography* entry both Ronalds and Faraday attended. He also made and sold electrical equipment. He had broad and deep knowledge of the subject but was not an original thinker. In 1814, he published the book *Elements of Electricity and Electro-chemistry*, which was well received at the time and today gives a very useful picture of what was and was not known about electricity when Ronalds started in the field. In his writing however he comes across as a self-important man.

Ronalds ran some errands for Singer in Paris during his first recorded trip to the Continent in 1814 with his brother Hugh. Singer helped Ronalds in his early electrical endeavours but they fell out in 1815. Ronalds nonetheless collected 11 of his works for his library, including copies of his book in English, French, Italian and German and the paper in which Singer described Ronalds' first acknowledged invention as 'rather curious than useful'.[A12]

Sir James SOUTH FRS (1785–1867)

James South trained as a physician but, after his marriage, was able to spend his time on his favourite pastime of astronomy. He helped found the Royal Astronomical Society and, together with Sir John Herschel, catalogued many double stars at the large observatory at his home. He was an opinionated man, and had close allies (such as Charles Babbage and Thomas Romney Robinson) and various foes (like George Airy). Hostilities came to a head when he attempted to build a very large telescope in the early 1830s.

Ronalds was quite friendly with South. He arranged for the Nichols family to be given a demonstration of his equipment in 1834, where they 'saw a star about 3 o'clock in the afternoon'.[J33] Ronalds tried to assist South in completing his large telescope and later discussed his early photo-recording ideas with him. South in turn introduced Ronalds to one of his German friends in 1845 who was 'very anxious'[C162] to see the meteorological instruments at Kew.

Earl STANHOPE and his Family

Charles Stanhope FRS (1753–1816), the third Earl Stanhope, was a statesman and scientist. Politically, he had revolutionary sympathies, while his inventions included the so-called Stanhope printing press. Ronalds collected copies of his book *Principles of Electricity* (1779) in English and French. He also referred to 'Great Lord Stanhope'[C375] in his book on the electric telegraph and in his autobiographical letter. Ronalds is depicted sitting next to Stanhope in an engraving entitled 'The Distinguished Men of Science of Great Britain 1807–8'.

Charles' son Philip Stanhope FRS (1781–1855), the fourth Earl Stanhope, also had broad interests and was a seasoned traveller. He was a great fan of Ronalds' drawing instruments, and ended a letter to him:

They cannot be appreciated by any person more than by
your obedient, humble Servant
Stanhope [C93]

Five letters from the Earl to Ronalds regarding the instruments survive in the Ronalds Archive.

Charles Vincent WALKER FRS (1812–82)

Charles Walker was electrician to the South Eastern Railway for many years. He was one of the first to observe geomagnetically-induced currents in telegraph lines. In early 1849 he demonstrated submarine telegraph messaging from a ship in the English Channel, via insulated cable to shore at Folkestone and along the railway telegraph line to London. He also helped implement the transmission of accurate time signals from Greenwich Observatory through the railway infrastructure to Dover, London and beyond.

Walker was in addition a significant author, translator and publisher in electrical subjects. He visited Kew Observatory in 1844 while preparing an early article on Ronalds' atmospheric electricity apparatus. There are 13 of his papers in the Ronalds Library, along with a signed photographic portrait. He was President of the STE in 1876.

Professor Sir Charles WHEATSTONE FRS (1802–75)

Charles Wheatstone was a well-known Professor of Experimental Philosophy at King's College in London. He made important contributions in diverse areas, including acoustics (with his family's music instrument business), vision (where he explained binocular depth perception) and encryption, as well as electrical science. He is perhaps best known today for the Wheatstone bridge to measure an unknown electrical resistance, which he popularised but did not invent. Wheatstone additionally put considerable effort into science oversight. He served on the BAAS

Council for over 30 years and also long contributed to various committees of the Royal Society.

Wheatstone and Faraday were closer friends than Ronalds was with either of them. Wheatstone was apparently very shy in large groups and Faraday conducted his lectures for him. In private he had a self-assured style, described as 'irrepressible egotism'[B87] by one biased observer (Cooke's brother). He was involved in various disputes through his career concerning the origin of ideas, as we see regarding the Kew Observatory as well as the telegraph. He argued with the 'disputatious' Brewster (as he called him)[8] over his discovery of stereoscopy in a series of letters to *The Times*.

Wheatstone was a youth when he met Ronalds and the two had a rather complex relationship through much of their careers. Wheatstone exhibited the trait of playing on both sides. With Ronalds, Wheatstone at times helped (for example, in instigating his election to the Royal Society), and at other times hindered, efforts for him to obtain research funding and recognition. Difficulties in their interactions may have been rooted in the similarity of their scientific talents and interests and Wheatstone's apparent inclination to take some advantage of Ronalds' humility. Numerous letters from Wheatstone to Ronalds survive concerning the Kew Observatory; their tone is cool relative to much of Ronalds' correspondence but this is not atypical of Wheatstone's style. The Ronalds Library has seven of his works and the Catalogue notes several instances where his achievements may have been overstated.

Dr William WOLLASTON FRS (1766–1828) and his Family

William Wollaston was an English chemist and physicist who made important contributions across several fields. He discovered two chemical elements and developed a process for making platinum malleable. In optics, he invented the camera lucida and a lens tailored for use in the camera obscura. He also published works on electricity and magnetism of which the Ronalds Library has three, as well as his eulogy and portrait. Briefly President of the Royal Society, his bequest of £2,000 to the organisation enabled the establishment of a Donation Fund to support research — it still exists today. Ronalds benefited from an early grant from the Fund and augmented it on his death by much more than he had received. Wollaston's brother developed an early odometer that was inspiration for one of Ronalds' surveying instruments and his son was one of the first customers for Ronalds' tripod.

---

[8]*The Times* 15/11/1856 10e.

BIBLIOGRAPHY

Website

E1. Further detail on Sir Francis Ronalds. Available at http://www.sirfrancisronalds.co.uk.

Sir Francis Ronalds' Printed Works

R1. Ronalds. F., Experiments on the Variable Action of the Electric Column, *Phil. Mag.*, **XLIII** (1814), pp. 414–19; reprinted in: Sturgeon W, *Annals of Electricity, Magnetism, and Chemistry*, **9** (1842) pp. 305–9.

R2. Ronalds. F., On Electricity, *Phil. Mag.*, **XLIV** (1814), pp. 442–5.

R3. Ronalds. F., On Electro-galvanic Agency employed as a Moving Power; with a Description of a Galvanic Clock, *Phil. Mag.*, **XLV** (1815), pp. 261–4.

R4. Ronalds. F., On the Electric Column of Mr. De Luc, *Phil. Mag.*, **XLV** (1815), pp. 466–7.

R5. Ronalds. F., On Correcting the Rate of an Electric Clock by a Compensation for Changes of Temperature, *Phil. Mag.*, **XLVI** (1815), pp. 203–4.

R6. Ronalds. F., An Account of an Atmospheric Electrometer, *Quart. J. Sci. Arts*, **II** (1817), pp. 249–53.

R7. Ronalds. F., *Descriptions of an Electrical Telegraph and of some other Electrical Apparatus*, (1823), 83pp 8 plates.

R8. Ronalds. F., Observations on Atmospheric Electricity made on Vesuvius, in June and July, 1819, *Quart. J. Sci. Arts*, **XIV** (1823), pp. 333–4.

R9. Ronalds. F., *Descriptions of Mr Ronalds's Patent Perspective Tracing Instruments and Tripod Staff*, (1825?).

R10. Ronalds. F., *Mechanical Perspective*, 1ˢᵗ Ed. (1824?), 2ⁿᵈ Ed. (1828), 36pp 8/12 plates.

R11. Blair. A. and Ronalds. F., *Sketches at Carnac (Brittany) in 1834*, (1836), 24 plates.

R12. Ronalds. F., Report Concerning the Observatory of the British Association at Kew, *1843–4, Report of the British Association for 1844*, (1845), pp. 120–42.

R13. Ronalds. F., On the Kew Observations, *Report of the British Association for 1845* (1846), p. 341.

R14. Ronalds. F., On the Meteorological Observations at Kew, with an Account of the Photographic Self-registering Apparatus, *Report of the British Association for 1846*, (1847), pp. 10–11; reprinted in: *Athenæum*, (1846), 1000–1; reprinted as: Self-registering Meteorological Apparatus at Kew, *Year-Book of Facts in Science and Art*, (1847), pp. 125–7; reprinted in: *The Decorator's Assistant*, **I** (1847), pp. 55–6.

R15. Ronalds. F., On Photographic Self-registering Meteorological and Magnetic Instruments, *Phil. Trans. R. Soc. Lond.* Part I, **137** (1847), pp. 111–17.

R16. Ronalds. F., Experiment Made at the Kew Observatory on a New Kite-Apparatus for Meteorological Observations, or other purposes, *Phil. Mag.*, **XXXI** S3 (1847), pp. 191–2; reprinted as: Experiment Made at the Kew Observatory, *Year-Book of Facts in Science and Art*, (1848), p. 280; reprinted as: The Use of the Kite in Meteorology, *Monthly Weather Review*, **24** (1896), pp. 416–7.

R17. Ronalds. F., Notice of Observations Carried on at the Kew Observatory, *Report of the British Association for 1847*, (1848), pp. 30–31; reprinted in: *Literary Gazette*, (1847), p. 519; reprinted as: The Kew Observatory, *Year-Book of Facts in Science and Art*, (1848), pp. 279–80.

R18. Ronalds. F., *Epitome of the Electro-Meteorological and Magnetic Observations, Experiments, &c. made at the Kew Observatory*, (1848), 12pp.

R19. Ronalds. F., Report Concerning the Observatory of the British Association at Kew, *1848–9, Report of the British Association for 1849*, (1850), pp. 80–7.

R20. Ronalds. F., Report Concerning the Observatory of the British Association at Kew, *1849–50, Report of the British Association for 1850*, (1851), pp. 176–86.

R21. Ronalds. F., Report Concerning the Observatory of the British Association at Kew, *1850–1, Report of the British Association for 1851*, (1852), pp. 335–70.

R22. Ronalds. F., Great Britain and Ireland, *A Walk through the Universal Exhibition of 1855*, Galignani A. & W. (Ed.), (1855), pp. 153–60.

R23. Ronalds. F., *British Association Reports 1844, 1849, 1850, 1851*, reprinted (1855).

R24. Ronalds. F., *Descriptions de quelques Instruments Météorologiques et Magnétiques*, (1855), 68pp 14 plates.

R25. Ronalds. F., Photographie, *Cosmos Revue Encyclopédique Hebdomadaire*, **VIII** (1856), pp. 541–9.

R26. Ronalds. F., *Description of an Electrical Telegraph*, 2ⁿᵈ Ed. (1871), 25pp 4 plates.

R27. Ronalds. F., Correspondence, *J. Soc. Telegraph Engrs.*, **I** (1872), pp. 243–4.

R28. Frost. A. J., (Ed.), *Catalogue of Books and Papers Relating to Electricity, Magnetism, the Electric Telegraph, &c. including The Ronalds Library, compiled by Sir Francis Ronalds, F.R.S.*, (1880); reprinted (CUP 2013).

Ronalds Family Patents

P1. Martineau. P. & M., *Methods of Refining or Clarifying certain Vegetable Substances*, UK patent (later numbered 3912), 8/5/1815.

P2. Ronalds. H., *Manufacture of Leather*, UK patent (later numbered 4205), 23/1/1818.

P3. Ronalds. F., *Apparatus for Tracing from Nature*, UK patent (later numbered 5132), 23/3/1825.

P4. Ronalds. H., *Apparatus for Propelling Boats, &c., by means of Jets of Water Drawn in and Forced out by Pumps*, US Patent 1899, 14/12/1840. Available at https://www.google.us/patents/US1899. (Accessed on 16/7/2015).

Notices of Sir Francis Ronalds' Death

D1. The Late Sir Francis Ronalds, F.R.S., *The Times* 12/8/1873, 8f.

D2. *Pall Mall Gazette* 12/8/1873, 4b.

D3. Death of Sir Francis Ronalds, F.R.S., *Morning Post* 12/8/1873, 6a.

D4. *Daily News* 13/8/1873, 5c.

D5. *Glasgow Herald* 16/8/1873, 7a.

D6. Death of Sir Francis Ronalds, *Hastings and St Leonards Observer* 16/8/1873, 7d.

D7. *Athenæum* 23/8/1873, pp. 244–5.

D8. *Sydney Morning Herald* 21/10/1873, 5d.

D9. *Nature*, **8** (1873), p. 313.

D10. Sir Francis Ronalds, *J. Soc. Arts*, **21** (1873), p. 764.

D11. The Record, *Appletons' Journal: A Magazine of General Literature*, **X** (1873), p. 352.

D12. Sir Francis Ronalds, F.R.S., *J. Soc. Telegraph Engrs.*, **II** (1873), p. 168.

D13. Second Annual Meeting, *J. Soc. Telegraph Engrs.*, **II** (1873), pp. 411–23.

D14. Airy. G. B., Anniversary Meeting. *Proc. R. Soc. Lond.*, **22** (1873), pp. 1–20.

D15. *Dodsley's Annual Register for the Year 1873*, Part II (1874), p. 149.

Memoirs of Sir Francis Ronalds

M1. Walker. W., Francis Ronalds, F.R.S., *Memoirs of the Distinguished Men of Science of Great Britain Living in the Years 1807–8*, (1862), p. 154; 2nd Ed. (1864), pp. 99–101; reprinted in *Telegraphic J.* 2/7/1864, pp. 3–4.

M2. Sir Francis Ronalds, F.R.S., *Illustrated London News* 30/4/1870, pp. 455, 464.

M3. *Debrett's Illustrated Baronetage, with the Knightage, of the United Kingdom*, (1871).

M4. RONALDS, Knt. Bachel., *Dod's Peerage, Baronetage, Knightage of Great Britain*, (1872) p. 547.

M5. Frost. A. J., Biographical Memoir of Sir Francis Ronalds, F.R.S., *Catalogue of Books and Papers Relating to Electricity, Magnetism, the Electric Telegraph, &c including The Ronalds Library* (1880), vii-xxvii; reprinted (1880), 21pp.

M6. Hartog. P. J., Ronalds, Sir Francis, *Dictionary of National Biography*, Lee. S., (Ed.) **XLIX** (1897), pp. 201–4.

M7. Appleyard. R., Francis Ronalds, *Pioneers of Electrical Communication*, (1930) pp. 301–31.

M8. The Men who Made Telegrams Possible, *Age*, Melbourne 31/5/1935.

M9. Symons. L., Francis Ronalds, Electrician and Book Collector, *Proc. IEEE*, **91** (2003), pp. 1980–3.

Representative Articles pertaining to Sir Francis Ronalds' Telegraph

T1.   Ronalds' Electric Telegraph, *Encyclopædia Britannica*, 7ᵗʰ Ed. **VIII** (1842), p. 662; also 8ᵗʰ Ed. (1855), p. 627.

T2.   Vail. A., Ronald's Electric Telegraph, invented in 1816, *American Electro Magnetic Telegraph*, (1847), pp. 130–1.

T3.   Ueber Telegraphen, *Allgemeine Bauzeitung*, (1848), pp. 205–79.

T4.   Moigno. F-N., Télégraphe de M. Ronalds, Conçu en 1815, Exécuté et Décrit en 1823, *Traité de Télégraphie Electrique*, (1849), 5, pp. 294–8.

T5.   Highton. E., Ronalds's Telegraph, *The Electric Telegraph: Its History and Progress*, (1852), pp. 49–53.

T6.   Noad. H. M., The Electric Telegraph, *Manual of Electricity*, (1859), pp. 747–51.

T7.   Shaffner. T. P., Ronald's Electric Telegraph, *Telegraph Manual: A Complete History and Description* (1859), pp. 147–56.

T8.   Sabine. R., *The Electric Telegraph*, (1867).

T9.   Description of an Electrical Telegraph, *Nature*, **5** (1871), p. 59.

T10. Fifty Years' Progress, *The Times* 2/1/1875, p. 10b-c.

T11. Clark. L., Inaugural Address, *J. Soc. Telegraph Engrs.*, **IV** (1875), pp. 1–23.

T12. Zetzsche. K. E., Ronalds, *Geschichte der Elektrischen Telegraphie*, **I** (1877), pp. 33–40.

T13. Noad. H. M. (Preece. W. H. Ed.), Ronalds' Electric Telegraph, *The Student's Text-Book of Electricity*, (1879), pp. 408–10.

T14. Adams. W. G., The Scientific Principles involved in Electric Lighting, *J. Soc. Arts*, **XXIX** (1881), pp. 730–3.

T15. Fahie. J. J., Ronalds' Telegraph, *A History of Electric Telegraphy, to the Year 1837*, (1884), pp. 127–45.

T16. Who Invented the First Electric Telegraph? A Popular Misunderstanding, *Theleme* **I**(7), (1889), pp. 67–71.

T17. Sime. J., A Pioneer in Electricity, *Lightning: a Popular Review of Electricity*, (Sep 1892), pp. 1–12; reprinted as: *Sir Francis Ronalds, F.R.S. and his Work in Connection with Electric Telegraphy in 1816*, (1893), 14pp.

T18. Lockwood. T. D., History of the Electric Telegraph Part II, *Telegraphic Historical Society of North America* (1895), pp. 90–106.

T19. Mottelay. P. F., A.D. 1816, *Bibliographical History of Electricity & Magnetism*, (1922), pp. 438–40.

T20. Black. R. M., Sir Francis Ronalds (1788–1873), *The History Of Electric Wires and Cables*, (London: Peter Peregrinus 1983), pp. 4–5.

T21. Wheen. A., *Dot-Dash to Dot.Com*, (Springer 2011), pp. 8–9.

Representative Articles pertaining to Sir Francis Ronalds' Work at the Kew Observatory

K1. Kaemtz. L. F., (Walker. C. V. trans.), *Complete Course of Meteorology*, (1845), pp. 551–73.

K2. Daniell. J. F., An Essay on some of the Phenomena of Atmospheric Electricity, *Elements of Meteorology*, 3rd Ed. **2** (1845), pp. 369–89.

K3. Weale. J., The Kew Observatory, *The Pictorial Handbook of London* (1854), pp. 667–70.

K4. Noad. H. M., Atmospheric Electricity, *A Manual of Electricity* (1859), pp. 184–95.

K5. Drew. J., Electrical Apparatus at the Kew Observatory, *Practical Meteorology*, 2nd Ed. (1860), pp. 227–34.

K6. Daguin. P. A., Barométrographes, *Traité de Physique* I (1861), pp. 369–70.

K7. Tissandier. G., (Thomson. J. Ed.), Photographic Registering Instruments, *A History and Handbook of Photography*, (1876), pp. 268–86.

K8. Radau. R., Les Applications Scientifiques de la Photographie II, *Revue des Deux Mondes*, **XXVI** (1878) pp. 198–216.

K9. Scott. R. H., The History of the Kew Observatory, *Proc. R. Soc. Lond.*, **39** (1885), pp. 37–86.

K10. Kew Observatory under Mr. Ronalds, *Sketcher and Theleme*, **1**(8) (1889), p. 79.

K11. Whipple. G. M., Photography in Relation to Meteorological Work, *Quart. J. R. Meteorological Soc.*, **XVI** (1890), pp. 141–6.

K12. Inwards. R., Meteorological Observatories, *Quart. J. R. Meteorological Soc.*, **XXII** (1896), pp. 81–98.

K13. Scrase. F. J., Observations of Atmospheric Electricity at Kew Observatory 1843 to 1931, *Geophys. Mem.*, **60** (1934), pp. 3–27.

K14. Whipple. F. J., Some Aspects of the Early History of Kew Observatory, *Quart. J. R. Meteorological Soc.*, **LXIII** (1937), pp. 127–35.

K15. Drummond. A. J., A Century of Progress in Atmospheric Electricity at Kew Observatory, *Quart. J. R. Meteorological Soc.*, **LXX** (1944), pp. 49–60.

K16. Jacobs. L., The 200-Years' Story of Kew Observatory, *Met. Mag.*, **98** (1969), pp. 162–71.

K17. Bryden. D. J., Quality Control in the Making of Scientific Instruments, *Bull. Scientific Instrument Soc.*, No. 88 (2006), pp. 48–59.

K18. Macdonald. L. T., Making Kew Observatory: the Royal Society, the British Association and the Politics of Early Victorian Science, *BJHS*, **48** (2015).

Books, Printed Reports and Theses

B1.   Danti. E., *Due Regole della Prospettiva Practica*, (1583).
B2.   Mottley. J., *Joe Miller's Jests*, (1739).
B3.   Campbell. R., *The London Tradesman*, (1747).
B4.   Sterne. L., *A Sentimental Journey Through France and Italy*, (1768).
B5.   Blair. H., *Lectures on Rhetoric and Belles Lettres*, (1783).
B6.   Bennet. A., *New Experiments on Electricity*, (1789).
B7.   Adams. G., *Geometrical and Graphical Essays*, (1791).
B8.   *Universal British Directory of Trade, Commerce, and Manufacture*, **II** (c.1791).
B9.   Darwin. E., *Phytologia or the Philosophy of Agriculture and Gardening*, (1800).
B10. De Cambry. J., *Monumens Celtiques, ou Recherches sur le Culte des Pierres*, (1805).
B11. Ronalds. H., *Ventriculi Fabrica Vitiata*, Dissertation, Univ. Edinburgh, (1814).
B12. Singer. G. J., *Elements of Electricity and Electro-Chemistry*, (1814).
B13. Birkbeck. M., *Notes on a Journey in America*, (1818).
B14. De Gourbillon J A, *Travels in Sicily and to Mount Etna in 1819*, (1820).
B15. Belzoni. G., *Narrative of the Operations and Recent Discoveries within the Pyramids, Temples, Tombs and Excavations, in Egypt and Nubia*, (1820).
B16. De Quincey. T., *Confessions of an English Opium-Eater*, (1821).
B17. Flower. R., *Letters from the Illinois 1820, 1821*, (1822).
B18. Waddington. G. and Hanbury. B., *Journal of a Visit to some parts of Ethiopia*, (1822).
B19. *Biographie Nouvelle des Contemporains*, **VIII** (1822).
B20. Cohen. B., *Compendium of Finance*, (1822).
B21. Ferguson. J., *Essays and Treatises*, (1823).
B22. Blair. A., *Graphic Illustrations of Warwickshire*, (1823–9).
B23. Henniker. F., *Notes during a visit to Egypt, Nubia, The Oasis, Mount Sinai and Jerusalem*, (1823).
B24. Mangles. J. and Irby. C. L., *Travels in Egypt and Nubia, Syria, and Asia Minor, during the Years 1817 & 1818* (1823).
B25. Young. T., *An Account of Some Recent Discoveries in Hieroglyphical Literature, and Egyptian Antiquities*, (1823).
B26. Blane. W. N., *An Excursion through the United States and Canada during the years 1822–23*, (1824).
B27. Hullmandel. C., *The Art of Drawing on Stone*, (1824).
B28. Mahé. J., *Essai sur les Antiquités du département du Morbihan*, (1825).
B29. Partington. C. F., *The Century of Inventions of the Marquis of Worcester*, (1825).
B30. Mure. W., *The Historie and Descent of the House of Rowallane*, (1825).
B31. Rundell. M., *New System of Practical Domestic Economy*, (1825).
B32. National Repository, *Catalogue of the Specimens of New and Improved Productions*, (1828).
B33. Murray. J., *A Treatise on Atmospherical Electricity*, (1830).

B34. West. W., *History, Topography and Directory of Warwickshire*, (1830).

B35. Ronalds. H., *Pyrus Malus Brentfordiensis: or, a Concise Description of Selected Apples*, (1831).

B36. Herbert. W., *The History and Antiquities of the Parish and Church of St. Michael, Crooked Lane, London*, (1833).

B37. Ibbetson. J. H., *A Brief Account of Ibbetson's Geometric Chuck*, (1833).

B38. Cary. W., *Catalogue of Optical, Mathematical, and Philosophical Instruments, made and sold by W. Cary*, (after 1833).

B39. Holtzapffel & Co, *Manufacturers of Engines, Lathes, Mechanical and Edge Tools*, (1834).

B40. Cromwell. T., *Walks through Islington*, (1835).

B41. Hebert. L., *Engineer's and Mechanic's Encyclopædia*, (1836).

B42. Ronalds. A., *The Fly-fisher's Entomology* (1836), 2nd Ed. (1839), 3rd Ed. (1844), 4th Ed. (1849), 5th Ed. (1856), 6th Ed. (1862), 7th Ed. (1868), 8th Ed. (1877), 9th Ed. (1883), 10th Ed. (1901), 11th Ed. (1913), New Ed. (1921).

B43. Long. G., (Ed.), *The Penny Cyclopædia of the Society for the Diffusion of Useful Knowledge*, **VIII** (1837).

B44. *Repertory of Patent Inventions*, **XI** New Series, (1839).

B45. Wartmann. E., *Mémoire sur deux Balances a Réflexion*, (1841).

B46. Skey. R. S., *Report to the Committee of the Birmingham and Liverpool Junction Canal*, (1841)

B47. Sturgeon. W., *Lectures on Electricity*, (1842).

B48. Liebig. J., *Familiar Letters on Chemistry*, (1843).

B49. Finlaison. J., *An Account of some Remarkable Applications of the Electric Fluid by Mr. Alexander Bain; with a Vindication of his Claim to be the First Inventor of the Electro-Magnetic Printing Telegraph*, (1843).

B50. Holtzapffel, *Turning and Mechanical Manipulation*, **I** (1843) - **V** (1884).

B51. Dodd. G., *British Manufactures: Chemical*, (1844).

B52. Airy. G. B., *Account of the Northumberland Equatoreal and Dome*, (1844).

B53. Newman. J., *Catalogue of Philosophical Instruments*, (1845).

B54. Royal Polytechnic Institution, *Catalogue*, (1844), (1845).

B55. Faulkner. T., *History and Antiquities of Brentford, Ealing, & Chiswick*, (1845).

B56. Thomason. E., *Sir Edward Thomason's Memoirs during Half a Century*, (1845).

B57. Weld. C. R., *A History of the Royal Society*, **2** (1848).

B58. *A Treatise on Gold Discovery and Gold Washing, compiled from Ure and Hebert, with Description of a Patent Washing Machine for Economising Labour*, (1851), SLV.

B59. Great Exhibition, *Prospectuses of Exhibitors*, (1851).

B60. Hunt. R., (Ed.) *Hunt's Hand-book to the Official Catalogues of the Great Exhibition*, (1851).

B61. *Exhibition of the Works of Industry of All Nations, 1851. Reports by the Juries*, (1852).

B62. Turnbull. L., *Lectures on the Electro-Magnetic Telegraph*, (1852).

B63.   Hogg. J., *Elements of Experimental and Natural Philosophy*, (1853).

B64.   Snow Harris. W., *Rudimentary Electricity*, (1853).

B65.   Sheepshanks. R., *A Letter to the Board of Visitors of the Greenwich Royal Observatory*, (1854).

B66.   Cooke. W. F., *The Electric Telegraph: was it invented by Professor Wheatstone?*, (1854).

B67.   Wheatstone. C., *A Reply to Mr Cooke's Pamphlet*, (1855).

B68.   Smyth. C. P., *Description of New or Improved Instruments for Navigation and Astronomy*, (1855).

B69.   Ronalds E, Richardson T, Watts H, Knapp F L, *Chemical Technology*, 1st Ed. (1855).

B70.   Tresca. H., *Visite a L'Exposition Universelle de Paris, en 1855*, (1855).

B71.   Le Verrier. U., *Annales de L'Observatoire Impérial de Paris*, **I** (1855).

B72.   Cooke, W. F., *A Reply to Mr Wheatstone's Answer*, (1856).

B73.   Cooke. W. F., *The Electric Telegraph: was it invented by Professor Wheatstone?*, (1857).

B74.   FitzRoy. R., *Barometer and Weather: How to Foretell Weather*, (1858).

B75.   De la Rive. A. (Walker. C. V. Trans.) *A Treatise on Electricity in Theory and Practice*, **III** (1858).

B76.   Becquerel. A. C. and Becquerel. A. E., *Résumé de l'Histoire de l'Électricité*, (1858).

B77.   Noad. H. M., *A Manual of Electricity*, (1859).

B78.   Prescott. G. B., *History, Theory, and Practice of the Electric Telegraph*, (1860).

B79.   Hardcastle. E. C., *Memoir of Joseph Hardcastle, Esq., A Record of the Past*, (1860).

B80.   Walker. W., *Memoirs of the Distinguished Men of Science of Great Britain Living in the Years 1807–8*, (1862); 2nd Ed. (1864).

B81.   Gordon. M., *Christopher North: A Memoir of John Wilson*, (1862).

B82.   *Abstracts of Magnetical Observations made at the Magnetical Observatory, Toronto 1853–62*, (1863).

B83.   Brooke. C., *The Elements of Natural Philosophy*, 6th Ed. (1867).

B84.   *Kelly's Post Office Directory of Essex, Herts, Middlesex, Kent, Surrey and Sussex*, (1867).

B85.   *Report of the Meteorological Committee of the Royal Society 1867*, (1868).

B86.   Cole. H., *Reports on the Paris Universal Exhibition, 1867*, **I** (1868).

B87.   Cooke. T. F., *Authorship of the Practical Electric Telegraph of Great Britain*, (1868).

B88.   Cooke. T. F., *Invention of the Electric Telegraph: The Charge against Sir Charles Wheatstone of Tampering with the Press*, (1869).

B89.   Silliman. B., *Principles of Physics, or Natural Philosophy*, (1869).

B90.   Fergusson. J., *Rude Stone Monuments in all Countries*, (1872).

B91.   Tissandier. G., *La Photographie* (1873).

B92.   Rennie. J., *Autobiography*, (1875).

B93.   Burke. B., *Genealogical and Heraldic History of the Landed Gentry of Great Britain and Ireland*, (1875).

B94. Science and Art Department, *Catalogue of the Special Loan Collection of Scientific Apparatus at the South Kensington Museum*, (1876).

B95. South Kensington Museum, *Conferences held in connection with the Special Loan Collection of Scientific Apparatus*, (1876).

B96. Tissandier. G., (Thomson. J. Ed.), *A History and Handbook of Photography*, (1876).

B97. Martineau. H., *Autobiography, with Memorials by M W Chapman*, (1877).

B98. Flower. E. F., *Bits and Bearing-Reins: With Observations on Horses and Harness*, (1877).

B99. Kingzett. C. T., *The Alkali Trade*, (1877).

B100. *Minutes of Council of the Royal Society 1870–77*, **IV** (1878).

B101. Meteorological Society, *Modern Meteorology, A Series of Six Lectures*, (1879).

B102. Hicks. J. J., *Catalogue of Standard, Self-Recording and other Meteorological Instruments*, (c.1880).

B103. Flower. G., *History of the English Settlement in Edwards County Illinois*, (1882).

B104. Heathman. J. H., *On the Preservation of Life and Property from Fire*, (1882).

B105. Gibson. E., *Recollections of my Youth, Written at the Request of my Daughter*, (c.1885).

B106. Withers. W. B., *History of Ballarat*, (1887).

B107. Gray. J., *Electrical Influence Machines*, (1890).

B108. Munro. J., *Heroes of the Telegraph*, (1891).

B109. Scott. R. H., *Elementary Meteorology*, 6th Ed. (1893).

B110. Waldo. F., *Modern Meteorology*, (1893).

B111. Hill. C., *Frederic Hill, An Autobiography of Fifty Years in Times of Reform*, (1894).

B112. Lukin. J., *Turning Lathes*, 4th Ed. (1894).

B113. Moore. J. W., *Meteorology, Practical and Applied*, 1st Ed. (1894), 2nd Ed. (1910).

B114. Lefroy. J. H., (Lefroy. C. A. Ed.) *Autobiography*, (1895).

B115. Airy. G. B., *Autobiography of Sir George Biddell Airy*, (1896).

B116. Addison. W. I., *A Roll of the Graduates of the University of Glasgow*, (1898).

B117. Pierce. F. C., *Field Genealogy*, **I** (1901).

B118. Holtzapffel and Co, *Catalogues of Lathes and Tools*, (after 1901).

B119. Steuart. W. M., *Central Electric Light and Power Stations, 1902*, (1905).

B120. Smith. J., (Ed.) *The Cyclopedia of Victoria*, **III** (1905).

B121. Hendy. J. G., *The History of the Early Postmarks of the British Isles*, (1905).

B122. Shaw. W. A., *The Knights of England*, (1906).

B123. Podmore. F., *Robert Owen, A Biography*, (1906).

B124. *Diary of William Owen 1824–1825*, Indiana Historical Society (1906).

B125. Stokes. G. G., (Larmor. J., Ed.) *Memoir and Scientific Correspondence of the late Sir George Gabriel Stokes*, **II** (1907).

B126. Allibone. S. A., *A Critical Dictionary of English Literature and British and American Authors*, **II** (1908).

B127. Kolbe. B., (Skellon. J., Trans.), *An Introduction to Electricity*, (1908).

B128. Weaver. W. D., (Ed.) *Catalogue of the Wheeler Gift*, (1909).

B129. Galton. F., *Memories of My Life*, (1909).

B130. Pierce. R. M., *Dictionary of Aviation*, (1911).

B131. Titford. C., *History of Unity Church, Islington, London*, (1912).

B132. Bird. J. and Norman. P., (Eds.) *Survey of London*, **6** (1915).

B133. Schuster. A. and Shipley. A. E., *Britain's Heritage of Science*, (1917).

B134. Sotheran. H., *Catalogue of Rare and Standard Books*, (1918).

B135. Howarth. O. J., *The British Association for the Advancement of Science: A retro-spect 1831–1921*, (1922).

B136. *Catalogue of the Collections in the Science Museum — Meteorology*, (1922).

B137. Mottelay. P. F., *Bibliographical History of Electricity & Magnetism*, (1922).

B138. Sanford. F., *Bulletin of the Terrestrial Electric Observatory*, **II** (1925).

B139. Appleyard. R., *The History of the Institution of Electrical Engineers (1871–1931)*, (1939).

B140. Det Norske Institutt for Kosmisk Fysikk, *The Auroral Observatory at Tromsø*, (1932). Available at http://flux.phys.uit.no/Hist/NOBS_nr1.pdf (Accessed on 15/7/2015).

B141. Swann. E., *Christopher North*, (1934).

B142. *Washington: A Guide to the Evergreen State*, (1941).

B143. *The Diaries of Donald Macdonald 1824–1826*, Indiana Historical Society, (1942).

B144. Haynes. W., *The Stone that Burns*, New York, (1942).

B145. Black. G. F., *The Surnames of Scotland*, (1946).

B146. Wise. S. J., *Electric Clocks*, 2nd Ed. (1951).

B147. Clow. A. and Crow. N. L., *The Chemical Revolution*, (1952).

B148. Abraham Lincoln Association, *The Collected Works of Abraham Lincoln*, **VII** (1953).

B149. Brownhill. W. R., *The History of Geelong and Corio Bay*, (1955).

B150. Middleton. W. E. K., *The History of the Barometer*, (Baltimore: Johns Hopkins 1964).

B151. Flower. S., (Flower. E. F., Ed.) *Great Aunt Sarah's Diary, 1846–92*, (Printed Privately 1964).

B152. Hubbard. G., *Cooke and Wheatstone and the Invention of the Electric Telegraph*, (London: Routledge 1965).

B153. Middleton. W. E. K., *The History of the Thermometer*, (Baltimore: Johns Hopkins 1966).

B154. Middleton. W. E. K., *Invention of the Meteorological Instruments*, (Baltimore: Johns Hopkins 1969).

B155. Stephens. W. B., Ed. *A History of the County of Warwick*, **8** (London: Victoria County History 1969).

B156. Trevelyan. R., *Princes under the Volcano*, (New York: William Morrow 1973).

B157. Butler. J. M., *Convict by Choice*, (Melbourne: Hill of Content 1974).

B158. Gingrich. A., *Fishing In Print — A Guided Tour Through Five Centuries of Angling Literature*, (New York: Winchester 1974).

B159. Marinaro. V., *In the Ring of the Rise*, (New York: Lyons & Burford 1976).

B160. Elliott. J. M., (Ed.) *Robert Dale Owen's Travel Journal 1827*, Indiana Historical Society, (1977).

B161. Metcalf. P., *The Halls of the Fishmongers' Company*, (Chichester: Phillimore 1977).

B162. Darlington. B., (Ed.) *The Love letters of William and Mary Wordsworth*, (Univ. Chicago 1981).

B163. Morrell J & Thackray A, *Gentlemen of Science: Early Years of the British Association for the Advancement of Science*, (Oxford: Clarendon Press 1981).

B164. Baker. T. F. and Elrington. C. R., (Ed.) *A History of the County of Middlesex*, **7** (London: Victoria County History 1982).

B165. Ronalds. A. F., *The Ronalds Family of Australia*, (Printed Privately 1985).

B166. Reader. W. J., *A History of The Institution of Electrical Engineers 1871–1971*, (London: IEE 1987).

B167. Ogden. W. G., *Notes on the History and Provenance of Holtzapffel Lathes*, (N. Andover: Museum Ornamental Turning 1987).

B168. Hinman. M. J., *Men who Ruled Coventry 1725–1780*, Coventry and Warwickshire Pamphlet 14, (Coventry: Historical Association 1988).

B169. Darragh. T. A., *The Establishment and Development of Engraving and Lithography in Melbourne to the time of the Gold Rush*, (Garravembi 1990).

B170. Porter. M., *The Competitive Advantage of Nations*, (New York: Free Press 1990).

B171. *The Angling Library of Jock Grey*, (Cheltenham: P. Jowett 1991).

B172. James. F. A., *The Correspondence of Michael Faraday*, **I** (IET 1991) - **VI** (IET 2011).

B173. Walker. J. R. and Burkhardt. R. W., *Eliza Julia Flower: Letters of an English Gentlewoman*, (Muncie: Ball State Univ. 1991).

B174. Court. F. E., *Institutionalizing English Literature*, (Stanford Univ. 1992).

B175. Kemp. M., *The Science of Art*, (Yale Univ. 1992).

B176. Morris. C., *Fanny Wright: Rebel in America*, (Univ. Illinois 1992).

B177. Place. G., *The Rise and Fall of Parkgate: Passenger Port for Ireland*, (Manchester: Chetham Soc. 1994).

B178. Harris. R. S. and Harris T. G., *The Eldon House Diaries: Five Women's Views of the 19th Century*, (Toronto: Champlain Soc. 1994).

B179. Porter. F. and Macdonald. C., (Eds.) *What my Heart Dictates*, (Auckland: Auckland Univ. 1996).

B180. Latham. J. E., *Search for a New Eden: James Pierrepont Greaves (1777–1842)*, (Madison: Fairleigh Dickinson Univ. 1999).

B181. Read. A., (Ed.) *Architecturally Speaking*, (London: Routledge 2000).

B182. Herrin. S. J., *The development of printing in nineteenth-century Ballarat*, (Melbourne: Bibliographical Soc. ANZ 2000).

B183. Bowers. B., *Sir Charles Wheatstone FRS: 1802–1875*, (London: IEE 2001).

B184. Hook. D. H. and Norman. J. M., *Origins of Cyberspace* (Novato: historyofscience. com 2002).

B185. Silverman. K., *Lightning Man: The Accursed Life of Samuel F. B. Morse*, (New York: Alfred A. Knopf 2003).

B186. Swade. D. D., *Calculation and Tabulation in the Nineteenth Century: Airy versus Babbage*, PhD Thesis, UCL (2003). Available at http://ed-thelen.org/bab/Swade_PhD.pdf (Accessed on 15/7/2015).

B187. Burns. R. W., *Communications: An International History of the Formative Years*, (London: IEE 2004).

B188. Matthew. H. C. and Brian Harrison. H., (Eds.) *Oxford Dictionary of National Biography*, (Oxford Univ. Press 2004).

B189. Anderson. K., *Predicting the Weather: Victorians and the Science of Meteorology*, (Chicago: Univ. Chicago 2005).

B190. Gubbins. D. and Herrero-Bervera. E., (Eds.) *Encyclopedia of Geomagnetism and Paleomagnetism*, (Springer 2007).

B191. Kutney. G., *Sulfur: History, Technology, Applications & Industry*, (ChemTec 2007).

B192. Nagel. S., *Marie-Thérèse, Child of Terror*, (Bloomsbury 2008).

B193. Bailloud. G., Boujot. C., Cassen. S., Le Roux, C. T., *Carnac: Les Premières Architectures de Pierre*, (Paris: CNRS 2009).

B194. Brookes. L. C., *The Sheaf Catalogs of George John Spencer*, Master's Thesis, San Jose State Univ. (2009). Available at http://scholarworks.sjsu.edu/cgi/viewcontent.cgi?article=4635&context=etd_theses (Accessed on 15/7/2015).

B195. Walker. J. M., *History of the Meteorological Office*, (New York: Cambridge Univ. Press 2012).

Articles and Abstracts

A1.   Obituary — Mr John Field, *Gentleman's Mag.*, **LXVI** Part II (1796), p. 792.

A2.   De Saussure. M., Agenda, or a Collection of Observations and Researches the Results of which may serve as the Foundation for a Theory of the Earth, *Phil. Mag.*, **IV** (1799), pp. 259–65.

A3.   Cuthbertson. J., An Account of Improvements in Electrical Batteries; a Method of Augmenting their Power, with Experiments, Shewing the Proportional Lengths of Wire Fused by Different Quantities of Electricity, *J. Nat. Phil., Chem., Arts*, **II** (1799), pp. 525–35.

A4.   Volta. A., On the Electricity Excited by the Mere Contact of Conducting Substances of Different Kinds, *Phil. Mag.*, **VII** (1800), pp. 289–311.

A5.   Nicholson. W., Account of the new Electrical or Galvanic Apparatus of Sig. ALEX. VOLTA, *J. Nat. Phil., Chem., Arts*, **IV** (1801), pp. 179–87.

A6.   Nicholson. W., Scientific News, Accounts of Books, &c., *J. Nat. Phil., Chem., Arts*, **V** (1801), pp. 235–40.

A7.   Erman. P., Observations and Doubts concerning Atmospheric Electricity, *J. Nat. Phil., Chem., Arts*, **X** (1805), pp. 294–300, **XI** (1805), pp. 17–19.

A8.   De Luc. J. A., On the Electric Column and Aerial Electroscope, *J. Nat. Phil., Chem., Arts*, **XXVII** (1810) pp. 81–99.

A9.   Intelligence and Miscellaneous Articles, *Phil. Mag.*, **XXXVII** (1811), pp. 79–80.

A10.   Cary. J., Description of the Patent Reflecting Semicircle, invented by Sir Howard Douglas, *Phil. Mag.*, **XXXVIII** (1811), pp. 186–7.

A11.   De Luc, J. A., On the variable Action of the Electric Column, *Phil. Mag.*, **XLIV**, (1814), pp. 248–53.

A12.   Singer. G. J., The Electric Column considered as a maintaining Power, or First Mover for mechanical Purposes, *Phil. Mag.*, **XLV** (1815), pp. 359–63.

A13.   Singer. G. J., Correction of some Errors in Mr. Singer's Paper on the Mechanical Applications of the Electric Column, *Phil. Mag.*, **XLVI**, (1815), pp. 11–12.

A14.   Report of Chemistry, Natural Philosophy, &c., *Monthly Mag.*, **XXXIX** (1815), p. 447.

A15.   Proceedings of Learned Societies. Royal Society, *Phil. Mag.*, **XLV** (1815), pp. 65–7.

A16.   A List of Contributing Members to the Society, *Trans. Soc. Encouragement of Arts, Manufactures and Commerce*, **XXXIII** (1815), pp. 249–85.

A17.   Gilbert. L. W., Vorläufige Nachricht von grossen zu Stuttgard ausgeführten trocknen Säulen, und von einer sogenannten electrischen Uhr. *Annalen der Physik*, No. 3 (1815), pp. 187–90.

A18.   Thomson. T., Account of the Improvements in Physical Science during the Year 1816, *Annals of Philosophy*, **IX** (1817), pp. 1–89.

A19.   Electromètre Atmosphérique de Francis Ronalds, *Annales de Chimie et de Physique*, **IV** (1817), p. 104.

A20.   Ronalds. H., Description of the Different Varieties of Brocoli, with an Account of the Method of Cultivating them, *Trans. Hort. Soc. Lond.*, **III** (1820), pp. 161–9.

A21.   Occurrences in London and its Vicinity, *Gentleman's Mag.*, **XC** Part I (1820), pp. 79–82.

A22.   Account of Mr Ronalds' Pendulum Doubler of Electricity, *Edinburgh Phil. J.*, **IX** (1823), pp. 323–5.

A23.   Improvement on the Electrical Machine, *Edinburgh Phil. J.*, **IX** (1823), pp. 395–6.

A24.   On the Periodical reappearance of Thunder Storms, *Edinburgh Phil. J.*, **IX** (1823), p. 398.

A25.   Electrical Apparatus, *London J. Arts Sci.*, **VII** (1824), 152–3; reprinted in: *Glasgow Mechanics' Magazine*, **I** (1824), p. 367.

A26.   Herschel. J. F. and South. J., Observations of the apparent distances and positions of 380 double and triple Stars, *Phil. Trans. R. Soc. Lond.*, **114** (1824), pp. 1–412.

A27.   Invention of a new Tracing Apparatus, to facilitate Drawing from nature, *London J. Arts, Sci.*, **XII** (1826), pp. 21–3.

A28.   Instrument pour dessiner d'après nature; par M. Ronalds, *J. Hebdomadaire des Arts et Métiers*, **V** (1826), pp. 364–6.

A29.   National Repository, *Mechanics' Magazine*, **IX** (1828), pp. 307–9.

A30.   The National Repository, *La Belle Assemblée*, **VIII** (1828), p. 87.

A31.   Hebert. L., Patent Instrument for Drawing in Perspective, *Register of Arts*, **II** (1828), pp. 202–5.

A32.   Proceedings of the Royal Institution of Great Britain, *Quart. J. Sci. Arts*, (Jan–Jun 1828), pp. 417–41; reprinted in: *Arcana of Science* (1829), p. 26.

A33.   The London Nurseries, *Loudon's Gardener's Magazine*, **V** (1829), pp. 736–7.

A34.   The National Repository, *British Magazine, Monthly J. Lit., Sci., Art*, **I** (1830), pp. 42–3.

A35.   Pyrus Malus Brentfordiensis Review, *Loudon's Gardener's Magazine*, **VII** (1831), pp. 587–90.

A36.   Hebert. L., Patent Perspective Tracing Instrument, *Register of Arts*, **IV** (1830), pp. 296–7.

A37.   Hebert. L., Improved Four-Wheeled Carriage, *Register of Arts*, **IV** (1830), 299–300.

A38.   Hebert. L., An Improved Block for Rigging or Tackle, *Register of Arts*, **VI** (1832), p. 115.

A39.   Hebert. L., A newly-invented Sun-Dial, *Register of Arts*, **VI** (1832), pp. 115–16.

A40.   Hebert. L., An Improved Portable Stand and Plane for a Theodolite, or a Spirit Level, *Register of Arts*, **VI** (1832), p. 189.

A41.   Hebert. L., Drawing Machine, *Register of Arts*, **VI** (1832), pp. 236–7.

A42.   Brewster. D., On Thunder Rods, *Edinburgh Encyclopædia*, **VIII** (1832), pp. 341–2.

A43.   Observations on Dracontia, *Gentleman's Mag.*, **CIII** Part II (1833), pp. 143–4.

A44.   Obituary — Mr Hugh Ronalds, *Gentleman's Mag.*, **I** New Series (1834), pp. 337–8.

A45.   Obituary — Mr Hugh Ronalds, *Horticultural Register*, **III** (1834) p. 92.

A46.   Deane. J. B., Observations on Dracontia, *Archaeologia*, **XXV** (1834), pp. 188–229.

A47.   Deane. J. B., Remarks on certain Celtic Monuments at Locmariaker, in Britany, *Archaeologia*, **XXV** (1834), pp. 230–4.

A48.   Wheatstone. C., An Account of some Experiments to measure the Velocity of Electricity and the Duration of Electric Light, *Phil. Trans. R. Soc. Lond.*, **124** (1834), pp. 583–91.

A49.   Faraday. M., On the Electricity of the Voltaic Pile; its source, quantity, intensity and general characters, *Experimental Researches in Electricity — 8th Series*, (1834) pp. 425–70.

A50.   Forbes. J. D., On the Horary Oscillations of the Barometer near Edinburgh, deduced from 4410 Observations, *Trans. R. Soc. Edinburgh*, **XII** (1834) pp. 153–90.

A51.   Thompson. R., Report upon the Principal Varieties of the Cherry Cultivated in Garden of the Society, *Trans. Hort. Soc. Lond.*, **I** S2 (1835), pp. 248–94.

A52.   Review — The Fly-fisher's Entomology, *New Sporting Magazine*, **XI** (1836), pp. 343–5.

A53.   Macneill. J. B., Recent Canal-boat Experiments, *Trans. Inst. Civ. Eng.*, i (1836), pp. 237–82.

A54.   Faraday. M., On Induction, *Experimental Researches in Electricity — 12th Series*, (1838), pp. 83–123.

A55.   Jordan. T. B., On a New Mode of Registering the Indications of Meteorological Instruments, *R. Cornwall Polytechnic Soc. Sixth Annual Report*, (1838), pp. 184–9.

A56. Partington. C. F., (Ed.) Air Thermometer, in Electricity, *British Cyclopædia Arts Sci.*, **I** (1838), p. 43.

A57. Partington. C. F., (Ed.) Perspective Instruments, *British Cyclopædia Arts Sci.*, **II** (1838), pp. 170–2.

A58. Partington. C. F., (Ed.) Portraiture, *British Cyclopædia Arts Sci.*, **II** (1838), p. 333.

A59. Cooke and Wheatstone's Electric Telegraph, *Mechanics' Magazine*, **XXXIV** (1841), pp. 433–5.

A60. Poe. E. A., Review of New Books, *Graham's Magazine*, **XX** (1842), pp. 68–72.

A61. Goodwin. J., Progress of the Two Sicilies under the Spanish Bourbons, from the year 1734–1840, *J. Statistical Soc. Lond.*, **V** (1842), pp. 47–73, 177–207.

A62. Electricity, *Encyclopædia Britannica*, 7th Ed. **VIII** (1842), pp. 565–663; also 8th Ed. (1855), pp. 523–627.

A63. Singer's Improved Electrometer, *Encyclopædia Britannica*, 7th Ed. **VIII** (1842), pp. 654–6, also 8th Ed. (1855), pp. 617–19.

A64. Holtzapffel's Book on Turning, *Mechanics' Magazine*, **XXXVIII** (1843), pp. 37–41.

A65. The People's Hand-book to the Polytechnic Institution, *Punch*, **V** (1843), p. 91.

A66. Hebert. L., Perspective Instruments, *Engineer's and Mechanic's Encyclopædia*, **II** (1836), pp. 287–9; reprinted in: *Magazine of Science and School of Arts*, **V** (1844), pp. 305–6.

A67. Hennessy. H., On the Application of Photography to Registering the Thermometer and Barometer, *Phil. Mag.*, **XXVII** S3 (1845), pp. 273–6.

A68. Collen. H., On the Application of the Photographic Camera to Meteorological Registration, *Phil. Mag.*, **XXVIII** S3, (1846), pp. 73–5.

A69. Rennie. J., Address of Sir John Rennie, President to the Annual General Meeting, *Minutes of the Proceedings ICE*, **5** (1846), pp. 19–122.

A70. Self-Registering Instruments, *Practical Mechanic and Engineer's Magazine*, **II** S2 (Nov 1846), 33–4; reprint of: *Literary Gazette*, (1846), pp. 874–5.

A71. Brewster. D., Photography, *North British Review*, **VII** (1847), pp. 465–504.

A72. Brooke C, On the Automatic Registration of Magnetometers, and other Meteorological Instruments, by Photography, *Phil. Trans. R. Soc. Lond.*, **137** (1847), pp. 59–77.

A73. Glaisher. J., Remarks on the Weather during the Quarter ending March 31, 1848, *Phil. Mag.*, **XXXII** S3 (1848), pp. 506–18.

A74. Hebert. L., Air Thermometer, in Electricity, *Engineer's and Mechanic's Encyclopædia*, **I** (1848), p. 53.

A75. Karsten. G., (Ed.) *Fortschritte der Physik im Jahre 1846*, **II** (1848), pp. 240–1.

A76. Barlow. W. H., On the Spontaneous Electrical Currents Observed in the Wires of the Electric Telegraph, *Phil. Trans. R. Soc. Lond.*, **139** (1849), pp. 61–72.

A77. Birt. W. R., Report on the Discussion of the Electrical Observations at Kew, *Report of the British Association for 1849*, (1850), pp. 113–99.

A78. Birt. W. R., On the Connexion of Atmospheric Electricity with the Condensation of Vapour, *Phil. Mag.*, **XXXVI** S3 (1850), pp. 161–71.

A79. Quetelet. M., On Atmospheric Electricity, especially in 1849, *Phil. Mag.*, **I** S4 (1851), pp. 329–32.

A80. Institution of Civil Engineers, *London J.*, **XL** (1852), pp. 305–11.

A81. Sabine. E., On Periodical Laws Discoverable in the Mean Effects of the Larger Magnetic Disturbances — No. II, *Phil. Trans. R. Soc. Lond.* **142** (1852), pp. 103–24.

A82. Sabine. E., Report on the Kew Magnetographs, *Report of the British Association for 1851*, (1852), pp. 325–7.

A83. Welsh. J., Report to Francis Ronalds, Esq., on the Performance of his three Magnetographs during the Experimental Trial at the Kew Observatory, *Report of the British Association for 1851*, (1852), pp. 328–35.

A84. Weale. J., The Observatories of London and its Vicinity, *Pictorial Handbook of London*, (1854), pp. 631–699.

A85. Relation des Observations météorologiques exécutées dans quatre ascensions aérostatiques, *Annales de Chimie et de Physique*, **XLI** S3 (1854), pp. 503–8.

A86. The Electric Telegraph, *London Quart. Review*, **XCV** (1854), pp. 62–85.

A87. Faraday. M., On Electric Induction — Associated cases of Current and Static Effects, *Phil. Mag.*, **VII** S4 (1854), pp. 197–208.

A88. Faraday. M., On Subterraneous Electro-telegraph Wires, *Phil. Mag.*, **VII** S4 (1854), pp. 396–8.

A89. Glaisher. J., Philosophical Instruments and Processes, as represented in the Great Exhibition, *Lectures on the Progress of Arts and Science resulting from the Great Exhibition*, (1854), pp. 243–301.

A90. Royal Society — the Morality of its Members, *Mechanics' Mag.*, **LX**, (1854) pp. 386–91, 413–18

A91. Annual General Meeting, *Monthly Notices of the Royal Astronomical Society*, **XV** (1855), pp. 101–44.

A92. Académie des Sciences, *Cosmos Revue Encyclopédique Hebdomadaire*, **VII** (1855), pp. 660–2.

A93. Nouvelles et Faits Divers, *Cosmos Revue Encyclopédique Hebdomadaire*, **VII** (1855), pp. 705–11.

A94. Nouvelles et Faits Divers, *Cosmos Revue Encyclopédique Hebdomadaire*, **VIII** (1856), pp. 29–33.

A95. Académie des Sciences, *Cosmos Revue Encyclopédique Hebdomadaire*, **VIII** (1856), pp. 371–79.

A96. Porro. I., Oscillations Diurnes du Pendule, *Cosmos Revue Encyclopédique Hebdomadaire*, **VIII** (1856), pp. 578–9.

A97. Association Britannique, *Cosmos Revue Encyclopédique Hebdomadaire*, **IX** (1856), pp. 185–90.

A98. The Weather and its Prognostics, *North British Review*, **XXV** (1856), pp. 173–205.

A99. Johnson. M. J., On the Detection and Measurement of Atmospheric Electricity by the Photo-Barograph and Thermograph, *Report of the British Association for 1855*, (1856), p.40.

A100. Du Moncel. T., Lumière Electrique, *Exposé des Applications de l'Electricité*, 2ⁿᵈ Ed. **III** (1857), pp. 214–81.

A101. Poggendorff. J. C., Ronalds, Francis, *Biographisch-literarisches Handwörterbuch der exakten Wissenschaften*, (1858), pp. 684–5.

A102. Sinobas. M. R., Discurso que Sobre los Fenómenos de la Electricidad Atmosférica, *Real Academia de Ciencias*, Madrid, (1858).

A103. Stewart. B., Construction of the Self-Recording Magnetographs at the Kew Observatory, *Report of the British Association for 1859*, (1860), pp. 201–28.

A104. Stewart. B., On the Great Magnetic Disturbance which extended from August 28 to September 7 1859, as Recorded by Photography at the Kew Observatory, *Phil. Trans. R. Soc. Lond.*, **151** (1861), pp. 423–30.

A105. Walker. C. V., On Magnetic Storms and Earth-Currents, *Phil. Trans. R. Soc. Lond.*, **151** (1861), pp. 89–131.

A106. Clark. L., On Electrical Quantity and Intensity, *Proc. R. Instit. Great Britain*, **III** (1862), pp. 337–41.

A107. Salleron. M., *Comptes rendus hebdomadaires des séances de l'Académie des Sciences*, **59** (1864), p. 916.

A108. The British Association Observatory at Kew, *The Engineer*, (1866), pp. 137–8, 159–60.

A109. Sabine. E., Note on Meteorological Correspondence, *Proc. R. Soc. Lond.*, **15** (1867), pp. 29–38.

A110. Anniversary Meeting, *Proc. R. Soc. Lond.* **15** (1867), pp. 268–88.

A111. Stewart. B., A Comparison between some of the Simultaneous Records of the Barographs at Oxford and at Kew, *Proc. R. Soc. Lond.*, **15** (1867), pp. 413–14.

A112. Thomson. W., Report on Electrometers and Electrostatic Measurements, *Report of the British Association for 1867*, (1868), pp. 489–512.

A113. Everett. J. D., Results of Observations of Atmospheric Electricity at Kew Observatory, and at King's College, Windsor, Nova Scotia, *Phil. Trans. R. Soc. Lond.*, **158** (1868), pp. 347–61.

A114. Yates. J., Remarks on the origin of the Megalithic Structures of Carnac in Brittany, *Archaeological J.*, **28** (1871), pp. 68–9.

A115. Rules and Regulations, *J. Soc. Telegraph Engrs.*, **I** (1872), pp. 10–18.

A116. The Society of Telegraph Engineers, *J. Soc. Telegraph Engrs.*, **I** (1872), pp. 19–39.

A117. Knight. E. H., Disk Telegraph, *Knight's American Mechanical Dictionary*, **I** (1874), p. 708.

A118. Society of Telegraph Engineers, *Telegraphic J. and Electrical Review*, **II** (1874), p. 369.

A119. Thomson. W., Inaugural Address, *J. Soc. Telegraph Engrs.*, **III** (1874), pp. 1–17.

A120. Scientific Prophecy, *Popular Science Monthly*, **VII** (May 1875), pp. 17–19.

A121. Obituary, Sir Charles Wheatstone, *Proc. R. Soc. Lond.*, **24** (1875), xvi–xxvii.

A122. Frost. A. J., The Ronalds' Library and Catalogue, *J. Soc. Telegraph Engrs.*, **VIII** (1879), pp. 12–15.

A123. Obituary — Mr Robert Ronalds, *Gardeners' Chronicle*, **XIV** (1880), p. 346.

A124. Siemens. C. W., The Dynamo-Electric Current and its Application, *J. Soc. Telegraph Engrs.*, **IX** (1880), pp. 278–307.

A125. Electricity at the Crystal Palace, *Nature*, **25** (1882), pp. 515–17.

A126. Knight. E. H., Rain-gage, *Knight's American Mechanical Dictionary*, **II** (1882), pp. 1871–4.

A127. The History of the Electric Telegraph, *Sci. American Supp.* No. 384, (1883), pp. 19–27.

A128. Remarks on Underground Telegraphs, *J. Soc. Telegraph Engrs.*, **XVI** (1887), pp. 421–37.

A129. Thompson. S. P., The Influence Machine, from 1788 to 1888, *J. Soc. Telegraph Engrs.*, **XVII** (1888), pp. 569–628.

A130. Chronological Table, *Encyclopædia Britannica*, 9ᵗʰ Ed. **V** (1889), pp. 720–54.

A131. Ancient House in Brentford, *Sketcher and Theleme*, **1**(13) (1889), pp. 138–9.

A132. Symons. G. J., On Barometric Oscillations during Thunderstorms, and on the Brontometer, and Instrument designed to facilitate their Study, *Proc. R. Soc. Lond.*, **XLVIII** (1890), pp. 59–68.

A133. De Closmadeuc. G., Deux Archéologues anglais à Carnac en 1834, *Bulletin de la Société polymathique du Morbihan*, (1891), pp. 15–48.

A134. List of Portraits and Busts in the Apartments of the Royal Society at Burlington House, *Proc. R. Soc. Lond.*, **50** (1891), pp. 516–523.

A135. Obituary — Sir George Airy, *Proc. R. Soc. Lond.*, **51** (1892), i–xxi.

A136. Anniversary Meeting, *Proc. R. Soc. Lond.*, **52** (1892), pp. 299–325.

A137. Chree. C., Observations on Atmospheric Electricity at the Kew Observatory, *Proc. R. Soc. Lond.*, **60** (1896), pp. 96–132.

A138. Chree. C., Magnetic Work at the Kew Observatory, Richmond, Surrey, *Terrestrial Magnetism Quart. J.*, **I** (1896), pp. 23–6.

A139. Chree. C., Description of the Kew Observatory, *Record R. Soc. Lond.*, (1897), pp. 137–53.

A140. Royal Meteorological Society Jubilee Celebration, *Quart. J. R. Meteorological Soc.*, **XXVI** (1900), pp. 173–202.

A141. James. M. S., The Progress of the Modern Card Catalog Principle, *Public Libraries*, **VII** (1902), pp. 185–9; reprinted in: Rowland. A. R. (Ed.), *The Catalog and Cataloging*, (1969), pp. 56–63.

A142. Library Dinner, *Trans. AIEE*, **XXI** (1903), pp. 97–128.

A143. Atmospheric Electricity, *Encyclopædia Britannica*, 11ᵗʰ Ed. (1911).

A144. Carter. J., The Electric Telegraph, *Nature*, **108** (1921), p. 568.

A145. Calendar of Industrial Pioneers, *Nature*, **110** (1922), p. 199.

A146. Fussell. G. E., The London Cheesemongers of the Eighteenth Century, *Economic J.*, **3** (1928), pp. 394–8.

A147. Patterson. J., A Century of Canadian Meteorology, *Quart. J. R. Meteorological Soc.*, Suppl. **LXVI** (1940), pp. 16–33.

A148. Bush. V., Arthur Edwin Kennelly 1861–1939, *Nat. Acad. Sci. USA Biographical Memoirs*, **XXII** (1940), pp. 83–119.

A149. Thiessen. A. D., Her Majesty's Magnetical and Meteorological Observatory, Toronto, *J. R. Astronomical Soc. Canada*, **39** (1945), pp. 221–30, 267–78, 311–19, 355–69, 394–408; **40** (1946), pp. 221–6, 256–64, 365–72.

A150. Monaghan. J., (Ed.) From England to Illinois in 1821: The Journal of William Hall, *J. Illinois State Historical Soc.*, **39**, (1946), No. 1, pp. 21–67, No. 2, pp. 208–53.

A151. Smith. E. C., Britain's Scientific Shrines (2), *Nature*, **167** (1951), pp. 790–2.

A152. Meteorological Office, Memorial Tablets at Kew Observatory, *Met. Mag.*, **83** (1954), pp. 321–2.

A153. Strout. A. L., Writers on German Literature in *Blackwood's Magazine*, *Library*, **9** (1954), pp. 35–44.

A154. Multhauf. R. P., The Introduction of Self-Registering Meteorological Instruments, *Bull. 228: Contributions from the Museum of History and Tech.*, (1963), pp. 95–116.

A155. King. W. J., The Development of Electrical Technology in the 19th Century: 3. The Early Arc Light and Generator, *Bulletin 228: Contributions from the Museum of History and Tech.*, (1963), pp. 334–406.

A156. Blanchard. D. C. and Bjornsson. S., Water and the Generation of Volcanic Electricity, *Monthly Weather Review*, **95** (1967), pp. 895–8.

A157. Brown. P. L. and Batman, J., (1801–1839), *Australian Dictionary of Biography*, **I** (1966), pp. 67–71.

A158. Twyman. M., The Lithographic Hand Press 1796–1850, *J. Printing Historical Soc.*, **3** (1967), pp. 3–50.

A159. Hart. C., Meteorological Kites, *Kites, An Historical Survey* (1967), pp. 104–13.

A160. Jarvis. R. C., Eighteenth Century London Shipping, in Hollaender. A. E. *et al.*, (Eds.), *Studies in London History presented to Philip Edmund Jones*, (1969).

A161. Scrase. F. J., Some Reminiscences of Kew Observatory in the Twenties, *Met. Mag.* **98** (1969), pp. 180–6.

A162. Aked. C. K., The First Electric Clock, *Antiquarian Horology*, **VIII**, (1973), pp. 276–89.

A163. Payne-Gaposchkin. C. H., The Nashoba Plan for Removing the Evil of Slavery: Letters of Frances & Camilla Wright 1820–29, *Harvard Library Bulletin*, **23** (1975), pp. 221–51, 429–61.

A164. Hackmann. W. D., Eighteenth Century Electrostatic Measuring Devices, *Annali dell'Istituto e Museo di Storia della Scienza di Firenze*, **III**(2) (1978), pp. 3–58.

A165. Stern. W. M., Where, oh where, are the Cheesemongers of London?, *London J.*, **5** (1979), pp. 228–48.

A166. Cawood. J., The Magnetic Crusade: Science and Politics in Early Victorian Britain, *ISIS*, **70** (1979), pp. 492–518.

A167. Salter. M. A., Quarreling in the English Settlement: The Flowers in Court, *J Illinois State Historical Soc.*, **75**(1) (1982), pp. 101–14.

A168. Croft. A. J., The Oxford Electric Bell, *Eur. J. Phys.*, **V** (1984), pp. 193–4.

A169. Miller. D. P., The Revival of the Physical Sciences in Britain, 1815–40, *Osiris*, **2** S2 (1986), pp. 107–34.

A170. Alfred Ronalds and his Victorian Separation Medal, *J. Numismatic Assoc., Australia*, **9** (1998), pp. 59–66.

A171. Hoskin. M., Astronomers at War: South v. Sheepshanks, *J. History of Astronomy*, **XX** (1989), pp. 175–212.

A172. Barraclough. D. R. *et al.*, 150 Years of Magnetic Observatories: Recent Researches on World Data, *Surv. Geophys.*, **13** (1992), pp. 47–88.

A173. Pooley. J., The diary of Mary Anne Nichols 1823–34, a publisher's daughter in Hammersmith, *Trans. LAMAS*, **44** (1993), pp. 171–196.

A174. Hackmann. W., Sir Francis Ronalds' Electric Observatory, *Bull. Scientific Instrument Soc.*, No. 43 (1994), pp. 27–8.

A175. Malin. S. R., Geomagnetism at the Royal Observatory, Greenwich, *Quart. J. R. Astr. Soc.*, **37** (1996), pp. 65–74.

A176. Hunt. J. L., James Glaisher FRS (1809–1903) Astronomer, Meteorologist and Pioneer of Weather Forecasting: 'A Venturesome Victorian', *Quart. J. R Astr. Soc.*, **37** (1996), pp. 315–47

A177. Tinazzi. M., Perpetual Electromotive Of Giuseppe Zamboni, *Atti del XVI Congresso Nazionale di Storia della Fisica e dell'Astronomia*, Como (1996).

A178. Rycroft. M. J. *et al.*, The Global Atmospheric Electric Circuit, Solar Activity and Climate Change. *J. Atmospheric & Solar-Terrestrial Phys.*, **62** (2000), pp. 1563–76.

A179. Hackmann. W., The Enigma of Volta's "Contact Tension" and the Development of the "Dry Pile", in Bevilacqua. F. & Frenonese. L. (Eds.), *Nuova Voltiana: Studies on Volta and his Times*, **3** (2000), pp. 103–19.

A180. Brooks. S., A History of the Card Catalog, in Eberhart. G. M. (Ed.), *The Whole Library Handbook*, **3** (ALA 2000), pp. 311–4.

A181. Elliott. B., The Landscape of Kensal Green Cemetery, in Curl. J. S. (Ed.), *Kensal Green Cemetery: the origins and development of the General Cemetery of All Souls, Kensal Green, London, 1824–2001*, (Chichester: Phillimore 2001), pp. 287–96.

A182. Sherratt. A., and Roughley. C., The Equipment of Scientific Prehistory in Victorian Times, European Assoc. of Archaeologists Annual Meeting, EAA 2002, Thessaloniki (2002). Available at http://www.area-archives.org/files/AREA_WWW_Thessaloniki_session_EAA%20session.pdf (Accessed on 15/7/2015).

A183. Harrison. R. G., The Global Atmospheric Electrical Circuit and Climate, *Surv. Geophys.*, **25** (2004), pp. 441–84.

A184. Wallace. G., Meteorological Observation at the Radcliffe Observatory, in Burley. J. & Plenderleith. K., (Eds.) *A History of the Radcliffe Observatory*, (2005), pp. 103–28.

A185. Mather. T. A., and Harrison. R. G., Electrification of Volcanic Plumes, *Surv. Geophys.*, **27** (2006), pp. 387–432.

A186. Kresek. R., History of the Osborne Firefinder, Forest Ranger Archives (2007). Available at http://nysforestrangers.com/archives/osborne%20firefinder%20by%20kresek.pdf (Accessed on 15/7/2015).

A187. Patanè. A., Viaggiatori in Sicilia: Il viaggio nel Regno delle Due Sicilie (1819–1820) di Gourbillon, *la Freccia Verde*, **XV11**(131) (2008).

A188. Bott. V., Some Brentford Nursery Gardeners, *Brentford & Chiswick Local History J.*, **17** (2008), pp. 3–8.

A189. Elliott. B., English Fruit Illustration in the Early Nineteenth Century, Part 1, *RHS Lindley Library Occasional Papers*, **4** (2010), pp. 37–71.

A190. Aubin. D., A History of Observatory Sciences and Techniques, in Lasota. J. P. (Ed.), *Astronomy at the Frontiers of Science*, (Springer 2011), pp. 109–21.

A191. Peres. I. M. *et al.*, The Photographic Self-Recording of Natural Phenomena in the Nineteenth Century, in Roca-Rosell. A. (Ed.), *Circulation of Science and Technology*, Proc. 4th Int. Conf. ESHS, Barcelona, Nov 2010 (2012), pp. 462–76.

A192. Aplin. K. L. and Harrison. R. G., Atmospheric Electric Fields during the Carrington Flare, *Astronomy & Geophysics*, **55** (2014), 5.32–5.37.

A193. Ronalds. B. F., Remembering the First Battery-Operated Clock, *Antiquarian Horology*, **36** (2015), pp. 244–48.

A194. Ronalds. B. F., Sir Francis Ronalds and the Electric Telegraph. *Int. J. History of Eng. & Tech.*, **86** (2016), pp. 42–55.

A195. Ronalds. B. F., The Bicentennial of Francis Ronalds's Electric Telegraph. *Phys. Today*, **69**(2) (2016), pp. 26–31.

A196. Ronalds B. F., The Ronalds Family in Brentford and Ontario, *Brentford & Chiswick Local History J.*, **25** (2016), pp. 14–17.

A197. Ronalds B. F., Ronalds Nurserymen in Brentford and Beyond, Submitted.

A198. Ronalds B. F., Betsey Ronalds (1788–1854): Horticultural Illustrator, Submitted.

A199. A Union Home for Engineering Societies. Available at http://ethw.org/images/8/88/History%2C_A_Union_Home_for_Engineering_Societies.pdf (Accessed on 28/4/2016).

Multi-year Series

S1.   *Report of the British Association for the Advancement of Science*, (1832)–(1871).

S2.   United Kingdom Census, (1841)–(1911).

S3.   *Report of the Astronomer Royal to the Board of Visitors*, (1841)–(1870).

S4.   Royal Observatory Greenwich, *Results of Magnetical and Meteorological Observations*, (1845)–(1880).

S5.   *Astronomical and Meteorological Observations made at the Radcliffe Observatory*, (1854)–(1864).

S6.   *Kew Committee Report*, (1853)–(1871) (BAAS); (1872)–(1899) (Royal Society).

S7.   *Report of the Meteorological Committee of the Royal Society*, (1867)–(1875).

S8.   *Journal of the Society of Telegraph Engineers [and Electricians]*, (1872)–(1888).

S9.   *Journal of the Institution of Electrical Engineers*, (1889)–(1960).

S10. *Literary Gazette and Journal of Belles Lettres, Arts, Sciences, &c*, (1825)–(1855).

S11. *Athenæum*, (1840)–(1875).

Correspondence

C1.   B. Ronalds (Brentford) to R. Ronalds (Uxbridge) 10/8/1813, WU B2284.

C2.   C. Ronalds (Blackheath) to R. Ronalds (Uxbridge) 16/8/1813, WU B2284.

C3.   H. Ronalds (Brentford) to R. Ronalds 14/9/1813, WU B2284.

C4.   J-A. de Luc (Windsor) to F. Ronalds (Hammersmith) ?/1/1815, IET 1.6.56.
C5.   J. Henniker to F. Ronalds (Hammersmith) 24/3/1815, IET 1.6.56.
C6.   J. Banks (Soho) to F. Ronalds (Hammersmith) 18/4/1815, IET 1.6.56.
C7.   W. Brande (London) to F. Ronalds (Hammersmith) ?/5/1815, IET 1.6.56.
C8.   J-A. de Luc (Windsor) to F. Ronalds (Hammersmith) 4/5/1815, IET 1.6.56.
C9.   F. Ronalds (Hammersmith) to J-A de Luc (Windsor) draft 9/5/1815, IET 1.6.56.
C10.   J-A. de Luc (Windsor) to F. Ronalds (Hammersmith) 11/5/1815, IET 1.6.56.
C11.   J-A. de Luc (Windsor) to F. Ronalds (Hammersmith) 16/6/1815, IET 1.6.56.
C12.   F. Ronalds (Hammersmith) to J-A. de Luc (Windsor) draft ?/8/1815, IET 1.6.56.
C13.   G. Singer (London) to F. Ronalds (Hammersmith) 28/9/1815, IET 1.6.56.
C14.   F. Ronalds (Hammersmith) to ? (Paris?) draft 3/11/1815, IET 1.6.56.
C15.   F. Ronalds (Hammersmith) to Melville copy 11/7/1816, IET 1.9.2.
C16.   R. Hay (Admiralty) to F. Ronalds (Hammersmith) 29/7/1816, IET 1.9.2.
C17.   J. Barrow (Admiralty) to F. Ronalds (Hammersmith) 5/8/1816, IET 1.9.2.
C18.   F. Ronalds (Paris) to J. Ronalds (Hammersmith) 8/9/1818, IET 1.1.10.
C19.   F. Ronalds (Milan) to Family (Hammersmith) 13/10/1818, IET 1.1.11.
C20.   F. Ronalds (Rome) to Family (Hammersmith) 25/11/1818, IET 1.1.11.
C21.   F. Ronalds (Naples) to Family (Hammersmith) 15/1/1819, IET 1.1.11.
C22.   F. Ronalds (Naples) to J. Ronalds (Hammersmith) 21/2/1819, IET 1.1.11.
C23.   F. Ronalds (Naples) to C. Ronalds (Hammersmith) 4/4/1819, IET 1.1.11.
C24.   F. Ronalds (Naples) to C. Ronalds (Hammersmith) 9/4/1819, IET 1.1.11.
C25.   F. Ronalds (Messina) to J. Ronalds (Hammersmith) 6/8/1819, IET 1.1.11.
C26.   F. Henniker (Valletta) to F. Ronalds (Catania) 29/9/1819, IET 1.1.74.
C27.   F. Ronalds (Catania) to C. Ronalds (Hammersmith) 4/11/1819, IET 1.1.11.
C28.   F. Ronalds (Valetta) to C. Ronalds (Hammersmith) 6/12/1819, IET 1.1.11.
C29.   F. Ronalds (Larnaca) to E. Ronalds (Upper Thames St) copy 21/4/1820, IET 1.1.11.
C30.   P. Lee (Alexandria) to F. Ronalds (Cyprus) 27/4/1820, IET 1.1.13.
C31.   F. Ronalds (Limassol) to E. Ronalds (Upper Thames St) 2/5/1820, IET 1.1.14.
C32.   F. Ronalds (Limassol) to Briggs & Lee (Alexandria) copy 2/5/1820, IET 1.1.21.
C33.   P. Lee (Alexandria) to F. Ronalds (Smyrna) 24/5/1820, IET 1.1.15.
C34.   P. Lee (Alexandria) to F. Ronalds (Smyrna) 29/5/1820, IET 1.1.16.
C35.   F. Ronalds (Smyrna) to E. Ronalds (Upper Thames St) ?/6/1820, IET 1.1.17.
C36.   J. Wilson (Edinburgh) to A. Blair (Birmingham) 20/6/1820, Leeds.
C37.   J.Wilson to A. Blair (Birmingham) 11/8/1820, Leeds.
C38.   F. Ronalds (Venice) to Family (Queen Sq) 30/9/1820, IET 1.1.11.
C39.   W. Davidson (Milan) to A. Ferrier (Rotterdam) 22/10/1820, IET 1.1.18.
C40.   F. Ronalds (Geneva) to E. Ronalds (Upper Thames St) 31/10/1820, IET 1.1.11.
C41.   J. Wilson to A. Blair (London) 19?/11/1821, Leeds.
C42.   F. Ronalds (Strasbourg) to E. Ronalds (Queen Sq) draft 29/11/1820, IET 1.1.11.
C43.   F. Ronalds (Carlsruhe) to E. Ronalds (Queen Sq) 1/12/1820, IET 1.1.11.
C44.   J. Wilson (Edinburgh) to A. Blair (Birmingham) 6/2/1822, Leeds.
C45.   F. Ronalds (Queen Sq) to F. Henniker (London) draft 1822, IET 1.1.82.
C46.   F. Henniker (London) to F. Ronalds (Queen Sq) various 1822, IET 1.1.76, 1.1.79,
       1.6.61–62.

C47. F. Ronalds to F. Henniker draft 1822, IET 1.1.81.

C48. F. Ronalds (Brighton) to F. Henniker draft 4/11/1822, IET 1.6.63.

C49. F. Ronalds to F. Henniker draft, IET 1.6.64.

C50. H. Ronalds (Albion) to F. Ronalds 31/3/1823, IET 1.5.22.

C51. J. Cooper (Messina) to F. Ronalds (Messina) 4/9/1823, IET 1.1.68.

C52. A. Blair (Birmingham) to J. Wilson (Kendal) 23/10/1823, Leeds.

C53. J. Wilson (Edinburgh) to A. Blair (London) 7/11/1825, Leeds.

C54. H. Ronalds (Albion) to E. Flower (England) 15/2/1826, Chicago.

C55. J. P. Greaves to C. F. Wurm (Brighton), 12/7/1826, Staats- und Universitätsbibliothek Hamburg, N. L. Wurm Archive, 50, 102.

C56. J. Wilson to A. Blair (London) 14/12/1826, Leeds.

C57. K. Ronalds (Albion) to E. Flower (Birmingham) 14/7/1827, Chicago.

C58. A. Blair (Fulham) to J. Wilson (London) 11/9/1827, Leeds.

C59. F. Wright (Ship) to E. Ronalds (Croydon), 16/12/1827, Massachusetts Historical Soc, C. E. French Papers.

C60. R. Owen (London) to E. Ronalds (Croydon) 16/11/1828, Indiana Historical Soc M0219 Box 1.

C61. E. Ronalds (Brixton) to R. Owen (London) 7/9/1829?, Co-op ROC/17/31/1.

C62. J. Wilson (Kendal) to A. Blair (Birmingham) 15/9/1829, Leeds.

C63. F. Ronalds (Hastings) to C. Holtzapffel? draft 9/2/1830, IET 1.6.100.

C64. A. Blair (London) to L. Horner (UCL) 8/1/1831 UCL College Correspondence.

C65. A. Blair (London) to J. Wilson (Edinburgh) 12/1/1831, Leeds.

C66. F. Wright (Paris) to R. Owen (London) 16/3/1831, Co-op ROC/22/65/2.

C67. F. Ronalds to J. South draft 16/5/1833, IET 1.5.4.

C68. F. Ronalds to Simpson draft 1833?, IET 1.6.29.

C69. C. Holtzapffel to F. Ronalds 10/6/1834, IET 1.1.87.

C70. C. Holtzapffel to F. Ronalds 14/6/1834, IET 1.1.87.

C71. F. Ronalds (Chiswick) to J. Nichols 3/9/1834, Nichols Archive Project, Private Collection PC1/17/185 NAD2943.

C72. J. Lawe (Seaforth) to B. Ronalds (Brentford) 24/10/1834, WU B1450.

C73. J. B. Deane to F. Ronalds 29/1/1836?, IET 1.2.17.

C74. J. Wilson (Edinburgh) to A. Blair (London) 31/7/1836, Leeds.

C75. A. Blair (Bristol) to J. Wilson (Edinburgh) 24/4/1837, Leeds.

C76. S. Carter (London) to F. Ronalds 20/10/1837, IET 1.6.77.

C77. F. Ronalds to S. Carter draft ?/10/1837, IET 1.6.73.

C78. I. Gifford (London) to F. Ronalds (Chiswick) 27/10/1837, IET 1.6.94.

C79. H. Gem to F. Ronalds ?/12/1837, IET 1.6.87.

C80. F. Ronalds to S. Carter draft ?/12/1837, IET 1.6.69.

C81. J. Rennie and G. Rennie to F. Ronalds 1/2/1838, IET 1.6.81.

C82. F. Ronalds to ? draft 1838, IET 1.6.71.

C83. J. Hawkins (Chancery) to F. Ronalds (Chiswick) 2/2/1839, IET 1.6.99.

C84. K. Ronalds (Albion) to E. Flower (Stratford) 25/2/1839, Chicago.

C85. F. Ronalds to E Bayley draft 1839?, IET 1.5.41.

C86. E. Bayley (St Petersburg) to C. Field (St Petersburg) 30/8/1839, IET 1.5.42.

C87.   Stanhope (Chevening) to J. Cary (London) 31/10/1839, IET 1.6.19.
C88.   Stanhope (Chevening) to F. Ronalds (Chiswick) 5/11/1839, IET 1.6.20.
C89.   F. Ronalds (Chiswick) to Stanhope (London) draft ?/11/1839, IET 1.6.21.
C90.   Stanhope (Chevening) to F. Ronalds (Chiswick) 16/11/1839, IET 1.6.22.
C91.   Stanhope (Chevening) to F Ronalds (Chiswick) 6/12/1839, IET 1.6.23.
C92.   F. Ronalds (Chiswick) to Stanhope (Chevening) draft 23/12/1839, IET 1.6.25.
C93.   Stanhope (Chevening) to F. Ronalds (Chiswick) 28/12/1839, IET 1.6.26.
C94.   F. Ronalds (Chiswick) to *The Times* draft 24/5/1840, IET 1.5.17.
C95.   M. Ronalds (Brentford) to B. Ronalds (Seaforth) 29/6/1840, WU B2284.
C96.   F. Ronalds to J. Cary draft 1840?, IET 1.6.34.
C97.   J. Cary to F. Ronalds 1840?, IET 1.5.3.
C98.   S. Sharpe to F. Ronalds 29/07/1840, IET 1.2.13.
C99.   F. Ronalds (Chiswick) to Lord Mayor (London) draft 5/11/1840, IET 1.5.14.
C100.   W. Cooke to G. Cooper (Brentford) 4/12/1840, IET 7.1.072.
C101.   G. Cooper (Brentford) to F. Ronalds (Chiswick) 5/12/1840, IET 1.6.41.
C102.   E. Sabine to J Herschel 13/1/1842, R Soc HS.15.136.
C103.   A. Blair (Fontainebleau) to J. Wilson 28/4/1841, Leeds.
C104.   C. Wheatstone (London) to E. Sabine 24/6/1842, R Soc Sa.1779.
C105.   A. Blair (Fontainebleau) to F. Ronalds (Chiswick) 9/8/1842, IET 1.2.12.
C106.   M. Ronalds (Brentford) to H. Ronalds Jnr (Raleigh) 12/10/1842, WU B2284.
C107.   C. Wheatstone (London) to F. Ronalds 11/11/1842, IET 1.3.30.
C108.   F. Ronalds to C Wheatstone 16/11/1842 copy, IET 1.4.17b.
C109.   L. Ronalds (Coventry) to B. Ronalds (Brentford) 15/12/1842, WU B1450.
C110.   M. Ronalds (Brentford) to H. Ronalds Jnr (Raleigh) 31/1/1843, WU B558.
C111.   E. Ronalds (London) to H. Ronalds Jnr (Raleigh) 3/4/1843, WU B558.
C112.   F. Ronalds to C Wheatstone draft 24/6/1843, IET 1.4.17.
C113.   C. Wheatstone (London) to F. Ronalds 30/6/1843, IET 1.3.39.
C114.   F. Ronalds (Chiswick) to W. Aiton (Kew) draft 13/7/1843, IET 1.4.17.
C115.   J. Daniell (London) to F. Ronalds 14/7/1843, IET 1.3.41.
C116.   C. Wheatstone (London) to F. Ronalds 7/8/1843, IET 1.4.43.
C117.   F. Ronalds (Kew) to C. Wheatstone copy 18/8/1843, IET 1.4.17.
C118.   M. Ronalds (Brentford) to H. Ronalds Jnr (Raleigh) 30/8/1843, WU B2284.
C119.   G. Airy (Greenwich) to C. Wheatstone 7/9/1843, Cambridge RGO 6/701.
C120.   C. Wheatstone (London) to G. Airy 9/9/1843, Cambridge RGO 6/701
C121.   C. Wheatstone (London) to F. Ronalds 11/9/1843, IET 1.3.45.
C122.   B. Ronalds (Brentford) to H. Ronalds Jnr (Raleigh) 15/9/1843, WU B558.
C123.   G. Airy (Greenwich) to M. Faraday 20/9/1843, Cambridge RGO 6/701.
C124.   M. Faraday (Royal Institution) to G. Airy 21/9/1843, Cambridge RGO 6/701.
C125.   G. Airy (Greenwich) to M. Faraday 21/9/1843, Cambridge RGO 6/701.
C126.   F. Ronalds to G. Airy copy 28/9/1843, IET 1.4.17b.
C127.   G. Airy (Greenwich) to F. Ronalds 9/11/1843, IET 1.3.49.
C128.   F. Ronalds (Chiswick) to C. Wheatstone draft 28/11/1843, IET 1.4.17.
C129.   F. Ronalds (Chiswick) to C. Wheatstone draft 22/12/1843, IET 1.4.17.

C130. G. Airy (Greenwich) to F. Ronalds 23/12/1843, IET 1.3.52.

C131. F. Ronalds to C. Wheatstone draft ?/1/1844, IET 1.4.17.

C132. M. Faraday (Royal Institution) to G. Airy 13/1/1844, Cambridge RGO 6/701

C133. F. Ronalds (London) to G. Airy 29/4/1844, Cambridge RGO 6/701.

C134. G. Airy (Greenwich) to F. Ronalds 27/5/1844, IET 1.3.11.

C135. J. Daniell (London) to F. Ronalds 18/6/1844, IET 1.3.13.

C136. C. Wheatstone (London) to F. Ronalds 27/6/1844, IET 1.3.12.

C137. C. Wheatstone (York) to F. Ronalds 30/9/1844, IET 1.3.15.

C138. F. Ronalds to H. Collen 14/10/1844 copy, IET 1.4.17b.

C139. C. Walker (London) to F. Ronalds ?/10/1844, IET 1.3.16.

C140. J. Phillips (Dublin) to F. Ronalds 6/11/1844, IET 1.3.20.

C141. J. Phillips (Dublin) to F. Ronalds 18/11/1844, IET 1.3.22.

C142. F. Ronalds to H. Collen copy 19/12/1844, IET 1.4.17b.

C143. J. Daniell (Norwood) to F. Ronalds 2/1/1845, IET 1.3.55.

C144. B. Ronalds (Brentford) to H. Ronalds Jnr (Raleigh) 15/1/1845, WU B2284.

C145. C. Wheatstone (London) to F. Ronalds 10/2/1845, IET 1.3.59.

C146. F. Ronalds to E. Sabine copy ?/2/1845, IET 1.4.17b.

C147. E. Sabine (Woolwich) to F Ronalds 12/2/1845, IET 1.3.60.

C148. J. Herschel (Collingwood) to C. Wheatstone 6/4/1845, R Soc HS.25.11.6.

C149. G. Airy to C. Brooke 25/10/1845, Cambridge RGO 6/675.

C150. G. Airy (Greenwich) to F. Ronalds 31/10/1845, IET 1.3.69.

C151. F. Ronalds (Kew) to G. Airy 4/11/1845, Cambridge RGO 6/701.

C152. G. Airy (Greenwich) to F. Ronalds 5/11/1845, IET 1.3.70.

C153. F. Ronalds (Kew) to G. Airy 6/11/1845, Cambridge RGO 6/701.

C154. G. Airy (Greenwich) to F. Ronalds 8/11/1845, IET 1.3.72

C155. J. Glaisher to G. Airy 10/11/1845, Cambridge RGO 6/701.

C156. G. Airy (Greenwich) to F. Ronalds 12/11/1845, IET 1.3.73.

C157. J. Glaisher to G. Airy 27/11/1845, Cambridge RGO 6/701.

C158. J. Glaisher (Greenwich) to F. Ronalds 2/12/1845, IET 1.3.87.

C159. G. Airy (Greenwich) to F. Ronalds 12/12/1845, IET 1.3.77.

C160. F. Ronalds (Chiswick) to G. Airy 14/12/1845, Cambridge RGO 6/701.

C161. G. Airy (Greenwich) to F. Ronalds 15/12/1845, IET 1.3.79.

C162. J. South (Kensington) to F. Ronalds 31/12/1845, IET 1.3.86.

C163. J. Galloway (Kew) to F. Ronalds 8/2/1846, IET 1.3.97.

C164. F. Ronalds to J. Glaisher copy ?/2/1846, IET 1.4.17b.

C165. F. Ronalds to E. Sabine copy 24/3/1846, IET 1.4.17b.

C166. E. Sabine (Woolwich) to F. Ronalds 31/3/1846, IET 1.3. 102.

C167. C. Wheatstone (London) to F. Ronalds 1/4/1846, IET 1.3.103.

C168. F. Ronalds to E. Sabine copy 4/4/1846, IET 1.4.17b.

C169. J. Egerton (London) to F. Ronalds 7/4/1846, IET 1.3.106.

C170. G. Airy (Greenwich) to F. Ronalds 21/7/1846, IET 1.3.110.

C171. F. Ronalds to J. Glaisher copy 17/8/1846, IET 1.4.17b.

C172. J. Glaisher (Blackheath) to F. Ronalds 27/8/1846, IET 1.3.115.

C173. J. Glaisher (Greenwich) to F. Ronalds 29/8/1846, IET 1.3.116.

C174. J. Glaisher (Greenwich) to F. Ronalds 4/9/1846, IET 1.3.119.

C175. E. Sabine (London) to F. Ronalds 4/9/1846, IET 1.3.118.

C176. E. Sabine (Woolwich) to F. Ronalds 22/9/1846, IET 1.3.122.

C177. F. Ronalds to E. Sabine copy 28/9/1846, IET 1.4.17b.

C178. J. Lefroy (Knightsbridge) to F. Ronalds 30/9/1846, IET 1.3.123.

C179. F. Ronalds (Kew) to J. Lefroy copy 2/10/1846, IET 1.4.17b.

C180. J. Lefroy (Knightsbridge) to F. Ronalds 2/10/1846, IET 1.3.124.

C181. J. Glaisher (Blackheath) to F. Ronalds 3/10/1846, IET 1.3.126.

C182. E. Sabine (Woolwich) to F. Ronalds 5/10/1846, IET 1.3.127.

C183. S. Banks (Newmarket) to F. Ronalds 5/10/1846, IET 1.3.128.

C184. J. Glaisher (Blackheath) to F. Ronalds (Chiswick) 7/10/1846, IET 1.3.129.

C185. F. Ronalds (Chiswick) to G. Airy 8/10/1846, Cambridge RGO 6/676.

C186. J. Glaisher to G. Airy 9/10/1846, Cambridge RGO 6/676.

C187. G. Airy (Greenwich) to F Ronalds 14/10/1846, IET 1.3.131.

C188. F. Ronalds (Chiswick) to G. Airy 16/10/1846, Cambridge RGO 6/42.

C189. J. Newman (London) to F. Ronalds 9/11/1846, IET 1.3.135.

C190. G. Airy (Greenwich) to F. Ronalds 26/1/1847, IET 1.3.141.

C191. G. Airy (Greenwich) to F. Ronalds 12/2/1847, IET 1.3.144.

C192. G. Airy to C. Brooke 7/4/1847, Cambridge RGO 6/675.

C193. F. Ronalds (Kew) to G. Airy 24/5/1847, Cambridge RGO 6/701.

C194. J. Glaisher (Blackheath) to F. Ronalds 5/6/1847, IET 1.3.156.

C195. J. Ronalds (Brighton) to F. Ronalds (Oxford) ?/6/1847?, IET 1.4.56.

C196. G. Airy (Greenwich) to F. Ronalds 10/8/1847, IET 1.3.163.

C197. W. Birt (Bethnal Green) to F. Ronalds 21/8/1847, IET 1.3.168.

C198. M. Ronalds (Brentford) to H. Ronalds Jnr 4/10/1847, WU B558.

C199. J. Herschel (Collingwood) to F. Ronalds 7/10/1847, IET 1.3.175.

C200. T. Malone (London) to F. Ronalds 15/10/1847, IET 1.3.178.

C201. J. Lee (Hartwell) to F. Ronalds 22/10/1847, IET 1.3.179.

C202. T. Malone (London) to F. Ronalds 10/11/1847, IET 1.3.183.

C203. H. Ronalds Jnr (Raleigh) to E. Ronalds (Brentford) 8/3/1848, WU B2284.

C204. G. Airy (Greenwich) to F. Ronalds 10/3/1848, IET 1.3.184.

C205. J. Glaisher (Blackheath) to F. Ronalds 13/3/1848, IET 1.3.185.

C206. F. Ronalds (Chiswick) to S. Sharpe (Highbury) 17/5/1848, UCL GB 0103 MS ADD 206.

C207. W. Birt (Bethnal Green) to J. Herschel 19/5/1848, IET 1.3.190.

C208. J. Glaisher to F. Ronalds 19/5/1848, IET 1.3.191.

C209. W. Birt (Bethnal Green) to E. Sabine 24/5/1848, R Soc Sa.163.

C210. J. Herschel (Collingwood) to E. Sabine 1/6/1848, R Soc Sa.656.

C211. G. Humphreys (Greenwich) to F. Ronalds 12/6/1848, IET 1.3.196.

C212. E. Sabine (Woolwich) to J. Herschel 17/6/1848, R Soc HS.15.222.

C213. F. Ronalds (Chiswick) to E. Sabine 30/6/1848, R Soc Sa.1087.

C214. E. Sabine (Woolwich) to J. Herschel 3/7/1848, R Soc HS.15.224.

C215. A. Weld (Stonyhurst) to F. Ronalds 9/7/1848, IET 1.3.202.

C216. S. Sharpe (Highbury) to F. Ronalds (Chiswick) 21/7/1848, IET 1.3.208.

C217. J. Herschel (Collingwood) to E. Sabine 21/7/1848, R Soc Sa.657.

C218. T. Malone (London) to F. Ronalds 26/7/1848, IET 1.3.209.

C219. E. Sabine to J. Herschel 30/7/1848, R Soc HS.15.332.

C220. L. Jones (Paddington) to F. Ronalds 12/8/1848, IET 1.3.211.

C221. G. Airy (Greenwich) to H. Ward (Admiralty) 17/8/1848, Cambridge RGO 6/675.

C222. P. Siljeström (London) to F. Ronalds 27/8/1848, IET 1.3.216.

C223. E. Sabine (Parsonstown) to F. Ronalds 27/8/1848, IET 1.3.217.

C224. C. Younghusband (Charlton) to F. Ronalds 7/9/1848, IET 1.3.222.

C225. E. Sabine (Ireland) to F. Ronalds 9/9/1848, IET 1.9.1.

C226. E. Sabine (Ireland) to J. Herschel 9/9/1848, R Soc HS.15.227.

C227. C. Field (London) to F. Ronalds (Chiswick) 14/9/1848, IET 1.3.224.

C228. J. Herschel to G Airy 24/9/1848, Cambridge RGO 6/675.

C229. G. Airy (Greenwich) to J. Herschel 25/9/1848, Cambridge RGO 6/675.

C230. E. Sabine (Dublin) to F. Ronalds 26/9/1848, IET 1.9.1.

C231. J. Lee (Hartwell) to F. Ronalds 6/10/1848, IET 1.3.230.

C232. H. Lloyd (Dublin) to F. Ronalds 13/10/1848, IET 1.3.234.

C233. E. Sabine (Woolwich) to Northampton 21/10/1848, R Soc Sa.1177.

C234. J. Herschel (Collingwood) to G. Airy 26/10/1848, Cambridge RGO 6/675.

C235. J. Herschel (Collingwood) to Northampton 27/10/1848, R Soc Sa.659.

C236. J. Herschel (Collingwood) to G. Airy 28/10/1848, Cambridge RGO 6/675.

C237. G. Airy (Greenwich) to J. Herschel 30/10/1848, Cambridge RGO 6/675.

C238. E. Sabine to F. Ronalds 8/11/1848, IET 1.3.245.

C239. F. Ronalds (Chiswick) to E. Sabine 10/11/1848, R Soc Sa.1088.

C240. J. Herschel (Collingwood) to E. Sabine 15/11/1848, R Soc Sa.658.

C241. E. Sabine (Woolwich) to F. Ronalds 27/11/1848, IET 1.9.1.

C242. E. Sabine to F. Ronalds 5/12/1848, IET 1.3.251.

C243. E. Sabine to F. Ronalds 7/12/1848, IET 1.9.1.

C244. C. Brooke (London) to G. Airy 22/12/1848, Cambridge RGO 6/676

C245. E. Sabine (Woolwich) to J. Herschel 3/1/1849, R Soc HS.15.228.

C246. E. Sabine (Woolwich) to F. Ronalds ?/1/1849, IET 1.3.254.

C247. T. Malone to H. Talbot 17/2/1849, British Library Add MS 88942/2/80; also Correspondence of William Henry Fox Talbot, http://foxtalbot.dmu.ac.uk/ (Accessed 9/7/2015).

C248. G. Field to F. Ronalds 28/2/1849, IET 1.3.262.

C249. G. Field to F. Ronalds ?/3/1849, IET 1.3.263.

C250. W. Hooker (Kew) to F. Ronalds 27/4/1849, IET 1.3.275.

C251. Northampton (London) to E. Sabine 27/4/1849, R Soc Sa.947.

C252. E. Sabine to F. Ronalds 28/4/1849, IET 1.9.1.

C253. J. Herschel (Collingwood) to E. Sabine 2/5/1849, IET 1.9.1.

C254. W. Hooker (Kew) to F. Ronalds 2/5/1849, IET 1.3.277.

C255. A. Weld (Stonyhurst) to F. Ronalds 13/5/1849, IET 1.3.279.

C256. J. Herschel (Collingwood) to F. Ronalds 13/5/1849, IET 1.9.1.

C257. E. Sabine (Woolwich) to F. Ronalds 15/6/1849, IET 1.3.286.

C258. C. Younghusband (Charlton) to F. Ronalds 16/6/1849, IET 1.3.287.

C259. C. Brooke (Sandgate) to E. Sabine 28/7/1848, R Soc Sa.238.

C260. J. Herschel (Collingwood) to E. Sabine 1/8/1849, R Soc MM.15.59.

C261. C. Younghusband (Charlton) to F. Ronalds 13/8/1849, IET 1.3.310.

C262. E. Sabine (Woolwich) to F. Ronalds 30/8/1849, IET 1.3.312.

C263. A. Blair to F. Ronalds ?/9/1849, IET 1.1.95.

C264. E. Sabine to F. Ronalds 24/9/1849, IET 1.3.320.

C265. E. Sabine to F. Ronalds 25/9/1849, IET 1.3.321.

C266. W. Henshaw (Birmingham) to F. Ronalds 27/9/1849, IET 1.3.322.

C267. J. Lee (Hartwell) to F. Ronalds 29/9/1849, IET 1.3.323.

C268. C. Younghusband to F. Ronalds 1/10/1849, IET 1.3.326.

C269. E. Sabine to F. Ronalds 2/10/1849, IET 1.3.327.

C270. J. Lee (Hartwell) to F. Ronalds 15/10/1849, IET 1.3.33.

C271. E. Sabine to F. Ronalds 17/10/1849, IET 1.3.335.

C272. J. Lefroy (Toronto) to F. Ronalds 1/11/1849, IET 1.3.340.

C273. F. Ronalds (Kew) to W. White 4/11/1849, Wellcome Library MS7545/18.

C274. W. Birt (Kew) to J. Phillips 15/11/1849, R Soc Sa.169.

C275. W. Birt (Kew) to R. Hunt 16/11/1849, R Soc Sa.170.

C276. W. Birt (Kew) to J. Phillips 21/12/1849, R Soc Sa.172.

C277. W. Birt (Kew) to J. Herschel 15/1/1850, R Soc HS.4.137.

C278. E. Sabine (St Leonards) to J. Herschel 15/2/1850, R Soc HS.15.336.

C279. W. Birt (Kew) to E. Sabine 2/3/1850, R Soc Sa.173.

C280. E. Sabine (St Leonards) to J. Herschel 6/3/1850, R Soc HS.15.230.

C281. C. Younghusband (Woolwich) to F. Ronalds 6/3/1850, IET 1.3.371.

C282. J. Herschel (Collingwood) to E. Sabine 10/3/1850, R Soc Sa.661–2.

C283. C. Younghusband (Woolwich) to F. Ronalds 12/3/1850, IET 1.3.372.

C284. A. Blair (London) to Royal Archaeological Institute 15/3/1850, Society of Antiquaries Library.

C285. W. Birt (Kew) to E. Sabine 16/3/1850, R Soc Sa.175.

C286. T. Malone (London) to F. Ronalds 16/3/1850, IET 1.3.375.

C287. C. Younghusband (Woolwich) to F. Ronalds 27/3/1850, IET 1.3.379.

C288. E. Sabine (Teignmouth) to F. Ronalds 19/4/1850, IET 1.3.390.

C289. E. Sabine (Woolwich) to F. Ronalds 11/5/1850, IET 1.3.397.

C290. E. Sabine (Woolwich) to F. Ronalds 5/6/1850, IET 1.3.404.

C291. W. Birt (Kew) to E. Sabine 5/6/1850, R Soc Sa.179.

C292. J. Lefroy (Toronto) to F. Ronalds 26/6/1850, IET 1.3.417.

C293. J. Herschel (Collingwood) to E. Sabine 9/7/1850, IET 1.3.423.

C294. W. Birt (Kew) to F. Royle 11/7/1850, R Soc Sa.180.

C295. J. Lefroy (Toronto) to F. Ronalds 17/7/1850, IET 1.3.425.

C296. W. Birt (Kew) to G. Airy 29/7/1850, Cambridge RGO 6/403.

C297. F. Royle (Edinburgh) to J. Gassiot 8/8/1850, R Soc Sa.1126.

C298. E. Sabine (Lucerne) to J. Lefroy 9/8/1850, IET 1.3.441.

C299. E. Sabine (Lucerne) to F. Ronalds 11/8/1850, IET 1.3.440.

C300. G. Airy (Taymouth) to W. Sykes 16/8/1850, Cambridge RGO 6/403.

C301. W. Birt (Kew) to F. Royle 22/8/1850, R Soc Sa.181.

C302. F. Ronalds (Kew) to E. Sabine 23/8/1850, R Soc Sa.1092.

C303. F. Royle (Acton) to F. Ronalds 24/8/1850, IET 1.3.446.

C304. W. Sykes (Inverness) to G. Airy 24/8/1850, Cambridge RGO 6/403.

C305. J. Welsh (Edinburgh) to F. Ronalds 26/8/1850, IET 1.3.447.

C306. G. Airy to J. Forbes 27/8/1850, Cambridge RGO 6/403.

C307. E. Sabine (Geneva) to F. Ronalds 8/9/1850, IET 1.3.450.

C308. J. Gassiot (London) to W. Sykes (London) 19/9/1850, R Soc Sa.591.

C309. E. Sabine (Amiens) to F. Ronalds 6/10/1850, IET 1.3.460.

C310. J. Lefroy (Toronto) to F. Ronalds 8/10/1850, IET 1.3.464.

C311. W. Sykes (London) to F. Ronalds 12/10/1850, IET 1.3.467.

C312. E. Sabine (London) to F. Ronalds 14/10/1850, IET 1.3.469.

C313. J. Lefroy (Toronto) to C. Younghusband part ?/10/1850, IET 1.3.479.

C314. M. Faraday (Royal Institution) to H. Acland 18/10/1850, James. F. A., (Ed.), *The Correspondence of Michael Faraday*, **IV** (1999), pp. 192–3.

C315. E. Sabine (London) to F. Ronalds 25/10/1850, IET 1.3.477.

C316. E. Sabine (London) to J. Herschel 16/11/1850, Herschel Collection, Harry Ransom Center, Univ. Texas at Austin, Reel 1093 H/M-0524.

C317. E. Sabine (London) to F. Ronalds 17/11/1850, IET 1.3.485.

C318. E. Sabine (London) to F. Ronalds 26/11/1850, IET 1.3.490.

C319. G. Airy (Greenwich) to F. Ronalds 26/11/1850, IET 1.3.489.

C320. E. Sabine (London) to F. Ronalds 9/12/1850, IET 1.3.495.

C321. E. Sabine (London) to F. Ronalds 14/12/1850, IET 1.3.496.

C322. J. Egerton (London) to F. Ronalds 5/3/1851, IET 1.3.521.

C323. W. Sykes (London) to F. Ronalds 6/5/1851, IET 1.3.542.

C324. J. Gassiot (London) to F. Ronalds 28/6/1851, IET 1.3.569.

C325. W. Birt (London) to E. Sabine 1/7/1851, R Soc Sa.182.

C326. B. Ronalds (Brentford) to H. Ronalds Jnr 8/8/1851, WU B558.

C327. F. Ronalds (Chiswick) to G. Airy 9/9/1851, Cambridge RGO 6/677.

C328. F. Ronalds (Kew) to G. Airy (Greenwich) 29/10/1851, IET 1.9.1.

C329. G. Airy (Greenwich) to F. Ronalds 6/12/1851, IET 1.9.1.

C330. F. Ronalds (Kew) to J. Chavarri copy 19/3/1852, IET 1.9.1.

C331. F. Ronalds (Chiswick) to J. Chavarri copy 6/5/1852, IET 1.9.1.

C332. J. Herschel to E. Sabine 22/5/1852, R Soc HS.15.238.

C333. W. Sykes to F. Ronalds 6/6/1852, IET 1.9.1.

C334. J. Cheyne (HMS Assistance) to F. Ronalds 12/7/1852, IET 1.9.1

C335. F. Ronalds (Chiswick) to J. Chavarri copy 29/10/1852, IET 1.9.1.

C336. F. Ronalds to M. Johnson 17/1/1853, IET 1.9.1.

C337. G. Airy (Greenwich) to F. Ronalds 27/1/1853, Cambridge RGO 6/677.

C338. F. Ronalds (London) to G. Airy 31/1/1853, Cambridge RGO 6/677.

C339. M. Johnson (Oxford) to F. Ronalds 7/2/1853, IET 1.9.1.

C340. F. Ronalds (London) to E. Faà di Bruno copy 20/3/1853, IET 1.9.1.

C341. E. di Bruno (Turin) to F. Ronalds 12/4/1853, IET 1.9.1.

C342. F. Ronalds (London) to M. Johnson copy 27/4/1853, IET 1.9.1.

C343. F. Ronalds (London) to J. Chavarri copy 9/5/1853, IET 1.9.1.

C344. M. Johnson (Oxford) to F. Ronalds 17/6/1853, IET 1.9.1.

C345. J. Gassiot (London) to F. Ronalds 5/8/1853, IET 1.9.1.

C346. F. Ronalds (Oxford) to H. Noad (London) copy 25/8/1853, IET 1.9.1.

C347. F. Ronalds (Folkestone) to M. Johnson (Oxford) copy 27/10/1853, IET 1.9.1.

C348. M. Johnson (Oxford) to F. Ronalds 30/10/1853, IET 1.9.1

C349. M. Johnson (Oxford) to F. Ronalds 8/11/1853, IET 1.9.1

C350. M. Johnson (Oxford) to F. Ronalds 14/11/1853, IET 1.9.1.

C351. F. Ronalds (Paris) to J. Gassiot copy 15/12/1853, IET 1.9.1.

C352. J. Gassiot (London) to F. Ronalds (Paris) 17/12/1853, IET 1.9.1.

C353. F. Ronalds (Paris) to M. Johnson copy 18/12/1853, IET 1.9.1.

C354. M. Johnson (Oxford) to F. Ronalds 21/12/1853, IET 1.9.1.

C355. W. Burder (Bristol) to F. Ronalds 7/3/1854, IET 1.1.9.

C356. F. Ronalds (Paris) to W. Burder (Bristol) copy 20/3/1854, IET 1.9.1.

C357. M. Johnson (Oxford) to F. Ronalds 30/6/1854, IET 1.9.1.

C358. F. Ronalds (Paris) to J. Gassiot copy 15/2/1855, IET 1.9.1.

C359. C. Weld (London) to F. Ronalds (Paris) 8/3/1855, IET 1.9.2.

C360. M. Sinobas (Madrid) to F. Ronalds 17/4/1855, IET 1.9.1.

C361. F. Ronalds (Paris) to M. Johnson copy 3/6/1855, IET 1.9.1.

C362. M. Johnson (Oxford) to F. Ronalds 16/6/1855, IET 1.9.1.

C363. M. Johnson (Oxford) to F. Ronalds 24/7/1855, IET 1.9.1.

C364. F. Ronalds (Paris) to C. Piazzi Smyth (Edinburgh) copy 26/8/1855, IET 1.9.1.

C365. F. Ronalds (Paris) to T. Romney Robinson (Armagh) copy 18/2/1856, IET 1.9.1.

C366. M. Johnson (Oxford) to F. Ronalds 5/6/1856, IET 1.9.1.

C367. F. Ronalds (Bordeaux) to M. Sinobas copy 7/7/1856, IET 1.9.1.

C368. F. Ronalds (Montpellier) to H. Lloyd copy 20/5/1857, IET 1.9.1.

C369. H. James (Southampton) to F. Ronalds 15/12/1857, IET 1.9.1.

C370. E. Ronalds (Dr) (Edinburgh) to F. Ronalds 30/3/1858, IET 1.9.1.

C371. E. Ronalds (Dr) (Edinburgh) to F. Ronalds 19/6/1858, IET 1.9.1.

C372. F. Ronalds (Geneva) to I. Tremblay copy 7/7/1858, IET 1.9.1.

C373. H. Vaughan (Venice) to F. Ronalds (Padua) 6/5/1859, IET 1.9.1.

C374. H. Vaughan (Venice) to F. Ronalds (Padua) 10/7/1859, IET 1.9.1.

C375. F. Ronalds (Padua) to S. Carter (Battle?) 21/2/1860, UCL GB 0103 MS ADD 206.

C376. F. Ronalds (Milan) to C. Matteucci draft 19/2/1861, IET 1.9.1.

C377. L. Clark (Westminster) to F. Ronalds (Bellagio) 24/6/1861, IET 1.9.2.

C378. F. Ronalds (Bellagio) to L. Clark (Westminster) copy 12/9/1861, IET 1.9.2.

C379. F. Ronalds (Milan) to E. Sabine copy 18/4/1862, IET 1.9.1.

C380. E. Sabine (London) to F. Ronalds (Milan) 24/4/1862, IET 1.9.1.

C381. F. Ronalds (Milan) to *The Times* copy 17/5/1862, IET 1.9.1.

C382. F. Ronalds (Milan) to C. Matteucci copy 30/5/1862, IET 1.9.1.

C383. Lusignan (Milan) to F. Ronalds (Milan) 10/6/1862, IET 1.9.1.

C384. G. Airy (Greenwich) to G. Stokes (Royal Society) 31/1/1866, Cambridge RGO 6/393.

C385. G. Stokes (Royal Society) to G. Airy 15/2/1866, Cambridge RGO 6/393.

C386. F. Ronalds (Battle) to Derby copy 1/11/1866, IET 1.9.2.

C387. L. Clark (Westminster) to F. Ronalds (Battle) 13/11/1866, IET 1.9.2.

C388. F. Ronalds (Battle) to L. Clark 9/12/1866, Bakken.

C389. W. Cooke (Carnarvon) to F. Ronalds (Battle) 11/12/1866, IET 1.9.2.

C390. G. Barrington to S. Carter 15/12/1866, IET 1.9.2.

C391. F. Ronalds (Battle) to Derby copy 24/12/1866, IET 1.9.2.

C392. F. Ronalds (Battle) to W. Cooke draft 3/1/1867, IET 1.9.2

C393. J. Peacock (Hammersmith) to F. Ronalds (Battle) 10/2/1868, IET 1.9.2

C394. L. Clark (Westminster) to F. Ronalds 23/4/1869, Bakken.

C395. F. Ronalds (Battle) to G. Airy 23/10/1869, Cambridge RGO 6/485.

C396. F. Ronalds (Battle) to S. Carter 25/11/1869, IET 1.9.2.

C397. S. Carter (Battle) to G. Glyn 24/2/1870, IET 1.9.2.

C398. S. Carter (Westminster) to G. Glyn 2/3/1870, IET 1.9.2.

C399. W. Gladstone (London) to F. Ronalds 28/3/1870, IET 1.9.2.

C400. F. Ronalds (Battle) to W. Gladstone 29/3/1870, British Library Add MS 44426 f36r.

C401. S. Carter (Rome) to F. Ronalds 30/3/1870, IET 1.9.2.

C402. M. Carter (Rome) to F. Ronalds 31/3/1870, IET 1.9.2.

C403. T. Gibson (Tunbridge Wells) to F. Ronalds 1/4/1870, IET 1.9.2.

C404. F. Ronalds (Battle) to L. Clark 11/4/1870, Jeremy Norman's HistoryofScience.com

C405. W. Cooke (Tooting) to F. Ronalds (Battle) 18/4/1870, IET 1.9.2.

C406. S. Carter (Battle) to Royal Society 14/1/1871, R Soc MC.9.152.

C407. J. Peacock (Hammersmith) to F. Ronalds (Battle) 6/12/1871, IET 1.9.2.

C408. F. Ronalds (Battle) to L. Clark ?/12/1871, Bakken.

C409. W. Morris to J. Morris 18/3/1878, British Library Add MS 45338; also *The Collected Letters of William Morris* (1984).

C410. L. Harris (London) to Ruston (Brentford) 30/8/1880, WU B1458.

C411. E. Ronalds (Dr) (Edinburgh) to J. Fahie 26/4/1882, IET 1.9.2.119.

C412. Thompson & Debenham (London) to G. Harris (London) 6/3/1882, WU B1458.

C413. E. Ronalds (Brighton) to L. Harris 18/9/1884, WU B2284.

C414. L. Montgomrey (Twickenham) to L. Harris (London) 14/6/1886, WU B2279.

C415. L. Clark (Westminster) to W. Preece 30/5/1892, Science Museum, T/1892–158.

C416. L. Montgomrey (Twickenham) to L. Harris (London) 11/8/1896, WU B1462.

C417. H. Carter (Notting Hill) to L. Cust (London) 26/5/1897, NPG 46/11/97, RP 1095.

C418. A. Ronalds (Buninyong) to C. Taylor (Ballarat) 23/6/1913, Ronalds Family papers.

C419. E. Ronalds (Great Malvern) to J Greg (London) 7/10/1928, The Ronalds Family 1726–1875.

C420. R. Orr (London) to T. Rowatt (Edinburgh) 23/7/1931, National Museum of Scotland

C421. E. Ronalds (Great Malvern) to J. Greg (London) 21/6/1932, The Ronalds Family 1726–1875.

C422. D. Miller (Brighton) to A. Ronalds (Balwyn) 6/9/1968, Ronalds Family papers.

Journals and Manuscripts

J1.   Court of the Lord Lyon Register of Arms and Bearings in Scotland, Edinburgh.

J2.   Old Parish Records for Kilmarnock, Ayrshire, with kind permission of Registrar General for Scotland.

J3.   *Brentford Congregational Church collection,* LMA N/C/34.

J4.   Boyd's Roll, Register of Apprentices and Freemen of the Drapers' Company of London.

J5.   Royal and Sun Alliance Insurance, Hardcastle & Field (1791), LMA CLC/B/192/F/001/MS11936/375.

J6.   Corporation of London, *Drapers' Admission by Redemption — Francis Ronalds* (1802), LMA COL/CHD/FR/02/1260/009.

J7.   W. Robertson, *Estate Legal Documents* (1807–16), WU B558.

J8.   R. Dent, *Plan of the Parish of Saint Mary Islington* (1805), Islington Library.

J9.   St Paul's Cathedral, *Islington Prebendal Manor Court Roll,* LMA CLC/313/N/008/MS14225.

J10.   Northampton, *Lease Agreement, 6 Canonbury Place,* LMA E/NOR/L/04/176.

J11.   *Estate Catalogues for 6 Canonbury Place* (1812), Columbia Univ., Nichols Family Papers.

J12.   Fishmongers' Company, *Title Deed, 109 Upper Thames Street* (1863), LMA CLC/L/FE/G/162/MS06276.

J13.   H. Crabb Robinson, *Diaries* (1811–67), Dr Williams's Library, with permission from the Trustees.

J14.   F. Ronalds, *Ideas Book* (*c.*1814–29), IET 1.6.136.

J15.   F. Ronalds, *Dry Piles and Electric Clock* (1814–23), IET 1.5.56, 1.1.7.2.

J16.   F. Ronalds, *Update of Priestley's History and Present State of Electricity* (1817), IET 1.2.37, 1.2.39.

J17.   F. Ronalds, *Arm Chair Journal* (1/9/1818–21/10/1818), IET 1.1.1.

J18.   F. Ronalds, *Journal: Italy, Malta* (16/1/1819–4/1/1820), IET 1.1.2.

J19.   F. Ronalds, *Journal: Malta* (31/12/1819), IET 1.1.3.

J20.   F. Ronalds, *Journal: Egypt, Holy Land* (23/1/1820–3/4/1820), IET 1.1.4.

J21.   F. Ronalds, *Journal: Cyprus, Turkey, Greece, Balkans, Venice* (2/5/1820–10/9/1820), IET 1.1.5.

J22.   F. Ronalds, *Journal: Italy, Switzerland, Germany, France* (10/9/1820–12/11/1820), IET 1.1.6.

J23.   F. Ronalds, *Letter Summary* (4/8/1819–18/11/1820), IET 1.1.19.

J24.   F. Ronalds, *Travel Notes* (1820), IET 1.1.73.

J25.   F. Ronalds, *Summary of Holy Land* (*c.*1821), IET 1.1.7.

J26.   F. Ronalds, *Expenses* (1818–20), IET 1.1.40.

J27. F. Ronalds, *Note re Preserving Papyrus* (1820), IET 1.1.48.

J28. F. Ronalds, Passports (1820–34), IET 1.1.22, 1.1.29–30, 1.1.34–5.

J29. Royal and Sun Alliance Insurance, Garratt and Ronalds (1821), LMA CLC/B/192/F/001/MS11936/486/974913.

J30. *Notes on Trade in Marseille* (1823), IET 1.1.65.

J31. F. Ronalds, *Travel Memorandums* (1823–4), IET 1.1.58.

J32. F. Ronalds, *Expenses Comparison Messina and London* (1823), IET 1.1.37.

J33. M. A. Nichols, *Diary* (1823–34), Nichols Archive Project, Private Collection PC9/1–9 NAD3223–3231.

J34. F. Ronalds, *Thermometric Ventilator* (*c.*1825), IET 1.6.127.

J35. *Drawings of Perspective Tracing Instruments* (*c.*1825), IET 1.6.16.

J36. F. Ronalds, *Perspective Sketches* (*c.*1829), IET 1.6.118.

J37. *Description de la première espèce d'Instrument Perspectif et Patent pour Esquisser d'après Nature*, IET 1.6.15.

J38. B. Ronalds and Family, *Art Album*, WU X1593.

J39. F. Ronalds, *Pocket Surveying Instrument* (*c.*1829), IET 1.6.108.

J40. F. Ronalds, *Anamorphosis*, IET 1.6.27.

J41. F. Ronalds, *Drawing and Engraving Instruments*, (*c.*1829), IET 1.6.113.

J42. F. Ronalds, *Origin & Progress of Engraving in Copper & Wood* (*c.*1830), IET 1.6.134.

J43. F. Ronalds, *Specimens Executed by means of the Turner's Pen* (*c.*1830), IET 1.5.51.

J44. F. Ronalds, *Worksheet comparing Motions of various Lathe Attachments* (*c.*1830), IET 1.6.109

J45. F. Ronalds, *Diary* (May 1833), IET 1.5.4.

J46. J. South, *Correspondence on Equatorial Mount*, R Astronomical Soc MS South 3.

J47. A. Blair & F. Ronalds, *Journal at Carnac* (1834), IET 1.2.31.

J48. F. Ronalds, *Memorandums of Bearings, Measurements &c at Carnac* (1834), IET 1.2.29.

J49. F. Ronalds, *Carnac Monuments* (*c.*1835), IET 1.2.18.

J50. F. Ronalds, *Dracontia* (*c.*1836), IET 1.2.14.

J51. A. Blair, *Celtic and Germanic Ancient Geographies* (*c.*1837–42), IET 1.2.6.

J52. A. Blair, *Ancient Monuments in Sweden* (*c.*1837–42), IET 1.2.10.

J53. F. Ronalds, *Drawing of Vase made with Large Perspective Tracing Instrument* (*c.*1837), IET 1.7.52.

J54. F. Ronalds, *Turner's Manual* (*c.*1837), IET 1.2.1–2, 1.5.39–40, 1.6.74.

J55. F. Ronalds, *Canal Propulsion* (1837–8), IET 1.2.27, 1.4.51, 1.4.54, 1.6.79.

J56. F. Ronalds, *Pamphlet on Preservation of Life & Property from Fire* (*c.*1840), IET 1.5.15.

J57. F. Ronalds, *Fire Finder* (*c.*1840), IET 1.5.9, 1.5.11–13.

J58. F. Ronalds, *Rain and Vapour Gauge* (*c.*1842), IET 1.2.4.

J59. F. Ronalds, *Ancient Monuments in England* (1842–3), IET 1.2.25.

J60. *Journal of Meteorological Observations at Kew Observatory*, 1842–3, Met Office Y54.F-Y52.D.

J61.   *Journal of Meteorological Observations at Kew Observatory*, 1843–4, Met Office Y54.F-Y52.D.

J62.   *Journal of Meteorological Observations at Kew Observatory*, 1844–5, Met Office Y54.F-Y52.D.

J63.   *Journal of Meteorological Observations at Kew Observatory*, 1845–6, Met Office Y54.F-Y52.D.

J64.   *Journal of Meteorological Observations at Kew Observatory*, 1849–50, Met Office Y54.F-Y52.D.

J65.   F. Ronalds, *Visit to Greenwich Observatory* (27/12/1843), IET 1.4.17.

J66.   *Certificate of a Candidate for Election — Francis Ronalds* (1844), R Soc.

J67.   F. Ronalds, *Draft Note for the Athenæum and Literary Gazette*, (1846), IET 1.4.6.

J68.   Annotated segment of first print of Murchison's BAAS Presidential Address (1846), IET 1.3.500.

J69.   F. Ronalds, *Notes for 1847 BAAS Conference* (1847), IET 1.4.8.

J70.   F. Ronalds, *Punch-lists for Camera and Photography Rooms* (c.1847), IET 1.4.47.

J71.   *Unused Photographic Paper, Gelatine Sheet Tracings and Prints* (1848–50), IET 1.7.4.

J72.   Note regarding Talbot (19/4/1848), IET 1.3.187.

J73.   F. Ronalds, *Memorial on Self-registering Instruments*, draft (30/9/1848), IET 1.9.1.

J74.   F. Ronalds, *Directions & Precautions for Toronto Observatory*, draft (1849), IET 1.4.23.

J75.   F. Ronalds, *Directions & Precautions for Toronto Observatory* (1849), IET 1.4.26 also Met Office Y16.J3.

J76.   F. Ronalds, *Notes of BAAS Council Meeting* (25/10/1849), IET 1.3.336.

J77.   F. Ronalds, *Report to Kew Committee* (22/3/1850), R Soc Sa.1091.

J78.   *Doctor's certificate for W. Birt* (3/6/1850), R Soc Sa.178.

J79.   F. Ronalds, *Notes in Edinburgh* (1850), IET 1.5.28.

J80.   F. Ronalds, *Diary of 1850 BAAS Conference in Edinburgh* (1850), IET 1.5.30.

J81.   F. Ronalds, *Notes of Barograph, Magnetograph and Electrograph in Operation* (1851), IET 1.5.38.7.

J82.   F. Ronalds, *Correspondence and Memorial on the Electric Telegraph*, IET 1.9.2.

J83.   *Kew Committee Minutes* (1850–55), Met Office.

J84.   *Kew Observatory Diary* (28/8/1850–31/10/1851), Met Office.

J85.   F. Ronalds, *Diary Excerpts* (1851–62), IET 1.9.1.

J86.   G. Airy, *Address to the Visitors of the Royal Observatory*, Greenwich (Jan 1863), Cambridge RGO 55/3.

J87.   *Kew Observatory Barograms* (1862–1981), Met Office Z26.A-Z27.J.

J88.   *Ronalds Family Letters from New Zealand* (1853–60), Alexander Turnbull Library qms-1719.

J89.   B. Ronalds, *Travel Journals*, WU B2285.

J90.   H. Ronalds Jnr, *Diary* (1851–4) WU B1462, (1854–62) WU B4186.

J91.   H. Ronalds Jnr, *Genealogical Notes*, WU B1454.

J92.   H. Ronalds Jnr, *Family Tree*, WU B2285.

J93. H. Ronalds, *Ronalds' Genealogy*, UCL GB 0103 MS ADD 206.

J94. A. Harris, *Diary*, WU B4185.

J95. L. Harris, *Diary* (1867–95), WU B1453.

J96. F. Ronalds, *Memorial on the Invention of the Electric Telegraph* (12/2/1870), IET 1.9.2.

J97. *Trust Deeds of the Ronalds Library* (1875), (1908), (1922), (1949), IET.

J98. *Record of the Ronalds Family,* Ronalds Family Papers, USA Branch.

J99. *The Ronalds Family 1726–1875*, **1–3**, Ronalds Family Papers, NZ Branch.

J100. H. J. Woodhouse, *Victorian Pioneers of Litho-drawing and Engraving* (1889), SLV.

J101. A. F. Ronalds, *Cousins by the Dozens* (1966).

J102. E. M. Day, Ronalds family history documentation, Melbourne.

J103. B. O. Rayner, Extended Ronalds family tree.

J104. D. Shailes, *The Ronalds of Brentford* 1753–1880 (2009), Chiswick Library.

J105. J. Cloake, *The King's Observatory, Old Deer Park Richmond, Historical Report* (c.2010).

J106. *Papers of Sir Joseph Banks*, Section 5, Series 19, State Library of NSW. Available at http://www2.sl.nsw.gov.au/banks/series_19/19_31.cfm (Accessed on 28/8/2015).

Family Wills

W1. Hugh Ronald, 1719.

W2. Francis Clarke (Snr), 1731.

W3. Nathaniel Field, 1755.

W4. Nathaniel Hardcastle, 1786.

W5. (Old) Hugh Ronalds, 1787.

W6. Francis Ronalds, 1806.

W7. William Field (Snr), 1811.

W8. Emily Venning, 1834.

W9. Jane Ronalds, 1848.

W10. Charles Field, 1850.

W11. (Sir) Francis Ronalds, 1866, with four later codicils.

Archive Abbreviations

Bakken	Bound with *Descriptions of an Electrical Telegraph* (1823), Collections of Bakken Library and Museum.
Chicago	George Flower Family Papers, Chicago History Museum.
IET	Institution of Engineering and Technology Archives, London.
Cambridge	Royal Greenwich Observatory Archive, by permission of Science and Technology Facilities Council and Syndics of Cambridge Univ. Library.
Leeds	GB 206 Brotherton Collection MS 19c Wilson, Univ. Leeds.
LMA	London Metropolitan Archives, City of London.

Co-op   Robert Owen Collection, National Co-operative Archive, Manchester.
Met Office   National Meteorological Library and Archive, Exeter.
R Soc   Herschel and Sabine Collections, Royal Society, London.
SLV   State Library of Victoria, Melbourne.
UCL   University College London Library Services, Special Collections.
WU   Harris Family Fonds, Ronalds Family Papers, Western Archives, Western
             Univ., London Ontario.

INDEX